Springer Tracts in Mechanical Engineering

Springer Tracts in Mechanical Engineering (STME) publishes the latest developments in Mechanical Engineering - quickly, informally and with high quality. The intent is to cover all the main branches of mechanical engineering, both theoretical and applied, including:

- Engineering Design
- Machinery and Machine Elements
- Mechanical Structures and Stress Analysis
- Automotive Engineering
- Engine Technology
- Aerospace Technology and Astronautics
- Nanotechnology and Microengineering
- Control, Robotics, Mechatronics
- MEMS
- Theoretical and Applied Mechanics
- Dynamical Systems, Control
- Fluids Mechanics
- Engineering Thermodynamics, Heat and Mass Transfer
- Manufacturing
- Precision Engineering, Instrumentation, Measurement
- Materials Engineering
- Tribology and Surface Technology

Within the scope of the series are monographs, professional books or graduate textbooks, edited volumes as well as outstanding PhD theses and books purposely devoted to support education in mechanical engineering at graduate and post-graduate levels.

Indexed by SCOPUS, zbMATH, SCImago.

Please check our Lecture Notes in Mechanical Engineering at http://www.springer.com/series/11236 if you are interested in conference proceedings.

To submit a proposal or for further inquiries, please contact the Springer Editor **in your region**:

Ms. Ella Zhang (China)
Email: ella.zhang@springernature.com
Priya Vyas (India)
Email: priya.vyas@springer.com
Dr. Leontina Di Cecco (All other countries)
Email: leontina.dicecco@springer.com

All books published in the series are submitted for consideration in Web of Science.

More information about this series at http://www.springer.com/series/11693

Peter F. Pelz · Peter Groche ·
Marc E. Pfetsch · Maximilian Schaeffner
Editors

Mastering Uncertainty in Mechanical Engineering

Springer

Editors
Peter F. Pelz
Department of Mechanical Engineering
Technische Universität Darmstadt
Darmstadt, Hessen, Germany

Peter Groche
Department of Mechanical Engineering
Technische Universität Darmstadt
Darmstadt, Hessen, Germany

Marc E. Pfetsch
Department of Mathematics
Technische Universität Darmstadt
Darmstadt, Hessen, Germany

Maximilian Schaeffner
Department of Mechanical Engineering
Technische Universität Darmstadt
Darmstadt, Hessen, Germany

ISSN 2195-9862 ISSN 2195-9870 (electronic)
Springer Tracts in Mechanical Engineering
ISBN 978-3-030-78356-3 ISBN 978-3-030-78354-9 (eBook)
https://doi.org/10.1007/978-3-030-78354-9

This Springer imprint is published by the registered company Springer Nature Switzerland AG
The registered company address is: Gewerbestrasse 11, 6330 Cham, Switzerland

Preface

End of Certainty

The great success of classical physics and the closely related engineering mechanics, such as the precision of Kepler's laws verified by Tycho Brahe's data, has trapped us in deterministic thinking: although engineers deal with data uncertainty. In practice, we see little quantification of model uncertainty or even mastering structural uncertainty in design and usage of technical systems today.

In this book, we fully recognise "The End of Certainty" as Prigogine puts it[1] in socio-technical systems. The ratio, but also the intuition of engineers, is based on reliable data and a reliable imagination of the relevant reality. The design of safe and sustainable systems needs exploring the complete design space and not only a small step into it. It needs a holistic view starting from the resources or component and ending at the socio-technical systems and their evolution in time.

This insight and the primacy of safety and sustainability including costs, energy and material consumption force us to master uncertainty and especially ignorance as the most important manifestation of uncertainty. Safety and sustainability are the main objectives in solving the unknown-unknowns: mastering uncertainty is one of today's fundamental methodological, technological, economic and sociologic tasks for science, industry and society. This contributes to coining the current paradigm of science.

"Science finds, industry applies, man adapts"[2] is an outdated paradigm. Today, it is replaced either by Responsible Research and Innovation (RRI) or even Open Science.[3] Therefore, the question whether "science push" or "society pull" is adequate, is in fact answered by agreeing on the latter with broad consent of most scientists. Considering engineering design and engineering science, "society pull" has always been the motivation, as solving social needs sustainably is the task of any

[1]Prigogine, I., Stengers, I.: The End of Certainty: Time's Flow and the Laws of Nature. Free Press, New York (1997).

[2]A Century of Progress: Official Guide—Book of the Fair, 1933. A Century of Progress, Chicago (1933).

[3]Nordmann, A.: Vorlesung Ingenieurwissenschaft und Gesellschaft, TU Darmstadt (2019).

engineering work. This book contributes to this task by presenting innovative methods, technologies and strategies for mastering uncertainty in technical systems.

Story Behind This Book

In 2007 a group of six engineers and mathematicians, Reiner Anderl, Herbert Birkhofer, Peter Groche, Holger Hanselka, Alexander Martin and Stefan Ulbrich, all members of Technische Universität Darmstadt at that time, recognised that uncertainty often propagates un-managed through the product life cycle. Even though tolerance synthesis and, already at that time, uncertainty quantification was a large field of science, it was recognised that a holistic strategy to master uncertainty throughout the product life cycle was missing. In other words, mastering uncertainty in the synthesis and usage was identified to have a much broader implication than quantification of uncertainty in the phases of the product life cycle. In fact, mastering uncertainty includes uncertainty quantification.

From the year 1950 on, Genichi Taguchi pioneered in mastering uncertainty by establishing his Robust Design methodology using applied mathematical methods, such as stochastics and optimisation applied to the design and production process. Thus, right from the beginning, it was obvious that expertise from engineering and applied mathematics needed to be combined. The cooperation of engineers and mathematicians has proven to be most fruitful for both sides. In the recent research phase since the year 2016, perspectives from law, history and linguistics, which are all present in this book, contributed to mastering uncertainty.

From January 2009 till March 2021 research on the topic "Mastering Uncertainty in Mechanical Engineering" has been conducted. Throughout, more than twelve research groups from the faculty of mechanical engineering, mathematics, law, linguistics and history collaborated. All involved researchers are indicated following this preface. They sum up to an impressive number of nearly 100 researchers. In addition, there are many students guided by those researchers who have dealt with the main topic of this book in their research work. All test rigs and mechatronic systems that serve as demonstrators to validate the findings of this book, such as the Modular Active Spring-Damper System, the Active Air Spring and the 3D Servo Press, were built in the workshops of the research groups. By that more than 20 workshop employees of Technische Universität Darmstadt supported the research work.

During this long period of time, the research has been continuously funded by the Deutsche Forschungsgemeinschaft (DFG, German Research Foundation)—project number 57157498—Sonderforschungsbereich (SFB, Collaborative Research Centre) 805. We all are grateful to the colleagues from other German universities for their support and interest. They served as reviewers in three main review sessions and two further research sessions at the very beginning of the project, as well as at the very end of the project. We would further like to thank the DFG for funding our research, namely Ursula von Gliscynski, Wieland Biedermann and Ferdinand Hollmann for their interest and support. They accompanied the research during that long period.

Openness, Friction, Intellectual Environment

A funded research group of up to 100 researchers, Principal Investigators (PI) and students, is a not to be overseen structure within any university. The SFB 805 "Mastering Uncertainty in load-bearing Systems of Mechanical Engineering" emerged from a discussion of six colleagues. As such, it is a typical example of bottom-up research. In an analogon to non-linear thermodynamics the structure may be called a dissipative structure. As it may be known, the prerequisites for its formation are, first, an open system boundary and, second, friction. For a university with a technical background, such as the Technische Universität Darmstadt, the openness to society including industry is given right from its foundation in the 19th century. We are grateful to our university for enabling the costly environment for experimental research including the care for research data being part of Open Science. The founder of cybernetics, Norbert Wiener, puts the task of a university into words: "The intellectual and technological environment must be right before the idea can blossom".[4]

This demanded and given openness also enables the spreading of people (and ideas). Holger Hanselka and Roland Platz successfully shaped the research as spokesperson and scientific manager to the SFB 805 for the first five years. Daniela Wagner has done the project controlling from the very beginning. When Holger Hanselka was appointed as President of the Karlsruhe Institute of Technology (KIT) in 2013, Peter Pelz took on the task of shaping and leading the research for the following eight years until 2021. He has been assisted by his co-workers Philipp Hedrich, Ingo Dietrich and Manuel Rexer. In preparing the last funding period Peter Pelz was advised by his colleagues Ulrich Konigorski and Jürgen Rödel.

Most of the former Ph.D. and master students are now transferring ideas to industry. Professors are super spreaders of ideas by profession. Hence, it is to be rated positively that the former PIs Ulf Lorenz is now at the University of Siegen, Lena Altherr at the University of Applied Sciences, Münster. The former Ph.D. students Tobias Eifler and Kai Mecke are now teaching and performing research at the Technical University of Denmark and the Jade Hochschule Wilhelmshaven. Furthermore, three spin offs emerged from the SFB 805, making transfer of ideas accessible to industry.

The editors would like to thank all authors for their contributions from very different perspectives, which hopefully make this book worth reading. Finally, we would like to thank in particular Nicolas Brötz, Jakob Hartig, Dorothea Saur, and Christoph Eyrich for a year of intensive work on this book. They completed our team in the discussion with the authors, in the management of the big book project

[4]Wiener, N.: Invention: The Care and Feeding of Ideas. MIT Press, Cambridge, Mass (1994)

and—very importantly—for the many quality controls and improvements that make a book valuable.

We are pleased that we were able to facilitate an open access book in the spirit of Open Science.

Darmstadt, Germany

Peter F. Pelz
Peter Groche
Marc E. Pfetsch
Maximilian Schaeffner

Acknowledgements

We would like to thank the following people who contributed to the Sonderforschungsbereich (SFB, Collaborative Research Centre) 805.

Nassr Al-Baradoni	Thomas Hauer
Jörg Avemann	Michael Haydn
Thomas Bedarff	Fuzhang He
Matthias Berger	Philipp Hedrich
Christian Bölling	Felix Heimrich
Matthias Brenneis	Marlene Helfert
Andreas Bretz	Benjamin Hess
Nicolas Brötz	Daniel Hesse
Stefan Calmano	Florian Hoppe
Ingo Dietrich	Laura Joggerst
Thorsten Ederer	Sebastian Kersting
Tobias Eifler	Eckhard Kirchner
Roland Engelhardt	Maximilian Knoll
Georg Enss	Jan Koenen
Tina Felber	Christina König
Paul Felber	Philip Kolvenbach
Robert Feldmann	Ulrich Konigorski
David Fischer	Matthias Kraft
Tillmann Freund	Martin Krech
Tristan Gally	Lisa Kristl
Christopher Gehb	Anja Kuttich-Meinlschmidt
Thiemo Germann	Philipp Leise
Felix Geßner	Eric Lenz
Mehmet Görtan	Jonathan Lenz
Benedict Götz	Sushan Li

(continued)

(continued)

Sebastian Güth	Sophie Loidolt
Kai Habermehl	Julian Lotz
Jakob Hartig	Gerhard Ludwig
Shashidhar Mallapur	Pia Schlemmer
Sonja Mars	Philipp Schmidt
Daniel Martin	Johann Schmitt
Alexander Matei	Andreas Schmitt
Johannes Mathias	Sebastian Schmitt
Michael Maurer	Fiona Schulte
Kai Mecke	Astrid Seekatz
Christiane Melzer	Adrian Sichau
Dirk Molitor	Nicolai Simon
Lucia Mosch	Julian Sinz
Thomas Müller	Britta Späh
Marius Oberle	André Sprenger
Steffen Ochs	Georg Staudter
Serge Parfait Ondoua	Markus Türk
Tuğrul Öztürk	Angela Vergé
Sebastian Pokutta	Daniela Wagner
Nils Preuß	Moritz Weber
Manuel Rexer	Matthias Weigold
Majid Rezaei	Marion Wiebel
José Rios	Heinrich Wiener
Stefan Rothenbücher	René Winterstein
Marlene Schaffland-Utz	Jan Wolf
Maximilian Schäffner	Jan Würtenberger
Matthias Scheitza	Maximilian Zocholl

Between 2008 and 2021, the following Principal Investigators (PI) have been or were part of the SFB 805.

Eberhard Abele	Ulf Lorenz
Lena Altherr	Alexander Martin
Reiner Anderl	Tobias Melz
Herbert Birkhofer	Jürgen Nuffer
Andrea Bohn	Peter Pelz
Ralph Bruder	Marc Pfetsch
Peter Groche	Roland Platz
Holger Hanselka	Stefan Ulbrich
Hermann Kloberdanz	Janine Wendt
Michael Kohler	Kai Wolf

Contents

Chapter 1
Introduction

Peter F. Pelz

Abstract In this chapter, the motivation for this book is given. The analysis process of socio-technical systems based on data and models is examined from the perspective of uncertainty. The synthesis process of systems based on models and/or intuition leads to the important concepts of function and quality as well as data, model, and structural uncertainty. This forms both the foundation and the introduction to the following chapters. It is shown that the mastering of uncertainty is the key to Sustainable Systems Design. Thus, the societal need for safety and sustainability is met.

"Ha, that will be a merry-go-round!
The bridge must sink into the ground."
"And with the train what shall we do
That crosses the bridge at seven?"
"That too."
"That must go too!"
"A bauble, a naught,
What the hand of man hath wrought!"
„Hei, das gibt einen Ringelreihn,
Und die Brücke muß in den Grund hinein."
„Und der Zug, der in die Brücke tritt
Um die siebente Stund?"
„Ei, der muß mit"
„Muß mit."
„Tand, Tand
Ist das Gebilde von Menschenhand!"

The Bridge by the Tay / Die Brück' am Tay
by Theodor Fontane (1880) [27]

P. F. Pelz (✉)
Department of Mechanical Engineering, TU Darmstadt, Darmstadt, Germany
e-mail: peter.pelz@fst.tu-darmstadt.de

P. Pelz et al. (eds.), *Mastering Uncertainty in Mechanical Engineering*,
Springer Tracts in Mechanical Engineering,
https://doi.org/10.1007/978-3-030-78354-9_1

How can we ensure product safety in a world of products with ever increasing complexity? This question arises when designing lightweight structures and sustainable systems. The question also comes up when implementing methods and technologies for controlled production quality. Mastering uncertainty is central to all these topics and requires contributions from engineering, mathematics and law. This book provides answers on how to master uncertainty in the life cycle of products from the design phase via the production phase to the usage phase. These answers are consolidated in strategies to master the uncertainty of a possible product usage, even if partly unknown at the beginning of a new engineering design.

Invitation to visit the building devoted to mastering uncertainty

We do not intend to represent a definition, a method, or a technology for their own sake. On the contrary, the building presented here, consisting of the fundamental floor, middle and top floor, inspires the visitor how to master uncertainty in his or her specific task. The craftsmen who built this house come from the fields of engineering, mathematics and law. Together they have pursued the goal of further developing systematic engineering design. To master uncertainty, we always focus on the function and quality of the product or system, i.e. its essence from the application perspective.

On the fundamental floor we submit data, models and structures. Here we lay the conceptual basis and define consistent uncertainty classes. On the middle floor we introduce methods and technologies to identify, evaluate and counteract uncertainty. On the top floor we introduce the strategies (i) robustness, (ii) flexibility, (iii) resilience. All three strategies contribute to mastering uncertainty.

In order not to develop a method for its own sake, we have tested all tools, i.e. definitions, technologies and strategies on the three technical systems that we have developed, manufactured and used over the last twelve years. The systems are active and semi-active systems. So, flexibility is achieved by the smart systems Active Air Spring and 3D Servo Press. All research and its presentation focus on a load-bearing example system, which is a lightweight structure. We invite you as our readers to be guests in our house and hope that you will profit from your visit.

The chapter's structure

Section 1.1 outlines the motivation and Sect. 1.2 the concept of holistic control of uncertainty over the product life phases. In Sect. 1.3, the focus is on the source and quality of models. Section 1.4 provides reflections on the sources and quality of data. Section 1.5 deals with the structures composed out of components. In Sect. 1.7, a broad motivation for mastering uncertainty is presented. The chapter closes with an overview of the book's chapters and the three demonstrator systems designed, manufactured and tested at the Technische Universität (TU) Darmstadt during the last twelve years.

1.1 Motivation

Back in the year 2008, an interdisciplinary group of about ten researchers designed a research program on the topic of this book: *Mastering Uncertainty in Design, Production and Usage of Load-Bearing Structures in Mechanical Engineering*. This led to the Sonderforschungsbereich 805 (SFB, Collaborative Research Centre), which was funded by the Deutsche Forschungsgemeinschaft (DFG, German Research Foundation) in three phases of four years each, from 2008 to 2021. About 60 doctoral students have completed their research work during this time. The researchers, all members of the TU Darmstadt, have come from fields as diverse as production engineering, structural mechanics, fluid power, applied mathematics including nonlinear and discrete optimisation, statistics, and law. The research topic as such is truly interdisciplinary, which is also reflected in the topics of this book.

The topic from a society point of view is motivated by an increasing number of product recalls in the automotive industry. In the era from 1990 to 1995, the number of vehicles recalled annually in the US market rose from 5 million to 20 million. In the year 2014, 64 million vehicles were recalled contrasting 17 million vehicles sold, see Fig. 1.1. Hence, for every vehicle that entered the US market in 2014, four vehicles were recalled for lack of safety [10, 28]. In the same year, 1.5 million vehicles were recalled and 3 million vehicles sold within Germany [41, 46].

Recalls are made on the basis of the Product Safety Act [7]: a recall is required if the product causes a sudden and for the user unforeseen serious danger. The decision is based on the likelihood of failure during the product's lifetime combined with the severity of possible personal injury [11]. In 2014, the recall of vehicles on the German market was in 70% of the cases due to mechanical safety problems and in 20% due to faults in the mechatronic system, including servo-hydraulics [46].

Product safety is equally a strong motivation for mastering uncertainty in the capital goods industry, in mechanical and plant engineering, and in the aerospace industry. Mainly the following three reasons led to the recalls mentioned:

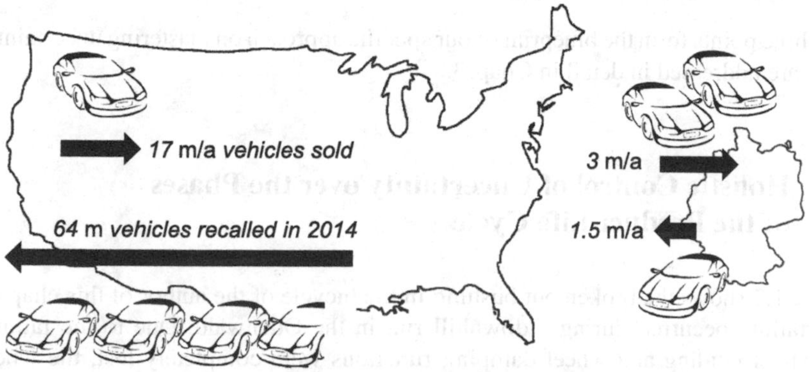

Fig. 1.1 Vehicle recalls in the US (left) and Germany (right) in 2014 [28]

(i) A conflict of objectives between effort and availability, while, at the same time, the future product usage is still uncertain, i.e. the design target is moving.
(ii) Increased demands on cross-company quality assurance due to the shift in value creation to globally developing and producing suppliers with the difficulties of communication and interfaces.
(iii) Increased development speed as a result of global competition.

As a reaction to the increasing speed of development, systems are more and more being developed virtually. This increases the demands on mastering the uncertainty in the models during the product life cycle. All of the above-mentioned points form the current boundary conditions under which safety-relevant load-bearing structures—whether passive, semi-active or active—are developed, produced and used today. At the same time, the importance of product safety law is growing. It is to be expected that complexity will increase even further, as self-adaptive systems gain in importance in the future.

After one decade of research within the SFB 805, the Boeing 737 MAX accident shows that today the mastering of uncertainty in all product life phases is more relevant than ever: on 29 October 2018, a Boeing 737 MAX airliner crashed because of a newly introduced pitch control system. In retrospect, the crash had five causes: firstly, insufficient testing of the newly introduced autonomous pitch control system; secondly, insufficient training of the pilots; thirdly, sensor failure; fourthly, the override control of the pilots by the software; fifthly, the lack of visual feedback to the pilots [40]. The crash of the Boeing 737 MAX in its consequence is an extreme but at the same time typical example of unmastered uncertainty.

Hence, there is a growing need to master uncertainty in all phases of the product life cycle by

(i) laying a solid foundation of *classification, definitions and metrics* of uncertainty;
(ii) assessing and developing *methods and technologies* for quantification, evaluation and master uncertainty;
(iii) developing and validating *strategies* to master uncertainty.

The three points form the blueprint of our specific approach on mastering uncertainty. They are addressed in detail in Chap. 3.

1.2 Holistic Control of Uncertainty over the Phases of the Product Life Cycle

Figure 1.2 shows the broken out bushing in the bicycle of the author of this chapter. The failure occurred during a downhill run in the Odenwald. Due to the failure, the wheel guiding and wheel damping functions were completely lost, the wheel being blocked. The rider remained unharmed. The bicycle's usage can be described

Riding down this slope of a bike trail in the Odenwald ended up with the broken bushing support shown on the right.
slope
broken bushing support
ROCK SHOX

Fig. 1.2 Broken bushing support of the author's mountain bike. The failure occurred during a downhill run in the Odenwald

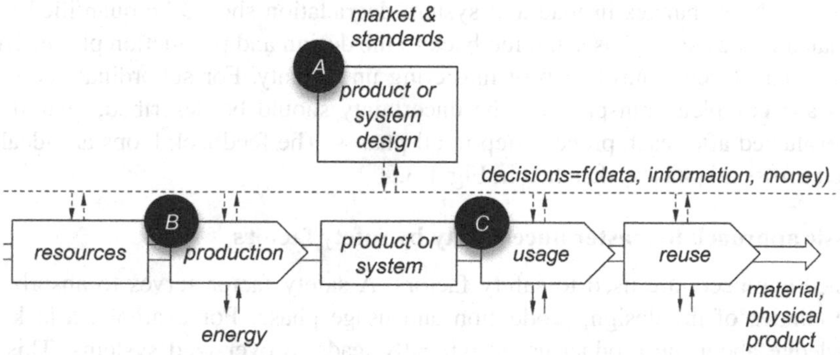

Fig. 1.3 **A** product or system design, **B** production, **C** usage; all phases are interconnected by the flow of physical goods and data, information and money

by factors, such as geography, speed, damper setting, rider's weight, maintenance condition and others.

But not only the usage has to be evaluated: in order to avoid such a failure, the uncertainty over all phases of the product life cycle including product design must be viewed holistically. The failure of the load-bearing structure can have its causes in unmastered uncertainty in one, two or all three phases (A) product design, (B) production or (C) usage, cf. Fig. 1.3. Within this book, we exclude the phases resources and reuse. We are aware that sourcing and recycling are important topics but they are not what we want to focus on.

The phases of the product life cycle are, on the one hand, interconnected by the flow of material or physical products. On the other hand, the phases are interconnected by the flow of data, as well as information including the flow of costs and profits [21]. Although the separation of the product life cycle including product design in phases is common [8, 29, 42], methods and strategies for the holistic, cross-phase mastering of uncertainty have not yet been developed and validated. The following hypothesis can therefore be formulated:

> *Uncertainty can be mastered, if uncertainty is described, quantified and evaluated in all phases of the product life cycle; further, if it is reacted to and learnt from experience, and if follow-up processes are anticipated.*

A process is seen here in a general sense. It may be a production process with an input and output of a physical material flow. It may also be the usage of a component of a load-bearing structure, such as a suspension strut, or a system being composed of many such components.

Following the chain from sourcing to production, to usage and reuse, it is evident, that the uncertainty of a specific product property propagates downstream. Provided this process is unmastered, an accumulation of uncertainty from process step to process step can occur. The task is to master a possible accumulation of uncertainty or even reduce the uncertainty along the process chain. Therefore, the product stress and strength or changes in load and system degradation should be quantified and evaluated in the usage phase, and fed back to the design and production phase. This is the outer closed control loop of mastering uncertainty. For subordinate control loops and complete transparency, the uncertainty should be described, quantified and evaluated after each process step in all phases. The feedback loops are ideally closed across all phases, sketched in Fig. 1.3.

Classic approach to master uncertainty by safety factors

Trained engineers are used to safety factors. A safety factor serves to absorb all uncertainties of the design, production and usage phase. For example, a lack of knowledge about the product usage typically leads to oversized systems. This is understandable, since the function of the product is of primary importance for its use. How the "quality" of this function is fulfilled ranks second. Oversizing may not necessarily be a shortcoming for the customer. However, it leads to the fact that design, production and usage are not sustainable. That this is quite serious can be seen from a simple number. In order to operate fluid systems in Europe in the year 2014, the estimated energy amount of 900 TWh was required [32].

The spatial separation of "generation" and use of (electrical) energy was pushed forward in the 19th century by Werner von Siemens. We will concentrate here only on the consumption side: the electrical energy that drives the fluid systems in use is provided by the output of about 100 large thermal power plants. It is estimated that 40 power plants alone could be saved by sustainable planning and operation of the fluid systems [32]. The driven machines on the consumption side serve the functions of cooling, heating, ventilation, transport, mixing, dosing as well as the power transmission from and through liquids and gases. The example drastically

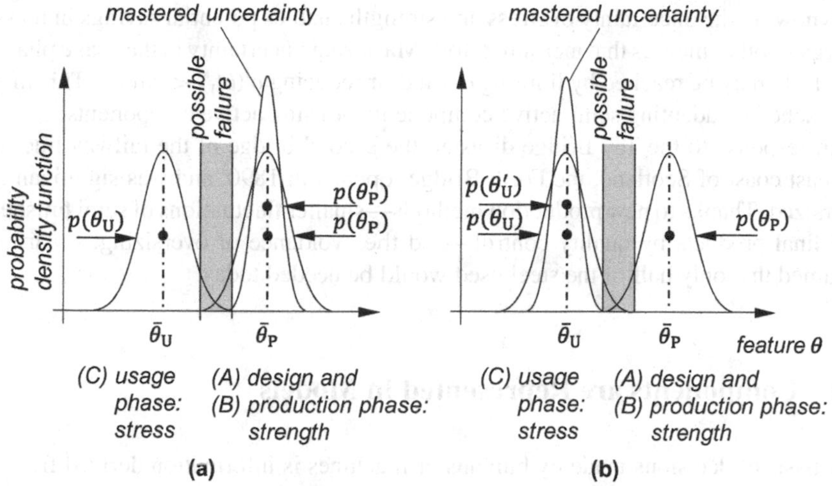

Fig. 1.4 Generic probability density function of properties θ of production $p(\theta_P)$ of the strength θ_P and usage $p(\theta_U)$ of the stress θ_U with mastered uncertainty in (**a**) design (A) and production (B) and (**b**) usage (C) [15]

illustrates the effect of oversizing. In the Anthropocene, saving energy in the use of energy consuming systems should be our priority. The good news is that sustainable systems design is promoted by the methods presented in this book, among others.

Accounting for uncertainty by safety factors is illustrated in Fig. 1.4. If we think of a load-bearing system, the function is described by a load history resulting in the system's stress. Here, the stress of the system shall be smaller than the strength of the system; otherwise there would be a failure. This happened to the Tay Rail Bridge on the night of the 28 December 1879 in a strong winter storm, only 19 months after its opening. Theodor Fontane then wrote his ballad 'The Bridge by the Tay' with the line "A bauble, a nought, what the hand of man hath wrought!" (in German "'Tand, Tand ist das Gebilde von Menschenhand'"), cf. quotation at the beginning of this chapter. Fontane, as a representative of society, criticises the unrestrained uncertainty in this poem. In fact, the wind load and, thus, the stress during the usage phase was underestimated in the planning [22].

This is indicated in Fig. 1.4a where there is an overlap between the probability density function $p(\theta_U)$ of the stress θ_U in the usage phase with the probability density function $p(\theta_P)$ of the strength θ_P in production. Both are influenced by the system's design and production.

In the framework of *stochastic uncertainty*, cf. Chap. 2, the density function of the feature θ has mean $\overline{\theta}$ and standard deviation $\sigma(\theta)$. Hence, mastering uncertainty in the design and production phase, Fig. 1.4a, may be reached by increasing $\overline{\theta}_P$ and/or reducing $\sigma(\theta'_P) < \sigma(\theta_P)$.

Knowing the uncertainty of stress and strength enables potential savings in mass, energy or other metrics that measure effort. Mastering uncertainty in the usage phase, Fig. 1.4b, may be reached by limiting $\overline{\theta}_{\mathrm{U}}$ and/or reducing $\sigma(\theta'_{\mathrm{U}}) < \sigma(\theta_{\mathrm{U}})$. This may be reached by adapting semi-active components or using active components.

In response to the Tay Bridge disaster, the second bridge of the railway line on the east coast of Scotland, the Forth Bridge, opened in 1890, and was significantly oversized. Thanks to new production methods—smaller fluctuations of semi finished and final products by quality control—and the avoidance of oversizing, it can be assumed that only half of the steel used would be needed today.

1.3 Components are Represented in Models

The basis of decisions made by humans or machines is information derived from a representation of a process, i.e. a model [19, 25]. Each model serves the purpose to represent the relevant part of reality and derive specific information out of the model. Hence, there are no general, purpose-free models. Since models are the prerequisite for evaluating the propagation of uncertainty in process chains, designing and optimising robust systems and selecting suitable process chains or structures from the solution space, a careful inspection of models is needed. This is even more important, as models connect data and structures, as will be seen, cf. Chap. 2.

The object to be represented by a model is a component or process of a technical system. In mechanical engineering, we distinguish between physical and software components. Each fulfils a sub-function of a system. Functions can be combined to form a module, an assembly, a sub-process chain or a single process. In the following, we use the representative term *component*.

A model represents only a part of the relevant reality. The model may even cover a part of the unreality. The data are embedded into the models. This is illustrated by the schematic Fig. 1.5: Data are linked to the models. Therefore, they are represented as a subset of the model. The boundary between relevant reality and the model is called *model horizon* [18]. The part of relevant reality not represented by the model is *ignorance*.

The model horizon is concisely described by the trained engineer and later philosopher Wittgenstein "The limits of my language mean the limits of my world" [47]. This does not mean, that every model has to be written in the mathematical language. Experience and implicit knowledge, which are often the basis of intuition, can also be regarded as a model. In fact, engineering design and production is often based on intuition. Intuition should therefore not be confused with ignorance.

The physicist and philosopher Heinrich Hertz, judged models with respect to their conciseness and simplicity. Hertz [19] demands that a model should be

(i) *consistent*, in a logical sense;
(ii) *correct*, i.e. the model implementation is done properly, and the model provides an appropriate map of the technical system;

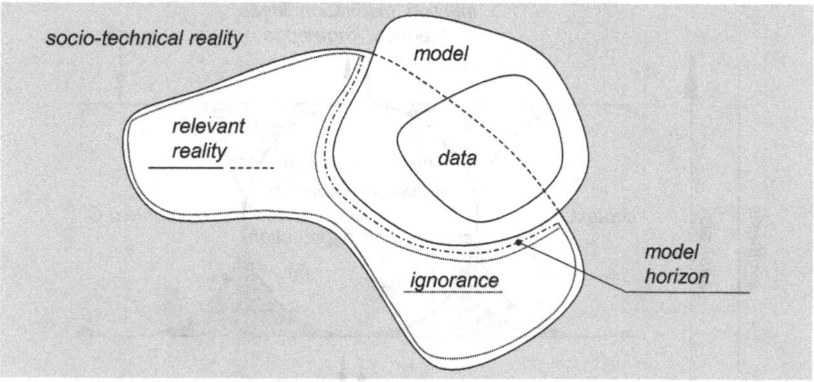

Fig. 1.5 Euler diagram to clarify relevance, ignorance, model and model horizon: socio-technical reality is separated in relevant and irrelevant reality. This separation is task-dependent. Humans or machines generate representations of this relevant reality, i.e. they model the relevant reality. It is not possible to completely model the relevant reality. The uncovered part of the relevant reality is named ignorance. The boundary between the model and the relevant reality is named model horizon. Data are formed by the values of parameters and process variables. Modelling only a part of the relevant reality is summarised by "there's more to the picture than meets the eye" [48]

(iii) *concise*, i.e. it should contain as few empty relations and assumptions as possible.

The latter requirement is known as the principle of simplicity. In the following, we shed some light on the three requested features of a model: consistency, correctness and conciseness.

Firstly, a model is considered (i) *consistent* with a theory framework if the model is free of contradictions to the knowledge represented by that framework. Simple examples illustrate the demand for consistency and the difference between consistency and correctness: a polynomial model or a neural network, both data-driven models, can certainly represent measurement data, such as a stress-strain relation of an elastomer or an adsorption isotherm of a gas and its adsorption material. Therefore, most engineers refer to the two models as correct or verified because they represent reality in a sufficient precision. The correctness may be quantified by the confidence and prediction levels of the model. In order for the models to be *consistent* with a more general theory framework, both models shall be consistent with the second law of thermodynamics [17]. If this demand for consistency is ignored, model uncertainty can be dramatic when the models are applied outside their calibration range. Vice versa, consistency reduces model uncertainty. Axiomatic i.e. deductively derived models are consistent per se, cf. Fig. 1.6. Consistency of data-driven models is improved by Bayesian inference using prior knowledge, see Fig. 1.6. A prominent example for this is Kalman filtering first used in the Apollo program 60 years ago for trajectory prediction [2, 23].

In today's language, consistency and verification are used synonymously. In the context of this book we follow this common usage in Chap. 2 and beyond, knowing

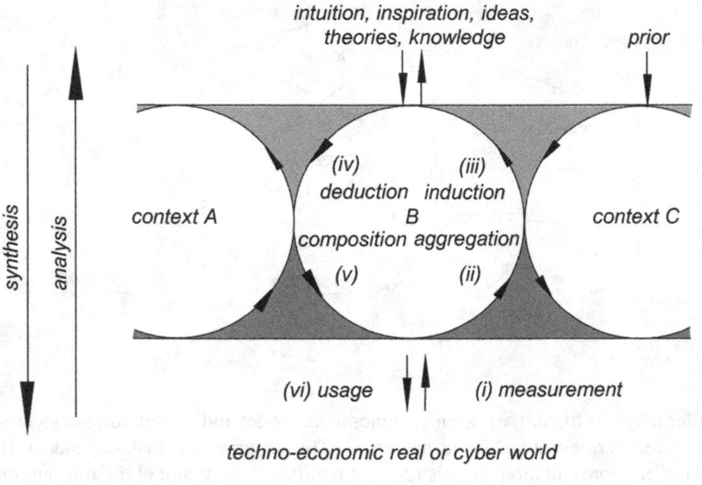

Fig. 1.6 Mapping of the real or cyber world to the world of intuition, inspiration, ideas, theories and knowledge by analysis and synthesis

that Hertz and empiricists like Hume or Popper understand the process of verification differently.

Secondly, it is important to stress, that from Hertz's point of view, widely used verification and validation processes address only the second of Hertz's demands, the (ii) *correctness*. Design engineers on the one hand and scientists on the other hand use the term *verification* in the above mentioned sense. If it is about the engineering design process of a product (physical or cyber), then verification is the examination of the specification-compliant implementation and work of models, methods and technologies.

For the empiricists it is about the '*truth*' of models. According to Karl Popper, the 'truth' of physical models is in principle not generally verifiable: the hypothesis-model, i.e., the model "all swans are white" is falsified by the proof that black swans actually exist.

Here, too, there is an ambiguity: when scientists speak of the validity of a model, they evaluate the 'truth' of the model by comparing the model prediction with reality. According to Popper, this can only be done for a limited empirical context. This narrower concept of truth has proved extremely successful in the natural and engineering sciences since Galileo Galilei. This concept of 'truth' is a very successful concept.

Therefore, natural scientists should rather stick to the concept of verification in order to have a clear language, but we will not change this. For designers the language is clearer. When design engineers talk about product *validation*, they have acceptance in mind. They ask: Does a product fulfill its purpose and is it accepted? This means the product is useful.

Thirdly, the requirement for the (iii) *conciseness* or simplicity of a model has so far been underestimated when dealing with uncertainty. Two models serving the same purpose may both be equally consistent and equally valid but may differ in the number of assumptions needed to form the model [19]. To quote the medieval philosopher William Occam [30], a father of modern epistemology: "frustra fit per plura, quod potest fieri per pauciora", i.e.

> *it is unnecessary to let something happen by several [factors], which can also happen with few [factors].*

This applies to axiomatic models, but also to data driven models [1]. The principle of simplicity is known as Occam's razor in philosophy of science [30].

Scientists tend to model more and more nuances. Engineers tend to get lost in the details of a design. That is because it's easy to be complex but it's difficult to be easy. By doing so, there is a danger of losing the essence of a technical system out of focus. This implies the function, the effort to gain this function, the availability of the system and the system's acceptability.

Why is simplicity so important in the context of mastering uncertainty? Each unnecessary assumption or relation is a source of uncertainty. This becomes relevant for forecasts or extrapolation, when a model is to be applied outside its calibration range. This is important in the context of resilience, as a strategy to master uncertainty, cf. Sect. 6.3. Resilient systems are capable to anticipate downstream processes.

Today Occam's razor serves as guiding principle for axiomatic and more and more for data driven models [24].

In summary, simplicity, i.e. conciseness and also consistency reduce uncertainty, whereas the correctness of a model does not per se reduce uncertainty. This is relevant if models are to be used for forecasting, forward control, model prediction or anticipation of downstream processes.

What are the sources of our models?

In the above, we have used the terms axiomatic and data-driven models. The former is seen here not only for first principles, but also as a synonym for the state of knowledge, ideas and theories, which are independent of a specific application or even context.

As Fig. 1.6 schematically shows, the sphere of intuition, inspiration, ideas, theories and knowledge is filled by an upstream pipe, the analysis, formed out of (i) measurements done in the real or cyber world, (ii) aggregation of the measurements, (iii) induction of general relations, by possibly using prior knowledge. This prior knowledge may be accessible by the Bayesian inference or other means deducted from the sphere of ideas, cf. Sect. 5.3. The philosophic problem of induction discussed above is in the engineering science only of minor relevance.

Today, this upstream pipe is very successfully used in data-driven modelling or black box modelling, e.g. in industrial image processing. The great success, ease of use and low threshold of expertise lead us to consider such models as a panacea. They work successfully when, for example, image data are available in abundance.

The focus here is on overcoming uncertainty. The question arises whether data driven models are sufficient to enable the design process. It is inherent in the innovation process that the technical system is only just emerging. Therefore, only a limited amount of data can be collected in the early phase. Consequently, the situation of small data instead of big data is typical for the composition and operation of innovative technical systems. The models required by the engineers are therefore initially white box models deducted or adopted from a general theory, knowledge or experience. In summary, the synthesis of physical or cyber products is always triggered by an intuition. In the context of Sustainable Systems Design, this idea is motivated from a society need. The consequent methodological system design has a deduction phase (iv) and a composition phase (v). In the deduction phase component models are derived. Those component models are composed i.e. connected forming a system fulfilling the society needs ideally sustainable.

The upper sphere shown in Fig. 1.6 is not homogeneous."I believe in intuition and inspiration. [...] Inspiration is more important than knowledge." [45]. Thus, Einstein is consistent with David Hume, who considered the interplay between ratio and inspiration: "Reason is, and ought only to be the slave of the passions." The engineer, the homo faber, needs both inspiration and knowledge in an outbalanced interplay.

Sometimes prior knowledge may be used. This may be supplied from another context or a physical model test, cf. Sect. 4.3.6. As often, there is no black or white. Today the combination of both, upstream and downstream modelling approaches is common. The results are so-called grey box models, where grey is a mixture of the white axiomatic models with the black, i.e. data driven models. It remains a task integrating implicit knowledge into grey box models.

"All models are wrong, some are helpful"

is a quote from Box [4]. Models are inherently uncertain as Fig. 1.5 indicates. With regard to model uncertainty, cf. Sect. 2.2, models may have either an unsuited structure to model the relevant part of the reality or model parameters may be uncertain. It is evident that model uncertainty, due to an unsuited model structure, cannot be mastered by mastering the uncertainty of the model parameters. This is indicated by the Euler diagram shown in Fig. 1.5. Nevertheless, many engineers still cling to their familiar models, even if a model is unsuitable. By calibrating the model parameters, originally axiomatic models are degenerated to data-driven models without being recognised as such by the user. A wrong model structure is not helpful.

For example, an unsuitable structure may be given when trying to model a diffusion problem with an elliptic partial differential equation, if a parabolic equation is suitable. An unsuitable structure may also be present when a journal bearing is modelled using the Reynolds' equation of lubrication theory, even if the product of Reynolds' number Re and relative clearance ψ is greater than one. In this case the inductance within the bearing itself is a relevant part of reality. This inductance is ignored in classical lubrication theory, which is part of most engineering education. Dimensionless model parameters, such as the product of Reynolds' number and relative clearance ψ Re, are often weights of the different terms of a model resulting from a dimensional analysis.

To sum up, it is often the engineer's experience and his or her ability to evaluate the applicability of a model.

1.4 Data and Data Sources

Data are connected to physical or cyber components, which in turn are mapped to the models. This is the one side of the system. The other side is the structure with its individual components.

The data addressed here can be the value of any model parameter or any measurement signals gained from a process. There are three main data sources:

(i) the process itself,
(ii) a representative process,
(iii) the archived data.

For sources (i) and (ii), the data may come from (a), a sensor in the real (physical sensor) or cyber world (simulation data), or (b), a soft sensor. A soft sensor combines a model with a physical sensor to derive data that are not physically accessible with limited effort [16]. Provided the process itself delivers signals by means of integrated sensors, cf. Sect. 4.2.2.

A representative process (ii) is firstly a sample test, where the sample's properties are assumed representative for all similar parts; it is secondly a physical model test, where the model may be a scaled prototype. Performing a sampling inspection is common in quality assurance. Testing a downscaled physical model is common in turbo-machinery, aerospace and marine industry [38]. The necessary scaling methods are based on the Bridgman postulate [5] and the Buckingham Pi-theorem [6]. In both cases, the data are gained offline of the relevant process. This might have the advantage that measurement uncertainty is reduced. But any offline test must take into account physical dissimilarity. This dissimilarity may be treated by scaling methods, which are a source of uncertainty [20, 43], cf. Sect. 4.3.6.

The archived data (iii) can be quality-assured, i.e. findable, accessible, interoperable, reusable (FAIR). This requires data governance and curation. The storage has to take place in such a way that the raw data are linked to their metadata in a machine-readable form [13]. Often archived data are not FAIR. Archived data are also fuzzy data remembered by an engineer or worker.

Data quality has two sides [13]: firstly, formal data quality achieved by following the FAIR-principles, and secondly, content quality. Since uncertainty is associated with trust in data [14, 15], formal data quality should not be ignored: the higher the formal data quality, the more the data is trusted. A detailed view on data quality is given in Sect. 2.1.

Two or more data sources can be used simultaneously to derive information. This data fusion will lead us to a concept called data-induced conflicts, which will be discussed in Sect. 4.2. It is a concept that allows to assess confidence in data sources but also model uncertainty.

1.5 Component Structures

Structures consist of components, physical components and/or cyber components, i.e. algorithms in the form of software. Having treated component models and data, we come to the system level represented by the term *structure*.

In the classical engineering design [29], the system's function is usually the starting point from where a system's functional structure with related sub-functions is derived, cf. Sect. 3.3. The system's function structure is independent of a product, process or system realisation. After the decision on the integration or separation of the functions into individual physical or cyber components, the functional structure of the system is mapped to the components. These form the real system.

The decision about the integration or separation of the functions is guided by the mastering of the internal and external complexity. This decision process is the foundation of modular design, which allows an economic scaling. An illustrative example of modular design obtained by intelligent function integration and function separation is shown in Fig. 3.19.

With respect to the system's function and quality, a quantitative evaluation of the system's uncertainty is only possible at the system level; we evaluate the system's uncertainty with respect to effort, availability and acceptability, frequently being only possible at the component level. *Structural uncertainty*, cf. Sect. 2.3, therefore results from the fact that a multitude of possible functional structures can be found for a system's function that is still subject to uncertainty; and in turn a multitude of component structures can be realised for each functional structure. This results in a combinatorial explosion of the solution space [39], which is only partially comprehensible and assessable for humans. The unnoticed part of the solution space remains in the area of ignorance due to this structural uncertainty [33].

For example, the difference between data, model and structure is exemplified by the design task of a hydrostatic transmission sketched in Fig. 1.7. Figure 1.7a shows a double-acting piston, whose force-displacement curve has to be controlled by a structure or system formed out of the sketched components, i.e. the hydraulic valves. Figure 1.7b shows the load history, which may be uncertain. The system's function is described by such a load history.

Each valve is a component being described by a functional relation of input u, output z and model parameters m: $f(u, z, m, \dots) + \delta f = 0$. Here, the model f of the valve arises from a differential-algebraic system of equations, and δf is the residuum between model and reality. The operational inputs u determine the valve position, density, pressure difference and particle concentration. The parameters m include the maximal valve opening and diameter. The output is given by the wear history. Thus, the time-varying flow-characteristic and at the same time, the evolution or wear due to particle erosion are described, cf. Sect. 3.3. Hence, the wear for an arbitrary load history and structure is given [44].

The system is composed of different admissible components schematically collected in the design space as sketched in Fig. 1.7. The design space with admissible structures all fulfilling the demanded function is so large that it cannot be explored manually. The different design solutions S all differ in the system's degeneration due to particle wear.

If only one solution out of the design space is selected and the countless other solutions are ignored, we call this form of uncertainty firstly *structural uncertainty* and secondly *ignorance*. Only if an optimal structure S_{opt} is selected, here Fig. 1.8b, with regard to minimal particle erosion,

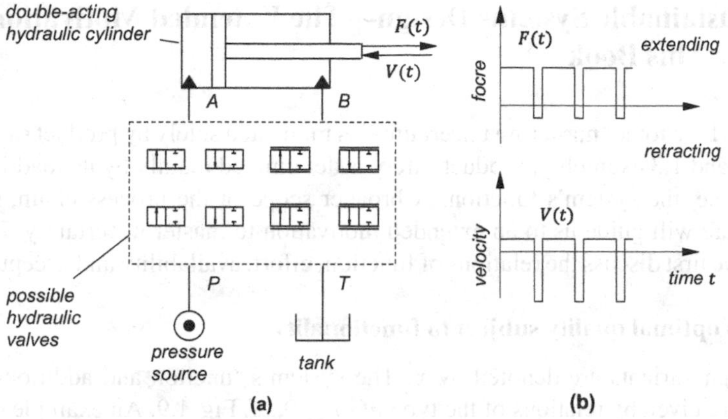

Fig. 1.7 **a** Design task of a hydrostatic transmission with minimal particle wear. The pressure source of a pump (high pressure) and the tank (atmospheric pressure) is to be connected by a so far unknown structure of hydraulic valves with a double-acting hydraulic cylinder; when the pump is connected to the left volume and the tank to the right volume, the cylinder extends; **b** the function is described by a load history; the control valves shall be selected from a field of possible hydraulic valves; the right half of the valves allow the pressure drop to be adjusted. The representation of the possible hydraulic valves implies that any structural solutions S are possible [44]

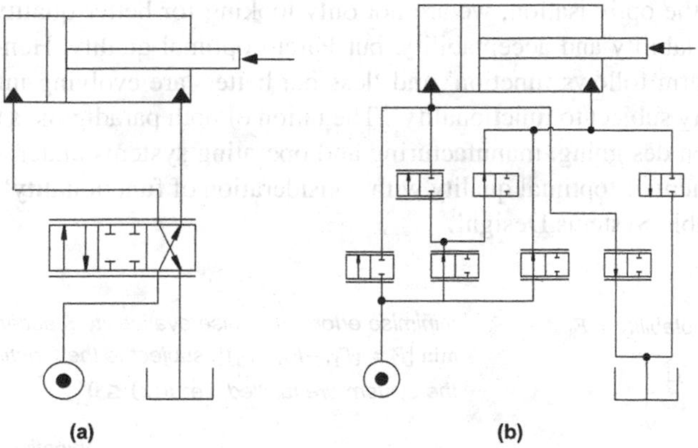

Fig. 1.8 Design for **a** ignored structural uncertainty, **b** minimal wear; the availability with respect to wear due to particle erosion is increased by a factor of 16 [44]

we speak of mastered structural uncertainty. Figure 1.8a shows the usual design using a standard 4/3 directional control valve with the optimal structure S_{opt} showing minimal wear as in Fig. 1.8b.

Data uncertainty $\theta = \overline{\theta} + \delta\theta$ and *model uncertainty* $f(u, z, m, \ldots) + \delta f = 0$ have to be encountered in the structural uncertainty. They propagate into the structure S. There are some examples in this book how this is achieved by means of robust optimisation, see for example Sect. 6.1.1.

1.6 Sustainable Systems Design—The Extended Motivation for This Book

In Sect. 1.1 the topic 'mastering uncertainty' is motivated solely by product safety. As Figs. 1.2 and 1.3 exemplify, product safety is determined mainly by its load-bearing capacity, i.e. the system's function. A broader scope of the process chain, system or structure will guide us to an extended motivation to master uncertainty. For this reason, we first discuss the relations of function, effort, availability and acceptability.

Towards optimal quality subject to functionality

The design variants are denoted by x. The system's function and additional constraints are given by relations of the type $g(x) \leq 0$, cf. Fig. 1.9. An example of such a constraint is e.g. seen in Fig. 1.7b. At this point, the discussion about structural uncertainty shows that the design variants x differ for each structure S. Hence, the paradigm 'form follows function' created by the American architect Louis Sullivan, at the beginning of the 20th century, is not an objective, it is a constraint. The missing objective is 'less but better' created by Rams and Klatt [35]. The renowned German designer Dieter Rams having worked many years for the company of Braun, demanded in the mid of the 20th century: 'Weniger, aber besser'! This is the missing objective. In the optimisation, we are not only looking for better quality measured in effort, availability and acceptability, but Pareto optimal quality. Hence, the two paradigms 'form follows function' and 'less but better' are evolving into 'towards optimal quality subject to functionality'. The union of both paradigms is the guiding principle when designing, manufacturing and operating systems under uncertainty. The achievement of 'optimal quality with consideration of functionality' is what we call 'Sustainable Systems Design'.

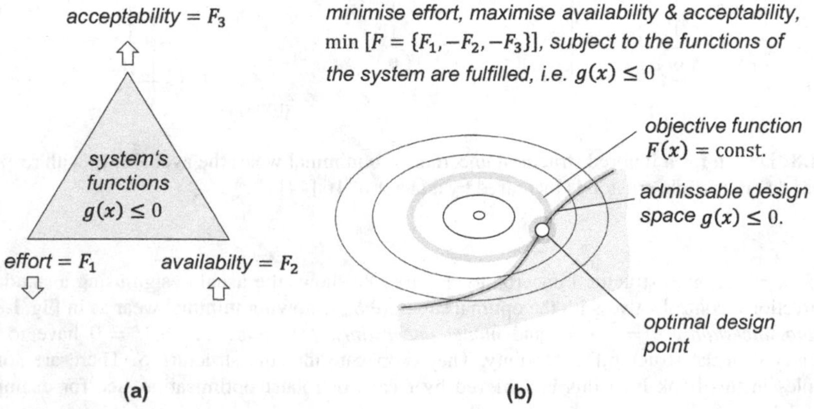

Fig. 1.9 Equivalence of **a** the Sustainable Systems Design and **b** the constrained optimisation problem

What is our understanding of function and quality?

The objectives are (i) minimise effort F_1, (ii) maximise availability F_2, and (iii) maximise acceptability F_3. The three objectives are often conflicting. Hence, the multi-criterial decision problem min $[F = \{F_1, -F_2, -F_3\}]$ leads to a Pareto set of optimal solutions [12]. The selected optimal solution always depends on the ranking of the three objectives (i) effort, (ii) availability, (iii) acceptability.

Linguistically, the system's function is described by verbs, such as to carry, store or transport. The function is mostly further specified by a load spectrum or load history. The objective function is determined by the quality of how the function is fulfilled. Here, quality symbolises the adverb to a verb, namely a function, like for example efficient transport. The adverb, i.e. the quality, characterises the three aspects of effort, availability and acceptability.

(i) *Effort* is measured, for example, by the total cost of ownership. Sometimes only the material or energy consumption are measured. In the usage phase of lightweight structures, the weight is the determining factor for the effort.
(ii) *Availability* can be measured, for example, by the sum of the mean time between two failures and the repair time relative to the total time. Alternatively, the anticipated remaining service life can also be specified. For this purpose, an assumption regarding future usage and an ageing model are necessary. A general ageing model is presented in Chap. 3.
(iii) From the three measures of the objective function, the *acceptability* is the most difficult measure. Acceptability has two sides, a formal and an informal side: A formal aspect of acceptability, presented in Sect. 5.1 lies in the conformity with regulations, such as the Product Safety Act [7]. Formally, acceptability can also be achieved through a regulation. For example, an ordinance can be the function of an electronic stability control system (ESP) mandatory for vehicles. For formal acceptability, the politically consented society needs are cast into regulations. Either products have to meet the regulations or the regulations demand defined technologies.
The counterpart to the formal side of acceptability is the *informal acceptability* gained through *positive user experiences*. The user may be a consumer in the consumer goods market, but also a company in a business-to-business market. This *user experience* has many facets and it would go far beyond the scope of this book to fully immerse into this field. Schmitt coined the term perceived quality in this context [37]. Instead, we focus on the facet product quality being important for informal acceptability.

It is obvious that the higher the experienced quality of a product and the lower the effort measured against the costs, the higher the acceptability. The quality is measured on the one hand by the expected *functional performance* given by the deviation $\delta g = g_s - g$ from the expectation, cf. the 3rd case study in Chap. 3, and on the other hand by the expected effort F_1 and by the expected availability F_2. As Fig. 1.10 shows, g_s is the specified function and g is the realised function.

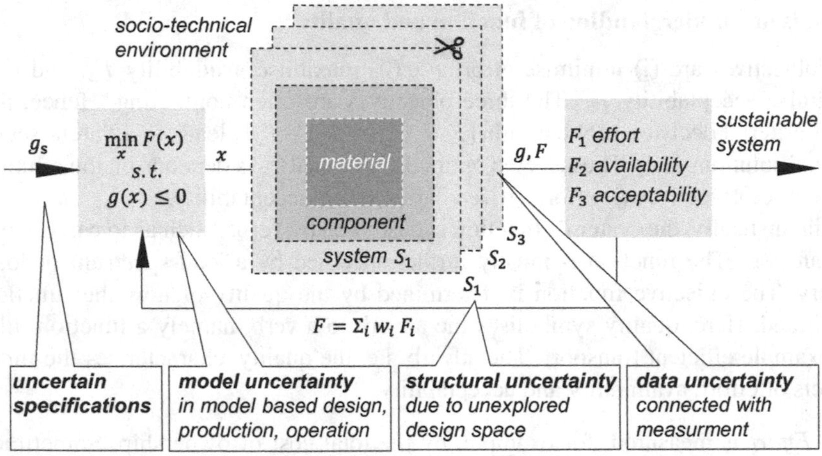

Fig. 1.10 Sustainable Systems Design presented as a closed loop, indicating the localisation of model uncertainty, structural uncertainty and data uncertainty, which is dealt with in Chap. 2. The process of system specification between different stakeholders is a source of uncertainty, which is discussed in Sect. 5.1.1

Customer expectations must match the quality promise. This is either explicitly given by the manufacturer or it must be consistent with the usual market quality. If necessary, the quality is also defined in regulations, see Sect. 5.1. Here, too, it can be seen that the various aspects of a product depend on each other: formal and informal acceptability overlap in parts.

The schematic representation of the constrained optimisation problem as a closed control loop helps identifying the different uncertainty sources, Fig. 1.10: model uncertainty, structural uncertainty and data uncertainty, which is dealt with in Chap. 2. The dynamic process of system specification between different stakeholders is a source of uncertainty, which is discussed in Sect. 5.1.1.

Sustainable Systems Design is model-based: the system function g and system quality F is evaluated on the basis of models. The recognition, evaluation and mastering of model uncertainty, cf. Sect. 2.2, is thus one core of this book. By integration of functions or separation of functions, by combination of materials and components often more than seven competing systems fulfil the same specific function g_s. The number seven is known to be the limit on human capacity for processing information [26]. Roughly speaking, all other possible variants remain in the field of ignorance for people in system design. This structural uncertainty can only be controlled by algorithms discussed in this book. For this, rules of the game and system boundaries have to be set. This must be recovered by the stakeholders. In order to quantify the system quality in the evaluation step, metrics for effort, availability and acceptability are necessary. The evaluation of the function and quality requires models, see Chap. 3. Secondly, weighting factors w_i are necessary. In the evaluation step, Pareto surfaces can be presented.

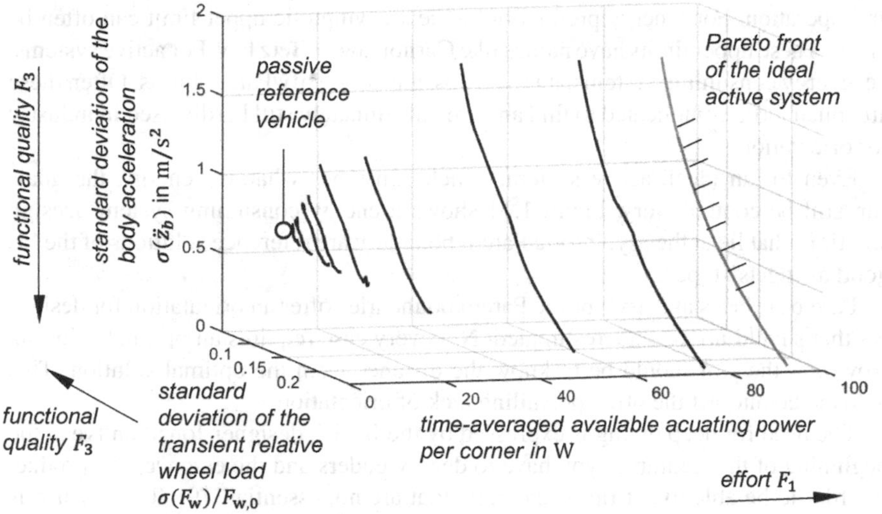

Fig. 1.11 Improvement of system performance by an active component, here the Active Air Spring introduced in Sect. 3.6.2. A compact car is driving over a country road at a speed of 70 km/h. The standard deviation of the body acceleration is plotted versus the time-averaged actuating power in *W* and the standard deviation of the relative wheel load. The white circle is the reference for the vehicle with a passive suspension system [36]

As good as it gets—orientation helps mastering uncertainty

The demand to improve quality beyond an existing Pareto surface requires an extended playing field or altered rules of the game. This is achieved by new technologies. A Pareto line for a chassis design using an Active Air Spring as component is shown in Fig. 1.11. The effort F_1 in the example of an active chassis may be defined by the power consumption. The acceptability in the example F_3 is given by the functional quality of the suspension system. The sub-functions are isolating the body and reducing wheel load fluctuation, i.e. to foster driving safety. As seen in the figure, the position of the Pareto line is determined by the available power of the active component. However, there is often a technology-independent, i.e. asymptotic Pareto boundary. The question 'what can be achieved in the optimal case, if there is no limitation, for example to the power?' can often be answered.

In engineering sciences, this asymptotic Pareto line or surface is determined by physical laws. The most prominent Pareto surface is the Carnot efficiency of a thermal power plant. Due to the second law of thermodynamics only the fraction $1 - T_1/T_2$ of the input heat flux $\dot{Q}$ may be transferred into mechanical power P_S. The knowledge of this asymptote i.e. Pareto surface motivated engineers to increase the combustion temperature T_2 more and more (T_1 is the ambient or cooling temperature). This triggered the development of high temperature material. For wind power [3] and water power [31, 34] we have similar upper limits independent of the system design and operation. For 'energy production' a clear asymptotic upper limit can often be given. These upper limits have names like Carnot law or Betz law. For active systems, i.e. energy consuming systems, it is also possible to specify Pareto limits. Often these are much more complicated to find and are unfortunately still hardly used in industry for orientation.

Even for an ideal, active system, which consumes whatever energy, the goals can still be contradictory. Figure 1.11 shows a energy consuming system. Design solutions that lie at the asymptotic Pareto boundary are reference solutions of the 'as good as it gets'-type.

Pareto surfaces and asymptotic Pareto boundaries offer an orientation for designers that should not be underestimated. Not every case requires an optimal solution. However, the aim should be to know the distance from the optimal solution. This helps to counteract the often prevailing lack of orientation.

The need for deep diving is expressed by the British designer Jonathan Ive at the beginning of this century: "you have to deeply understand the essence of a product in order to be able to get rid of the parts that are not essential" [9]. The essence is the system's function g and the system quality F seen from the user's perspective. The way to sustainability is cleared by optimal quality subject to functionality.

1.7 Outlook on the Following Book Structure

Mastering uncertainty in the phases design, production and usage does not only refer to the system's function but also to (i) effort, (ii) availability, and (iii) acceptability, as depicted in Fig. 1.9. Hence, product safety stands next to other motivations all covered in this book from a specific point of view:

(i) Ensuring product safety,
(ii) realising lightweight structure and Sustainable Systems Design,
(iii) controlling production quality.

The schematic Fig. 1.11 shows that mastering uncertainty may lead to resource savings. This is immanently important for lightweight structures where the weight is to be minimised for a given load-bearing function. The example sketched in Fig. 1.8 is an example of a Sustainable Systems Design under uncertainty, where the wear was minimised. In production, the control of uncertainty can save costs by making processes more flexible and adaptive. In Sustainable Systems Design, the control of uncertainty leads to robust or even resilient systems.

The three floors of mastering uncertainty in mechanical engineering

We organise this book with a picture of a truss structure, shown in Fig. 1.12. The truss structure has three floors. These are, firstly, the fundamental floor built from terms and definitions, secondly, approaches to uncertainty quantification on the one hand

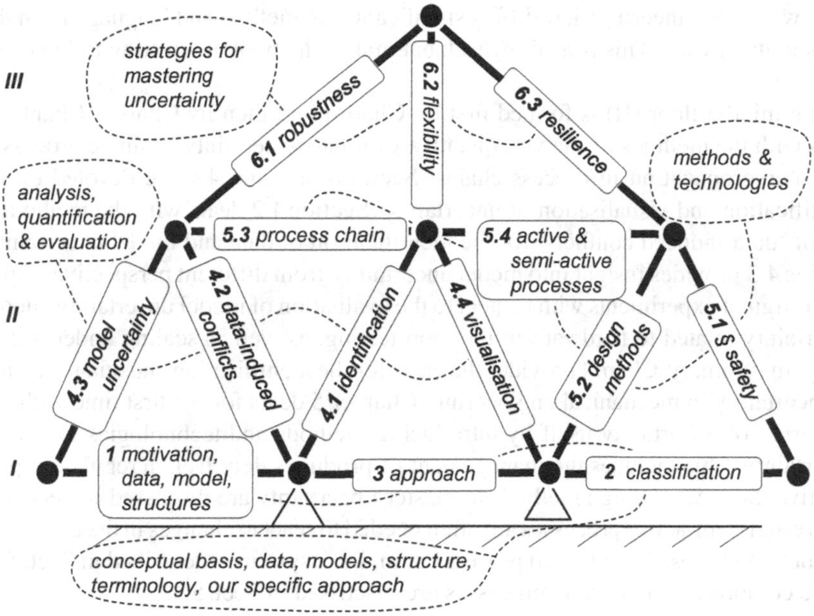

Fig. 1.12 Framework of mastering uncertainty presented in this book mapped on a truss structure

and methods and technologies on the other hand, and thirdly, strategies to master uncertainty.

The fundamental floor (I) is formed by the motivation as well as the reflection on data and models given in this chapter, the lower left bar in Fig. 1.12. At the beginning of our research more than ten years ago, it became clear that for mastering uncertainty a definition of uncertainty classes is important. Only when things are defined by name do they become tangible. The motivation and discussions in this chapter and Chap. 2 describe the classification of uncertainty into stochastic uncertainty, incertitude and ignorance by the first classifier and into data, model and structural uncertainty by the second classifier. This results in the matrix of uncertainty classes, shown in Fig. 2.2. With the first three bars and chapters the foundation is given for a solid middle floor. Chapter 3 provides our specific approach on mastering uncertainty. Within Chap. 3, we introduce three technical systems created, tested, and verified in the context of mastering uncertainty. The first system is a load-bearing structure representing a generic light weight structure called Modular Active Spring-Damper System; the second system is the Active Air Spring – a technology which is ideal to prevent kinetosis when driving autonomously; the third system is the 3D Servo Press allowing flexible production and a closed-loop control of the product properties. These three systems, all developed manufactured and validated from scratch at TU Darmstadt during the previous decade, form the heart of the book. They will be highlighted from different perspectives. The central bar Chap. 3 is connected via the two supports to the state of the art mechanical engineering, applied mathematics and law. The engineering view differs significantly in method and language from the mathematical view. This is no drawback but makes the book interesting to read—so we hope.

The middle floor (II) is formed first by Chap. 4 and then by Chap. 5. Chapter 4 deals with the methods to analyse, quantify, evaluate uncertainty in single processes and their propagation in process chains. Sections 4.1 and 4.4 are devoted to the identification and visualisation of uncertainty. Section 4.2 deals with the methodology of 'data-induced conflicts' for the identification of data and model uncertainty. Section 4.3 provides insight into model uncertainty from different perspectives: optimal design of experiments with respect to the evaluation of model uncertainty, model uncertainty related to hardware-in-the-loop testing, as well as scaling under uncertainty. In summary, Chap. 4 provides the basis for the identification and quantification of uncertainty in mechanical engineering. Chapter 5 deals for the first time with the mastering of uncertainty itself by introducing methods and technologies to master uncertainty. This includes the management of product safety from a regulatory perspective, Sect. 5.1. Design methods to master uncertainty are discussed in Sect. 5.2. Active and semi-active processes are often needed to react to changes in the usage and production phases. A controlled process chain, i.e. a system is described in Sect. 5.3, active components and single processes are discussed in Sect. 5.4.

The top floor (III) is devoted to strategies of uncertainty mastering. This floor builds on floors (I) and (II). We discuss three strategies: robustness Sect. 6.1, flexibility Sect. 6.2, and resilience Sect. 6.3. Progress in discrete and nonlinear robust optimisation methods is presented together with robust production processes and development methods for a robust system.

References

1. Bailer-Jones CAL (2017) Practical Bayesian inference: a primer for physical scientists. Cambridge University Press, Cambridge. https://doi.org/10.1017/9781108123891
2. Becker A (2018) Introduction to Kalman filter. https://www.kalmanfilter.net
3. Betz A (1920) The maximum of the theoretically possible exploitation of wind by means of a wind motor. Wind Engineering 37(4):441–446. https://doi.org/10.1260/0309-524X.37.4.441
4. Box G (1979) Robustness in the strategy of scientific model building. In: Launer RL, Wilkinson GN (eds) Robustness in statistics. Academic Press, pp 201–236. https://doi.org/10.1016/B978-0-12-438150-6.50018-2
5. Bridgman PW (1922) Dimensional analysis. Yale University Press
6. Buckingham E (1915) The principle of similitude. Nature 96(2406):396-397. https://doi.org/10.1038/096396d0
7. Bundesrepublik Deutschland (2011) Gesetz über die Bereitstellung von Produkten auf dem Markt: ProdSG, 08 Nov 2011
8. Bundschuh M (2007) Modellgestützte strategische Planung von Produktionssystemen in der Automobilindustrie: Ein flexibler Planungsansatz für die Fahrzeughauptmodule Motor, Fahrwerk und Antriebsstrang. Dissertation, University of Augsburg. http://www.gbv.de/dms/zbw/562092978.pdf
9. Castelluccio M (2019) Jony Ive exits Apple. Strategic Finance, pp 69–70

10. Doll N (2015) Verschlusssache Flensburg. Wie gut sind deutsche Autos? Die Welt, 30 July 2015. https://www.welt.de/wirtschaft/article144594870/Verschlusssache-Flensburg-Wie-gut-sind-deutsche-Autos.html
11. Europäische Union (2002) Richtlinie 2001/95/EG über die allgemeine Produktsicherheit, 15 Jan 2002
12. Feldman A, Serrano R (2010) Welfare economics and social choice theory. Springer, New York. Reprint
13. German Council for Scientific Information Infrastructures (2020) The data quality challenge: recommendations for sustainable research in the digital turn. http://www.rfii.de/?wpdmdl=4203. Accessed 21 Dec 2020
14. Hanselka H, Engelhardt R, Koenen JF, Enss GC, Sichau A, Platz R, Kloberdanz H, Birkhofer H (2010) A model to categorise uncertainty in load-carrying systems. In: Heisig P (ed) Proceedings of the 1st international conference on modelling and management engineering processes, pp 53–64
15. Hanselka H, Platz R: Ansätze und Maßnahmen zur Beherrschung von Unsicherheit in lasttragenden Systemen des Maschinenbaus. Konstruktion 2010(11–12), 55–62 (2010)
16. Hartig J, Schänzle C, Pelz PF (2019) Concept validation of a soft sensor network for wear detection in positive displacement pumps. In: 4th international rotating equipment conference
17. Haupt P (2010) Continuum Mechanics and Theory of Materials, 2nd edn. Springer, Berlin
18. Hedrich P, Cloos FJ, Würtenberger J, Pelz PF (2015) Comparison of a new passive and active technology for vibration reduction of a vehicle under uncertain load. In: Pelz PF, Groche P (eds) Uncertainty in mechanical engineering II, vol 807. Applied mechanics and materials. Trans Tech Publications, pp 57–66. https://doi.org/10.4028/www.scientific.net/AMM.807.57
19. Hertz H (1894) Die Prinzipien der Mechanik in neuem Zusammenhange dargestellt. Gesammelte Werke, Band 3. J.A. Barth, Leipzig
20. Holl M, Pelz PF (2017) Towards robust sustainable system design: an engineering inspired approach. In: Barthorpe R, Platz R, Lopez I, Moaveni B, Papadimitriou C (eds) Model validation and uncertainty quantification, vol 3. Conference proceedings of the society for experimental mechanics series, vol 136. Springer, Cham, pp 85–101. https://doi.org/10.1007/978-3-319-54858-6_10
21. Holl M, Rausch L, Pelz PF (2017) New methods for new systems—how to find the techno-economically optimal hydrogen conversion system. International Journal of Hydrogen Energy 42(36):22641–22654. https://doi.org/10.1016/j.ijhydene.2017.07.061
22. Jennings A (2014) Structures: From Theory to Practice, 1st edn. CRC Press, Boca Raton, FL
23. Kalman RE (1960) A new approach to linear filtering and prediction problems. Journal of Basic Engineering 82(1):35–45. https://doi.org/10.1115/1.3662552
24. MacKay DJC (2011) Information theory, inference, and learning algorithms. Cambridge University Press, Cambridge. Reprint
25. Magka D, Krötzsch M, Horrocks I (2014) A rule-based ontological framework for the classification of molecules. Journal of biomedical semantics 5:17. https://doi.org/10.1186/2041-1480-5-17
26. Miller GA (1956) The magical number seven, plus or minus two: Some limits on our capacity for processing information. Psychological Review 63(2):81–97. https://doi.org/10.1037/h0043158
27. Münsterberg M (1916) A Harvest of German Verse. D. Appleton and Co., New York
28. NHTSA National Highway Traffic Safety Administration (2020) National recall report. https://www.safercar.gov/Vehicle+Owners/vehicle-recalls-historic-recap. Accessed 21 Dec 2020
29. Pahl G, Beitz W, Feldhusen J, Grote KH (2007) Engineering design: a systematic approach, third edn. Springer, London. https://doi.org/10.1007/978-1-84628-319-2
30. Paqué R (1970) Das Pariser Nominalistenstatut. Walter de Gruyter, Berlin. https://doi.org/10.1515/9783110816785
31. Pelz PF (2011) Upper limit for hydropower in an open-channel flow. Journal of Hydraulic Engineering 137(11):1536–1542. https://doi.org/10.1061/(ASCE)HY.1943-7900.0000393
32. Pelz PF (2014) 250 Jahre Energienutzung: Algorithmen übernehmen Synthese, Planung und Betrieb von Energiesystemen: Fachvortrag anlässlich der Ehrenpromotion von Hans-Ulrich Banzhaf

33. Pelz FP, Pfetsch ME (2015) Der Unsicherheit von Anfang an auf der Spur. PT-Magazin für Wirtschaft und Gesellschaft, 30 July 2015
34. Pelz PF, Metzler M, Schmitz C, Müller TM (2020) Upper limit for tidal power with lateral bypass. Journal of Fluid Mechanics 889:A32. https://doi.org/10.1017/jfm.2020.99
35. Rams D, Klatt J (1995) Weniger, aber besser: Skizze für Phonokombination TP 1. Jo Klatt Design und Design Verlag, Hamburg
36. Rexer M, Brötz N, Pelz PF (2020) Much does not help much: 3D pareto front of safety, comfort and energy consumption for an active pneumatic suspension strut. In: 12th international fluid power conference, pp 37–42. https://doi.org/10.25368/2020.91
37. Schmitt R (2014) Perceived quality: Subjektive Kundenwahrnehmungen in der Produktentwicklung nutzen, 1st edn. Symposion, Düsseldorf
38. Spurk JH (1992) Dimensionsanalyse in der Strömungslehre. Springer, Berlin. https://doi.org/10.1007/978-3-662-01581-0
39. Suhl L, Mellouli T (2013) Optimierungssysteme: Modelle, Verfahren, Software, Anwendungen, third edn. Springer-Lehrbuch. Springer Gabler, Berlin. https://doi.org/10.1007/978-3-642-38937-5
40. Vartabedian R (2019) How a 50-year-old design came back to haunt Boeing with its troubled 737 Max jet. Los Angeles Times, 15 Mar 2019
41. Verband der Automobilindustrie (VDA) (2020) Neuzulassungen. https://www.vda.de/de/services/zahlen-und-daten/jahreszahlen/neuzulassungen.html. Accessed 21 Dec 2020
42. Verein Deutscher Ingenieure (1993) VDI 2221:1993–05 Methodik zum Entwickeln und Konstruieren technischer Systeme und Produkte [Systematic approach to the development and design of technical systems and products]. Beuth, Berlin
43. Vergé A, Lotz J, Kloberdanz H, Pelz PF (2015) Uncertainty scaling – motivation, method and example application to aload carrying structure. In: Pelz PF, Groche P (eds) Uncertainty in mechanical engineering II, vol 807. Applied mechanics and materials. Trans Tech Publications, pp 99–108. https://doi.org/10.4028/www.scientific.net/AMM.807.99
44. Vergé A, Pöttgen P, Altherr LC, Ederer T, Pelz PF (2016) Lebensdauer als Optimierungsziel – Algorithmische Struktursynthese am Beispiel eines hydrostatischen Getriebes. O+P – Ölhydraulik und Pneumatik 60(1–2):114–121
45. Viereck G (1929) What life means to Einstein. An interview with Albert Einstein. The Saturday Evening Post, pp 17, 110, 113, 114, 117, 26 Oct 1929
46. Weingartner M (2016) Millionen Rückrufe von Autos in Deutschland. Academic Press, 12 Feb 2016
47. Wittgenstein L (1922) Tractatus Logico-Philosophicus. Trubner & Co, London, Kegan Paul, Trench
48. Young N (1979) Crazy horse: rust never sleeps. Reprise Records

Chapter 2
Types of Uncertainty

Peter F. Pelz, Marc E. Pfetsch, Sebastian Kersting, Michael Kohler, Alexander Matei, Tobias Melz, Roland Platz, Maximilian Schaeffner, and Stefan Ulbrich

Abstract The goal of this chapter is to define different types of uncertainty in technical systems and to provide a unified terminology for this book. Indeed, uncertainty comes in different disguises. The first distinction is made with respect to the knowledge on the source of uncertainty: stochastic uncertainty, incertitude or ignorance. Then three main occurrences of uncertainty are discussed: data, model and structural uncertainty.

In this book we focus on physical and cyber-physical systems that are designed, manufactured and used. Hence, our context is that of engineering design, production and usage, in combination with applied mathematics providing methods and strategies as well as law providing a social and judicial framework. Uncertainty occurs in every step of system design, production and usage and needs to be anticipated in the design phase. Supporting the analysis, this chapter is concerned with different types of uncertainty and their quantification.

Indeed, before mastering uncertainty, uncertainty has to be identified. In order to do so, it is helpful to define individual uncertainty types. We classify uncertainty using two independent classifiers identifying its appearance and effect. The first classifier captures the effect of the uncertainty on the system at its core. It distinguishes between *stochastic uncertainty*, *incertitude* and *ignorance*. The resulting decision diagram is shown in Fig. 2.1. The second classifier distinguishes data, components and structures. Together they lead to the 3×3 matrix shown in Fig. 2.2.

P. F. Pelz (✉) · T. Melz · R. Platz · M. Schaeffner
Department of Mechanical Engineering, TU Darmstadt, Darmstadt, Germany
e-mail: peter.pelz@fst.tu-darmstadt.de

M. E. Pfetsch · S. Kersting · M. Kohler · A. Matei · S. Ulbrich
Department of Mathematics, TU Darmstadt, Darmstadt, Germany

T. Melz
Fraunhofer Institute for Structural Durability and System Reliability LBF, Darmstadt, Germany

P. F. Pelz et al. (eds.), *Mastering Uncertainty in Mechanical Engineering*,
Springer Tracts in Mechanical Engineering,
https://doi.org/10.1007/978-3-030-78354-9_2

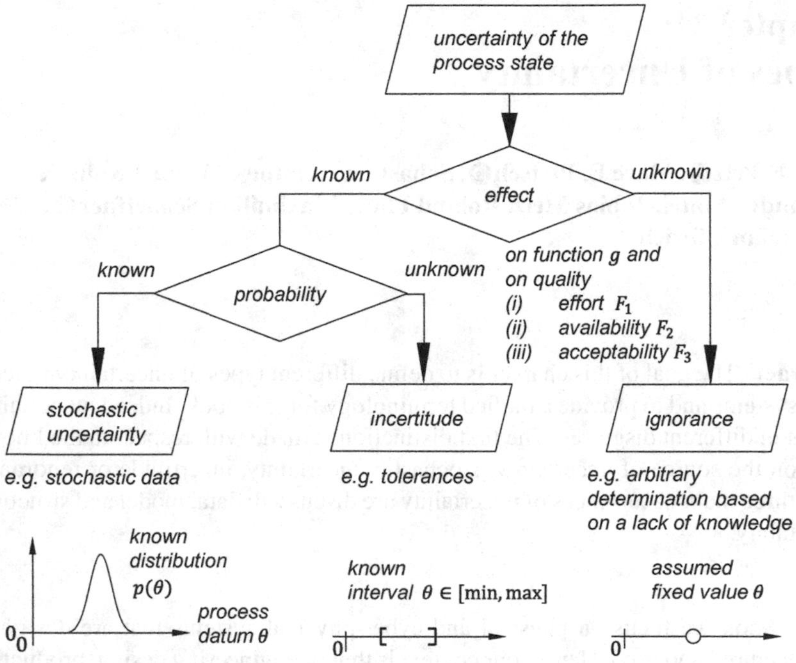

Fig. 2.1 Classifier using effect and probability to separate into stochastic uncertainty, incertitude and ignorance [8, 18]

This matrix can be applied in all phases of the product life cycle, i.e. design, production and usage. This is even more important, since uncertainty quantification has become a thriving research field in engineering, computer science and mathematics over the last twenty years [20, 39].

Classification by effect and probability

Fig. 2.1 shows the first classifier as a decision diagram. The first decision is whether the *effect* of an uncertain process property on the process or the structure's function is known or unknown. This includes the decision of whether the effect on the system function and quality is known or unknown; recall that the system function is usually represented as a constraint $g(x) \leq 0$ and quality is measured using effort F_1, availability F_2 and acceptability F_3, see Sect. 1.6.

If the effect is unknown, we speak of *ignorance*. If the effect is known, then we speak of *probability*. The second question is whether the probability of the effect is known or unknown. If the probability of the effect is only partially quantified, we speak of *incertitude*. If the probability of the effect is sufficiently quantified, as shown schematically in Fig. 1.4, then we speak of *stochastic uncertainty* [8, 18].

The category *stochastic uncertainty* implies that a probability density function of the process state as sketched in Fig. 1.4 is known. In this case, it is possible to describe, quantify and evaluate uncertainty. The category *incertitude* is very common in mechanical engineering. It is used for many processes in mass production and, for example, in the *Austauschbau* described by Franz Relaux in the year 1899 [47], i.e. a part A fits to a part B even though different people manufacture the two parts on different machines. The incertitude of the *Austauschbau* is mastered by measurement data from systematically drawn samples or by experience manifested in tolerance classes. The two categories stochastic uncertainty and incertitude lead to a non-deterministic system design whereas *ignorance*, i.e. disregarded uncertainty, implies a deterministic system design.

Classification by data, component and structure

The first classifier distinguishes between the effect and quantification of uncertainty. The second classifier is motivated by system design [42]. This classifier can best be understood by keeping in mind a physical system, such as one of the three demonstrators presented in Sect. 3.6, i.e. the lightweight structure MAFDS, the Active Air Spring and the 3D Servo Press, or the hydrostatic transmission as depicted in Fig. 1.7. A process chain, a system and a structure consist of components or individual processes that fulfil the sub-functions of a system. In the following, when mentioning a model and model uncertainty, we may refer to the model as the individual process or component of the system. However, we may also refer to the composed system satisfying one or more specific system functions g_s. As pointed out in Chap. 1, different systems may satisfy the same function. Some possible systems may not be evaluated. We call this nescience *structural uncertainty*.

Applying the second classifier yields (a) data, (b) model and (c) structural uncertainty. These classes form the columns of Fig. 2.2. The first classifier shown in Fig. 2.1 leads to the rows (i) stochastic uncertainty, (ii) incertitude and (iii) ignorance of the 3×3 matrix shown.

From top to bottom, the confidence in data, models and structures is decreasing. This distinction into different types of uncertainty in data, i.e. *data uncertainty*, is then as follows. Data $\theta = \bar{\theta} + \delta\theta$ is subject to stochastic uncertainty if it can be modelled as realisations of a random variable with a distribution $P(\theta)$ or density $p(\theta)$ and expected value/mean $\bar{\theta}$. Incertitude appears if the data is only known to lie within a given fuzzy set or interval. If uncertainty of the data is not considered and thus ignored in the problem analysis, we speak of ignorance. For a more detailed discussion, see Sect. 2.1.

For *model uncertainty* the classification is as follows: a validated and verified model is subject to quantified stochastic uncertainty. There is incertitude, as long as the model is only suspected, i.e. assumed without experimental evidence. If the model is unknown, this is called ignorance, consistent with the scheme shown in Fig. 1.5. The presented classification is supplemented by the required model characteristics given in Sect. 1.3; a model has the following three qualities here: consistency, correctness and conciseness, see Sect. 1.3.

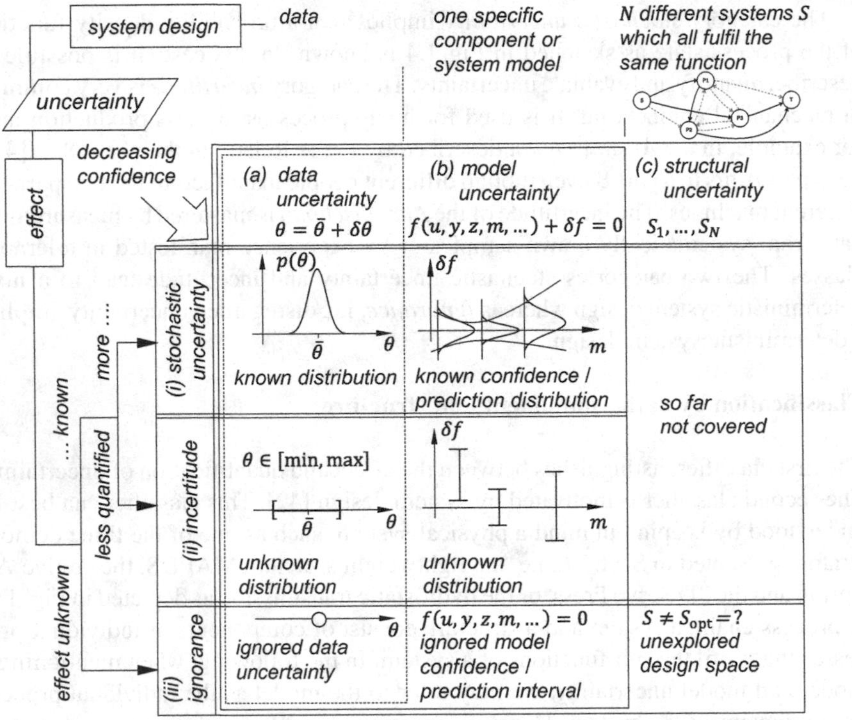

Fig. 2.2 Classification of uncertainty into (i) stochastic uncertainty, (ii) incertitude, (iii) ignorance; and into **a** data, **b** model, **c** structural uncertainty

To represent specific components or systems, a model in *implicit* form

$$f(u, y, z, m, \ldots) = 0, \tag{2.1}$$

is used, where f is the model function, u are inputs, like control or boundary values, y are internal variables, such as states, z is the model's output, i.e. the quantities of interest, and m are the model parameters which need to be calibrated. Sometimes u is split into binary design or other decision variables. In many cases, the above equation can be solved for given u and m such that y and z are uniquely determined. The model is then reduced to an *explicit* form. This often occurs if the model represents an input-output relation.

As mentioned in Chap. 1, in engineering and natural sciences, exact models do not exist. Consequently, to represent reality, one can use a model discrepancy function $\delta f(\ldots)$ that captures the difference between reality and the model given by f. This leads to the "real" model

$$f(u, y, z, m, \ldots) + \delta f(\ldots) = 0. \tag{2.2}$$

Hence, δf is in most cases different from zero and its analytical expression is in general unknown. If δf can be modelled as a random variable, we consider it as stochastic model uncertainty. In this case, the distribution of δf depends on the parameters m and on additional prior assumptions. If a non-probabilistic approach is adopted to model the discrepancy function, then the min-max values of δf have a functional relation to the parameters m. If δf cannot be quantified, non-exactness is assumed and ignorance prevails, as pointed out in Chap. 1. For a more detailed discussion, we refer to Sect. 2.2.

With respect to *structural uncertainty*, there are $N \gg 1$ competing structures S_i, $i = 1, \ldots, N$, all satisfying the same specific system function g_s within an accuracy interval $g_s - g = \pm\delta g$. But the structures may differ in quality F. If the complete design specification is not explored, i.e. $S \neq S_{opt}$ we speak of ignored structural uncertainty. Structural uncertainty is discussed in more detail in Sect. 2.3.

2.1 Data Uncertainty

Sebastian Kersting, Roland Platz, Michael Kohler, and Tobias Melz

In engineering sciences, generating and evaluating data for and from numerical simulations, experimental tests of technical systems with high safety requirements, a representative process or archived data play an important role to adequately predict and evaluate the system's performance, cf. Sect. 1.4. Data uncertainty is present, if the amount, type and distribution of required data, such as model parameters, is incomplete, unknown or insufficient; and in this context data quality as discussed in Sect. 1.4 is an important factor. This section clarifies the expression *data uncertainty* and classifies various approaches to describe different forms thereof. For the latter, a brief overview of probabilistic with further differentiation between frequentist and Bayesian inferences, and non-probabilistic, as well as parametric and non-parametric approaches to analyse data uncertainty is given. Following these approaches, the classification illustrated in Fig. 2.1 into *stochastic data uncertainty* and *incertitude* is discussed in this section.

2.1.1 Introduction

Within this book, we distinguish between two general types of data: model parameters m and state variables u and z that have a quantifiable value. Model parameters describe the technical system's characteristics, such as geometrical and material properties for a mathematical or computer model, e.g. length or width of a beam element and mass, Young's modulus, etc. Aggregating geometrical and material properties of one or multiple components that effect processes in a system or structure may lead to new model parameter expressions like stiffness or damping as important quantities in the system's computer models for load-bearing systems. The models relate to them in a mathematical way and *model uncertainty* may become relevant, see Sect. 2.2. State variables describe the input and output conditions, such as mechanical loading, stress and strain, displacement, velocity, acceleration etc. They are mathematically related to the model parameters via the models.

On the one hand, model parameters m and state variables u are the input data for numerical simulations with computer models for predicting the system's behaviour, for example the dynamic behaviour of a structure due to vibrational excitation. They are also the input quantities for the experimental tests to validate the prediction. On the other hand, the system's behaviour as the output z of both simulation and test are mostly state variables or other measured quantities that can be inferred from model parameters by mathematical conversions. For example, the measurement of a lateral force and a resulting lateral deflection of a beam element leads to its stiffness.

Data in form of model parameters or state variables may have a single or a distributed value. Both are subject to uncertainty. Single or distributed values may vary in many possible ways depending on the designer's knowledge about the data and other conditions that make them uncertain. For example, a model parameter, such as the length of a beam element, may vary due to production tolerances, which influences the output in computer simulations or experimental tests.

In general, two basic types of data uncertainty occur: *aleatoric* or *epistemic* data uncertainty. *Aleatoric* uncertainty [67] is also known as irreducible uncertainty [62] or variability [67]. It is objective [32], and mostly characterised by a probabilistic distribution function. In [7], it is presumed to be the intrinsic randomness of a phenomenon. *Epistemic* uncertainty is also known as reducible uncertainty [50, 62], ignorance uncertainty [50], or simply uncertainty [67]. It is reducible [62], subjective [32], and occurs due to a lack of knowledge [62], insufficient or incomplete data [7]. Both types of uncertainty may be described via probabilistic and non-probabilistic approaches, cf. [33, 38, 44]. The probabilistic approaches can be further divided into *parametric*, with *frequentist* or *Bayesisan* inference approaches, or *non-parametric*.

Taking into account *aleatoric* and *epistemic* data uncertainty, this book distinguishes between *stochastic data uncertainty* and *incertitude*. They depend on the knowledge and assumptions about the data distribution and are explained in the following.

2.1.2 Stochastic Data Uncertainty

We assume that data is subject to stochastic uncertainty, if it can be modelled as realisations of a random variable Θ with a distribution $P(\theta)$. In this case, a *parametric* or a *non-parametric* approach as well as a *frequentist* or a *Bayesian* inference approach may be used to approximate the distribution for further uncertainty analysis. The approaches highly depend on the knowledge about the data, e.g. if a sample of measured data exists and, if applicable, its sample size. All approaches to analyse stochastic data uncertainty are conducted under the assumption that an event is possible with a given probability [33].

In the *parametric* approach, one assumes that the underlying distribution depends on finite dimensional parameters that fully describe the distribution $P(\theta)$. In case of measuring errors or production tolerances, a typical choice would be a normal distribution. Here the distribution of a random variable Θ is uniquely identified by the parameters mean $\bar{\theta}$ and standard deviation $\sigma(\theta)$. If the distribution is incomplete or no samples are available, engineering or physical knowledge can be used to approximate the parameters. Otherwise, if an adequate sample is on hand, a maximum likelihood estimator can be applied to estimate the distribution parameters, cf. [21].

In the *non-parametric* approach, the description of distributions is based solely on observations, cf. [57, Chap. 4.8]. The underlying distribution does not need specific finite dimensional parameters. Instead, e.g. a kernel density estimator can be applied to estimate the corresponding probability density function $p(\theta)$ of $P(\theta)$, cf. [43, 49]. In most cases, if only a relatively small sample with less than 50 data points is available, the maximum-likelihood approach yields more adequate results with respect to high convergence than the kernel density estimator—if the distribution assumption is correct. However, the actual sample size needed to achieve sufficient results may vary and depends on the specific application.

If the distribution of input data is known or estimated as described above, several methods may be used for a probabilistic computer simulation for obtaining its probabilistic output prediction, e.g. Monte Carlo Simulation (MCS) methods as described in [5, 10, 15, 39, 41, 51, 54].

MCS methods depend on the selected inference approach, meaning that output data distributions may vary from a *frequentist* or *Bayesian* perspective. They differ in the underlying assumptions made regarding the nature of data distributions [57]. In the *frequentist* view, probabilities are defined as the frequency that an event occurs if an experiment is repeated a large number of times. The *Bayesian* perspective treats probabilities as a distribution of subjective values based on prior knowledge and assumptions. They are constructed or updated as data is observed; algorithms are, e.g. Markov Chain Monte Carlo Techniques, Metropolis-Hastings Algorithms etc., see [57]. *Bayesian* inference-based approaches are mainly used for model parameter calibration and to determine model uncertainty, see Sects. 2.2, 4.1 and 4.3. However, they are computationally demanding because of the need to infer the posterior distributions of each parameter [9].

2.1.3 Incertitude

We say that data is subject to incertitude, if it *can not* be modelled as realisations of Θ with a distribution $P(\theta)$. The distribution of the input data is unknown. Instead, the analysis may be conducted based on *fuzzy set theory* or direct interval analysis that provide information about the possibility that a certain data value lies between a minimum and a maximum. These *possibilistic* approaches, in contrast to *probabilistic* approaches to assess stochastic data uncertainty, basically analyse whether an event is possible or impossible [33]. They are explained briefly in the following.

Fuzzy data uncertainty

The fuzzy set theory was introduced by Zadeh in 1965 [66]. Since then, numerous subdomains like fuzzy logic, fuzzy modelling, fuzzy arithmetic etc. have emerged [19]. Within the Fuzzy framework, data is expressed as member elements of a set $\mathcal{A}$. They can be defined by using a characteristic or, respectively, membership function $\mu_A\colon \mathcal{A} \to [0, 1]$. A membership of an element x is given if $\mu_A(x) = 1$, non-membership if $\mu_A(x) = 0$, [19]. The membership functions vary in form, they can be triangular, Gaussian, exponential etc. A practical way is to use so called α-cuts that divide the membership function into intervals with the aim to determine possibilities that a value is inside the interval, [33]. An overview of fuzzy methods for data uncertainty analysis in engineering applications is given in [48], applied fuzzy arithmetic can be found in [19] and fuzzy set theory based on fuzzy arithmetic is discussed in [66]. However, due to the absence of a structured and systematic elaboration of the theory, only a few practical approaches have been conducted [19].

Interval based data uncertainty

If neither distribution nor membership functions are available to describe the occurrence of data, an interval based approach may be useful. In this case, it is commonly assumed that the uncertain data lies between a minimum and a maximum value. Using a pessimistic perspective, a worst-case scenario is then the object of further investigations. Each parameter interval consists of a pair of min/max values, the four basic computation rules for adding, subtracting, multiplying, and dividing are valid for each interval parameter. However, the quality of the direct interval arithmetic evaluation depends on how often the interval parameters are present in a governing function of the deterministic computer model, i.e. the intervals become larger when they are propagated. Moreover, building a computer model with interval parameters that include paired min/max values can be demanding and time-consuming. Also, the analysis tends to overestimate uncertainty by using only extreme values that occur only rarely within the interval arithmetic [1].

Intervals can be stochastically motivated, e.g. $\bar{\theta} \pm 3\,\sigma(\theta)^2$-intervals specify minimum and maximum values, if a normal distribution of data is assumed. Eventually, possibilistic methods can be applied as shown in [1, 40].

In higher dimensions and to avoid using extreme values that lead to overestimation, the limiting intervals are typically replaced by ellipsoids for which the worst-case analysis becomes more complicated. For example and as shown in [11, 28–30, 55], sophisticated optimisation techniques introducing uncertainty sets are necessary to master data uncertainty in this pessimistic setting, see also Sect. 6.1.

2.2 Model Uncertainty

Alexander Matei, Roland Platz, Stefan Ulbrich, and Maximilian Schaeffner

In science and technology, mathematical models are frequently employed for the explanation of natural phenomena and for the description, quantification and control of engineering processes. We specifically focus on *mathematical* models and their accuracy, for other types of models we refer to Chap. 1. We do so because mathematical models enable numerical simulations to *predict* the behaviour, outcome or result of real technical products, systems and processes along their life cycle, see Sect. 1.2. However, it is a common observation that the usage of these models is affected by uncertainty, which can be traced back to the system design phase. Several causes can be identified for this uncertainty in early stage product development and, particularly, in the mathematical modelling. Our ignorance about the physical behaviour of a technical system leads to models that are only approximations of reality and may only be valid for a particular range of inputs and parameters, cf. Fig. 1.5.

There are two categories of ignorance which we briefly want to mention. The first is called *lack of knowledge* and stands for objects or processes which are unknown, unfamiliar and nameless to us. Examples for lack of knowledge arise whenever a novel material is exposed to new circumstances. Its reaction to the environment, its behaviour under load or pressure and its wear, all of which needs to be observed, evaluated and generalised. This is a challenge to scientists and engineers alike. The second category comprises effects that are known to us, but they are neglected, ignored and kept out of consideration in the modelling. We call it *disregard of knowledge*. To give an example, one may think of a linear elastic spring under load where the deformation of the spring is proportional to the loading force, only if the latter stays below a certain threshold. For loads above this limit, the spring material shows nonlinear, plastic or hysteresis-type behaviour which is difficult to model and thus it is often neglected.

Another source of model uncertainty arises from the numerical approach used to discretise such equations that are impossible to solve analytically. In engineering applications, the finite element method is a common numerical approximation scheme. In most cases, uncertainty caused by the numerical discretisation with finite elements can not only be quantified but also mastered by standard approaches, e.g. by developing error estimates [46]. In most cases, numerical errors happen on a relatively small scale, whereas the more severe sources of model uncertainty are missing or incomplete physical or empirical relations. In addition, human factors also contribute to model uncertainty. The methods and technologies to detect, quantify and master model uncertainty, which are presented in this book, see Sect. 4.3, are generally applicable, irrespective of the above mentioned causes.

Model uncertainty exists, if the functional relations between input and output, model parameters and other internal variables, as well as the scope and complexity of the model, are unknown, incomplete, inadequate or unreasonable. The dilemma the designer encounters is that in early stage design, before calibration, verification and validation processes start, the extent of uncertainty is difficult to detect. Even in the usage phase, after the final product has been assembled, a degree of uncertainty remains, since the full-scale product does not exactly match the small-scale prototype, neither does the computer model. Various mathematical approaches using different prior knowledge about functional relations, scope and complexity as well as the required data, which can be uncertain as well, have been developed to deal with model uncertainty.

It is one aim of this book to understand and to evaluate the uncertainty especially in load-bearing mechanical systems with high safety requirements. This section gives a brief overview of different approaches to describe, quantify and master model uncertainty. In this context, the term 'master' means to be aware and to be able to quantify model uncertainty in verification and validation processes, e.g. by data-based training. This leads to an adapted mathematical model that, eventually, adequately describes and predicts the system and process behaviour observed in reality.

2.2.1 Functional Relations, Scope and Complexity of Mathematical Models

In this chapter, we consider *models* as images of reality in the domain of mathematical abstraction. These mappings describe or represent knowledge about a system in the language of functional relations given in implicit form

$$f(u, y, z, m) = 0 \tag{2.3}$$

between input u and output data z, model parameters m, such as material properties, and internal variables y, like states. Accordingly, we mean by *model uncertainty* that these images of reality are imperfect, i.e. the governing *physical* relations f between inputs, outputs, parameters and internal variables are unknown or incomplete, or they are partly reduced to *physically-inconsistent* approximations (Sect. 1.3) of more complex, but expensive, expressions. Thus, in the presence of model uncertainty, Eq. (2.3) does not reflect reality. A common representation of model uncertainty introduces a *model discrepancy function* δf which accounts for lack or disregard of knowledge, numerical errors and human factors in the mathematical modelling as mentioned before. The "real" functional relation is then considered to be implicitly given by

$$f(u, y, z, m) + \delta f(\ldots) = 0. \tag{2.4}$$

In many cases, for given u and m, the implicit equation (2.3) is solved to obtain the model's output z explicitly. If the model function f is differentiable and its derivative is invertible with respect to the internal variables y, then the implicit function theorem yields a *reduced model* η that represents an input-output relation to the quantity of interest z. The model equation can now be written in explicit form

$$\eta(u, m) = z, \tag{2.5}$$

and likewise for the case where the model discrepancy function is added, cf. Eq. (2.4). In the sequel, we follow the main literature on this topic that considers only computer models η, i.e. reduced models, which describe an input-output relation.

Evidently, different assumptions about functional relations like axiomatic or empirical, linear or non-linear, and time-invariant or time-variant alter the output of the model significantly [37]. Moreover, the model's scope and complexity can be varied. Thus, any computer simulation that is based upon uncertain models leads to erroneous predictions of the quantity of interest that affect the verification and validation process necessary to prove the model's consistency and correctness, cf. Sect. 1.3.

The designer has several options to choose from possible modelling assumptions about functional relations, as well as scope and complexity, as mentioned above. When a mathematical model has been built, the functional relations are subject to a verification and calibration process to prove and to update the numerical simulation, the computer code and, if applicable, the model parameters [2]. Eventually, the computer model is validated against experimental tests of a real system, like the Modular Active Spring-Damper System, see Sect. 3.6.1. As for the scope and complexity, the number of degrees of freedom can be high and costly depending on the form and discretisation of the model, e.g. analytical, finite elements or multi-body-models that also influence the results of verification, calibration and validation processes. Furthermore, Occam's razor can be used as a guiding principle to keep models as simple as possible, see Sect. 1.3, because more complex models often tend to be more susceptible to uncertainty.

The functional relations as well as the scope and complexity of models are considered as being independent from how model parameters and input data are present—as a simple value, randomly distributed or as intervals. These are subject to data uncertainty, which is covered in Sect. 2.1.

2.2.2 Approaches to Detect, Quantify, and Master Model Uncertainty

Basically, two different approaches give information about detectable, quantifiable and masterable model uncertainty: a *deterministic* analysis and a *probabilistic frequentist* or *probabilistic Bayesian inference-based* perspective. The latter needs subjective prior data distribution information as discussed in Sect. 2.1. The deterministic and Bayesian inference-based approach allow data-based training, which is an important criterion for the verification and validation processes. However, finding the adequate correction terms or prior information, especially for models with high complexity, remains a challenge [9]. Using a probabilistic frequentist approach, see Sect. 2.1, however, does not take into account prior information other than randomness of data.

In accordance to this book's notation, deterministic approaches do not master stochastic uncertainty in a mathematical model. However, they may be suited to quantify and master incertitude, e.g. when only extreme values like minima and maxima are known or assumed, or when parts of a model are missing to detect ignorance, cf. Fig. 2.2. Probabilistic approaches, however, describe and master all three types of uncertainty. In the following, we give an overview of the two approaches.

Deterministic framework

On the one hand, a *deterministic* analysis can be used, for example, in fault diagnosis [56] to quantify uncertainty in the model equations. To do so, the model is adjusted by a model discrepancy function $\delta\eta$ which is assumed to stay within a bounded uncertainty set $\mathcal{U}_\eta$:

$$\eta(u, m) + \delta\eta(u) = z, \quad \delta\eta \in \mathcal{U}_\eta. \tag{2.6}$$

A standard residual analysis then enables the engineer to distinguish component failures from the effects of model uncertainty. However, this method strongly relies on the assumption that the model discrepancy function $\delta\eta$ stays within a bounded uncertainty set $\mathcal{U}_\eta$.

On the other hand, a deterministic analysis may be used in model verification and validation processes. In [9, 57], the possibility to approximate the residual between the model output and the observed quantity of interest via a polynomial p is mentioned. In this case, the polynomial p takes the role of the discrepancy function:

$$\eta(u, m) + p(u, m, \theta) = z. \tag{2.7}$$

This necessarily leads to an augmented parameter set (m, θ) consisting of the original physical axiomatic and empiric parameters m as well as the non-physical polynomial parameters θ, which do not give an enhanced physics-based understanding of the model's uncertainty or shortcoming in predicting reality [9]. The augmented parameter set needs to be calibrated, which is usually performed by an optimisation scheme.

Probabilistic framework

Within the probabilistic framework, *Bayesian inference-based* approaches are frequently used to assess the prediction quality of a mathematical model under given experimental data [9, 16, 52, 61]. In [34] or [37], Bayesian calibration techniques and a plausibility prior argument are used for model selection. Another important approach introduces a stochastic process δ for the discrepancy function [22]:

$$\eta(u, m) + \delta(u, m, \phi) + \varepsilon = z, \quad (2.8)$$

where ε is stochastic noise, usually produced by the measurement error, and ϕ are hyperparameters, which need to be tuned by an optimisation scheme. Based on this setting, [64] introduce a technique for model validation. A method designed for time-dependent systems is proposed in [3]. Combinations of a high- and a low-fidelity model and experimental data are used to construct a multi-fidelity model, which is proposed in [13] as an improved predictor. A scaled Gaussian Process is used in [17] for model calibration and prediction. It is claimed that the method bridges the gap between the least-squares calibration and the Gaussian Process calibration and, thus, is an improvement of the method introduced in [22]. Furthermore, a Bayesian interval hypotheses-based approach [36] and a Bayesian inference-based approach [37] are used to compare different models based on their internal functional relations from axiomatic or empiric assumptions, see also Sect. 4.3.3 for assessing model uncertainty in the Modular Active Spring-Damper System, see Sect. 3.6.1.

In order to apply the Bayesian methods, it is necessary to select prior distributions. Often, this is subjective and it is unclear how to choose them, which may lead to unrealistic assumptions. Furthermore, it is pointed out in [60] that the approach in [22] may lead to inadequate results in case of an imperfect computer model. As a consequence, further verification of the computer codes needs to be conducted.

From the *frequentist's* perspective, a simple approach to quantify model uncertainty is to use validation metrics, such as the area validation metric [35] or the Mahalanobis distance [68]. They give a quantitative measure of disagreement between the model output and the observed quantity of interest based upon observed measurements. In order to select the most adequate model, an arbitrary threshold on the metric is imposed, or classical hypothesis testing is performed. Surrogate models are often used to quantify the uncertainty in a technical system. Usually, these methods use computer simulations and a small sample of experimental data to estimate properties of probability distributions, such as quantiles [25–27] or densities [14, 23, 24]. A detailed description for a method based on an imperfect computer model is given in Sect. 4.3.8. A case where computer models are assumed to fit the reality is shown in Sect. 5.2.6. In [12], another method to detect model uncertainty is proposed which is based on optimum experimental design and hypothesis testing, see also Sect. 4.3.1. In [65], another approach to quantify the model error is developed. Here, the model discrepancy function is estimated on bootstrap samples via a regression estimation, e.g. smoothing splines or artificial neural networks.

However, the application of frequentist methods often demands a larger sample size, which can become a problem, since generating data is an expensive and time-consuming process.

2.3 Structural Uncertainty

Peter F. Pelz and Marc E. Pfetsch

Besides data and model uncertainty introduced in Sects. 2.1 and 2.2, structural uncertainty forms the third important pillar of uncertainty under consideration in this book. Structural uncertainty refers to the fact that only part of the possible solutions are evaluated with respect to uncertainty. In this sense, the model of the system is incomplete. However, the focus is on the system and not on the ignorance of the models of the system components, as in model uncertainty.

Let us start with the viewpoint of classical product design. Given a requested system function, the designer plans the system structure. The systems are assembled by components and modules, which then form the system. In production, this arises when combining single processes to process chains. When a given system function can be generated by a multitude of different function structures, and each function structure can be generated by a multitude of elements, this results in a "combinatorial explosion" of possibilities that, in general, cannot be evaluated by humans anymore. This lack of knowledge on other possibilities is then called *structural uncertainty*.

The consideration of this type of uncertainty seems to be new, but the term "structural uncertainty" is sometimes used in the sense of "model structure uncertainty", i.e. the structure of the model is uncertain. Some examples from different disciplines can be found in [4, 6, 58, 59]. In our book, the latter meaning is captured by the term "model uncertainty", see Sect. 2.2.

In comparison to data and model uncertainty, structural uncertainty has the advantage that its presence has no direct negative effect on product safety. However, economically better solutions might be lost.

Structural uncertainty can be tackled by using (discrete) mathematical optimisation methods that allow to consider all possibilities and select the best system choice with respect to a predefined objective function. These techniques require a combination of domain knowledge in order to set up a physical model and define the allowed elements. This is then integrated into a mathematical optimisation model, which is solved using optimisation software. As usual, one needs to balance the exactness of the model with the effort to obtain optimal solutions; see Sect. 1.3 for a general discussion of this balance. Often tailored solution methods need to be developed in order to achieve practically feasible solution times. The method for the quantification of structural uncertainty is therefore highly context-dependent.

This book contains examples that illustrate this approach, see, for example, Sects. 1.5 and 6.3.5. Many more examples can be found in the literature, e.g. [31, 45, 53, 63]. These examples show the flexibility of mathematical programming to deal with different manifestations of structural uncertainty. Nevertheless, currently, expert knowledge is needed to derive and efficiently solve appropriate models of reality.

Structural uncertainty has two other aspects that we want to briefly mention.

1. When using models in order to compose a system, the model predetermines the possible components that can be chosen, i.e. the model can only be optimised over its model horizon, see Sect. 1.3. Again, possible interesting solutions might be lost due to the choice of the model. One needs to be aware of this restriction, similar to the fact that models are always an approximation of reality.
2. The first aspect arises from the fact that the system is built from smaller elements. However, a quantitative evaluation of uncertainty usually only takes place on the level of the single elements, since an analysis for the complete system would be too complex or taking measurements would be too expensive. This book discusses several methods to deal with this uncertainty. For instance, a common way to handle the corresponding uncertainty is by flexibility, see Sect. 6.2. Another method is to make the system robust or to consider the effect of uncertain parameters already in the mathematical model and perform a robust optimisation, see Sect. 6.1.

References

1. Alefeld G, Mayer G (2000) Interval analysis: Theory and applications. J Comput Appl Math 121(1–2):421–464
2. American Society of Mechanical Engineers (ASME) Standards Committee on Verification and Validation in Computational Solid Mechanics (PTC 60/V&V 10) (2007) Guide for verification and validation in computational solid mechanics. ASME publications
3. Bayarri MJ, Berger JO, Cafeo J, Garcia-Donato G, Liu F, Palomo J, Parthasarathy RJ, Paulo R, Sacks J, Walsh D: Computer model validation with functional output. Annals of Statistics 35(5), 1874–1906 (2007). https://doi.org/10.1214/009053607000000163
4. Bojke L, Claxton K, Sculpher M, Palmer S (2009) Characterizing structural uncertainty in decision analytic models: A review and application of methods. Value in Health 12(5):739–749. https://doi.org/10.1111/j.1524-4733.2008.00502.x
5. Braaten E, Weller G: An improved low-discrepancy sequence for multidimensional quasi-Monte Carlo integration. Journal of Computational Physics 33(2), 249–258 (1979)
6. Briggs AH, Weinstein MC, Fenwick EA, Karnon J, Sculpher MJ, Paltiel AD (2012) Model parameter estimation and uncertainty: A report of the ISPOR-SMDM Modeling Good Research Practices Task Force-6. Value in Health 15(6):835–842. https://doi.org/10.1016/j.jval.2012.04.014
7. Der Kiureghian A, Ditlevsen O: Aleatory or epistemic? Does it matter? Structural Safety 31(2), 105–112 (2009)
8. Engelhardt RA, Koenen JF, Enss GC, Sichau A, Platz R, Kloberdanz H, Birkhofer H (2010) A model to categorise uncertainty in load-carrying systems. In: 1st MMEP international conference on modelling and management engineering processes, pp 53–64
9. Farajpour I, Atamturktur S: Error and uncertainty analysis of inexact and imprecise computer models. Journal of Computing in Civil Engineering 27(4), 407–418 (2012)
10. Felber T, Kohler M, Krzyżak A (2015) Adaptive density estimation from data with small measurement errors. IEEE Trans. Inform. Theory 61(6):3446–3456. https://doi.org/10.1109/TIT.2015.2421297
11. Gally T, Gehb CM, Kolvenbach P, Kuttich A, Pfetsch ME, Ulbrich S (2015) Robust truss topology design with beam elements via mixed integer nonlinear semidefinite programming. In: Pelz PF, Groche P (eds) Uncertainty in mechanical engineering II, vol 807. Applied mechanics and materials. Trans Tech Publications, pp 229–238

12. Gally T, Groche P, Hoppe F, Kuttich A, Matei A, Pfetsch ME, Rakowitsch M, Ulbrich S (2021) Identification of model uncertainty via optimal design of experiments applied to a mechanical press. Optim Eng, to appear
13. Goh J, Bingham D, Holloway JP, Grosskopf MJ, Kuranz CC, Rutter E (2013) Prediction and computer model calibration using outputs from multifidelity simulators. Technometrics 55(4):501–512. https://doi.org/10.1080/00401706.2013.838910
14. Götz B, Kersting S, Kohler M (2020) Estimation of an improved surrogate model in uncertainty quantification by neural networks. Ann Inst Stat Math. To appear
15. Graham C, Talay D (2013) Stochastic Simulation and Monte Carlo Methods: Mathematical Foundations of Stochastic Simulation, vol 68. Springer
16. Green PL: Bayesian system identification of a nonlinear dynamical system using a novel variant of simulated annealing. Mechanical Systems and Signal Processing 52, 133–146 (2015)
17. Gu M, Wang L (2018) Scaled Gaussian stochastic process for computer model calibration and prediction. SIAM/ASA Journal on Uncertainty Quantification 6(4):1555–1583. https://doi.org/10.1137/17M1159890
18. Hanselka H, Platz R (2010) Ansätze und Maßnahmen zur Beherrschung von Unsicherheit in lasttragenden Systemen des Maschinenbaus. Konstruktion (11/12):55–62
19. Hanss M (2005) Applied Fuzzy Arithmetic: An Introduction With Engineering Applications. Springer
20. Imholz M, Vandepitte D, Moens D (2015) Analysis of the effect of uncertainty clamping stiffness on the dynamical behaviour of structures using interval field methods. In: Pelz PF, Groche P (eds) Uncertainty in mechanical engineering II, vol 807. Applied mechanics and materials. Trans Tech Publications, pp 195–204
21. Kalbfleisch JG (1979) Probability and Statistical Inference. II. Springer, New York
22. Kennedy MC, O'Hagan A (2001) Bayesian calibration of computer models. Journal of the Royal Statistical Society: Series B (Statistical Methodology) 63(3):425–464. https://doi.org/10.1111/1467-9868.00294
23. Kersting S, Kohler M (2019) Uncertainty quantification based on (imperfect) simulation models with estimated input distributions. Submitted for publication
24. Kohler M, Krzyżak A (2017) Improving a surrogate model in uncertainty quantification by real data. Submitted for publication
25. Kohler M, Krzyzak A (2018) Estimating quantiles in imperfect simulation models using conditional density estimation. Ann Inst Stat Math. https://doi.org/10.1007/s10463-018-0683-8
26. Kohler M, Krzyżak A (2019) Estimation of extreme quantiles in a simulation model. Journal of Nonparametric Statistics 31(2):393–419. https://doi.org/10.1080/10485252.2019.1567727
27. Kohler M, Tent R (2019) Nonparametric quantile estimation using surrogatemodels and importance sampling. Metrika 83:141-169.https://doi.org/10.1007/s00184-019-00736-3
28. Kolvenbach P, Lass O, Ulbrich S (2018) An approach for robust PDE-constrained optimization with application to shape optimization of electrical engines and of dynamic elastic structures under uncertainty. Optim Eng 697–731. https://doi.org/10.1007/s11081-018-9388-3
29. Kuttich A, Ulbrich S (2017) Feedback controller design and topology optimization for truss structures under uncertain dynamic loads. In: von Scheven M, Keip MA, Karajan N (eds) 7th GACM colloquium on computational mechanics for young scientists from academia and industry
30. Lass O, Ulbrich S (2017) Model order reduction techniques with a posteriori error control for nonlinear robust optimization governed by partial differential equations. SIAM J Sci Comput 39(5):S112–S139. https://doi.org/10.1137/16M108269X
31. Leise P, Altherr LC (2018) Optimizing the design and control of decentralized water supply systems – a case-study of a hotel building. In: International conference on engineering optimization. Springer, pp 1241–1252
32. Lemaire M (2014) Mechanics and uncertainty. Wiley Online Library
33. Li S, Platz R (2017) Observations by evaluating the uncertainty of stress distribution in truss structures based on probabilistic and possibilistic methods. Journal of Verification, Validation and Uncertainty Quantification 2(3):031006. https://doi.org/10.1115/1.4038486

34. Lima E, Oden JT, Wohlmuth B, Shahmoradi A, Hormuth II DA, Yankeelov TE, Scarabosio L, Horger T: Selection and validation of predictive models of radiation effects on tumor growth based on noninvasive imaging data. Computer Methods in Applied Mechanics and Engineering 327, 277–305 (2017)
35. Liu Y, Chen W, Arendt P, Huang HZ: Toward a better understanding of model validation metrics. Journal of Mechanical Design 133(7), 071005 (2011)
36. Mallapur S, Platz R (2018) Quantification of uncertainty in the mathematical modelling of a multivariable suspension strut using Bayesian interval hypothesis-based approach. In: Pelz PF, Groche P (eds) Uncertainty in mechanical engineering III, vol 885. Applied mechanics and materials. Trans Tech Publications, pp 3–17
37. Mallapur, S., Platz, R.: Uncertainty quantification in the mathematical modelling of a suspension strut using Bayesian inference. Mechanical Systems and Signal Processing 118, 158–170 (2019)
38. Melzer C, Platz R, Melz T (2015) Comparison of methodical approaches to describe and evaluate uncertainty in the load-bearing capacity of a truss structure. In: Fourth international conference on soft computing technology in civil, structural and environmental engineering. Paper 26
39. Melzer CM, Krech M, Kristl L, Freund T, Kuttich A, Zocholl M, Groche P, Kohler M, Platz R (2015) Methodical approaches to describe and evaluate uncertainty in the transmission behavior of a sensory rod. In: Pelz PF, Groche P (eds) Uncertainty in mechanical engineering II, vol 807. Applied mechanics and materials. Trans Tech Publications, pp 205–217. https://doi.org/10.4028/www.scientific.net/AMM.807.205
40. Moore, R., Lodwick, W.: Interval analysis and fuzzy set theory. Fuzzy Sets and Systems 135(1), 5–9 (2003)
41. Müller-Gronbach T, Novak E, Ritter K (2012) Monte Carlo-Algorithmen. Springer
42. Pahl G, Beitz W, Feldhusen J, Grote KH (2007) Engineering design: a systematic approachD, 3rd edn. Springer, London. https://doi.org/10.1007/978-1-84628-319-2
43. Parzen E (1962) On estimation of a probability density function and mode. Ann Math Stat 33:1065–1076. https://doi.org/10.1214/aoms/1177704472
44. Platz R, Ondoua S, Habermehl K, Bedarff T, Hauer T, Schmitt S, Hanselka H (2010) Approach to validate the influences of uncertainties in manufacturing on using load-carrying structures. In: Proceedings of USD2010 international conference on uncertainty in structural dynamics, pp 20–22
45. Pöttgen P, Ederer T, Altherr L, Lorenz U, Pelz PF (2016) Examination and optimization of a heating circuit for energy-efficient buildings. Energ Technol 4(1):136–144
46. Repin SI, Sauter SA (2020) Accuracy of mathematical models. EMS tracts in mathematics, vol 33. European Mathematical Society Publishing House. https://doi.org/10.4171/206
47. Reuleaux F (1899) Der Konstrukteur: Ein Handbuch zum Gebrauch beim Maschinen-Entwerfen. Vieweg
48. Reuter U, Möller B (2007) Uncertainty Forecasting in Engineering. Springer
49. Rosenblatt M (1956) Remarks on some nonparametric estimates of a density function. Ann Math Stat 27:832–837. https://doi.org/10.1214/aoms/1177728190
50. Roy CJ, Oberkampf WL (2011) A comprehensive framework for verification, validation, and uncertainty quantification in scientific computing. Comput. Methods Appl. Mech. Engrg. 200(25–28):2131–2144. https://doi.org/10.1016/j.cma.2011.03.016
51. Saltelli A, Ratto M, Andres T, Campolongo F, Cariboni J, Gatelli D, Saisana M, Tarantola S (2008) Global Sensitivity Analysis: The Primer. John Wiley & Sons
52. Sankararaman S, Mahadevan S (2011) Model validation under epistemic uncertainty. Reliability Engineering & System Safety 96(9):1232–1241
53. Schänzle C, Altherr LC, Ederer T, Lorenz U, Pelz PF (2015) As good as it can be – ventilation system design by a combined scaling and discrete optimization method. In: Proceedings of the FAN
54. Schueller G, Pradlwarter H: Uncertainty analysis of complex structural systems. International Journal for Numerical Methods in Engineering 80(6–7), 881–913 (2009)

55. Sichau A, Ulbrich S (2012) A second order approximation technique for robust shape optimization. In: Hanselka H, Groche P, Platz R (eds) Uncertainty in mechanical engineering, vol 104. Applied mechanics and materials. Trans Tech Publications, pp 13–22. https://doi.org/10.4028/www.scientific.net/AMM.104.13
56. Simani S, Fantuzzi C, Patton RJ (2003) Model-based fault diagnosis techniques. In: Model-based fault diagnosis in dynamic systems using identification techniques. Springer, pp 19–60
57. Smith RC (2014) Uncertainty Quantification; Theory, Implementation, and Applications, Computational Science & Engineering, vol 12. SIAM, Philadelphia, PA
58. Strong M, Oakley JE (2014) When is a model good enough? Deriving the expected value of model improvement via specifying internal model discrepancies. SIAM/ASA Journal on Uncertainty Quantification 2(1):106–125. https://doi.org/10.1137/120889563
59. Thore P, Shtuka A, Lecour M, Ait-Ettajer T, Cognot R (2002) Structural uncertainties: Determination, management, and applications. Geophysics 67(3):840–852. https://doi.org/10.1190/1.1484528
60. Tuo R, Wu CFJ (2015) Efficient calibration for imperfect computer models. Ann Stat 43(6):2331–2352. https://doi.org/10.1214/15-AOS1314
61. Tuomi M, Pinfield D, Jones HRA (2011) Application of Bayesian model inadequacy criterion for multiple data sets to radial velocity models of exoplanet systems. Astronomy & Astrophysics 532:A116
62. Vandepitte D, Moens D (2011) Quantification of uncertain and variable model parameters in non-deterministic analysis. In: IUTAM symposium on the vibration analysis of structures with uncertainties. Springer, pp 15–28
63. Vergé A, Pöttgen P, Altherr LC, Ederer T, Pelz PF (2016) Lebensdauer als Optimierungsziel – Algorithmische Struktursynthese am Beispiel eines hydrostatischen Getriebes. O+P – Ölhydraulik und Pneumatik 60(1–2)
64. Wang S, Chen W, Tsui KL (2009) Bayesian validation of computer models. Technometrics 51(4):439–451. https://doi.org/10.1198/TECH.2009.07011
65. Wong RKW, Storlie CB, Lee TCM (2017) A frequentist approach to computer model calibration. J R Stat Soc Ser B Stat Methodol 79(2):635–648. https://doi.org/10.1111/rssb.12182
66. Zadeh LA (1965) Fuzzy sets. Inf Control 8(3):338–353
67. Zang TA, Hemsch MJ, Hilburger MW, Kenny SP, Luckring JM, Maghami P, Padula SL, Stroud WJ (2002) Needs and opportunities for uncertainty-based multidisciplinary design methods for aerospace vehicles. Technical report, TM-2002-211462, NASA
68. Zhao L, Lu Z, Yun W, Wang W (2017) Validation metric based on Mahalanobis distance for models with multiple correlated responses. Reliability Engineering & System Safety 159:80–89

Chapter 3
Our Specific Approach on Mastering Uncertainty

Peter F. Pelz, Robert Feldmann, Christopher M. Gehb, Peter Groche, Florian Hoppe, Maximilian Knoll, Jonathan Lenz, Tobias Melz, Marc E. Pfetsch, Manuel Rexer, and Maximilian Schaeffner

Abstract This chapter serves as an introduction to the main topic of this book, namely to master uncertainty in technical systems. First, the difference of our approach to previous ones is highlighted. We then discuss process chains as an important type of technical systems, in which uncertainty propagates along the chain. Five different approaches to master uncertainty in process chains are presented: uncertainty identification, uncertainty propagation, robust optimisation, sensitivity analysis and model adaption. The influence of the process on uncertainty and methods depends on whether it is dynamic/time-varying and/or active. This brings us to the main strategies for mastering uncertainty: robustness, flexibility and resilience. Finally, three different concrete technical systems that are used to demonstrate our methods are presented.

How can we ensure product safety, realise lightweight structures, enable sustainable systems or control production quality? All these questions lead to the issue of how to master uncertainty, cf. Chap. 1. The answer to this core issue has many different facets, and there is a need to structure the topic and demonstrate solution methods.

This chapter is dedicated to structuring the facets, but also to introducing our specific approach to mastering uncertainty. For this purpose, we introduce process chains and temporal characteristics of processes. Often, active processes seem to be a smart way to cope with uncertainty. We compare the cost benefit associated with the selection of active processes.

The original version of this chapter was revised with the missed out corrections. An erratum to this chapter can be found at https://doi.org/0.1007/978-3-030-78354-9_8

P. F. Pelz (✉) · R. Feldmann · Christopher M. Gehb · P. Groche · F. Hoppe · M. Knoll · J. Lenz · T. Melz · M. Rexer · M. Schaeffner
Department of Mechanical Engineering, TU Darmstadt, Darmstadt, Germany
e-mail: peter.pelz@fst.tu-darmstadt.de

T. Melz
Fraunhofer Institute for Structural Durability and System Reliability LBF, Darmstadt, Germany

Marc E. Pfetsch
Department of Mathematics, TU Darmstadt, Darmstadt, Germany

P. Pelz et al. (eds.), *Mastering Uncertainty in Mechanical Engineering*, Springer Tracts in Mechanical Engineering,
https://doi.org/10.1007/978-3-030-78354-9_3

The three strategies, (i) robustness, (ii) flexibility, (iii) resilience, are of particular interest when it comes to mastering uncertainty. In this chapter we introduce these and distinguish between the three strategies. Finally, to demonstrate and validate our specific approach, we define three demonstrator systems in the last section of this chapter.

3.1 Beyond Existing Approaches

Peter F. Pelz

Uncertainty quantification is a large research field today. It is the basis for machine learning and inference in the context of data-driven, i.e. black box models, cf. Chap. 1. Uncertainty quantification inevitably deals with model uncertainty and data uncertainty. Grey box models become more and more important in mechanical engineering, where prior knowledge is available.

Our research from the perspective of mechanical engineering has three aspects that go beyond the field of uncertainty quantification:

(i) Firstly, we focus on physical systems that are designed, manufactured and used. The materiality of the systems and the physicality of their use inevitably lead to new aspects in the mastering of uncertainty. One example is the mastering of uncertainty across all product life phases and the concept of closed loops between phases, Fig. 3.1. Another example is the consideration of the functional safety of physical systems.

(ii) Secondly, the decomposition into sub-functions and the assignment of real processes or components to the sub-functions, typical for mechanical engineering [36], is a guiding principle for us, which is reflected in the uncertainty classification given in Chap. 2. It leads us to the concept of structural uncertainty.

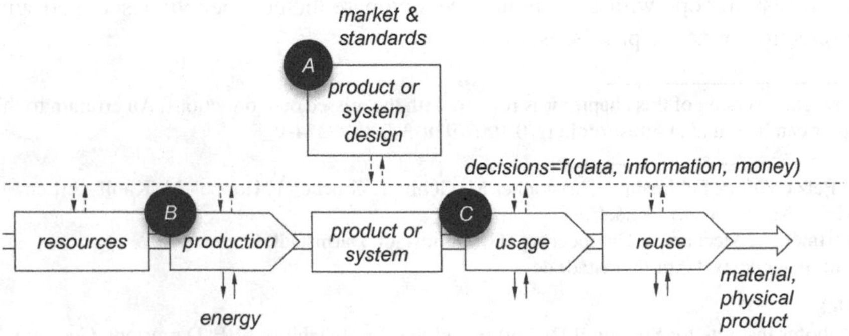

Fig. 3.1 **A** product or system design, **B** production, **C** usage; all phases are interconnected by the flow of material and physical products as well as data, information and money

But the systematic engineering design view also leads us to the conflicting objectives minimum effort, maximum availability and maximum acceptability. This view results in the broad motivation of mastering uncertainty. It is a basic approach to gain product safety, lightweight structures, sustainable systems and closed-loop production quality control, Chap. 1.

(iii) Thirdly, and most importantly in our view, we focus on mastering uncertainty; quantifying it may be a prerequisite. As a result, we put strategies for uncertainty management on the top level. Some of these strategies require new active or semi-active technologies that enable adaptation to changes in production or usage. Here, too, we extend the current state of science and technology by our contribution.

View from different disciplines

In the twelve years of research that form the foundation of this book, there has been continuous inspiration between the various disciplines. Thus, the framework structure outlined in Fig. 1.12 can be viewed equally from the perspectives of mechanical engineering, mathematics or law. As shown in the introduction, for example, Sustainable Systems Design can be complementarily presented from the perspectives of systematic engineering design or mathematics. We regard the different forms of representation in 'languages' and types rather as a benefit than a disadvantage. This viewpoint is represented throughout this book.

3.2 Uncertainty Propagation Through Process Chains

Peter F. Pelz

We consider process chains, systems and structures in all product life phases in view, cf. Fig. 3.1. Therefore, it is obvious that we need an abstract representation in mind on how to treat those chains in a generic way. Figure 3.2 shows the model of a generic process chain being composed of single processes.

Representation of process chains

In Fig. 3.2 the physical product or system moves from left to right through the process chain, or the system or product is represented by the process chain itself. The evolution of the density function $p(\theta)$ of the product or system feature $\theta = \bar{\theta} + \delta\theta$ along the process chain is shown. In each process, energy, material and information are fed into or are withdrawn from the product. This is done by the work equipment, a generic term for man or machine. This incorporates tools such as sensors, actuators, controllers, and equally energy transformers and storage devices, i.e. capacities. Such a process may be active or semi-active, transient or quasi-stationary; it may

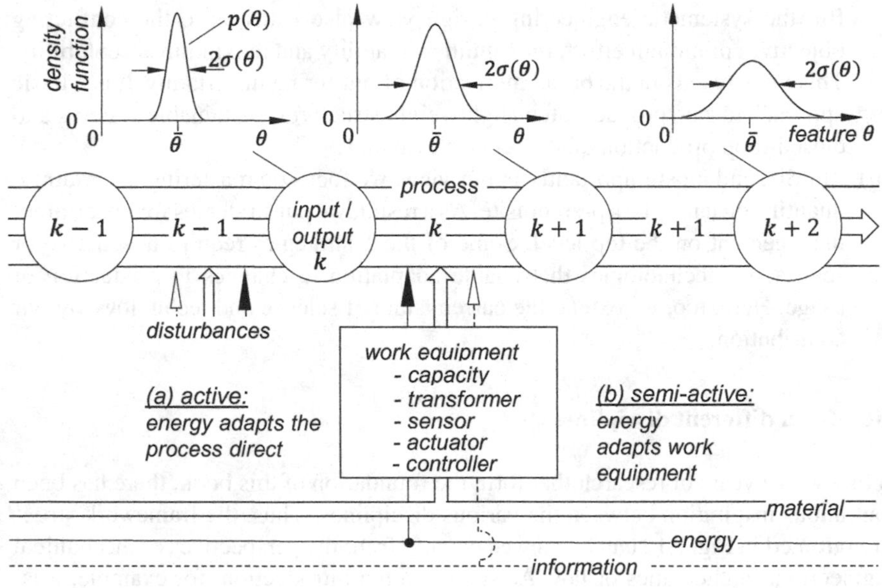

Fig. 3.2 Model of a generic process chain

be a static process, a relaxation process or a dynamic process. The different process characteristics will be discussed in Sect. 3.4.

In Chap. 1 we introduced engineering design thinking of structures built from components. This engineering design is represented by the generic process chain but also the branched structure or process chain depicted in Fig. 3.3. To model the general linkage of single processes, serial or parallel, the multi-pole presentation has spread [31]. The multi-pole representation is an object-oriented modelling approach. The representation is used in many domains, e.g. acoustics, electrical engineering, structural mechanics or fluid power. Other multi-domain model tools such as the bond graphs or Modelica are based on multi-pole representations.

Each component, i.e. each individual process, is easily modelled by a transfer matrix $A_1, \ldots, A_M$ with nonlinear algebraic entries. Using a multi-pole representation, the input data θ_k of the kth process are mapped to the output data θ_{k+1}: $\theta_k = A_k \theta_{k+1}$. Even though the reverse representation is often given, this representation is more concise when it comes to process chains. In a dynamical setting, the transfer matrices A_k may be given in the frequency domain. With a parallel arrangement of N components, the total transfer matrix is obtained by the sum of the individual matrices (i for input, o for output)

$$\theta_\mathrm{i} = \theta_\mathrm{o} \sum_{k=1}^{N} A_k .$$

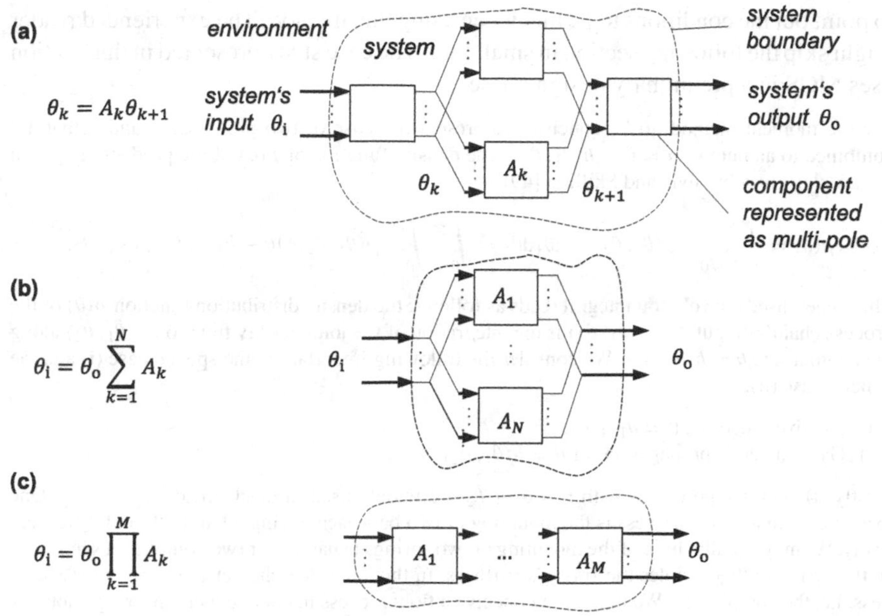

Fig. 3.3 **a** Multi-pole representation of a system or process chain; **b** parallel arrangement; **c** serial arrangement

With a serial arrangement of M components, the total transfer matrix is gained by the product of the individual matrices,

$$\theta_i = \theta_o \prod_{k=1}^{M} A_k .$$

Therefore, the model of the structure or the branched process chain can be easily assembled from the component models. Thus a model is available, which serves as a prerequisite for an uncertainty analysis or uncertainty propagation.

Uncertainty propagation

The process chains sketched in Figs. 3.2 and 3.3 help us to picture uncertainty propagation. The methods used in this book for uncertainty propagation are (i) moment methods and (ii) Monte Carlo simulations (MCS), cf. Sect. 2.1. The moment methods are the simplest and most commonly used methods. However, they are less powerful than the Monte Carlo methods. The latter are suitable in the preliminary design phase when compact, coarse granular models are used. In the later design stage, however, brute-force methods like MCS are often too expensive in terms of computing time.

In the following, we will shed some light on the Gaussian uncertainty propagation. This moment method is the most common one used by engineers, and it is worthwhile to point out the conditions to be met when using this method. The experienced reader might skip the following section in small font. The case study presented in this section uses MCS in a preliminary design phase.

For the moment method, to be concise, we present the case of two processes 1 and 2 that are combined to an output $\theta_3 = \theta = h(\theta_1, \theta_2)$. The density function of $p(\theta)$ of the product or system feature θ is given by Sivia and Skilling [49]

$$p(\theta) = \int_0^\infty \int_0^\infty p(\theta_1, \theta_2, \theta)\, \mathrm{d}\theta_1 \mathrm{d}\theta_2 = \int_0^\infty \int_0^\infty p(\theta_1, \theta_2)\delta\left[\theta - h(\theta_1, \theta_2)\right] \mathrm{d}\theta_1 \mathrm{d}\theta_2.$$

This generalised convolution integral reads as follows: the density distribution function $p(\theta)$ of the process chain's output $\theta = h(\theta_1, \theta_2)$ is the integration of the joint density function $p(\theta_1, \theta_2)$ along the contour line $\theta = h(\theta_1, \theta_2)$. We consider the following important in the special case (i) and the general case (ii):

(i) additive processes $\theta = \theta_1 + \theta_2$,
(ii) Gaussian error propagation for $\theta = h(\theta_1, \theta_2)$.

Firstly (i), additive processes with $\theta = \theta_1 + \theta_2$ are widely used in mechanical engineering. One example of an additive process is the mounting of two bars each having a length θ_1 and θ_2 respectively. We may equally think of the mounting of two springs in parallel or two compliances in series. In the first case $\theta_{1,2}$ denotes the spring's stiffness. In the latter $\theta_{1,2}$ characterises the inverse stiffness, i.e. the compliance. We may even conceive a flow process in chemical engineering, mobility research or logistics. In this case $\theta_{1,2}$ denotes the residence time within each individual process and θ the resulting residence time.

Inserting $h(\theta_1, \theta_2) = \theta_1 + \theta_2$ into the generalised convolution integral using the product rule $p(\theta_1, \theta_2) = p_1(\theta_1)p_2(\theta_2)$ for statistically independent processes, yields the convolution integral for the additive process chain

$$p(\theta) = \int_0^\infty p_1(\theta_1)p_2(\theta - \theta_1)\, \mathrm{d}\theta_1.$$

The density functions are normalised, i.e. the zero moment equals one. The first moment is the mean or expected value, the second central moment is the variance:

$$\overline{\theta} := \int_0^\infty \theta\, p(\theta)\, \mathrm{d}\theta, \quad \sigma^2 := \int_0^\infty (\theta - \overline{\theta})^2 p(\theta)\, \mathrm{d}\theta = \int_0^\infty (\delta\theta)^2 p(\theta)\, \mathrm{d}\theta.$$

Using the convolution to express $p(\theta)$ in the two definitions for the mean and variance yields for additive processes the well-known result

$$\overline{\theta} = \overline{\theta}_1 + \overline{\theta}_2, \quad \sigma^2 = \sigma_1^2 + \sigma_2^2.$$

The derivation does not rely on Gaussian density functions, i.e. the result is exact for arbitrary density functions. The generalisation for more than two individual processes reads $\overline{\theta} = \sum \overline{\theta}_k$, $\sigma^2 = \sum \sigma_k^2$.

Recall the example of two assembled bars; it is trivial that the mean value of the total length is the sum of the mean values of the components. But also the variance of the length of the overall structure is the sum of the variances of the bars. For two flow processes in serial connection, the mean value of the total residence time is the sum of the mean residence times within the individual processes. Diffusive mixing can cause a sharp front to blur. Due to the linearity of the diffusive transport, the variance of the signal at the output is equal to the sum of the variances of the two individual processes.

The mentioned flow process is a nice visual model for the propagation of uncertainty in additive processes. This is due to the fact that the density distribution at the outlet $p(\theta)$ can be measured as a change in concentration over time in a single experiment: assume we would add N objects (molecules, ...) in an arbitrarily short time at a time $t = 0$ into a constant flow Q. If the process has the volume V, then we measure the temporal change of the concentration given by $N/(V\bar{t})\ p(t)$ at the process output. Here, $\bar{t}$ is the mean residence time given by $\bar{t} = V/Q$. As always, things become clearer in a dimensionless representation: if we measure time in multiples of the mean residence time, i.e. $\theta := \frac{t}{\bar{t}}$, and concentration c in multiples of the typical concentration N/V, we measure $\delta(\theta) \to p(\theta)$, i.e. a Gaussian normal distribution for a purely diffusive process centred at $\theta = 1$. In a real experiment it is not possible to realise a Dirac delta function $\delta(\theta)$ but it is easy to realise a Heaviside function $\mathcal{H}(\theta)$. This is achieved by changing the concentration at the inlet at a time $\theta = 0$ suddenly from zero to one. In practice this means switching a valve. At the process outlet we then measure the cumulative density function $\mathcal{H}(\theta) \to P(\theta) = \int_0^\theta p(\tau)\mathrm{d}\tau$.

Secondly (ii), in the general case, the dependent variable θ is a nonlinear function of one, two or more independent variables $\theta = h(\theta_1, \theta_2)$. A perturbation of h at the operation point $(\theta_{10}, \theta_{20})$ marked by the index 0 is given by

$$h = h_0 + \left.\frac{\partial h}{\partial \theta_1}\right|_0 \delta\theta_1 + \left.\frac{\partial h}{\partial \theta_2}\right|_0 \delta\theta_2 + \mathrm{O}(\delta\theta^2).$$

Inserting this perturbation into the general convolution integral and neglecting terms of the order of magnitude $\mathcal{O}(\delta\theta^2)$ yields the Gaussian uncertainty propagation [29]

$$\sigma_{\mathrm{g}}^2 \approx \left[\frac{\partial h}{\partial \theta_1}\right]_0^2 \sigma_1^2 + \left[\frac{\partial h}{\partial \theta_1}\frac{\partial h}{\partial \theta_2}\right]_0 \sigma_{12}^2 + \left[\frac{\partial h}{\partial \theta_2}\right]_0^2 \sigma_2^2.$$

The index g indicates the Gaussian approximation of the variance. Here,

$$\sigma_{12}^2 := \int_0^\infty \int_0^\infty (\theta_1 - \bar{\theta}_1)(\theta_2 - \bar{\theta}_2) h(\theta_1, \theta_2)\, \mathrm{d}\theta_1 \mathrm{d}\theta_2 = \overline{\theta_1\theta_2} - \bar{\theta}_1\bar{\theta}_2$$

denotes the co-variance of the two independent variables.

The Gaussian uncertainty propagation is the simplest and most commonly used method of uncertainty propagation, but it is rarely critically questioned. From the derivation it is clear that the trust into the method should decrease with increasing nonlinearity in $h(\theta_1, \theta_2)$. This becomes clear for the simpler case $\theta = h(\theta_1)$. This coordinate transformation shown in Fig. 3.4b leads to a distortion of a symmetric input density function $p_1(\theta_1) \to p(\theta)$. As a result, the mean $\bar{\theta} = \int_0^\infty h(\theta_1)p_1(\theta_1)\,\mathrm{d}\theta_1$ differs from the approximation $\bar{\theta}_{\mathrm{g}} = h(\theta_{10})$ and further σ_{g}^2 is only an approximation of σ^2, Fig. 3.4b. In summary we recognise that the usage of the Gaussian approximation $(\delta\theta)^2 \approx (\delta\theta)_{\mathrm{g}}^2 = \sum (\partial h/\partial\theta_k)^2(\delta\theta_k)^2$ for uncorrelated input data θ_k needs some justification.

For the strongly nonlinear process chain being treated in the following Sect. 3.3, the moment method, i.e. the Gaussian uncertainty propagation is not adequate. In Sect. 3.3 we use the bootstrapping method [10]. Other methods and examples for uncertainty propagation are found in Sect. 6.1.6.

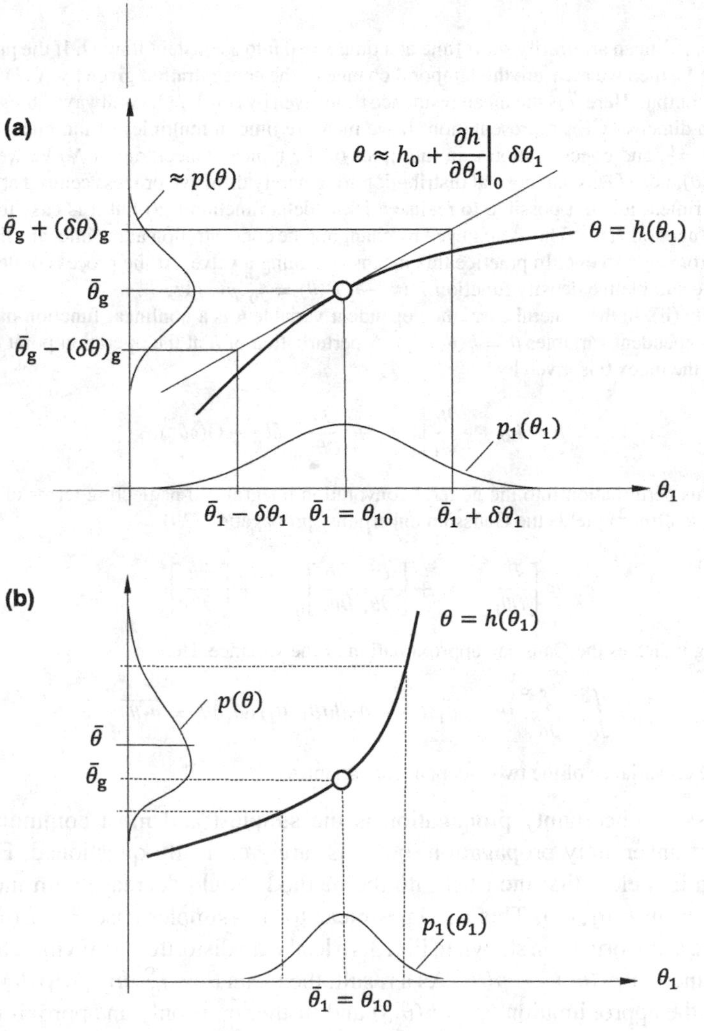

Fig. 3.4 **a** Gaussian uncertainty propagation shown to be associated to a coordinate transformation or mapping $\theta_1 \to \theta = h(\theta_1)$; **b** approximation $\overline{\theta}_g \approx \overline{\theta}$ due to the nonlinearity in $h(\theta_1)$

3.3 Five Complementary Methods for Mastering Uncertainty in Process Chains

Peter F. Pelz

In Sect. 3.2 we applied moment methods of uncertainty propagation to process chains. In several parts of the book we resume these and use more in-depth methods of uncertainty propagation, cf. Sect. 6.1.6. Uncertainty quantification includes uncertainty propagation [30]. Uncertainty propagation is often recognised as uncertainty

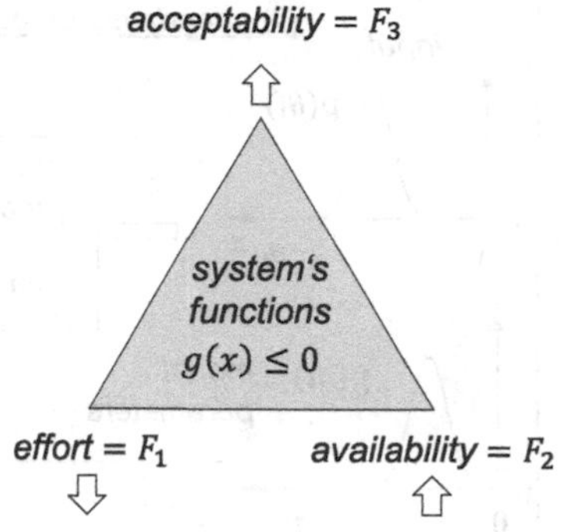

Fig. 3.5 Sustainable Systems Design: Optimal quality subject to functionality

analysis. However, it is only one of at least five methods that form a complementary toolbox for the mastering of uncertainty in mechanical engineering.

We will discuss the five methods of the toolbox from the perspective of Sustainable Systems Design, which was presented in Chap. 1: the shift to the new paradigm 'optimal quality subject to functionality' combines the two incomplete paradigms 'form follows function' (Sullivan) [51] and 'less but better' (Rams) [42]. The quality $F = \{F_1, F_2, F_3\}$ is represented by (i) the effort F_1 to manufacture and operate the system, (ii) the availability F_2 of the system, as well as (iii) the acceptability F_3 of the system by the user and the society. The functionality is represented by the system's functions, which are represented by a subset of the constraints, $g(x) \leq 0$. Here, functional quality is one aspect of acceptability and is measured by the deviation δg from the customer expectation given by the specifications g_s. The lower the loss in functional quality δg, the higher the acceptability.

In summary, this understanding of the interplay between function and quality as described here as well as in Sect. 1.7 and Fig. 3.5 is more general than the understanding in the concept of Robust Design [41]. With Robust Design, firstly, effort is measured in costs, secondly, availability is understood as part of the function, and thirdly, the only measure for acceptability is functional quality. As pointed out in Sect. 1.7, the more general understanding of function and quality allows the mapping of Sustainable Systems Design to a constrained optimisation problem, provided the metrics for effort F_1, availability F_2 and acceptability F_3 are well defined, cf. Fig. 1.10.

In this section, we will introduce the methodological building blocks and show the interaction of the methods by means of a use case. The five methods addressed here are

(i) uncertainty identification,
(ii) uncertainty propagation,
(iii) robust optimisation under uncertainty and/or multi-objective optimisation,
(iv) sensitivity analysis,
(v) model adaption.

Firstly (i), for uncertainty identification we developed the concept of "data-induced conflicts". Here, the trust in underlying data sources is identified, cf. Sect. 4.2. Sec-

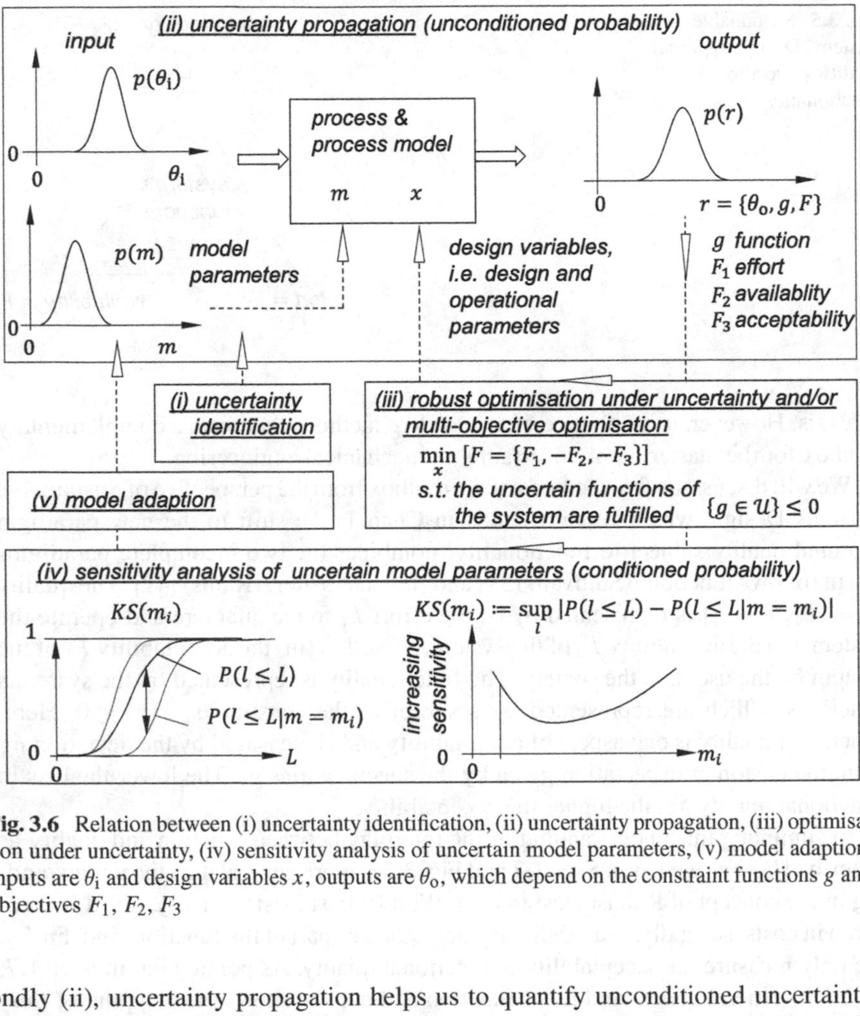

Fig. 3.6 Relation between (i) uncertainty identification, (ii) uncertainty propagation, (iii) optimisation under uncertainty, (iv) sensitivity analysis of uncertain model parameters, (v) model adaption. Inputs are θ_i and design variables x, outputs are θ_o, which depend on the constraint functions g and objectives F_1, F_2, F_3

ondly (ii), uncertainty propagation helps us to quantify unconditioned uncertainty of system output data z, system functions g and system objectives F, cf. Sect. 4.3. Thirdly (iii), robust optimisation under uncertainty combined with multi-objective optimisation helps us to select the design variants x robustly, i.e. in such a way that the system quality F is optimised subject to the uncertain system function g. In other words, the system has an optimal quality F subject to the condition that it has to satisfy not only one, but a set of constraints in the uncertainty set $g \in \mathcal{U}$. Fourthly (iv), sensitivity analysis allows us to indicate those model parameters m which are most relevant for model adaption (v).

The interaction of the five methods with a generic process and the data types introduced in Chap. 2 is best understood in a control loop or flow diagram with the generic input-output-process model, shown in Fig. 3.6 in its centre. At the boundary of the generic process, we recognise the data introduced in Chap. 2, i.e.

(i) functional requirements g and quality data $F = \{F_1, F_2, F_3\}$,

(ii) input u split into design variables x and other input variables,
(iii) output z,
(iv) model parameters m, e.g. mass flux, energy flux, information,
(v) internal variables y.

The overall data of the generic process subject to uncertainty are collected into $r = \{\theta, g, F\}$. The model parameters m and design variants x are inherent to the model as depicted in the flow diagram. Perturbations to the model are not shown but recognised as present. Note that in terms of uncertainty propagation in process chains of Sect. 3.2 the input would be $\theta_i = \theta$ (minus the design variables) and the output $\theta_o = z$. Moreover, the assignment of data to a certain type is not strict. For example, it is common in optimisation to formulate a sub-objective as a constraint, cf. Sect. 5.1.1. Depending on the context and the method, a model parameter can also become an output datum, for example.

To show the relations between the five methods (i)–(v) we will go clockwise through the control loop starting from the generic process. The reliability of underlying data sources (i) is evaluated by means of the concept of data-induced conflicts presented in Sect. 4.2.

The result of the generic process comprises the data z, given by the flows of mass, energy and information, further the system functions g and the system quality F. The density function $p(r)$ is the result of uncertainty propagation (ii) through the process. The propagation of incertitude or stochastic uncertainty through general nonlinear processes are easily applicable for engineers and Monte Carlo experiments form a practicable method. The interpretation of the Monte Carlo experiments is based on the "law of large numbers", see Chap. 2. That is, the signal to noise ratio of the experiment outcomes increases with the number of calculated samples. Hence, the downside of the brute force method in contrast to the moment methods can be the high computational effort.

The result of the robust optimisation under uncertainty combined with multi-objective optimisation (iii) are suggestions how to select the design variants x. Depending on the subjectively selected weights between the three different sub-objectives, the components are selected, their size is determined provided the selection is done from a size-ranged modular design and finally the operational conditions are determined. Hence, the optimal design and operational variants are determined by the robust optimisation under uncertainty and/or multi-objective optimisation.

Sensitivity analysis (iv) is the fourth method in the control loop, cf. Fig. 3.6. In addition to the input data w and the design variants x, there are predefined model parameters m. In axiomatic models, these are often material parameters of constitutive equations, such as the modulus of elasticity in Hooke's law. However, model parameters can also be dimensionless parameters as mentioned above.

Iooss et al. [30] point out that "sensitivity analysis provides users of mathematical and simulation models with tools to appreciate the dependency of the model output from the model input and to investigate how important each model input is in determining its output". There are many methods for sensitivity analysis. Saltelli et al. [44] provide a concise review paper on this topic. Also Ghanem et al. [15] give an overview on sensitivity analysis in Part IV of their "Handbook of Uncertainty Quantification": there are local and global sensitivity methods. The local methods are gradient based, whereas most of the global sensitivity methods are probabilistic. For the method considered here, in contrast to uncertainty propagation, probabilistic sensitivity analysis is based on Bayes theorem, cf. Sect. 4.1.2, i.e. on the concept of conditional probability. From the large number of local and global methods for sensitivity analysis we present only one. A very general, efficient and hence for engineers easily applicable global sensitivity method is the PAWN method used by Holl et al. [27].

From all model parameters m one is selected, i.e. m_i. The sensitivity with respect to the selected parameter is given by Kolmogorov–Smirnov metric as originally proposed by Baucells and Borgonovo [3]. This is the largest distance

$$\mathrm{KS}(m_i) := \sup_l |P(l \leq L) - P(l \leq L|m = m_i)|$$

between the unconditional cumulative probability function $P(l \leq L)$ and the large number of conditional cumulative probability functions $P(l \leq L|m = m_i)$. Here the model parameter $m_i \in [\min, \max]$ is selected arbitrarily from its interval. We have L numerical experiments. The lth experiment is identified by the numerator l.

The larger the sensitivity of m_i the more attention is to be paid to this parameter. The model adaption (v) is gained manually or again by means of optimisation methods. The model adjustment can have several stages: Firstly, model parameters can be adapted. Secondly, the model granularity can be adapted and thirdly, the model type or structure can be customised. The adaptation of model parameters mentioned here is mostly understood as calibration. The model granularity concerns the time and space resolution. The most profound change is the adaptation of the model structure. Here, model structure should not be mismatched with structural uncertainty introduced in Sect. 2.3.

In a data-driven model, the structure is given e.g. by the number of layers in an artificial neural network and the calibration determines the weights of each neuron's input. Obviously the change of structure is more profound than the calibration of the weights, and obviously the calibration is based on optimisation. The counterpart of the data-driven model is the axiomatic model. The structure of such a model must be adjusted if the model is to describe wave propagation, however, the model is represented by an elliptic differential equation. The calibration of data-driven models is widely used today. More interesting is the structural adaptation of axiomatic or grey-box models.

Four case studies are included in the chapter to illustrate the content. Depending on the reader's interests, the case studies may be read or skipped.

The first case study presented applies the concept of a multi-pole process chain, uncertainty propagation by MCS and sensitivity analysis. Furthermore, firstly the concept of parameter optimisation based on scalable modules is presented, and secondly the transformation of a transient investment into a time-invariant cost flow is shown.

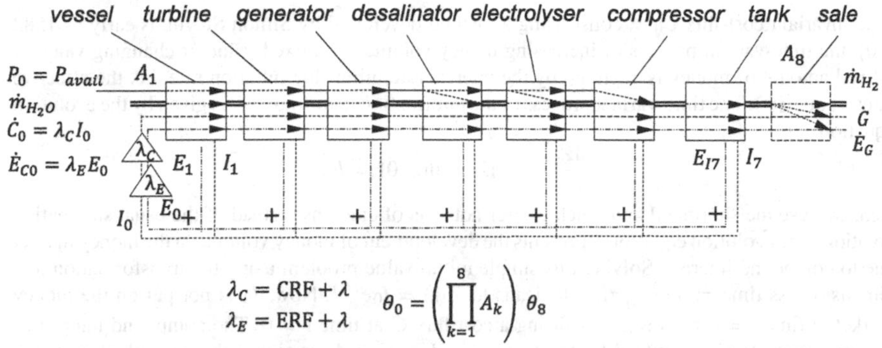

Fig. 3.7 Multi-pole representation of the energy ship; the structure is predefined; the size of the components, i.e. wind powered vessel, turbine, generator, desalinator, electrolyser, compressor and tank are design variants; the same is applicable to the operating parameters, such as the rotational speed of the turbine; to evaluate even ecological gains one information stream may equally be a flux of ecological costs $\dot{E}_C$

1st case study—Mastering uncertainty in the earliest design phase

By way of illustration, we show how the named methods can be applied to control uncertainties in the earliest design phase of a regenerative system. The concept of an 'energy ship' is to be designed and evaluated under uncertainty. This ship is supposed to convert wind energy into electric energy using a wind-powered vessel. A hydrokinetic turbine integrated into the vessel is used to convert kinetic energy of the water into electricity. In a further step, the electricity is used to produce hydrogen using sea water.

The obvious advantages of the 'energy ship' compared to far off-shore wind power are threefold: firstly, the energy supply system solves the transport problem of off-shore electricity production inherent in the system; secondly, availability is increased compared to existing solutions by allowing the ship to avoid a storm; thirdly, known modules, each with known uncertainty, are combined to form a new system. We will return to this point in Sect. 3.5 when speaking about flexibility enabled by combining modules.

The process chain needed as a basis for a decision for or against a technology made under uncertainty is a techno-economic model of the 'energy ship' using the multi-pole representation introduced in Fig. 3.7. The fluxes through the chain are the fluxes of energy, mass and money. The system model is formed by the process chain, the model uncertainty is given by the uncertain physical and economic parameters.

To shed some light on model uncertainty, here, we will only focus on the economic model. The physical model parts may be found in [40]. The simple balance reads profit=revenue-costs: $\dot{G} = \dot{R} - \dot{C}$. There was no periodic profit $\dot{G}$, if the mass specific production cost of hydrogen would equal the market price f_{H_2}. Hence, the system design should be oriented at the threshold $\dot{G} = 0$. If the production costs equal the market price, the production costs are named levelised cost of hydrogen (LCOH), i.e. $\dot{G} = 0 : \mathrm{LCOH} = f_{\mathrm{H}_2}$.

The main and first part of the costs are due to the investment costs I_0. The smaller second part of the costs are due to the maintenance; usually they are assumed to be proportional I_0. The factor λ is a rate, i.e. the dimension is an inverse time. Hence, the maintenance costs rate are λI_0.

Since I_0 usually has to be spent immediately at the beginning of the project, investment is inherently a time variant process given by Dirac's delta function, $I_0\delta(t)$. In order to make decisions under uncertainty, time-variant problems are much more difficult to treat than time-invariant problems, cf. Fig. 3.11. Fortunately, we can transform the transient investment $I_0\delta(t)$ into an equivalent time-invariant cost-flux $\dot{C}_0 = \text{const}$ using a method developed by Simon Stevin as early as 1582 [50]: the time-variant process of increasing money volume for a fixed value or changing value of a fixed amount of money is captured by the interest rate minus the inflation rate, i.e. the effective interest rate $z(t)$. The time-variant process of the information data "money" is given by the evolution equation

$$\frac{\mathrm{d}Z}{\mathrm{d}t} = z(t)Z, \text{with} Z(0) = I_0.$$

Here, we use the shorthand and much clearer notation of calculus instead of the usual summation notation. The evolution equation represents the development of money volume on the money market due to compound interest. Solving this simple initial value problem using the transformation to a dimensionless time $\tau(t) = \int_0^t z(t')\,\mathrm{d}t'$ leads to $Z(t) = I_0 \mathrm{e}^{\tau(t)}$. Now, I_0 is not put on the money market at time $t = 0$ but invested causing a cost flux $\dot{C}$ at time $t > 0$. The compound interest of the time span $[0, t]$ is covered by the factor e^{τ}. Hence, in determining the equivalent cost flux $\dot{C}$ we demand the equivalent $\mathrm{d}I_0 e^{\tau(t)} = \dot{C}$ dt. Integration over the time period $[0, T]$ leads to $I_0 = \int_0^T \dot{C}\mathrm{e}^{-\tau(t)}\mathrm{d}t$. For the special case of constant cost flux $\dot{C} = \dot{C}_0 = \text{const}$ we can define the ratio of cost flux C_0 and invest I_0 as the capital recovery factor.

$$\text{CRF} := \frac{\dot{C}_0}{I_0} = \frac{1}{\int_0^T \mathrm{e}^{-\tau(t)}\mathrm{d}t} = \frac{1}{T}\frac{1}{1 - \mathrm{e}^{-zT}}.$$

In analogy to λ, the dimension of CRF is a rate, i.e. 1/s. The right side of the above equation is gained for constant effective interest rate $z = \text{const}$. As a result, the cost flux is given by the investment costs I_0 multiplied by the sum of the capital recovery factor CRF and the maintenance rate λ: $\dot{C} = I_0(\text{CRF} + \lambda)$. Hence, by Stevin's "trick" we transformed a time-variant process into a time-invariant process.

On the other side, the revenue flux $\dot{R} = \delta T \dot{m}_{\mathrm{H}_2} f_{\mathrm{H}_2}$, gained by the sale of hydrogen, embodies the counterpart of the cost flux. Here, the capacity factor is denoted as δ. Hence, the dimensionless product δT measures the availability of the system introduced in Chap. 1.

From the balance $\dot{G} = \dot{R} - \dot{C} = \delta T \dot{m}_{\mathrm{H}_2} f_{\mathrm{H}_2} - I_0$ (CRF + λ) we derive the levelised cost of hydrogen (LCOH) to be

$$\dot{G} = 0: \ \text{LCOH} = \frac{I_0(\text{CRF} + \lambda)}{\delta T \dot{m}_{\mathrm{H}_2}}.$$

Here, $\dot{m}_{\mathrm{H}_2}$ is the hydrogen mass flow. Speaking again in system's function and quality: the function of the system being here in focus is to convert energy. The quality of the system is given by LCOH. This quality is increased, if the effort measured in investment cost $I_0 = \sum_{k=1}^{8} I_k$ is decreased, if the availability measured in δT is increased or if the maintenance cost rate λI_0 is reduced.

Hence, the design task is described by

$$\min_x \text{LCOH s.t. } \mathrm{g}(x) \leq 0.$$

The constrains $g(x)$ are given by the mulit-pole model of the system. There are binary and real decision variants. A binary decision is needed in deciding whether the hydrogen production on the ship uses salty seawater or stored fresh water. In the first case a desalinator is required as an additional module, in the latter case not; but using fresh water would require an additional tank with a limited capacity. The real design variants determine on the one hand the size of the modules. These are usually obtained as size ranged modules. On the other hand they determine the operational variant such as rotational speed of the turbine.

Here, as opposed to the example in Sect. 1.5 as well as further examples within this book, we ignore the binary decisions. Thus, in our example the structure of the system is predefined and sketched as the process chain in Fig. 3.7. The design variants are provided by the questions "what should be the length $x_1 = L$ of the vessel?"; "what should be the size of the turbine measured in a cross-sectional area $x_2 = A$?" etc.

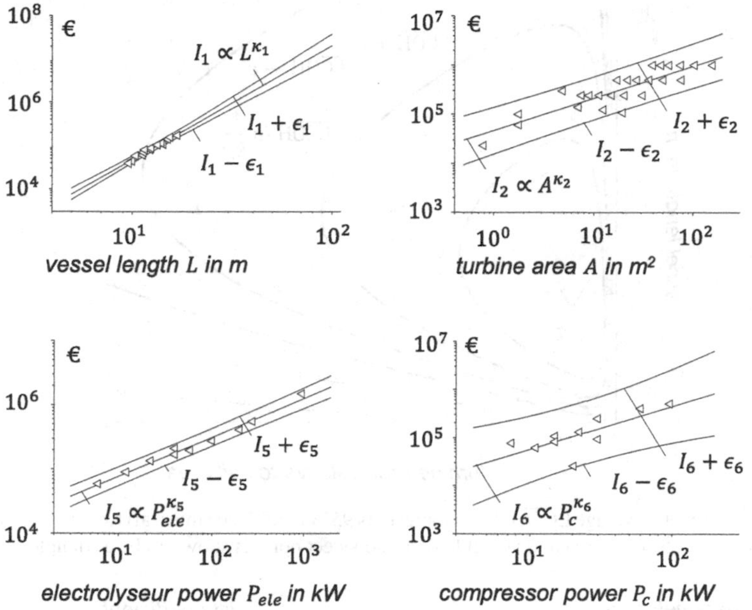

Fig. 3.8 Empirically derived investment cost models $I_k \propto x_k^{\kappa_k}$ for some of the modules needed to compose the energy ship; the 95% prediction interval $I_k \pm \epsilon_k$ is given [27]

The individual investment costs I_k scale with the design variants by means of power laws $I_k \propto x_k^{\kappa_k}$ as a market survey for size ranged components indeed shows, cf. Fig. 3.8.

Together with the physical model based on the energy, continuity and momentum equation, the techno-economical model is defined. The techno-economic-ecologic system model is a multi-pole model as described in the previous section. The model maps the input data $\theta_i = \theta_0$ to the output data $\theta_o = \theta_8$ by the matrix multiplication $\theta_i = A\theta_o$: $A = \prod_{k=1}^{8} A_k$. The entries of the 4×4 system matrix are the technical, economic and ecologic models being described by Holl et al. [27].

Figure 3.9 shows that the techno-economically optimal system analysis is defined by a vessel length of 37 m with the help of stochastic optimisation. At the optimum, the turbine area is 0.9% of the sail area. For this and all the other design variants not shown here, the levelised costs of hydrogen are 13.9 €/kg. Comparing this quality datum with the 2–3 €/kg for methane reformation shows that the production costs, even for the optimal system, are still high.

So far we have only addressed the prediction uncertainty of the component models, cf. Fig. 3.7. A global sensitivity analysis allows us to identify those model parameters $m_i \in m$ on which the optimisation result shows the largest sensitivity. In return, when adapting the model one should concentrate on mastering the uncertainty in the most sensitive model parameter.

Figure 3.10 shows the mean of $\text{KS}(m_i)$ [3] for some model parameter. The result is most sensitive with respect to uncertainty in the sail's lift parameter c_l. The system shows only a small sensitivity to the vessel displacement parameter relating the displacement volume to the cube of the ship's length, $\epsilon_D := \frac{V}{L^3}$. Interestingly, the system shows only an intermediate sensitivity to the availability δT.

In summary, this section reveals that the process chain model is at the heart of methods for mastering uncertainty. The methods discussed are complimentary and differ, for example, in terms of conditional and unconditional probability. Finally, we have used an example to show how a system can be evaluated in an early design phase. Here,

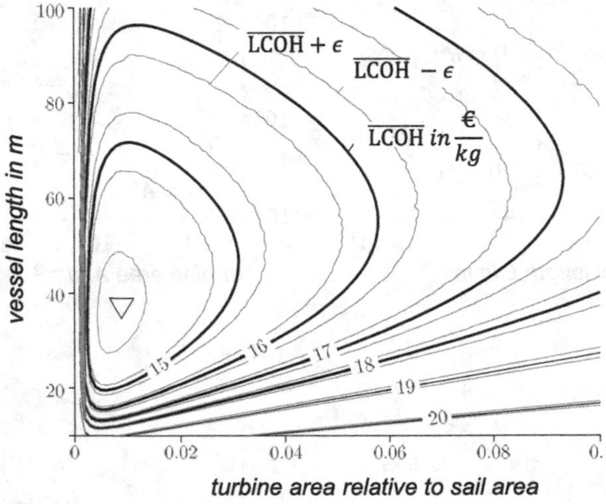

Fig. 3.9 Arithmetic average $\overline{\text{LCOH}} \pm \epsilon$ (ϵ represents 95%-confidence interval) for all combinations of dimensionless turbine area and vessel length; the sweet point is shown to be a triangle [27]

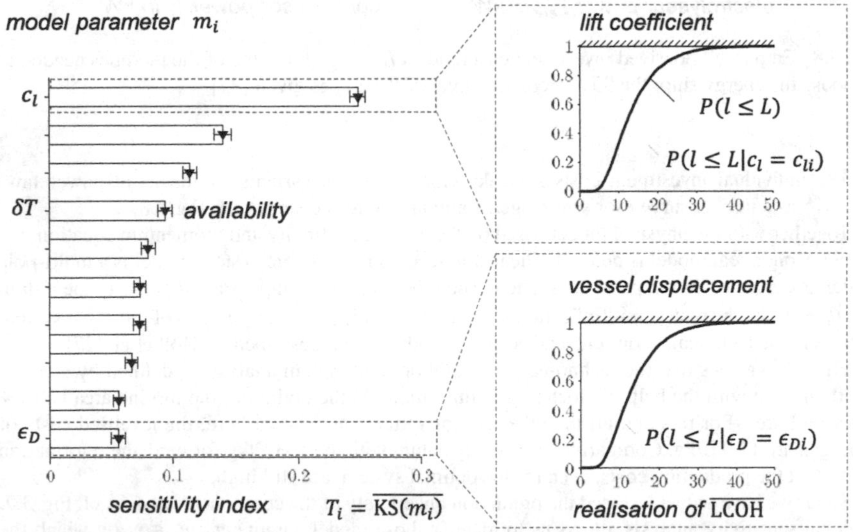

Fig. 3.10 Sensitivity analysis of the techno-economic optimal system (left) and graphical clarification on the example of the most sensitive and the less sensitive parameter [27]

it is necessary to map a time-variant process into a time-invariant process. Whether and how such a transformation succeeds is always problem- and model-specific. In the following section we will see how degradation can be treated in time-invariant terms, section by section. This leads us to the next section where we classify the time characteristics of process chains.

3.4 Time-Variant, Dynamic and Active Processes

Peter F. Pelz

Mechanical engineers are trained to solve problems in the field of dynamics, but they are much less familiar with the task of creating an optimal design of transient processes under uncertainty. Experience has shown that making decisions under uncertainty is much more difficult for time-variant processes than for time-invariant processes. Therefore, in the following criteria are to be compiled to determine when an actually time-variant process can be treated as a time-invariant process. Before doing this, it is worthwhile to classify the time-variant characteristics of the processes.

Figure 3.11 shows the Euler diagram that classifies the time characteristics of processes. Synonymous for time-invariant is stationary, synonymous for time-variant is transient.

A process is quasi-stationary if it reacts immediately to a temporal change in the boundary condition. In dynamics, a process is quasi-static and at the same time quasi-stationary when the system is excited by a frequency f far below the system's lowest natural frequency: $f \ll 1/\sqrt{LC}$. Here L is the inductance representing a solid or liquid body or an electric coil and C is the capacity or compliance associated with the storage of energy in a spring, in a gas volume or within an electric accumulator.

There is a slightly different case when the external load is in balance with frictional, i.e. dissipative $\propto R$, and elastic, i.e. conservative $\propto C^{-1}$, forces. The system then shows a relaxation time which is given by the product RC. The relaxation time is the typical time a system needs for relaxing to an equilibrium when disturbed. For $f \ll (RC)^{-1}$ the system behaves quasi-stationary. The model granularity with respect to temporal resolution is determined by the functional quality of the process

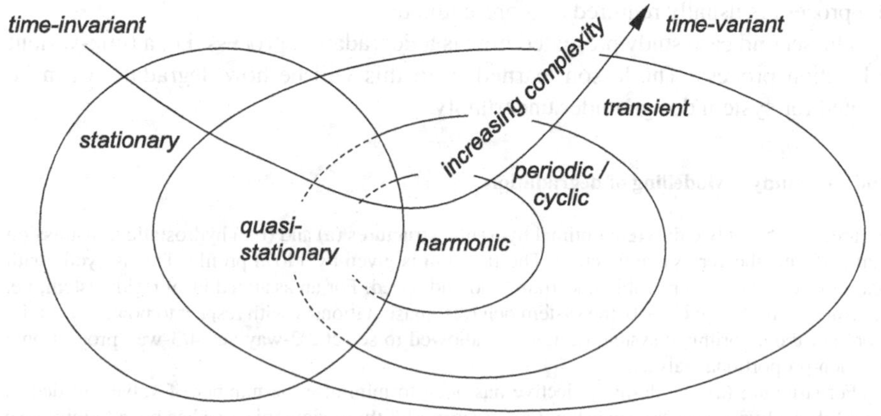

Fig. 3.11 Euler diagram for classification of time-variant and time-invariant processes

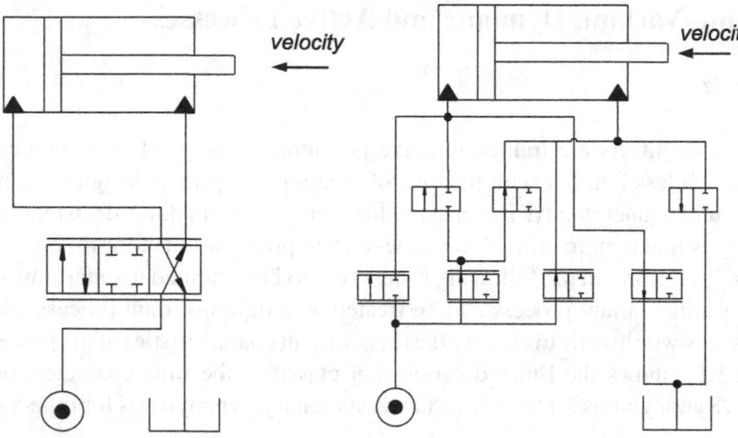

Fig. 3.12 Design of a hydrostatic transmission aiming for **(a)** the minimal number of valves; **(b)** minimal particle erosion over the life time of the hydrostatic transmission [53]

which shall be resolved in time. Thus the highest natural frequency being resolved is determined by the quality promise given explicitly or implicitly to the customer.

Special types of time-variant processes, i.e. transient processes, are periodic, i.e. cyclic processes. Again, a special type of cyclic processes are harmonic processes. Figure 3.11 indicates that the process complexity in terms of time behaviour is reduced if a time-variant process can be mapped to a time-invariant process or if a transient process can be mapped to a harmonic process. An example of the former was given in the first case study, in which the time-variant investment was mapped to the equivalent time-invariant cost flux, for the latter transforming the linearity of the process is usually required as a prerequisite.

The second case study presented now is a degradation process, i.e. a time-variant relaxation process. The lesson learned from this will be how degradation can be treated for system design under uncertainty.

2nd case study—Modelling of degradation

Figure 3.12 shows two different optimal hydraulic structures (a) and (b) a hydrostatic transmission that performs the very same function. The function is given by a load profile. This is cyclic with phases of constant and possible uncertain load and speed. For an assumed ideal rigid system, i.e. vanishing compliance $C \to 0$, the system behaves quasi-stationary with respect to power transmission. For the algorithmic system design it is allowed to select 2/2-way and 4/3-way proportional and non-proportional valves.

For structure (a), the design objective has been to minimise the number of valves needed to establish the hydrostatic transmission. For structure (b), the design objective has been to minimise the particle wear of the valves in each load cycle over the life time of the hydrostatic transmission. Mastering the structural uncertainty, cf. Sect. 2.3 in case (a) is simple: the design solution with two proportional and two non-proportional valves is obvious. The control of the structural uncertainty in case (b), on the other hand, is far from obvious. Although the system is dynamically time-invariant, due to the degradation process the system is time-variant in a long time frame. To master the structural uncertainty in the design process a degradation model has to be envisaged in the constrained optimisation.

In this case study, the degradation is depicted very generally as a relaxation process. This approach can be applied to many other ageing and wearing processes: particle erosion is one possible reason for degradation. Others are chemical degradation or hardening of polymer material due to temperature activated diffusion and/or chemical reaction. Also damage accumulation in cyclic loading of components may be treated as a degradation process.

For each of the mentioned processes, the degradation of a component property F may be described by an evolution equation $\frac{\mathrm{d}F}{\mathrm{d}t} = f(t, L(t), F)$. Here, t is the time and $L(t)$ the load history of the component. The initial condition is $F(t=0) = 1$ for the degradation process and $F(0) = 0$ for the complementary accumulation process. A separation approach of the right hand side is a good model of most degeneration or accumulation processes seen in nature and technology.

$$\frac{\mathrm{d}F}{\mathrm{d}t} = K\big(L(t)\big) f(F). \tag{3.1}$$

Here, the load history $L(t)$ and the time t enter the equation only implicitly by means of the degradation rate $K = K(L)$. The dependency is implicit, since only the rate depends on the load history. In fact we came across the evolution equation in the previous Sect. 3.3 where the rate is represented by the interest rate $z(t)$. The interest rate is determined by the global market and the central bank. The degradation rate is determined by the process intensity, i.e. the load. If we know $K(L)$ and the load history $L(t)$ we know the erosion process. The load history $L(t)$ is either determined by the measurement in the usage phase or by providing a model for the expected load history $L(t)$ based on our past experiences gained in the same or a similar process.

Transformation to a dimensionless time $\mathrm{d}\theta := K\ \mathrm{d}t,\ \theta(0) = 0$, yields $\frac{\mathrm{d}F}{\mathrm{d}\theta} = f(F)$. Thus, with an increasing load the rate K increases and the physical time t moves faster. Here, $F{:=}M/\widehat{M}$ is a product property M in relation to its initial value $\widehat{M}$ in the case of a degeneration or to the saturation value $\widehat{M}$ in the case of an accumulation influencing the product's function.

In technical systems often the initial degradation or accumulation phase is more relevant than the later phase. With a view to the asymptotic limit in the initial phase, the initial phase on the right-hand side of the evolution equation is independent of F. Hence, an approximation for the initial degradation or initial accumulation process is given by $\frac{\mathrm{d}F}{\mathrm{d}\theta} \approx \mp 1$. This leads to the damage accumulation.

$$\frac{M(t)}{\widehat{M}} = \theta(t) = \int_0^t K\left[L(t')\right] \mathrm{d}t', \tag{3.2}$$

known as Palmgren–Miner's rule [33, 37].

Figure 3.12b shows the optimal hydraulic structure with regard to the minimal particle erosion over the lifetime of the hydrostatic transmission. For particle erosion, the erosion rate is a function of the volume fraction c_v of solid particles within the fluid, the pressure drop history $\Delta p(t)$ across the valve, the history of the relative valve opening $0 \leq z_+(t) := \frac{z(t)}{z_{\max}} \leq 1$, the maximal cross sectional area $\pi D z_{\max}$ being proportional to the diameter D of the spool valve and the oil density ϱ, cf. Fig. 3.13. For dimension reasons we have the equivalent representation of the degradation rate $K(c_v, z_+, \Delta p, \varrho, \pi D z_{\max}) = K_+(c_v, z_+(t))\sqrt{\Delta p(t)/(\pi D z_{\max}\varrho)}$. For a load profile such as the one sketched in Fig. 1.7b the degradation process may be treated as stationary in each load phase of the load cycle.

The valve's function is given by the pressure amplification factor $V{:=}\frac{\mathrm{d}p}{\mathrm{d}z}$. Hence, the functional degradation Δz in each stationary phase of duration Δt is given by

$$\frac{\Delta z}{z} = \frac{\Delta z_+}{z_+} = \exp\left(-\frac{K_+}{3}\sqrt{\frac{\Delta p \Delta t^2}{\pi D z_{\max}\varrho}}\right) - 1. \tag{3.3}$$

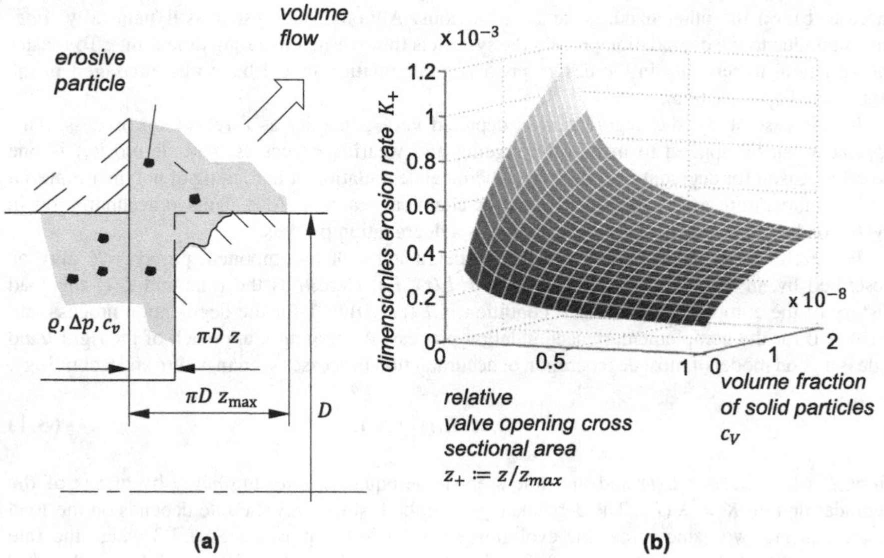

Fig. 3.13 Erosion model of the particle wear of a spool valve **a**. The experimental data **b** are generalised by means of a dimensional analysis [53]

With a view to finding an optimal structure in terms of minimal wear depicted in Fig. 3.12b a mixed integer nonlinear program (MINLP) is to be solved by

$$\min_{x} \max_{\text{valves}} \sum_{\text{load phases}} \Delta z \text{ s.t. } g(x) \leq 0\,. \tag{3.4}$$

In the MINLP the hydrostatic transmission is represented by a complete graph. Each possible valve, Fig. 1.7a, is mapped to an edge of the graph. The nodes are the hydraulic connections. The constraints $g(x)$ are given on the one side by the continuity and energy equations for each edge or node. On the other side they are given by the load profile sketched schematically in Fig. 1.7b.

As lessons learned, firstly, mastering structural uncertainty requires here the solution of a MINLP. Secondly, model uncertainty is relevant for the degradation model depicted in this case study. Thirdly and finally, the load profile sketched in Fig. 1.7b may be uncertain itself. Taking into account this uncertainty requires robust optimisation, cf. Chap. 6 techniques as presented in this book.

Costs and gains of active components

If one thinks of technologies to master uncertainty, one can think of active or "smart" technologies. Active technologies promise new degrees of freedom and the ability to adapt to unforeseen circumstances. Despite this promise, it is surprising to find that in practical applications far fewer active solutions are implemented than engineering research would suggest.

On the one hand, a reason for this is that the total costs for integrating active components can be considerable. On the other hand,—contrary to the actual intention—active components can also make a system vulnerable, because active components do not only include an actuator but also sensors, controllers as embedded software modules, software interfaces, cables, and connectors, all of which may fail. As seen in Chap. 1, training, legal conformity certification, commissioning, release, versioning, maintenance, procurement and much more increase the number of topics that influence the uncertainty of the system from a holistic point of view. If these are not mastered, the system may possibly be damaged.

It is therefore a conflict of objectives that has to be resolved. What are the costs and gains of the active components? Figure 1.11 helps with orientation: it is a matter of clearly formulating and weighing the function and the objectives and then deciding whether to use active components or not. What is also helpful is to answer the question: "What does an active component promise in an ideal case?". The answer to this question is an answer such as "as good as it gets" with respect to effort, availability, acceptability. The answer to the question "What does an active component promise in an ideal case?" usually leads to a Pareto boundary, Sect. 1.6.

Also here, mastering uncertainty is made easier if individual processes are defined. Hence, it is worthwhile to classify types of active processes. In the generic process chain shown in Fig. 3.1 we distinguish between active and semi-active processes. The classification is as follows [6]:

(i) for an active process, energy is transferred to the process and finally to the product or system directly;
(ii) by a semi-active process only the work equipment is adapted by the input energy;
(iii) a passive component or process is neither active nor semi-active.

A passive process may be a storage or a transport process. A passive component is e.g. a coil spring. An illustrating example of a semi-active process is the semi-active damper shown in Fig. 1.2. Manually, the damper setting may be changed from "hard" to "soft" by rotating the actuating lever. In the usage phase, no external energy is supplied for controlling the two connected sub-systems. This is different for the Active Air Spring being presented in Sect. 3.6.2 where the external energy is used to isolate two dynamic systems from each other. Thus, even the system's resonance when excited at the natural frequency may be suppressed by the active system.

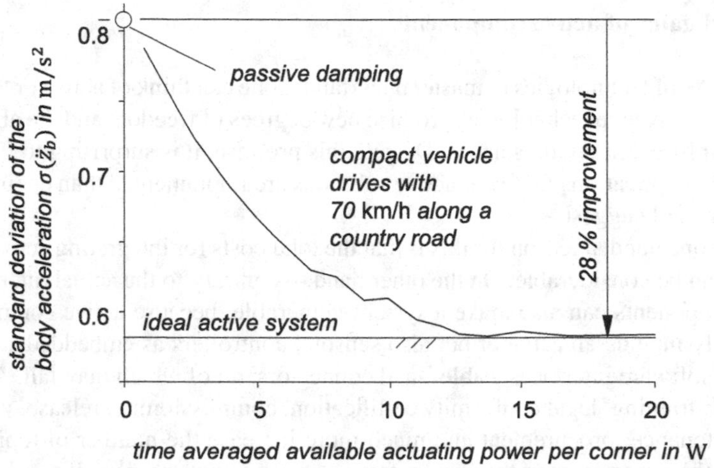

Fig. 3.14 Improvement of functional quality through an active component, here the Active Air Spring: A compact vehicle drives over a country road at a constant speed of 70 km/h. The standard deviation of the body acceleration is plotted against the time-averaged actuating power. The horizontal asymptote is determined analytically under the assumption of unlimited actuating power and actuating speed [43]

Example—Suspension system with integrated Active Air Spring

Figure 3.14 illustrates the improvement of the functional quality when replacing a passive component by an active one. Here, the benefit is an improved isolation function of a vehicle body from a country road's bumps. The picture results from a hardware-in-the-loop ride of a compact vehicle with a speed of 70 km/h along the road, cf. Fig. 3.14. The active component is the real hardware whereas the road and vehicle are virtual. For the passive system the standard deviation of the chassis acceleration is 0.81 m/s^2. The asymptote is given for the ideal active system by 0.58 m/s^2. This asymptote is formally derived by assuming an actuator with unlimited available actuating power and unlimited actuating speed. Figure 3.14 shows that, with the Active Air Spring, this level is sufficiently approached with a power consumption of roughly 10 W to 15 W for a compact vehicle such as the VW Golf. For reasons of dimension this power increases linearly with mass. The power consumption for a luxury vehicle having twice the body mass is 30 W for one vehicle corner and hence 120 W on average for the vehicle.

With regard to Figs. 3.5 and 3.14, the acceptability of the system arises by an increased quality of the functional performance. On the other hand, the energy and cost efforts are increased by the active components. Moreover, the additional components reduce availability due to their vulnerability as discussed above.

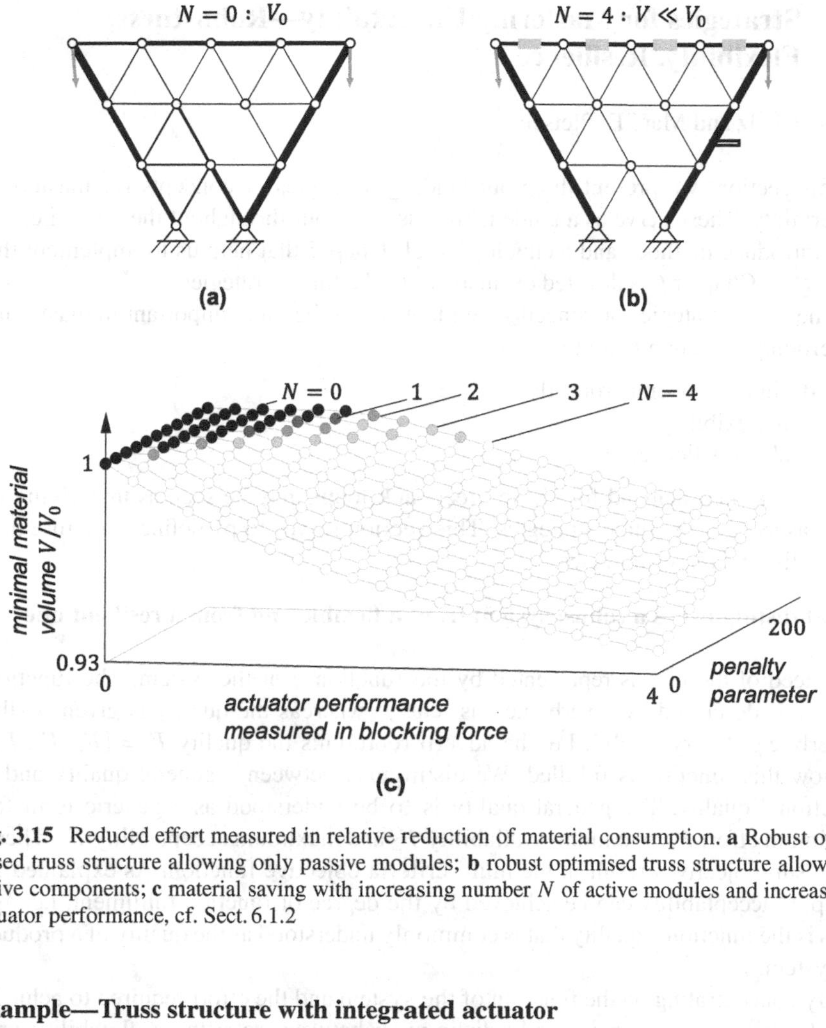

Fig. 3.15 Reduced effort measured in relative reduction of material consumption. **a** Robust optimised truss structure allowing only passive modules; **b** robust optimised truss structure allowing active components; **c** material saving with increasing number N of active modules and increasing actuator performance, cf. Sect. 6.1.2

Example—Truss structure with integrated actuator

In the above example, the active components serve to increase the functional quality of the system. From a different point of view, we may think of having the functionality unchanged and the active components serve to reduce the effort in material consumption.

This is exemplified by comparing two truss structures shown in Fig. 3.15a, b: Fig. 3.15a shows the optimised topology resulting from a robust optimisation of a passive, static system; Fig. 3.15b shows the structural optimisation for the same function, namely bearing defined load scenarios, when additional active components marked by a red box are integrated. Figure 3.15c shows the volume saving relative to the optimal passive system by replacing $N = 1, 2, 3, 4$ passive bars with active bars. The actuator performance is measured in the dimensionless actuating force ranging from 0 to 4.

3.5 Strategies for Mastering Uncertainty—Robustness, Flexibility, Resilience

Peter F. Pelz and Marc E. Pfetsch

In this section, we present three outstanding strategies or concepts for mastering uncertainty. These serve as a guide taking us on a tour throughout the book, i.e. we will introduce methods and technologies, cf. Chap. 5 that help us to implement the strategies. Chapter 6 is devoted exclusively to the three strategies.

The three strategies or concepts we identified to be most important in mastering uncertainty, cf. Chap. 6 are to

(i) design and operate robustly,
(ii) gain flexibility,
(iii) enable resilience.

The so far cited demand for closing feedback loops may be seen as underlying to most aspects of the three strategies. This section serves to predefine, structure and exemplify the three strategies.

What distinguishes a robust system from a flexible and from a resilient one?

The need of the user is represented by the function g of the system. The function is always described by a verb such as 'carry' whereas the quality is given by the adverb, e.g. 'carry safely'. I.e. the adverb represents the quality $F = \{F_1, F_2, F_3\}$ of how this function is fulfilled. We distinguish between a general quality and a functional quality. The general quality is to be understood as a generic term for minimum effort F_1, maximum availability F_2 and maximum acceptability F_3. These three sub-objectives result in the multi-criteria objective function. As explained in Chap. 1, acceptability can be achieved by the degree of function fulfilment, i.e. δg. This is the functional quality that is commonly understood as the quality of a product or system.

By concentrating on the function of the system and the effort required to achieve this function, we can now easily distinguish between robustness, flexibility and resilience. The differences are concisely collected in Fig. 3.16.

(a) A robust system is characterised by the fact that the system fulfils one predefined function g with accepted functional quality δg, even if the function and resources have been described uncertainly or are uncertain themselves—as long as they are part of the specified uncertainty set—or even if the system is disturbed by uncertain external influences. For the constrained optimisation problem the objective function is given by $\min\{F_1, -F_2, -F_3\}$ where the effort is given by F_1.
(b) A flexible system is characterised by the fact that the system fulfils $i = 1, \dots, N$ predefined functions g_i with accepted functional quality δg_i. For the constrained optimisation problem the objective function is given by $\min\{\Sigma_i F_{1,i}, -\Sigma_i F_{2,i}, -\Sigma_i F_{3,i}\}$. Usually the effort F_1 is the most important task; i.e. several functions should be possible with minimal effort: $\min \Sigma_i F_{1,i}$.

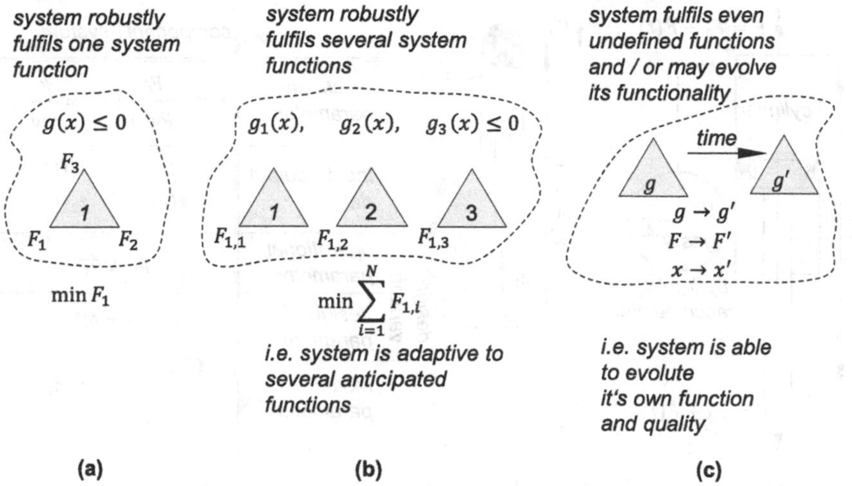

Fig. 3.16 Distinguishing robust, flexible and resilient systems from the perspective of the system's functions: **a** a robust system does not only fulfil its function at the design point, but also in the surrounding neighbourhood, the so-called uncertainty set; **b** a flexible system does not only fulfil one function but several functions i.e. several design points; **c** a resilient system fulfils its function at the design point like a robust system but still enables a residual function when disturbed; the system has the inherent ability to recover

(c) A resilient system is characterised firstly by the fact that it does not only fulfil a predefined function g such as a robust system, but also retains a residual function g_{res}, if the system is disturbed at time $t = 0$. A resilient system may secondly show the ability to recover: from a distorted state, the system may recover in a time $t = 0, \ldots, T$ to a function $g(x) \rightarrow g'(x')$ and/or a quality $F(x) \rightarrow F'(x')$. This function and quality may not be foreseen in the previous design and production phases. The design variants undergo transformation from x to x' during the evolution.

Robustness

Above we have defined the characteristics of a robust system. Here we would like to discuss how to establish a robust system. The comprehensive strategy is a methodology with many facets called robust design. Robust optimisation is a method needed in this methodology as we will demonstrate in a short example within this section and indeed in many examples throughout this book.

Robust Design goes back to Genichi Taguchi [14, 41]. In 1949, he started to develop the methodology of offline quality engineering, as Taguchi called it. The development of the method was accompanied by a project aimed at modernising the Japanese telecommunication system. The methodology proceeds from the system level to the product and process parameters. To reduce the influence of uncertain product and process parameters on the performance or functional parameters of

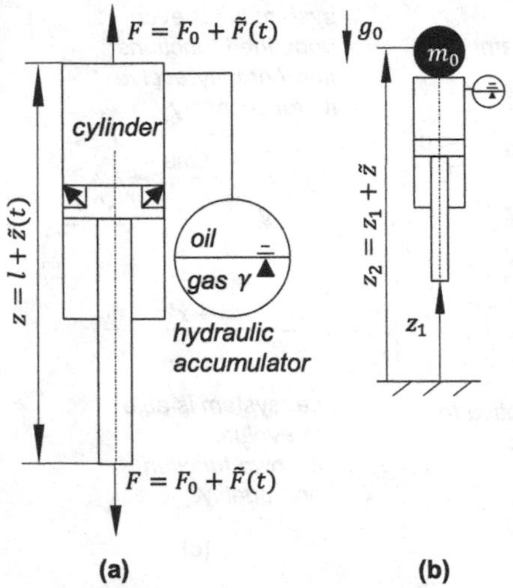

		component	system
	functional parameter	$k = \gamma \frac{F_0^2}{p_0 V_0}$	$\omega^2 = \frac{k}{m_0}$
	input / output data	$F(t)$ $z(t)$	$z_1(t)$ $z_2(t)$
design variants	operational parameter	$p_0 \pm \delta p$	
design variants	design parameter	$V_0 \pm \delta V$	
	model parameter	γ	γ, m_0

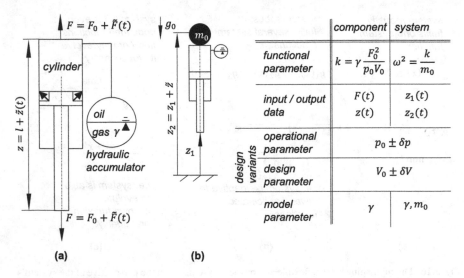

Fig. 3.17 Robust parameter design of a hydro-pneumatic suspension strut treated as a constrained optimisation problem; **a** shows the component and **b** the system; given is the list of functional parameters, input/output data, decision parameters and model parameters; note the change in parameters by broadening the view from the component to the system

the system, the sensitivity of design and process parameters are first determined by Design of Experiments (DoE). Following this first step called system design, the design variants are selected by robust optimisation. This second step is called parameter design. In the last step, called tolerance design, the tolerances for the design and operating parameters are defined.

Robust control in control theory is an approach to controller design. Hence, it has only one component of the system, the controller in focus. This controller shall provide its function for an incertitude range of system parameters. Thus a robust controller is static in contrast to an adaptive controller, adapting itself to system's variations [1].

In conclusion, the robustness of a system is gained by a methodology. Per se it does not need any special effort measured in investment or material. Of course, it is important for engineers today to know and apply the methods of Robust Design. They are constantly being further developed. Thus, the robust optimisation described in this book lays the foundation for Robust Design.

3rd case study—Parameter engineering in the methodology of Robust Design

The following case study firstly illustrates the parameter design as understood by Taguchi. Secondly, the case study further illustrates the benefits of inherently robust design and operational concepts when discussing the various physical effects that can be used in a suspension system.

Figure 3.17 shows the hydraulic scheme of a hydro-pneumatic suspension strut known from mobile hydraulics, forming presses or from the chassis of the Citroen DS passenger car. The function of the system is given by three sub-functions $g = \{g_{\text{i}}, g_{\text{ii}}, g_{\text{iii}}\}$:

(i) The first sub-function supports a chassis of mass m with the mass specific gravitational constant g_0 resulting in the time averaged force $F_0 = m_0 g_0$ at a distance l above the ground.
(ii) The second sub-function, the levelling known from the Citroen DS, is to enable a constant distance l independent of the loading $F_0 = m_0 g_0$.
(iii) The third sub-function is to isolate the chassis from vibrational excitation by the bumpy road.

This third sub-function is ensured by having a compliant system with the capability to store energy. Thus, the system has a natural frequency as discussed in Sect. 3.4. For an excitation frequency $\Omega^2/2 > \omega^2 := \frac{k}{m}$ the system is in isolation. Hence, the natural frequency ω_s is a predefined, i.e. specified value characterising the function of the system.

In Chap. 1 we discussed the objectives to be met when designing systems, cf. Fig. 1.9. All the three part-objectives, effort F_1, availability F_2 and acceptability F_3 form the general quality of the system. The acceptability is fostered when the special quality of the system, i.e. the degree of the functional fulfilment is improved. Every physical-technical system that fulfils the functional description given above competes in quality. The latter is given by the variance, i.e. the stochastic uncertainty $(\delta\omega)^2$ of the square of the system's natural frequency. The better the desired natural frequency is hit by a technical system the better is the system's functional quality.

In a mass production we anticipate the uncertainty of the functional parameters, here the square of the natural frequency ω^2 as $\omega^2 = \overline{\omega}^2 \pm (\delta\omega)^2$. While the discrete variants are assumed to be already selected when defining the structure, the task of the parameter design introduced by Taguchi is to select the continuous design variants u in such a way that the constrained optimisation problem is solved:

$$\min_u (\delta\omega)^2 \text{ s.t. } \overline{\omega}^2 = \omega_s^2. \tag{3.5}$$

Here, ω_s is specified and thus fixed, whereas $\overline{\omega}$ and $\delta\omega$ both depend on the design variants u. The task function is clear: we would like to have a minimal uncertainty in the functional parameter, i.e. the eigenfrequency. In the competition of physical modules we compare three systems, coil spring, hydro-pneumatic suspension and air spring. A coil spring alone would allow the first and the last sub-functions but not the levelling function. Hence, the coil spring fails the competition. An air spring allows all the three sub-functions; and air springs are indeed standard for load-bearing systems when there is a large variance in the mass $m_0 \in [m_{\min}, m_{\max}]$ as it is given in commercial or rail vehicles. But air springs require a compressed air system including compressor, air dryer and filter to level the system. This results in a constant time averaged gas volume V_0 in a loaded condition independent of the loading F_0 and hence mass m_0.

The needed effort, i.e. the investment in hydro-pneumatic systems is usually smaller. Here, the sum of gas and oil volume is constant within the suspension strut sketched in Fig. 3.17. With increased load F_0 oil is pumped by means of a pump into the system of constant volume. Hence, the gas within the hydraulic accumulator is compressed. The gas mass m_g does not change by the levelling.

For a minimal effort, we therefore would select in the first place a hydro-pneumatic suspension system. The design variants to be selected for the system are the absolute gas pressure p_0 of the accumulator and the gas volume V_0 of the accumulator in the unloaded condition, $u = (p_0, V_0)$. In our nomenclature p_0 is an operational variant for the manufacturing when the system is initially filled and V_0 is a design parameter. Both decision variants are uncertain. Presuming we know the incertitude in both decision variants, i.e. the operational variant of the production phase $p_0 = \overline{p}_0 \pm \delta p_0$ and the design variant $V_0 = \overline{V}_0 \pm \delta V_0$ we may solve the constrained optimisation problem.

Ignoring uncertainty of the use phase assumes a fixed suspended mass $m_0 = \text{const}$. The stiffness k of the suspension is the derivative of the force F compression z relation, i.e.

$$k := \left.\frac{\mathrm{d}F}{\mathrm{d}z}\right|_{F=F_0} = \gamma \frac{F_0^2}{p_0 V_0}. \tag{3.6}$$

Here γ is the isentropic exponent of the gas, i.e. a model parameter, cf. Fig. 3.17. With $k^2 = \bar{k}^2 + (\delta k)^2$ and the specified stiffness $k_s = \omega_s^2 m_0$ the constrained optimisation problem can be written as in an equivalent unconstrained problem by introducing the Lagrange parameter λ

$$\min_{p_0, V_0, \lambda} \left[(\delta k)^2 - \lambda(\bar{k} - k_s) \right] \tag{3.7}$$

(the two terms inside the square brackets sum up to the Lagrangian function). This problem is solved using Gaussian error propagation introduced in Sect. 3.2 for uncorrelated uncertainty

$$(\delta k)^2 = \left(\frac{\partial k}{\partial p_0}\right)^2_{(V_0, p_0)} (\delta p_0)^2 + \left(\frac{\partial k}{\partial V_0}\right)^2_{(V_0, p_0)} (\delta V_0)^2. \tag{3.8}$$

Hence, the parameter design, as Taguchi named it, yields the optimal value for the design variants:

$$\left(\frac{p_0}{\delta p}\right)_{\text{opt}} = \left(\frac{V_0}{\delta V}\right)_{\text{opt}} = \sqrt{\frac{\gamma}{k_s} \frac{F_0}{\delta p \, \delta V}}. \tag{3.9}$$

For this set of design variants, the product is said to be robust in the sense that the disturbances, i.e. the uncertainty in realising the pressure and volume, has the minimal impact on the uncertainty of the demanded functional property, i.e. the uncertainty of the stiffness.

So far Robust Design was applied and used on a component level; and as Fig. 3.17a indicates, it is the component for which the functional quality is indeed the stiffness. However, the quality the customer is interested in is not the quality of the component but the quality of the system. For the suspension system the functional quality is the natural frequency, Fig. 3.17b, and not the component's stiffness. Thus the sketched example is a typical example of how the system boundary, and hence the system view, cf. Fig. 3.3a, influence the design objective. For the—already disqualified—coil spring we have $\omega \propto 1/\sqrt{m_0}$. If we demand $(\delta\omega)^2$ to be small as a quality measure, storing energy by an elastic torsion of a beam—as it is done by storing energy in a coil spring—is not a physical effect ensuring robustness. The hydro-pneumatic suspension discussed above gives $\omega \propto \sqrt{m_0}$, i.e. the natural frequency increases with increasing load. This results directly from the given stiffness. Hence, both physical effects do now allow robustness with respect to mastering uncertainty in $m_0 \in [m_{\text{min}}, m_{\text{max}}]$.

There is a third principle to fulfil the three sub-functions. An air spring allows $\omega = \text{const}$ due to separating the sub-functions into (I) carrying and (III) isolating. In other words the air suspended system inherently uses three effects of functional separation. The load carrying sub-function is gained by the force balance $F_0 = (p_0 - p_a)A = m_0 g_0$ with the ambient pressure p_a and the springs cross-sectional area A. The levelling is done by adapting the pressure p_0 to the loading. The air spring's stiffness is

$$k := \left.\frac{\mathrm{d}F}{\mathrm{d}z}\right|_{F=F_0} = \gamma p_0 \frac{A^2}{V_0}. \tag{3.10}$$

Hence, the natural frequency of an air spring with $p_0 \gg p_a$ is

$$\omega^2 = \frac{k}{m_0} \approx \frac{1}{\gamma} \frac{g_0}{\frac{V_0}{A}}. \tag{3.11}$$

It is independent of the uncertain model parameter $m_0 \in [m_{\text{min}}, m_{\text{max}}]$ and the system is robust with respect to uncertainty in the usage phase at least in this one aspect. Typically, the height of a cylindrical spring $\frac{V_0}{A} \approx 350$ mm. This results in a natural frequency of 1 Hz. Indeed, if we compare springs in different applications, such as chassis of trains, commercial vehicles or passenger cars, they all show the same height of roughly 350 mm. The height is adapted to the sensitivity of adult humans regarding vibrations. The design is inherently robust with respect to the uncertainty in the usage phase.

Employing inherent robust design and operating concepts

We have learned from the above example that sometimes uncertainty can be overcome by choosing an inherently robust design and/or an inherently robust operating concept. In this example, energy storage in a gas volume is chosen as the physical effect. In the selected operating concept, the time-averaged gas volume (not the gas mass) is constant. This means that all three sub-functions mentioned above can be fulfilled independently of each other: carrying, levelling and isolating. In summary, the separation of functions was chosen as a concept to achieve robustness.

Separation of function is not the only concept inherently leading to robustness. Here we give a short, certainly incomplete list of design principles that are familiar to many engineers. Each concept fosters robustness and is hence a way how to master uncertainty. The seven design principles that inherently lead to robust construction solutions are to:

(i) enable overload protection,
(ii) enable overload capacity,
(iii) enable self-adaptation to increasing loads,
(iv) enable compensation of uncertainty,
(v) enable self-healing,
(vi) separate functions and
(vii) close feedback loops.

Each design concept and each concept of operation are illustrated by some short examples:

First (i), the *overload protection* is provided, for example, by a pressure relief valve or a split pin with a defined shear force. Cellular solids or foams [2] show an overload protection provided by the material itself. As long as the foam still has a compression margin, the compression force is basically determined by the buckling load of the cell walls. The foamed elastomer sole of jogging shoes enables—as desired—high energy absorption with a small force amplitude. The same principle, high energy adsorption with small force amplitude in the force-displacement diagram, is used for crash structures and railway buffers.

Second (ii), the *overload capacity* is made possible by additional energy storage. As to plastic materials, the energy storage is the irreversible deformation work with overload. As is well known, brittle materials tend to fail spontaneously. This principle applies in general. Supply and energy chains become robust if storage devices are integrated.

Third (iii), the *self-adaptation* to increasing loads is made possible e.g. by an O-ring: the higher the fluid pressure the higher the sealing contact pressure between the elastomer and the solid surface. The elasticity is a prerequisite for this elegant way of mastering uncertainty in the usage phase. This prerequisite was lost in the Challenger space shuttle disaster [52], because the elastomer of an O-ring tank seal had been frozen during a cold night. In the glass state, the material is frozen in a deformation state without the necessary contact stress. The leakage of fuel caused the explosion. The Challenger disaster is one more example of model uncertainty ignoring either the influence of the temperature on the function of the seal or ignoring the temperature as a model parameter.

The design concepts of overload protection, overload capacity and self-adaptation, especially as self-enhancement and self-repairing principles, are also used as a basis for the realisation of resilient system properties, cf. Sect. 6.3.2.

Forth (iv), the *compensation* of uncertainty is reached by integrating elastic elements at the interfaces of system components. Foil air bearings are typical for this. The integration of the bumpy elastic foil compensates uncertainty in the shaft and the journal diameter.

Fifth (v), *self-healing* materials or structures are the dream of mechanical engineers. Unfortunately these are very rare. A current example is a modern mountain bike tire. The air tube is replaced by a liquid sealant which is added to the inside of the tire. Every small puncture is self-healed by this sealant.

Sixth (vi), the *separation of functions* is the most basic concept to master uncertainty. A clear assignment of functions to components includes, for example, the avoidance of double fitting. The two functions sealing and load-bearing are known to be separate. A piston seal should not have a load-bearing function and a guide ring should not have a sealing function. In the above mentioned examples it is interesting to note that the separation of functions is a general concept to gain robustness of a system not only in the design and operation of technical systems: all modern forms of government use the principle of separation of powers [9]. Modern business organisations also identify roles, i.e. functions to gain robustness in many aspects including legal ones.

When designing load-bearing systems, the concept of separation of functions is applied to the flow of forces and is referred to as the design for clarity, cf. Sect. 6.1.6. The clarity of force flow significantly supports the mastering of accumulated uncertainty from the product life cycle processes.

Seventh (vii) *closing the feedback loops* is always a good advice: not always all facets of mastering uncertainty have to be considered: in nature, feedback control is a very successful concept to manage uncertainty; in Chap. 1 we pointed out the benefit to close feedback loops over the phases of the product life cycle. For feedback control, the model uncertainty discussed in Sect. 4.3 is only of minor importance. In fact, feedback control works even if there is no model of the system. On the other side, data uncertainty, i.e. the uncertainty of the measured or calculated data to be fed back, is most important when closing control loops. The situation is somewhat inverse in forward control. Here, in fact the model uncertainty and model quality are most important.

The given list of inherent robust design and robust operating concepts is not complete. The presented concepts are subliminally present in every design task. It is inherent to the optimisation and learning, not only in mechanical engineering.

Illustrating examples are e.g. the frequent weekly cooking of spaghetti and the rare roasting of a piece of meat. The latter may happen only for Christmas once a year. Only if the signal-to-noise-ratio when tasting salt is high enough, feedback will improve the cooking result of spaghetti from time to time. A typical feedforward control exists when the roast time of the roast has to be estimated in advance. This requires a model of the oven, temperature control and a model of the roast. The model of the roasting is parameterised by the weight of the meat piece. A good cook has such a model in mind. Again the model does not need to be in a mathematical form even though it might be so. If the model uncertainty is too large, the result will be of poor quality. The open control loop can be closed by a roasting thermometer, so that uncertainty is controlled without using a model at all.

Robustness, just like the mastering of uncertain loads, also requires insensitivity to disturbance parameters that affect the system externally. Knowledge of the influences on the system from its environment and vice versa is an essential prerequisite for development. The detailed analysis of disturbance influences by means of a process model is described in Sect. 5.2.3. In Sect. 5.1.2 it is shown how the documentation of desired functional properties and expected disturbance parameters positively influences the development process. The choice of physical effects that are insensitive to the expected disturbances can fundamentally contribute to the robustness in the early phases of the system design as shown in Sect. 6.1.5.

Flexibility

A flexible system is characterised by the fact that the system fulfils $i = 1, \ldots, N$ predefined functions g_i with accepted functional quality δg_i.

Recapturing the above given characterisation, the question arises how this flexibility can be achieved. There are several ways how to gain flexibility and there are many perspectives on the topic, cf. the review article about gaining flexibility in manufacturing [46]. As before, our perspective is a more general one being suitable to all phases of the product life cycle and also, as before, we have the systematic engineering design perspective [36]. From the functional view, we either can separate functions in different physical or software modules, or we can integrate functions in one physical module which may contain software modules.

Smart modularisation or smart modules, both concepts may enable flexibility

In the above explanations, we present the separation of functions as an inherent robust design and/or inherent robust operating concept. This applies at the level of a functional unit or a single component. Recognising that a module integrates some functional units, we have the opportunity to discuss the degree of functional integration in the module. Today the term "smart" module is ubiquitous. Therefore we should discuss the degree of "smartness". This degree is determined by the degree of integration or separation of functions.

In the following, we first differentiate between the two concepts to gain flexibility (i) modularisation by the smart separation of functions and (ii) smart modules by integration of function, see Fig. 3.18. This is followed by a more detailed look at the two contrasting concepts on how to gain flexibility.

Figure 3.18a shows flexibility gained by separation of functions into m subfunctions and the associated m physical or software modules. The matrix gives the utilisation n_j of the different modules. The effort F_1 is minimal if the total utilisation $\Sigma_j n_j$ is maximised and at the same time the number of modules m is minimised:

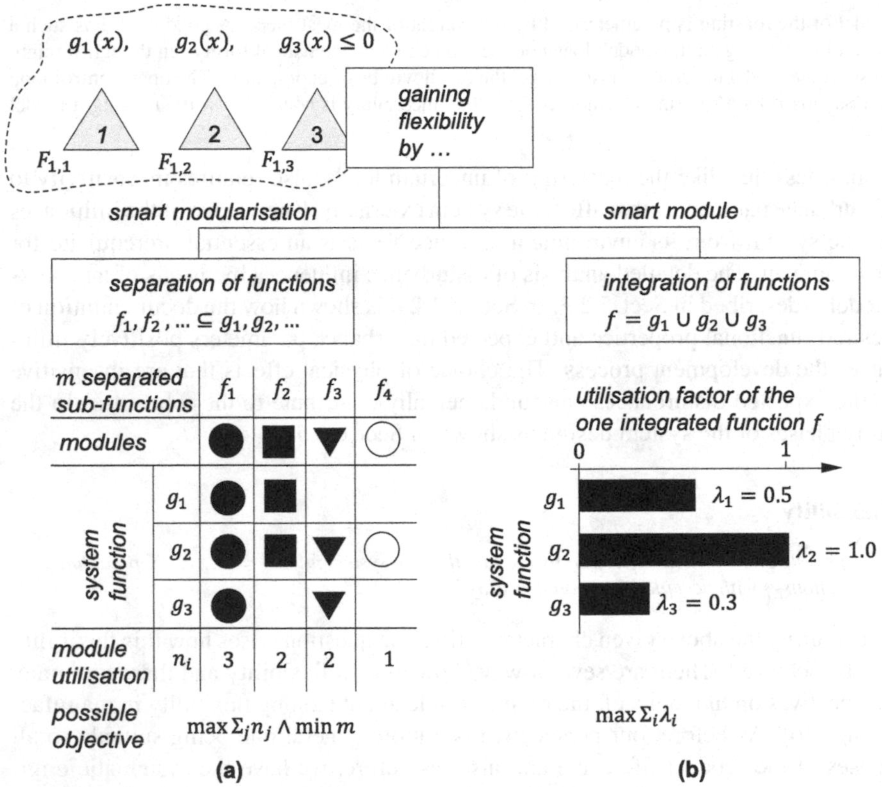

Fig. 3.18 **a** Flexibility gained by smart separation of functions into m sub-functions and the associated m physical or software modules; **b** flexibility gained by integration of function of the $g_1, g_2, \ldots$ user-defined individual functions into on master function realised by one smart module

$\max \Sigma_j n_j \wedge \min m$ subject to that in practise physical modules count more than software modules. The number of modules m is a measure of the internal complexity with regard to the design and the production phase. The external complexity comes from the user's view, e.g. the user would like to have common parts for maintenance. At the same time the specific user would like to have his or her needs matched fully by the system. Hence, the user experience should be kept in mind when designing a modular product.

Figure 3.18b shows how flexibility is gained by integration of function of N user defined individual functions into one master function realised by one "smart" module. For the average user the system shows too much functionality, which is of course not bad for the user but may be costly for the producer. The part of the functionality used by one user is denoted by $\lambda_i \leq 1$. The objective for the integration of function is hence $\max \Sigma_i \lambda_i$.

So far it has become clear that the two asymptotes of flexibility are first (i), "smart" modularisation and second (ii), "smart" modules. The first asymptote is a methodology, the second one is a technology, cf. Chap. 5.

Smart modularisation

We discuss modularisation as a smart separation and integration of functions into standardised modules as a strategy of flexibility and hence of mastering uncertainty. A module is a physical unit with several integrated functional units. The module can be used to fulfil different system functions by combining it with other modules. Thus, a component is formed out of several standardised segments. This concept of combining standardised modules was already used in the 13th century when standardised stone modules were prefabricated in a quarry to form structures of gothic cathedrals in medieval France. Today it is the basic principle to gain flexibility for example in ship building, process engineering and many other fields of engineering.

Modularisation is fostered with standardised interfaces for energy, forces, displacements and information; e.g. in the *Austauschbau*—being common in mass production, Chap. 1,—the uncertainty in geometric dimension is standardised by means of tolerance fields. Obviously by modularisation and standardisation the division of labour is fostered and the separation into the single elements of a supply chain is enabled. The development speed may be equally increased by integrating known modules into the system. The uncertainty of those modules is usually quantified. In some cases it is even possible to gain a legal conformity certification on the module level. If the module uncertainty is mastered together with the uncertainty on the basis of the interfaces, the system's functionality and the objective functions can be quantified.

It is self-evident that sources of uncertainty arise in the modularisation, in the first instance there are modules, or in the second instance there may be interfaces. Based on our experience the availability of a technical system is often limited by the failure of electrical power or signal connectors. Hence, not only the standardisation of the interfaces but also their robustness are most important when designing modular systems. The robustness of the interface is fostered by obeying the above five inherent robust design and operation concepts. So the concept (iv), compensation of uncertainty, may be realised by a flexible interface. As a rule, the interface, or more generally speaking, the module's boundaries should be selected in such a way that a module's sub-function and quality are only marginally influenced by a detailed location of the boundary. This rule is exemplified in the following case study.

4th case study—Smart modularisation of a control valve

In this case study we address the question of how to solve the constrained optimisation problem formulated in Fig. 1.9b: with minimised effort subject to the function g_i being fulfilled.

$$\min_x \sum_{i=1}^{N} F_{1,i}(x) \text{ s.t. } g_i(x) \leq 0, \quad i = 1, \ldots, N. \tag{3.12}$$

For this purpose, a product has to be evaluated first from the different stakeholder views. At least there is (a) the manufacturer and (b) the customer. In a business to business market there is often also (c) the planner working as a service provider and (d) the approval authority. All two or four stakeholders have different interests. Hence, the optimisation problem is more complex:

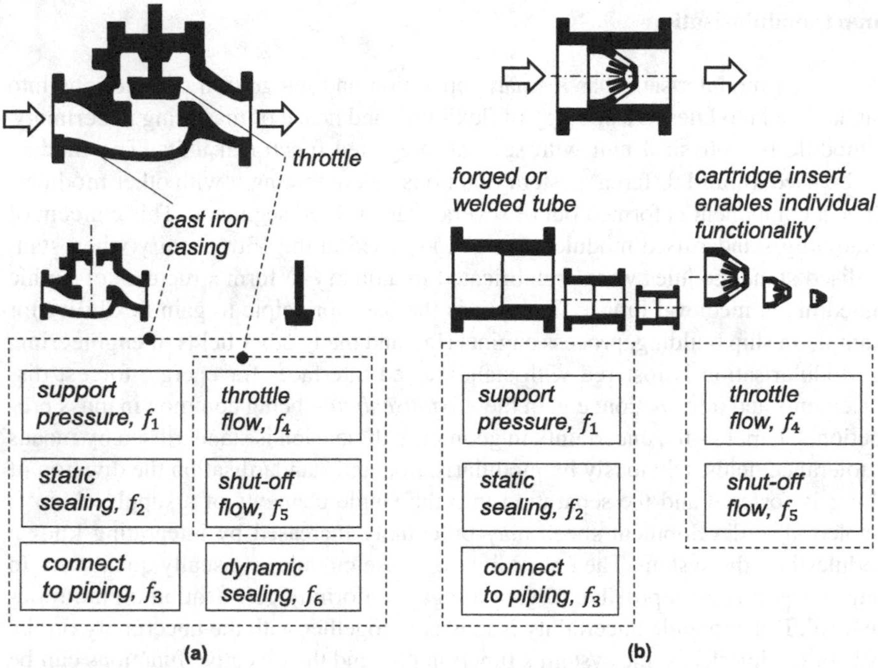

Fig. 3.19 Control valve design: **a** the integration of functions in a control valve hinders modularisation and increases internal and external complexity. Internal complexity may be measured by the number of cast moulds needed to meet the required functions. External complexity is measured by the different spare parts needed for maintenance. **b** Separation of functions enables a size ranged modular design of the valve reducing internal and external complexity

$$\min_x \sum_{i=1}^{N} \left[w_a F_{1,i}^a(x) + w_b F_{1,i}^b(x) + w_c F_{1,i}^c(x) + w_d F_{1,i}^d(x) \right] \text{ s.t. } g_i(x) \le 0,\ i = 1, \ldots, N. \tag{3.13}$$

Hereby, it becomes clear that the result of a modularisation strategy depends on the weights w_a, w_b, w_c, w_d of the individual stakeholder.

Figure 3.19 gives an example of a modularisation derived form a functional integrated control valve typically used in the process, energy or petrochemical industry. The functional separation as presented in Fig. 3.19b is obviously the prerequisite for the independent size ranged modular design of the forged or welded tube and the cartridge insert. The functional quality is determined by the insert allowing a detailed design to fulfil the primary prescribed function of the valve, i.e. throttle the flow. The tube segment fulfils the secondary functions only and this is made possible in a wide range independent from the insert. The interface between the module "tube" and the module "cartridge" allows large tolerances being important for successful modular designs, i.e. the interface is robust in the above prescribed sense. Looking at the tube, it is much simpler in design and manufacturing compared to the cast iron casing of the traditional design. This reduces the uncertainty in the supply chain and shortens delivery times. Overall the customer experience with the entire numerous facets can be improved by smart modularisation.

Mastering module uncertainty by smart test and development methods

The case study in Sect. 3.3, the 'energy ship', was motivated by the fact that combining modules with known uncertainty to build a new system enables mastering uncertainty. In fact that is what we observe today in the automotive industry were variants are derived from platforms and the platforms are based on modules. Hence, it is worthwhile shedding some light on smart test and development methods when talking about smart modularisation.

We focus on the following concepts to master the module uncertainty:

(i) early failure on the basis of a model test,
(ii) scaling of uncertainty in a size ranged modular design,
(iii) module in the loop.

The first concept is a very easy one being part of the agile development. Despite the overall simulation methods we have today, we should start very early in the development phase to have virtual and real mock ups but also – being even more important—physical functional models that allow the evaluation of the expected functions, functional quality and general quality. The second concept is based on the similarity principle to derive uncertainty of a scaled prototype from the model test, cf. Sect. 4.3.6. The third concept is widely used today. The uncertainty in function and quality of a module can be evaluated by integrating a real module into a virtual model of the system.

Figure 3.20 shows this concept known as hardware-in-the-loop (HiL). Instead of integrating a new or adapted module into a real system all at once, Fig. 3.20(*i*), the module is encapsulated by an active physical interface to the cyber world, thus simulating the overall system, Fig. 3.20(*ii*). This strategy evaluates the functional quality of new modular components with reduced effort and improved testing possibilities. The difficulty here is on the one hand the model uncertainty of the overall system and on the other hand the uncertainty in the flow of mass, energy and information between the test module and the active interface.

Smart modules

Instead of following the concept of smart modularisation we may think of satisfying a "power user" or a "power application" with only one smart module. Hypothetically, all the functions $g_1, g_2, \ldots$ shown in Fig. 3.18b could be represented in one system or functionally integrated module. Thus, the module is oversized for most applications; but in the overall picture this strategy can be advantageous with regard to the total effort $\Sigma_i F_{1,i}$. Such smart modules were already discussed in the past in the context of flexibility in manufacturing [46] but only now, with the help of control theory, these smart modules become reality: The ability of the smart module to adapt to a specific application g_i is necessary. This adaptivity often requires an actively controlled process. In the context of this book, the Active Air Spring presented in Sect. 3.6.2 or the 3D Servo Press presented in Sect. 3.6.3 are examples of smart modules. Since the integration of functions into smart modules will accompany us throughout the book, we have decided to keep this subsection short.

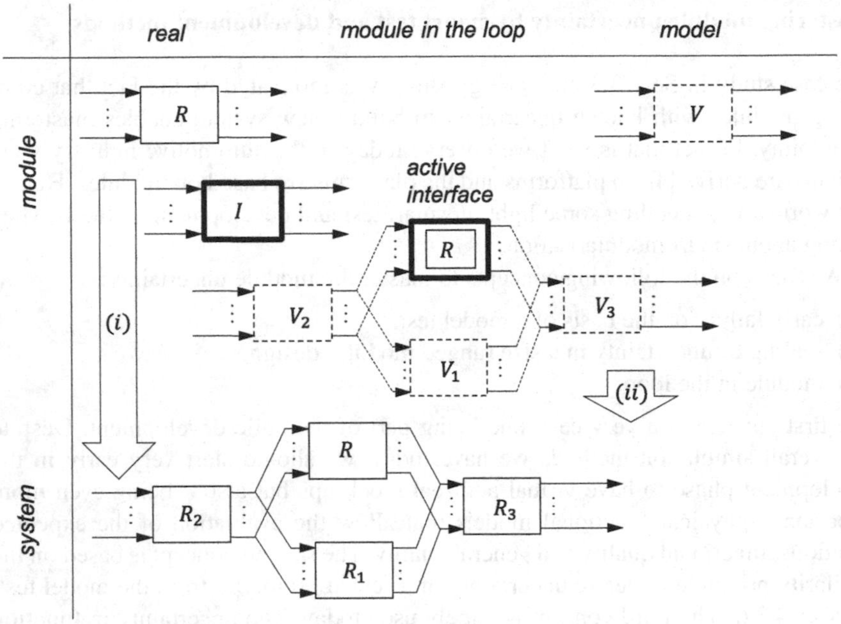

Fig. 3.20 Mastering uncertainty by stepwise integrating a module into a real system in combining the cyber world, with virtual modules V, and the real world, with real modules R by active interfaces. The interface I itself is an active component in the above defined sense. This concept is known as hardware- or module-in-the-loop (HiL, MiL)

Resilience

Again we recaptiulate the characterisation of a resilient system, cf. Sect. 6.3 as described above:

> *A resilient system is characterised firstly by the fact that it not only fulfils a predefined function g like a robust system, but also retains a residual function g_{res}, if the system is disturbed at time $t = 0$. A resilient system may secondly show the ability of recovery: from a distorted state, the system may recover in a time $t = 0, \ldots, T$ to a function $g(x) \to g'(x')$ and/or quality $F(x) \to F'(x')$. This function and quality may not be foreseen in the previous design and production phases. The design variants transform during the evolution from x to x'.*

In contrast, robust and flexible systems allow either the fulfilment of one function or several system functions. In both cases, i.e. for robustness and flexibility, the system's functions may be uncertain. Sometimes it occurs that the function, i.e. the usage phase, is in part unknown when designing and manufacturing the system. Sometimes a system only fulfils a partial function due to an accident or a catastrophe.

Seeing resilience as strategy to master ignorance or nescience and flexibility or robustness as a method to master incertitude or stochastic uncertainty is one possibility. But this classification shows a difficulty, as the result of the classification depends on the degree of information on the usage scenarios which allow us to distinguish between resilience and the other strategies. Hence, following this classification the difference between robustness and resilience is fuzzy at least in a strict scientific context: the transition between robustness and resilience is smooth and the use is determined by the context.

However, there is also another, namely a sharper distinction between the concepts of robustness and resilience. More precise is the characterisation from the perspectives of the system function and the system quality measured in effort, availability and acceptability: a resilient system is one that still shows a residual function when heavily disturbed. On top of this, a resilient system may have the ability to evolve its own function and quality. As pointed out in this chapter, evolution is a time-demanding process.

In fact, often this time-demanding process is commonly associated with resilience. Consistent with this understanding, we rather ask for the sub-functions or characteristics a system shall have to allow for the evolution of its functionality and quality from an initial function g, quality F associated to the design and operation parameters x at a time t to g', F', x' at a time t': $x \rightarrow x'$, $g(x) \rightarrow g'(x')$, $F(x) \rightarrow F'(x')$.

This evolution requires sub-functions that are called resilient in a system. It should be able to

(i) measure,
(ii) react,
(iii) learn,
(iv) anticipate.

The four functions can be represented by human or "smart" process chains. Depending on the process chain and system boundaries, man and machine can work together as partners or independent of each other.

There are some sequences in our mind set: For the evolution mentioned, flexibility may be a further prerequisite. Hence, we do have the understanding of resilience ⊃ flexibility ⊃ robustness. If we evaluate the four required sub-functions, there is again a sequence with (a) measure is easier than (b) react, is easier than (c) learn, is easier than (d) anticipate.

Chapter 6 is devoted to the in-depth introduction of the three strategies.

3.6 Exemplary Technical System Mastering Uncertainty

Maximilian Schaeffner

The theoretical considerations should equally be practically valid, verifiable and falsifiable. This applies especially to those methods and strategies outlined in the previous sections for our approach to master uncertainty in load-bearing structures of mechanical engineering. In particular, we have introduced methods and strategies for the analysing, quantifying and evaluating as well as finally mastering uncertainty along the product life cycle from the design via production to usage. In order to investigate and prove the effectiveness of our approach, it is crucial to apply the methods and technologies to exemplary technical systems. Therefore, technology demonstrators were designed and investigated both numerically and experimentally. More details are disclosed in the upcoming chapters.

Three of these technology demonstrators are introduced in greater detail in the following subsections. Section 3.6.1 provides an introduction of the Modular Active Spring-Damper System as a generic load-bearing system similar to an aircraft landing gear that captures a wide variety of applications. In Sect. 3.6.2 we describe the Active Air Spring as an active module to enable disturbance compensation in an automobile suspension. Finally, in Sect. 3.6.3 we introduce the 3D®Servo Press, which is a new press concept to increase flexibility and productivity in forming processes.

The three technology demonstrators shown in Sects. 3.6.1–3.6.3 stand as *pars pro toto* for all technologies presented within this book. They render the application of the methods and strategies possible to analyse, quantify, evaluate and finally master uncertainty across all phases of the life cycle. Examples are the identifying of data-induced conflicts in Sect. 4.2, quantifying model uncertainty in Sect. 4.3 and evaluating resilience in Sect. 6.3.

3.6.1 Modular Active Spring-Damper System

Christopher M. Gehb, Maximilian Schaeffner, Robert Feldmann, Jonathan Lenz, and Tobias Melz

The Modular Active Spring-Damper System (German acronym: MAFDS—Modulares aktives Feder-Dämpfer System) is a generic load-bearing system that serves as a platform for applying and testing methods and technologies to master uncertainty, compare Chap. 2. The motivating origin is based on an aircraft landing gear which reflects in particular the conflict between maintaining the main dynamic features, i.e. load bearing, load distribution, structural stabilisation and vibration control, combined with lightweight design. The investigations to describe, evaluate and master uncertainty are derived on a virtual MAFDS using phenomenological and mathematical models as well as on the physically realised structure of the MAFDS, see Fig. 3.21.

The MAFDS illustrates the possibilities of mastering uncertainty in a realistic and descriptive manner for scientific purposes. The mechanical requirements of the MAFDS are largely, but not exclusively, based on those of an aircraft landing gear and may also capture the quarter car dynamic behaviour of an automotive chassis. Nevertheless, the MAFDS and its test environment do not intent to replace, supplement or extend the existing industrial and product-oriented test procedure of commercial aircraft landing gears and automotive chassis or the aircraft landing gears and automotive chassis itself. In fact, the findings from applying and testing methods and technologies to master uncertainty in load-bearing systems are supposed to be transferable on scientific scales to many other load-bearing systems. Said findings serve to increase the acceptance and credibility of the investigated methods and tech-

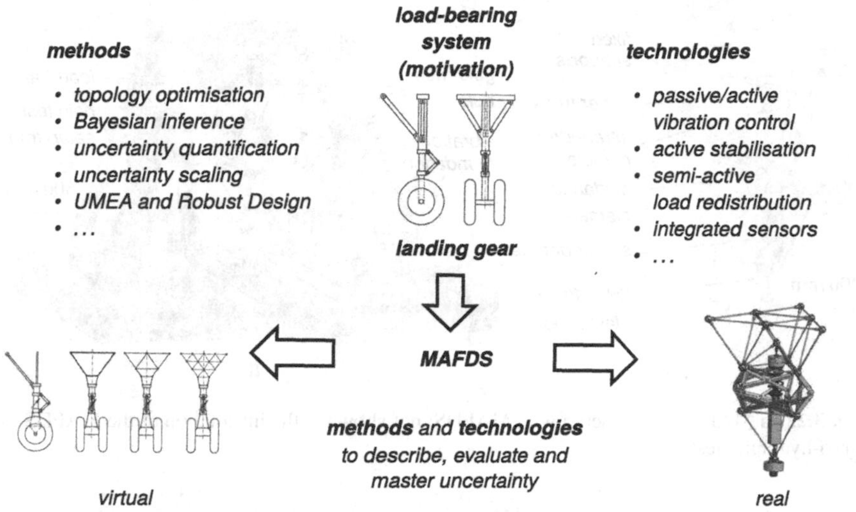

Fig. 3.21 Methods and technologies for mastering uncertainty in the generic load-bearing system MAFDS

nologies to master uncertainty. This uncertainty to be mastered includes all forms of uncertainty introduced in Chap. 2.

The virtual part comprises the methods and technologies to describe, quantify and evaluate the uncertainty, which means to examine the phenomenological and mathematical models and simulations to finally master the uncertainty [19]. These include e.g. investigation of topology variations in load-bearing structures that cannot be experimentally implemented due to the large number of possible variations. Furthermore, this part involves to apply and test methods, such as Robust Design, Uncertainty Mode and Effect Analysis (UMEA) in Sect. 5.2.1, process and uncertainty modelling, and model parameter calibration as in Chap. 4.

The physically realised part comprises components that are constructed, fully realised and experimentally tested, such as the truss structures, supports, beams and spring-damper, compare Fig. 3.22. Due to its modular structure, the virtual as well as the physically realised MAFDS enable the integration of different technologies to master uncertainty through e.g. load redistribution, stabilisation and vibration control. Therefore, various passive, sensory, semi-active and active components are applicable for each intended process manipulation. In the following, the possibilities of sensory, semi-active and active process manipulations for the MAFDS are presented after the test setup and the MAFDS are introduced in detail.

Test setup of MAFDS

The MAFDS is a load bearing system that can be used as a passive system and allows for the integration of semi-active and active components due to its modular setup [11]. The passive structure and its components are shown in Fig. 3.22a. It

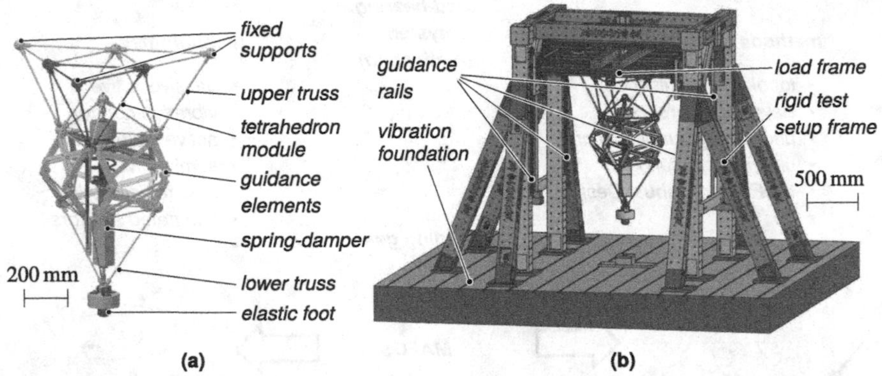

Fig. 3.22 **a** MAFDS, **b** test setup with MAFDS; not shown is the installation of the MAFDS in a servo-hydraulic test rig

consists of an upper truss structure, a lower truss structure with an elastic foot, guidance elements and a spring-damper. The three supports of the upper truss are fixed to a load frame that guides the MAFDS in vertical direction for experimental drop tests, see Fig. 3.22b. Additionally, it is possible to install the MAFDS in a servo-hydraulic test rig in order to apply base excitation, see Fig. 3.29 upper right picture. By this means, we realise a defined frequency excitation and e.g. road excitation. However, this section focuses on the drop test setup.

The upper truss consists of four tetrahedron modules with slender beams and solid nodes, of whom one is highlighted in Fig. 3.22a. The individual beams can be easily exchanged by sensory, semi-active or active beams, as presented in Sects. 5.3.6, 5.4.6 and 5.4.7. The four tetrahedron modules are linked to each other, yielding the upper truss structure. The lower truss of the MAFDS consists of one tetrahedron module. An elastic foot and an additional mass are attached to the lower node to represent the stiffness and the inertia of a wheel of an air plane landing gear or a car.

The dynamic behaviour of the MAFDS is governed by the properties of the spring-damper. For the application in the MAFDS, the passive suspension strut of a mid-range car serves as a spring-damper that is connected to the upper and the lower truss via a torque-free connection.

In order to enable the relative translation of the upper and lower truss structures in vertical direction, both are connected by three kinematic guidance elements. The design and position of the three guidance elements allow for simultaneous transmission of the transverse forces and bending moments between the lower and upper truss. In this way, the spring-damper is not charged with transverse forces or bending moments. Thus, the absorption and dissipation of impact energy and the low-frequency vibration reduction mostly take place via the spring-damper. A simple two degree freedom (2 DOF) model of the MAFDS to capture its dynamic behaviour will be presented in Sect. 4.3.3.

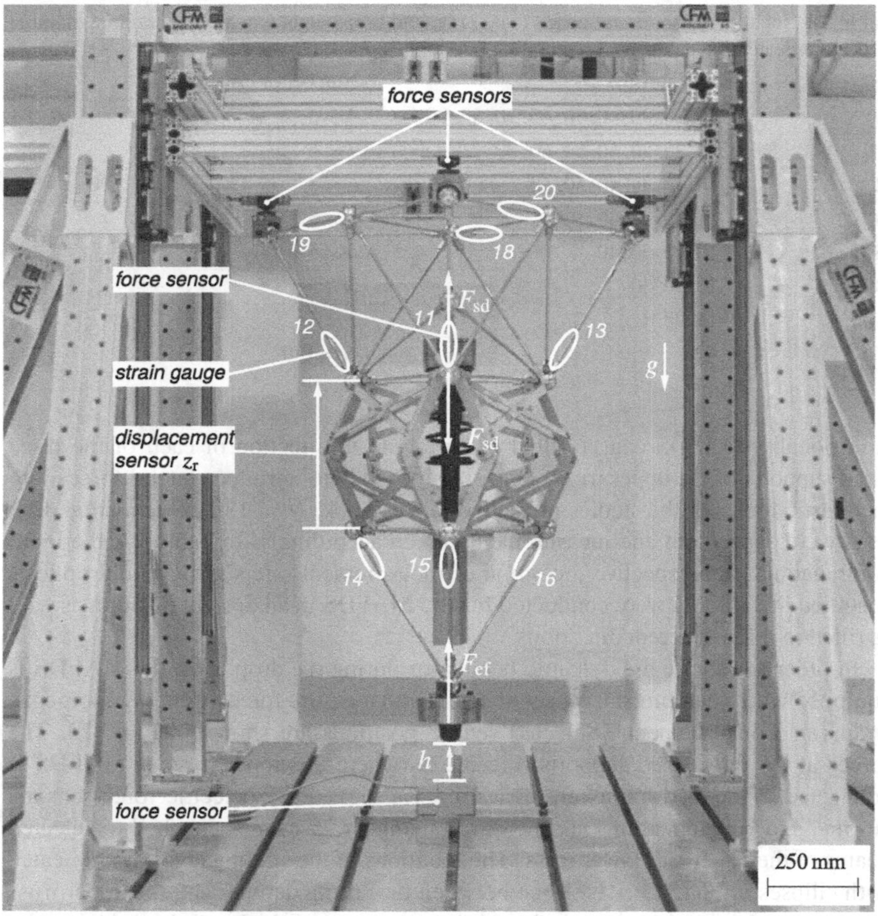

Fig. 3.23 MAFDS and its sensors and selected measured quantities [13]

The test setup used for the experimental testing of the MAFDS is depicted in Fig. 3.22b. It consists of a rigid test setup frame attached to a vibration foundation. Parallel guidance rails are mounted on the test setup frame and enable a low-friction vertical movement of the load frame that can translate along the guidance rails. The test setup enables the introduction of static as well as dynamic loads, the latter via drop tests of the MAFDS. These are carried out in a similar way as to landing gear tests. After the load frame with the MAFDS attached to it is lifted up to a desired drop height h as depicted in Fig. 3.23, the load frame is released for a drop test. Additional weights m_{add} can be added to the load frame, similar to varying loads of an airplane or a vehicle. The deterministic, but varying input quantities shall induce data uncertainty into the system, as defined in Sect. 2.1. The drop height h and the added mass m_{add} thus constitute the inputs to the experimental drop tests.

Table 3.1 Input combinations of added mass m_{add} and drop height h as well as resulting number of drop tests, upright and tilted configuration possible

m_{add} in kg	h in mm	Number of drop tests
0	10, 20, 30, 40, 50, 60, 70, 80, 90	9×5
10	10, 20, 30, 40, 50, 60, 70	7×5
20	10, 20, 30, 40, 50, 60	6×5
40	10, 20, 30, 40, 50	5×5
60	10, 20, 30, 40	4×5
80	10, 20, 30	3×5
100	10, 20	2×5
		$N = 180$

Additionally, MAFDS can be tilted by an angle in direction of each of the three fixed supports of the upper truss. Thus, it is possible to introduce lateral forces into the elastic foot. For the studies presented in this book, $N = 180$ experimental drop tests were carried out and measurements taken accordingly. In Table 3.1, the input combinations and respective repetition of measurements are shown. The drop tests specified in Table 3.1 were conducted for the MAFDS in an upright position as well as tilted in three different directions.

In order to capture the dynamic behaviour during the drop tests, the MAFDS is equipped with a comprehensive set of sensors to measure forces, bending moments, strains and displacements. Selected sensor positions are shown in Fig. 3.23. The forces at the three fixed supports, Fig. 3.22a, where the upper truss of MAFDS is connected to the load frame, are measured via triaxial piezoelectric force sensors. Normal and bending strains are measured using strain gauges attached to selected beams of the upper and lower truss. The locations of the strain gauges are indicated with ellipses in Fig. 3.23. The force between the spring damper and the upper truss is measured by a uniaxial strain-based force sensor and is denoted by the spring-damper force F_{sd}. The impact force that the elastic foot of the MAFDS exercises on the vibration foundation during the drop tests is measured by a triaxial strain-based force sensor with its vertical component being denoted by the elastic foot force F_{ef}. Furthermore, the relative displacement between the upper and the lower truss is measured using displacement sensors being denoted by z_r. Exemplarily, Fig. 3.24 shows measurements of the relative compression z_{r}, the spring-damper force F_{sd} and the elastic foot force F_{ef} for a drop test with zero additional weight $m_{\text{add}} = 0$ kg and drop height $h = 0.09$ m, where the peak values $z_{\text{r,peak}}$, $F_{\text{ef,peak}}$ and $F_{\text{sd,peak}}$ are marked.

Using the measurements of the MAFDS, the dynamic behaviour and aspects related to data uncertainty and model uncertainty (see Sects. 2.1 and 2.2.2) are further investigated in Sects. 4.2.3 and 4.3.3.

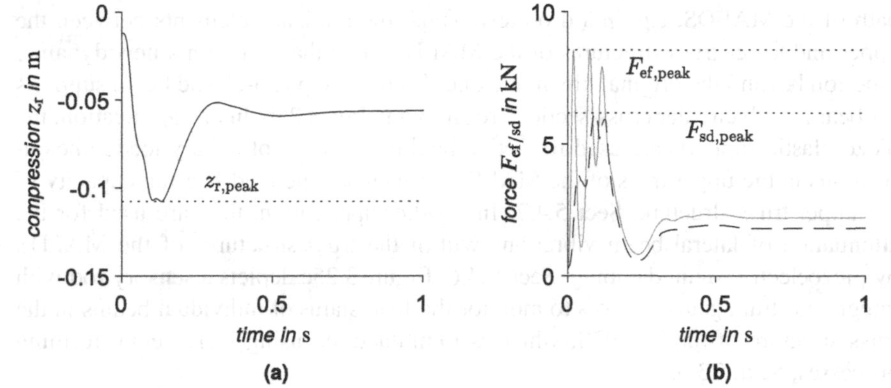

Fig. 3.24 Outputs from measurements of a drop test: **a** relative compression z_r (——), **b** force in the elastic foot F_{ef} (——) and force on the spring damper F_{sd} (– –)

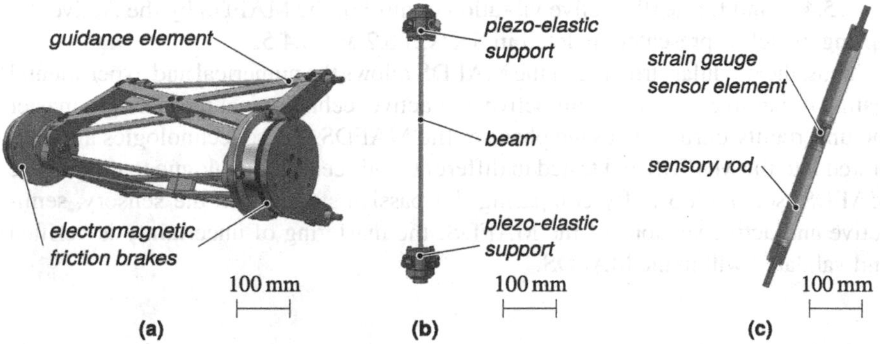

Fig. 3.25 Technologies for mastering uncertainty in the MAFDS by semi-active and active process manipulation: **a** semi-active guidance elements for load redistribution, Sect. 5.4.8, **b** beam with piezo-elastic supports for active buckling control, Sect. 5.4.7, and (semi-)active piezoelectric shunt damping, Sect. 5.4.6, **c** sensory rod for condition monitoring, Sect. 5.3.6; not shown is the Active Air Spring presented in Sect. 3.6.2

Semi-active and active process manipulation in the MAFDS

So far we have presented the design of the passive MAFDS that is used to quantify and evaluate uncertainty in a generic load-bearing system. In addition, the modular structure of the MAFDS allows for the possibility to modularly exchange sensory, semi-active and active components in order to master uncertainty within the usage phase, see Sect. 3.1. The technologies were developed as individual systems and validated in component tests, as described in detail in Sect. 5.4. Exemplarily, Fig. 3.25 shows three different technologies that can be integrated into the MAFDS to master different sources of uncertainty being present during the usage phase of the MAFDS.

In a new concept for load redistribution, compare Sect. 5.4.8, the semi-active kinematic guidance elements presented in Fig. 3.25a are used to influence the load path of the MAFDS, e.g. in a drop test. Thus, the guidance elements between the upper and lower truss structures of the MAFDS are enhanced with a new dynamic function beyond the original kinematic one. Innovative piezo-elastic beam supports for beams with circular cross-section are shown in Fig. 3.25b. In one application, the piezo-elastic supports are used for active buckling control of axially loaded beam-columns in the upper truss of the MAFDS to increase the load-bearing capacity of the upper truss structure, Sect. 5.4.7. In another application, they are used for the attenuation of lateral beam vibrations within the truss structures of the MAFDS by piezoelectric shunt damping, Sect. 5.4.6. Figure 3.25c depicts a sensory rod with integrated strain gauge sensors to monitor the load status of individual beams in the truss structure of the MAFDS, which is manufactured using incremental forming processes, Sect. 5.3.6.

Further technologies, which can be integrated into the MAFDS, but are not shown here, are the passive vibration control by a spring-damper with integrated hydraulic vibration absorber; this is an alternative to the passive spring-damper of the MAFDS, Sect. 5.4.4, and further the active vibration control of the MAFDS by the Active Air Spring, which is presented in detail in Sects. 3.6.2 and 5.4.5.

Thus, the modular structure of the MAFDS allows the numerical and experimental testing of passive, sensory, semi-active and active technological measures to master the uncertainty during the usage phase of the MAFDS. These technologies are integrated into the MAFDS and tested in different load scenarios analogue to the passive MAFDS, see Table 3.1. By comparing the passive structure to the sensory, semi-active and active versions of the MAFDS, the mastering of uncertainty is verified and validated within the MAFDS.

3.6.2 Active Air Spring

Manuel Rexer and Peter F. Pelz

Suspension systems in automobiles determine the driving comfort for the passengers as well as the driving safety of the vehicle. Currently, two trends are crucial for future suspension systems. Firstly, the trend towards autonomous driving is increasing the demands on driving comfort, as passengers are able to engage in other activities and thus may suffer more frequently from kinetosis [26, 48]. Secondly, the limited electrochemical energy storage of future vehicles requires energy-efficient passive or active suspension systems. This section discusses how both functional requirements can be met by the Active Air Spring, which has been developed and validated over the last 12 years.

It is helpful to understand the function of a suspension system before starting modelling or even designing. The suspension system of a vehicle performs four functions, i.e. (i) carrying the vehicle load, (ii) levelling the distance between the vehicle body and the road, (iii) isolating the body from road or vehicle dynamic excitations and (iv) limiting the dynamic force amplitude of the wheel. These four functions should be achieved with the least amount of packaging space, weight, energy and cost. Furthermore the four functions should be achieved robustly, i.e. even for uncertain loading or excitation or more general, even for uncertain customer expectations. There are two important quality measures: first, the functional quality as one measure of acceptability; second, the effort measured in energy consumption needed to achieve the functional requirement. Both are addressed in the following section.

Prevailing passive and semi-active systems

For economic reasons, the levelling function (ii) is usually not fulfilled in conventional passive suspension systems. A coil spring enables functions (i) and (iii) with minimal costs. Due to the periodic shift of potential energy from the coil spring with stiffness k_b to the kinetic energy of the chassis of mass m_b and vice versa, the system exhibits a natural frequency $\omega_\mathrm{b} = \sqrt{k_\mathrm{b}/m_\mathrm{b}}$. Above an excitation frequency $\Omega > \sqrt{2}\omega_\mathrm{b}$, the road excitations are isolated as desired. To limit resonant oscillations at $\Omega \approx \omega_\mathrm{b}$ and to fulfil the functional requirement (iv), namely to limit wheel load oscillation, a hydraulic damper is connected in parallel as a dissipative element. It is immediately apparent that the outlined and prevailing solution is not robust with respect to the uncertainty of the chassis mass m_b. This is due to the non-separate functions (i) and (iii).

Air suspension and hydro-pneumatics are more complex and costly suspension systems. Both enable the levelling function (ii). However, as discussed in Sect. 3.5, true separation of functions as one of seven inherently Robust Design principles is only realised in air suspension. Even with air suspension, the two functions of (iii) isolating the structure from road or vehicle dynamic excitations and (iv) limiting the dynamic force amplitude of the wheel are in conflict, as we show below using a dynamic vehicle model. Only the Active Air Spring, as one example of a smart module as discussed in Sect. 3.5 makes it possible to satisfy new demands resulting from the trends mentioned at the beginning of this section.

Pareto optimal passive and semi-active systems versus Pareto optimal active suspension system

In the case of conflicting tasks, Pareto-optimal solutions emerge. Indeed, we show that the Pareto line cannot be crossed by any active or semi-active state-of-the-art air or hydro-pneumatic suspension systems. In such systems, the spring force $k_\mathrm{b}(z, t)z$ and damper force $b_\mathrm{b}(\dot{z}, t)\dot{z}$ can be controlled by pneumatic or hydraulic valves. Here, $z = z_\mathrm{w} - z_\mathrm{b}$ denotes the compression of the suspension system and t the time, cf. Fig. 3.26.

Our research is carried out on three quarter car models, a virtual one, a hardware-in-the-loop system, cf. Sect. 4.3.4, and a real system. The latter is comparable to the MAFDS, Sect. 3.6.1, being a two-mass oscillator as well. The equation of motion of the system depicted in Fig. 3.26 is given by the following two equations of motions and the constitutive equation for the passive force change ΔF_p:

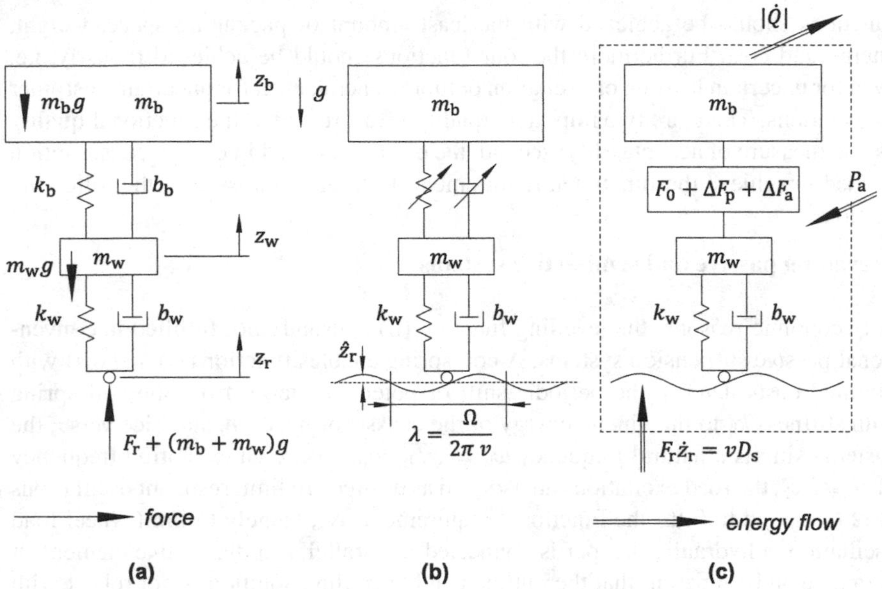

Fig. 3.26 Quarter car model for vertical dynamics of **a** passive, **b** semi-active and **c** active suspension systems with the energy flow across the system boundary

$$m_b(\ddot{z}_b + g) = F_0 + \Delta F_p + \Delta F_a, \tag{3.14}$$

$$m_w\ddot{z}_w + k_w(z_w - z_r) + b_w(\dot{z}_w - \dot{z}_r) = -\left(\Delta F_p + \Delta F_a\right), \tag{3.15}$$

$$\Delta F_p = k_b(z_w - z_b) + b_b(\dot{z}_w - \dot{z}_b)\,. \tag{3.16}$$

Here, ΔF_a is the active force between the body and wheel and $F_0 = m_b g$ is the static preload of the suspension system with the mass specific gravity force g. We assume linear springs (stiffness k) and dampers (damping constant b) and a foot point excitation z_r.

The passive, Fig. 3.26a, and semi-active system, Fig. 3.26b, consist of a spring and a damper connected in parallel between body mass, m_b and wheel mass m_w without any active force $\Delta F_a = 0$.

For a semi-active air spring being state-of-the-art today, the stiffness k_b is adapted by increasing the air spring volume V_0 by means of switching pneumatic valves. For semi-active damping systems the damping constant b_b is adapted by switching hydraulic valves.

Here, the focus is on the active force ΔF_a on the right side of the equations. Today there are no Active Air Springs in usage. Figure 3.26c shows the active suspension system. An active system can react to uncertainty from the usage phase and, for example, compensate the loss of comfort as we investigate in Sect. 5.4.5 using the example of the Active Air Spring presented here. When tuning a spring-damper system, the objectives of driving comfort and safety are in conflict because it is not

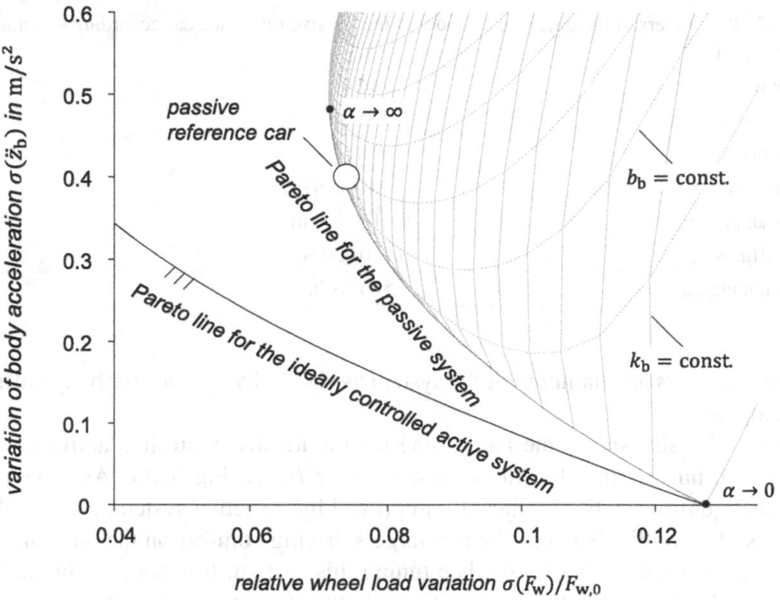

Fig. 3.27 Conflict diagram of driving safety and driving comfort with passive and active boundary lines as well as isolines for constant stiffness and constant damping of a passive suspension for a quarter car driving on a highway at 100 km/h [20]

possible to keep both constant at the same time, the force between the wheel and the ground and the body at rest. Hard-tuned sporty vehicles, for example, offer low driving comfort. As a result, a compromise must be found between the two target parameters during the tuning process. Mathematically, this conflict is described as a minimisation of body acceleration and wheel load fluctuation

$$\min\left(\sigma(\ddot{z}_b)^2 + \alpha^2\sigma(F_w)^2\right). \tag{3.17}$$

The two objectives driving comfort and driving safety—described by the standard deviation of the body acceleration $\sigma(\ddot{z}_b)$ and the standard deviation of the wheel load fluctuation $\sigma(F_w)$ – are weighted via the parameter α.

Figure 3.27 shows the conflict diagram for the quarter car where the standard deviation of body acceleration $\sigma(\ddot{z}_b)$ is plotted versus wheel load fluctuation $\sigma(F_w/F_{w,0})$ with the dynamic wheel load $F_w = k_w(z_w - z_r) + b_w(\dot{z}_w - \dot{z}_r)$ and the static wheel load $F_{w,0} = (m_b + m_w)g$ [34]. Table 3.2 lists the parameter of the passive reference car with a passive linear spring-damper system. The car is excited by a stochastic road signal z_r according to a ride on a highway at 100 km/h [34].

The Pareto line for the passive system, shown in Fig. 3.27 represents the optimal tuning of the spring-damper system of the passive suspension according to Eq. (3.17) with the curve parameter $0 < \alpha < \infty$. This Pareto line cannot be crossed by varying

Table 3.2 Parameters of the quarter car model of the passive reference car according to a middle-range car [20]

Parameter	Value
Body mass m_b	290 kg
Wheel mass m_w	40 kg
Body stiffness k_b	10 000 N/m
Body damping b_b	1140 Ns/m
Tyre stiffness k_w	200 000 N/m
Tyre damping b_w	566 Ns/m

the body stiffness and damping of the system as shown by isolines for body stiffness and damping.

Figure 3.27 also shows the Pareto line for the ideally controlled active system. This system has no limit for the actuating power P_a, cf. Fig. 3.26c. As shown, the functional quality can be significantly improved by an active system. Nevertheless, there is still a conflict between the two targets driving comfort on the one hand and driving safety on the other hand when tuning this system. It is not possible to keep body and wheel at rest at the same time while the system is externally excited. But still, the use of an active system makes it possible to improve both driving comfort and safety and thus to overcome the limits set by the Pareto line of the passive system. In addition, active systems have an extended working range compared to passive or semi-active systems. Due to their variability in force setting they are well suited in mastering the mentioned uncertainty.

Sustainable active suspension demands low viscous or Coulomb friction

From Newton's third law "actioni contrariam semper et aequalem esse reactionem" [35], known in short as 'actio est reactio', it follows that the axial compression force of the air spring is given by

$$F(t) = [p(t) - p_\mathrm{e}]A(t). \tag{3.18}$$

This equation applies, provided that the pressure distribution within the component is homogeneous, which is usually the case [38]. In any case, Eq. (3.18) applies if the static pressure $p(t)$ is the volume-averaged pressure in the component and p_e is the ambient pressure. For a piston or plunger, the load-bearing area $A(t)$ is equal to the plunger cross-sectional area. For a rolling lobe or bellow, the load-bearing area $A(t)$ is bounded by a closed curve along which the stress vector has no component in the compression direction, cf. Fig. 3.28. For the Active Air Spring in focus here, the load-bearing area does depend on time t. It is obvious that this can only be achieved by a compliant component. The innovative core of the Active Air Spring presented here is the active inner support of the bellow, cf. Fig. 3.28b.

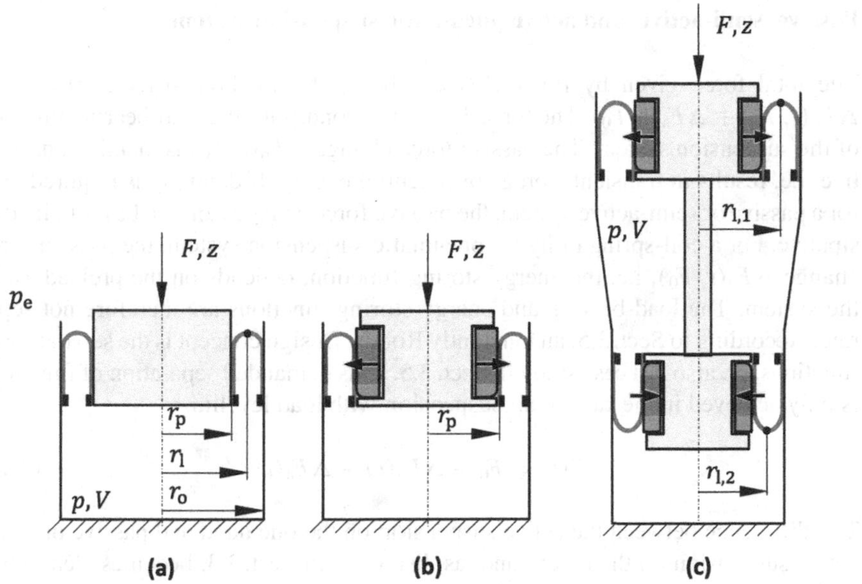

Fig. 3.28 Schematic sketch for **a** bellow sealed air spring with radius of the load-bearing area r_l, **b** actuated piston and **c**double bellow Active Air Spring [20]

For the system at rest at the design point, the spring force is in equilibrium with the gravitational force of the body mass $m_b g$: $F_0 = (p_0 - p_e)A = m_b g$. The index 0 marks the design point.

Transiently, the pressure $p(t)$ within the component can change due to a compliance C, a resistance R or an inductance L or a combination of those effects. In a pneumatic or hydro-pneumatic suspension, the above-mentioned load-bearing function including the levelling function is fulfilled by the fluidic component. The pressure p is adjusted quasi-stationarily for load levelling, cf. Sect. 3.4. This is made possible by a gas compressor in the case of an air spring, and by a displacement pump in case of hydro-pneumatics.

In the Active Air Spring presented here and in Sect. 5.4.5, the gas compliance C enables the second function mentioned above, the energy storage function. Inherent (passive) damping R makes every active system energetically inefficient since the actuator has not only to supply the required energy to the system, but also has to overcome internal frictional, i.e. damping forces. Thus, in the following, damping forces, viscous or Coulomb, are usually considered parasitic in the context of the Active Air Spring and Fluid Dynamic Vibration Absorber (FDVA) introduced in Sect. 5.4.4.

Passive, semi-active, and active pneumatic suspension system

The total force given by Eq. (3.18) can be split into three parts $F(t) = F_0 + \Delta F_p(t, F_0) + \Delta F_a(t, F_0)$. The force $F_0 = m_b g$ conditions the load-bearing function of the suspension system. The passive force change $\Delta F_p(t, F_0)$ is mainly conservative, i.e. results in transient storage of potential energy. If damping is required, e.g. for a passive or semi-active system, the passive force change can also be partially dissipative. For a coil-spring or hydro-pneumatic suspension system, the passive force change $\Delta F_p(t, F_0)$, i.e. the energy storing function, depends on the preload F_0 of the system. The load-bearing and energy-storing functions are therefore not separate. According to Sect. 3.5, an inherently Robust Design concept is the separation of functions, cf. also 3rd case study in Sect. 3.5. This demanded separation of functions is only achieved in the case of air suspension with load levelling:

$$F(t) = F_0 + \Delta F_p(t) + \Delta F_a(t). \tag{3.19}$$

The difference between the active suspension on the one hand and passive or semi-active suspension on the other hand, as discussed in Sect. 3.3, becomes clear when looking at the energy flux per unit time across the system boundary marked by a broken line in Fig. 3.26c.

Three energy flows must be taken into account. Firstly, the mechanical work $P_a = P_a(\Delta F_a)$, which acts on the suspension system per unit time through the actuator of the active component. The power P_a can be positive or negative, i.e. supplied or extracted. Secondly, the excitation work per unit time $P_r = P_r(F_r) = v D_s$, which results from the movement of the vehicle along a bumpy road (amplitude of the waves $\hat{z}_r$) with velocity v. In the latter case, the energy is taken from the drive system, which is needed for overcoming the additional drag force D_s. This force is additive to the rolling resistance of the wheels, which is essentially determined by viscoelastic-plastic deformation within the wheel elastomer and between the tyre and the road surface, and by the air resistance. Thirdly and finally, the energy dissipated in the damper per unit time is $P_d(F_d) = -|\dot{Q}|$. This power is always negative due to the second law of thermodynamics. It results in a heat flux $\dot{Q}$. Thus, the first law of thermodynamics for any load-bearing system reads

$$P_a(\Delta F_a) + P_r(F_r) = |\dot{Q}| \Rightarrow \begin{cases} P_r(F_r) = |\dot{Q}| & \text{passive/ semi-active,} \\ P_a(\Delta F_a) + P_r(F_r) = |\dot{Q}| & \text{active.} \end{cases} \tag{3.20}$$

Equation (3.20) is the foundation for classifying active systems on the one hand and passive or semi-active systems on the other hand, cf. Sect. 3.4. From the first energy balance $v D_s = P_r(F_r) = |\dot{Q}|$ it is easy to derive an upper limit for the energy consumption due to the (a) passive or (b) semi-active suspension system. Neglecting dissipation due to the tyre, for both cases the energy equation reads

$$vD_s = \int_0^{2\pi} k_w(z_r - z_w)\dot{z}_r \, d(\Omega t) = \int_0^{2\pi} b_b \dot{z}^2 \, d(\Omega t) . \qquad (3.21)$$

The angular excitation frequency $\Omega = (2\pi v)/\lambda$ is determined by the wave length λ of the road waviness and the driving velocity v. To simplify the calculation, we assume $m_w \ll m_b$. Therefore, the relative compression can be approximated by $\dot{z} = \dot{z}_w - \dot{z}_b \approx \dot{z}_w$. For this case, the last integral, i.e. the power loss is given by $|\dot{Q}| = \pi b_b \hat{z}_w^2 \Omega^2$. The first integral, i.e. the work exerted by the road on the suspension system, is in any case less than or equal to $\pi k_w \hat{z}_w \hat{z}_r \Omega$. The power needed for moving a vehicle along a bumpy road is

$$D = D_s + \text{air drag} + \text{rolling resistance}, \; D_s \leq \frac{\pi}{v} \frac{k_w^2 \hat{z}_r^2}{b_b} . \qquad (3.22)$$

Both air drag and rolling resistance increase with v, while the resistance due to the suspension system decreases with v. This is because the passive damper becomes dynamically stiffer with increasing speed. In the asymptotic limit of $\Omega \to \infty$, the suspension system becomes energetically conservative: energy storage takes place in the tyre alone.

If only the energy efficiency is taken into account, the damper would have to be replaced by a rigid bar. This is of course nonsensical, as the conflict diagram, Fig. 3.27, indicates. There, the two previously ignored functions of a suspension system are given as coordinates. These are the relative wheel load variation as a measure of driving safety and the body acceleration as a measure of driving comfort.

Before we arrive at that point, we discuss the difference between (a) a passive and (b) a semi-active system both sketched in the above Fig. 3.26. For both systems there is no active force, $\Delta F_a \equiv 0$. For (a) a passive system, the parameters b_b, k_b, b_w, k_w of the ordinary differential equations, Eqs. (3.14)–(3.16), are constant. For a (b) semi-active system, the parameters $b_b(t)$, $k_b(t)$ may vary with time. A semi-active damper or a semi-active air spring enables this. For the first case, the damping constant is changed by adapting the valve opening. For the second case, the spring constant is adapted by adapting the gas volume V_0 without performing mechanical work on the gas. The adaption is reached by opening or closing a pneumatic valve connecting adjacent volumes. For a cyclic excitation, the stiffness is $k_b(t) = k_b(t + T)$, were $T = 2\pi/\Omega$ is the cycle time. Hence, for the semi-active air spring, the suspension system is described by the Hill and Mathieu differential equation [18]

$$m_b \ddot{z} + k_b(t) z = 0 . \qquad (3.23)$$

It is important to emphasise that the body movement can only be ideally isolated from the movement of the wheel with (c) an active system. This may be achieved by dynamically balancing the spring force (and by balancing a parasitic damper force) with the controlled active force $\Delta F_a(t)$. From Eqs. (3.14) to (3.16) it follows that if the active force is controlled according to the rule $\Delta F_a(t) = k_b z = k_b(z_w - z_b)$, there will be an ideal isolation, i.e. $\ddot{z}_b = 0$. Ideal isolation becomes more and more important for future autonomous vehicles.

The additional drag D_s, present in the passive and semi-active system, can be reduced by an active system. In an ideal active system there is no damper. Hence, $|\dot{Q}| \to 0$ and $D_s \to 0$ for the ideal active system. In the following we analyse possibilities given by air suspension or hydro-pneumatic suspension.

Active and passive means of changing the compression force

To separate active and passive physical effects of changing the compression force, we take a look at the logarithmic derivative of the compression force F, Eq. (3.18),

$$\frac{\mathrm{d}F}{F} = \frac{\mathrm{d}p}{p - p_e} + \frac{\mathrm{d}A}{A}. \tag{3.24}$$

Replacing the increment d by the difference Δ at the design point, we get

$$\frac{\Delta F}{F_0} \approx \frac{\Delta p}{p_0 - p_e} + \frac{\Delta A}{A_0}. \tag{3.25}$$

In general, for the active system we have in Equation (3.24) $\mathrm{d}p = \mathrm{d}p_p + \mathrm{d}p_a$ and $\mathrm{d}A = \mathrm{d}A_p + \mathrm{d}A_a$. Only the active adjustments requires external energy. The passive adjustments for both pressure and load-bearing area are resulting from the relative compression $z = z_w - z_b$ of the spring. A contoured support of the rolling lobe either in form of contoured piston or a contoured outer guidance, results in $A(z)$, cf. Fig. 3.28. For instance, the load-bearing area is by predefined kinematics a function of the compression z. It follows that

$$\mathrm{d}A = \mathrm{d}A_p + \mathrm{d}A_a = \left.\frac{\partial A}{\partial z}\right|_a \mathrm{d}z + \mathrm{d}A_a = A'\mathrm{d}z + \mathrm{d}A_a. \tag{3.26}$$

Analogously, the change in gas pressure $\mathrm{d}p$ may have a passive and active part: $\mathrm{d}p = \mathrm{d}p_p + \mathrm{d}p_a$. In an asymptotic and hence simplified model, the change in pressure may be isentropic, $s =$ const, or isothermal, $\vartheta =$ const, cf. Sect. 4.3.5. The former is a good model for 'fast' processes, i.e. the cycle time T must be much shorter than the thermal relaxation time of the gas. The latter is a good model for 'slow' active systems. The load-levelling function of an air spring enabled by a compressor is such a slow active system. Here, there is sufficient time for thermal relaxation of the gas to the ambient temperature. It should be emphasised that only for preliminary design studies, the isentropic or isothermal asymptotes are needed. For observer models the thermal relaxation can easily be calculated, cf. Sect. 4.3.5.

For both, the 'fast' and 'slow' process, the change of the thermodynamic state is barotropic, because for both cases the pressure is only a function of the gas density: $p = p(\varrho, s = \text{const})$ or $p = p(\varrho, \vartheta = \text{const})$. The volume averaged pressure p is hence only a function of gas volume V and the gas mass m. The gas mass can only be changed (iii) actively by means of a compressor transporting gas from the ambient into the pressure chamber: $\mathrm{d}m = \mathrm{d}m_\mathrm{a}$. The gas volume is changed (i) passively when the air spring is compressed; the gas volume may be changed (ii) semi-actively by opening or closing a valve connecting the gas volume to an additional gas volume; finally, the gas volume may be changed (iii) actively by means of moving a piston against the gas pressure within the air spring. The passive and active means sum up to $\mathrm{d}V = \mathrm{d}V_\mathrm{p} + \mathrm{d}V_\mathrm{a}$. With the displacement area A_d defined as $A_\mathrm{d} := -\mathrm{d}V/\mathrm{d}z$ the pressure change is given by

$$\mathrm{d}p = \left.\frac{\partial p}{\partial V}\right|_{m,s} (-A_\mathrm{d}\mathrm{d}z + \mathrm{d}V_\mathrm{a}) + \left.\frac{\partial p}{\partial m}\right|_{V,\vartheta} \mathrm{d}m_\mathrm{a}. \tag{3.27}$$

From this, the logarithmic derivative, Eq. (3.24), of the compression force, Eq. (3.18), becomes

$$\begin{aligned}\frac{\mathrm{d}F}{F} = \left(-\frac{A_\mathrm{d}}{p - p_\mathrm{e}} \left.\frac{\partial p}{\partial V}\right|_{\mathrm{m,s}} + \frac{A'}{A}\right) \mathrm{d}z + \frac{1}{p - p_\mathrm{e}} \left.\frac{\partial p}{\partial V}\right|_{\mathrm{m,s}} \mathrm{d}V_\mathrm{a} \\ + \frac{1}{p - p_\mathrm{e}} \left.\frac{\partial p}{\partial m}\right|_{V,\vartheta} \mathrm{d}m_\mathrm{a} + \frac{\mathrm{d}A_\mathrm{a}}{A}.\end{aligned} \tag{3.28}$$

For the design point we have $A_\mathrm{d} \to A_{\mathrm{d,0}},\ A \to A_0,\ A' \to A'_0,\ p \to p_0,\ F \to F_0 = p_0 A_0,\ m \to m_0$. Hence, for the 'fast' change of the gas volume and the 'slow' change of the gas mass we have the asymptotic limits according to Sect. 4.3.5

$$\left.\frac{\partial p}{\partial V}\right|_{\mathrm{m,s}} \to -\gamma \frac{p_0}{V_0},\quad \left.\frac{\partial p}{\partial m}\right|_{V,\vartheta} \to \frac{p_0}{m_0}. \tag{3.29}$$

Here, the isentropic exponent is denoted by γ. Hence, the logarithmic derivative becomes

$$\frac{\Delta F}{F_0} = \frac{\Delta F_\mathrm{p}}{F_0} + \frac{\Delta F_\mathrm{a}}{F_0} = k_\mathrm{b} \frac{\Delta z}{F_0} + \frac{\Delta F_\mathrm{a}}{F_0}, \tag{3.30}$$

with the abbreviation for the stiffness

$$k_\mathrm{b} := \frac{\Delta F_\mathrm{p}}{\Delta z} = m_\mathrm{b} g \left(\gamma \frac{p_0}{p_0 - p_\mathrm{e}} \frac{A_{\mathrm{d,0}}}{V_0} + \frac{A'_0}{A_0}\right) \tag{3.31}$$

and the active force contribution

$$\frac{\Delta F_\mathrm{a}}{F_0} = \frac{p_0}{p_0 - p_\mathrm{e}} \left(\frac{\Delta m_\mathrm{a}}{m_0} - \gamma \frac{\Delta V_\mathrm{a}}{V_0}\right) + \frac{\Delta A_\mathrm{a}}{A_0}. \tag{3.32}$$

Inherent robust design concept of the Active Air Spring

From the stiffness, Eq. (3.31), we recognise the independence of the body natural frequency from the body mass m_b for $p_0 \gg p_\mathrm{e}$ (this condition is usually fulfilled) provided V_0 is controlled to be constant by means of a levelling function, i.e. a compressor as actuator,

$$\omega_b := \sqrt{\frac{k_b}{m_b}} = \sqrt{g\left(\gamma\frac{A_{d,0}}{V_0} + \frac{A'}{A_0}\right)}\,. \tag{3.33}$$

Thus, the system is inherently robust due to the separation of the two functions, first storage of potential energy and second levelling the body relative to the wheel to a predefined distance. As such, it is self-adaptive to load changes, meaning it compensates for uncertainties in the load. This is the reason why air suspension is standard in commercial vehicles and rail cars. The three principles of 'separation of functions', 'self-adaptation', and 'compensation' are three of the seven inherently Robust Design concepts discussed in Sect. 3.5.

From Eqs. (3.20) to (3.25) we recognise the semi-active nature of air springs with several volumes: by adapting the volume V_0, the stiffness and by that the dynamic behaviour is changed. But the semi-active control fails in preventing motion sickness. Thus, the following section discusses how to achieve an active force ΔF_a.

For the active system, the relevant cycle time is $T = 2\pi/\omega_b \sim 1$ s. Changing the gas mass Δm_a in a comparable time is not feasible. Hence, changing the gas mass is limited to the levelling function of the suspension system. Dynamically adapting the gas volume ΔV_a may be achieved by pumping oil, e.g. by a gear pump, as is done for hydro-pneumatic suspension systems. For an air suspension system to become an Active Air Spring, only the adaption of the load-bearing area is feasible [6]. Hence, for 'fast' cyclic forces $\Delta F_a(t) = \Delta F_a(t + T)$. Equation (3.32) reduces to

$$\Delta F_a = p_0 \Delta A_a\,. \tag{3.34}$$

Milestones in 12 years development of the Active Air Spring

Figure 3.29 shows the milestones in twelve years of development of the Active Air Spring as a time line. From the idea, the development led via a first prototype, which realises the adjustment of the load-bearing area via a hydraulically driven cam gear [4] to the second prototype with two connected hydraulic membrane actuators [20]. We were able to significantly reduce both the package and the weight of the prototype. The result is a hydraulic diaphragm actuator that is completely integrated into the piston of the Active Air Spring (Fig. 3.32). To ensure the relative force change $\Delta F_a/F_0$ is as large as possible, a double bellows air spring is used. Figure 3.28c shows the structure of two pistons connected by a rod. The load-bearing area thereby describes a circular ring.

Both pistons are equipped with an actuator allowing a total force change of $\Delta F_a \approx \pm 1$ kN at a static load of $F_0 = 2850$ N. A further advantage of the double bellow Active Air Spring is the hydraulic coupling of the two actuators via a double acting cylinder. As a result, the energy requirement of the air spring is reduced, as only differential forces have to be set, comparable to a mechanical rocker [43]. Fur-

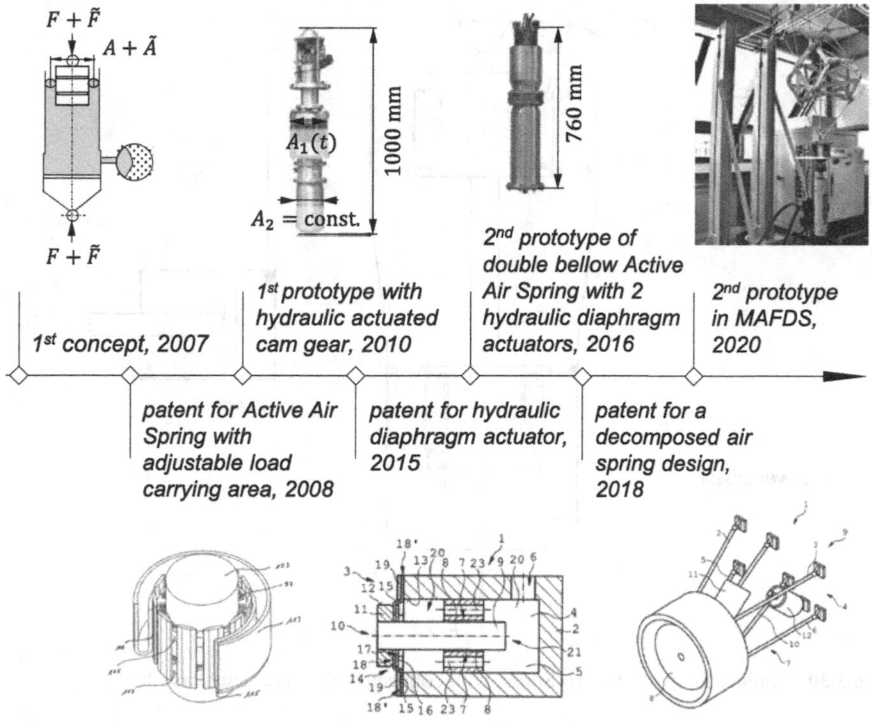

Fig. 3.29 Milestones in the development of the Active Air Spring [21, 22, 39]

thermore, energy can be recuperated by resetting the actuators through the bellows. Figure 3.30 shows the concept of the Active Air Spring including two actuator pistons and the hydraulic coupling. The actuator can be driven by different operating principles. A hydraulic drive is used for the technical prototype.

Table 3.3 gives the main characteristics of the developed Active Air Spring spring. Figure 3.31 shows the designed second prototype according to these parameters.

The two actuators are the core of the Active Air Spring. They work as single-acting hydraulic linear actuators. The reset is done via the bellows and the pressure in the air spring. In order to distribute the load on the bellows, the piston is divided into four segments [5], each moves radially up to ±3 mm [22]. These segments are guided by piston rods which are mounted in the piston via sliding bushes. Diaphragm cloths seal the actuators with low friction and without leakage [22]. Figure 3.32 shows the structure of the upper actuator.

Influencing factors on the performance of the active suspension system are

1. the maximum velocity at which the actuator forces can be adjusted,
2. the maximum actuator force $\Delta F_{a,max}$ and
3. the control concept and the selected controller.

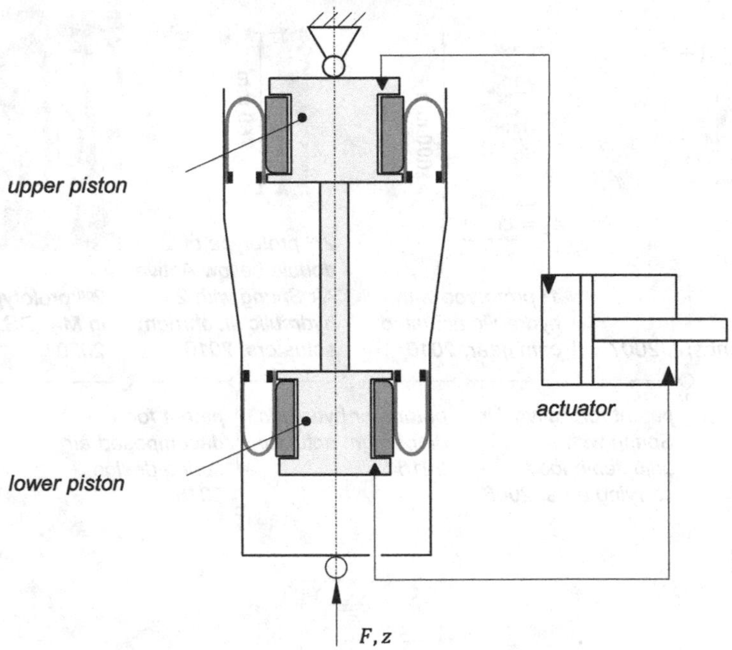

Fig. 3.30 Double bellow Active Air Spring with connected hydraulic actuators [20]

We considered all these factors in the development of the technical prototype. The uncertainty in the design parameters, such as the body mass for example, was taken into account in the sense of a resilient product development [25].

In the following, we present the results necessary for an optimal design and operation of the Active Air Spring.

1. *Actuator Velocity*. Active suspension systems can be divided in slow active systems working up to 5 Hz and fast active systems working up to 30 Hz. We showed that a slow active system is sufficient to improve driving comfort and avoid kinetosis [20]. The transfer function of the developed actuator corresponds approximately to a first order low pass filter with a cut-off frequency of 5 Hz [23].
2. *Actuator Force*. We showed that an actuator force of about 1 kN is sufficient to influence the vertical dynamics of a car with the properties shown in Table 3.2 [20]. With more available force, the driving comfort is not further improved.
3. *Controller*. The control concept consists of a primary controller that specifies the actuator force and a secondary controller that controls the realisation of the required actuator force via the hydraulic diaphragm actuators [20]. The structure and parameters of the primary control loop determine the performance of the active system. We designed the controller via an H2-optimisation [32]. The resulting controller structure includes a skyhook controller combined with a preview function. The preview function significantly increases the performance of

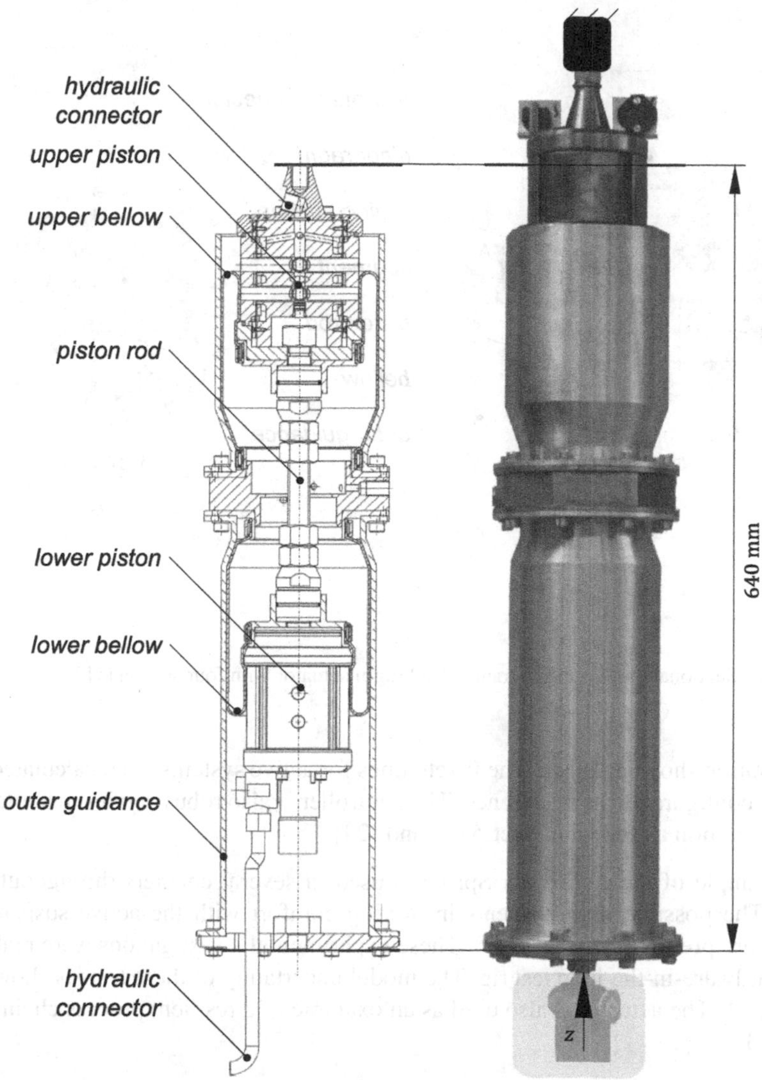

Fig. 3.31 Technical prototype of the Active Air Spring [20]

the active system. This results in almost the ideal active system [32]. Therefore the road excitation z_r with velocity $\dot{z}_r$ and acceleration $\ddot{z}_r$ as well as the body velocity $\dot{z}_b$ is fed back statically. More complex dynamic controllers are not used due to stability reasons [24]. Table 3.4 lists the implemented controller parameters.

The designed controller is robust against uncertain parameters such as varying body mass m_b or wheel stiffness k_w. Figure 3.33 shows that the results are always close to the active Pareto line. The controller is designed with the nominal con-

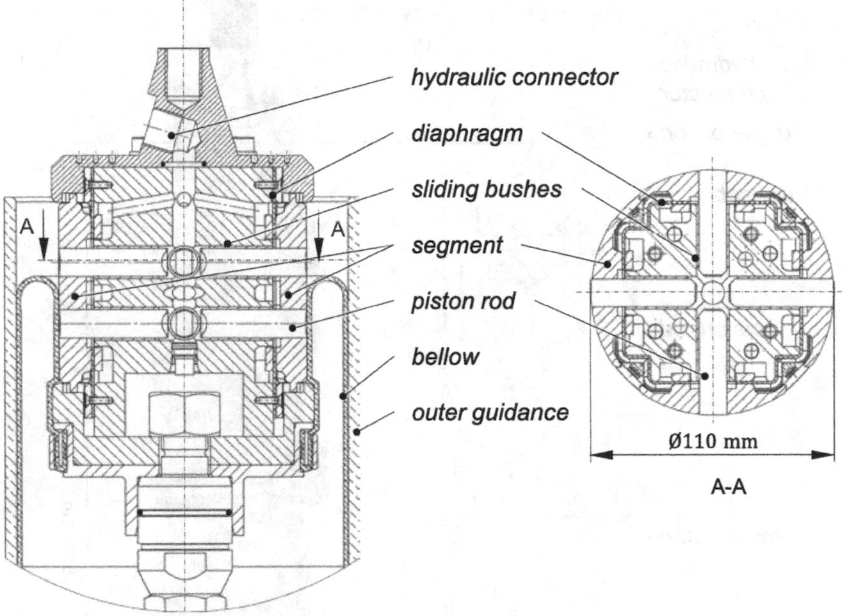

Fig. 3.32 Sectional view of the hydraulic diaphragm actuator with four segments [20]

figuration shown in black. The Pareto lines for active systems are recalculated for each configuration as a reference. The controller is also robust against uncertainty in excitation as shown in Sect. 5.4.5 and [24].

The example of the Active Air Spring is used in several chapters throughout this book. The possible improvements in driving comfort with the active suspension system are presented in Sect. 5.4.5. These experimental investigations were realised on a hardware-in-the-loop test rig. The model uncertainty of these tests is shown in Sect. 4.3.4. The actuator is also used as an example of a resilient process chain (see Sect. 6.3.7).

3.6.3 3D Servo Press

Maximilian Knoll, Florian Hoppe, and Peter Groche

Forming machines are used to provide forming forces and energies required to form and guide the tools. Depending on the requirements, individual types of machines are used. Due to fluctuations on the downstream market, the requirements for products and thus the forming processes can vary tremendously. As forming machines come with a high investment, it is important to predict future requirements and select

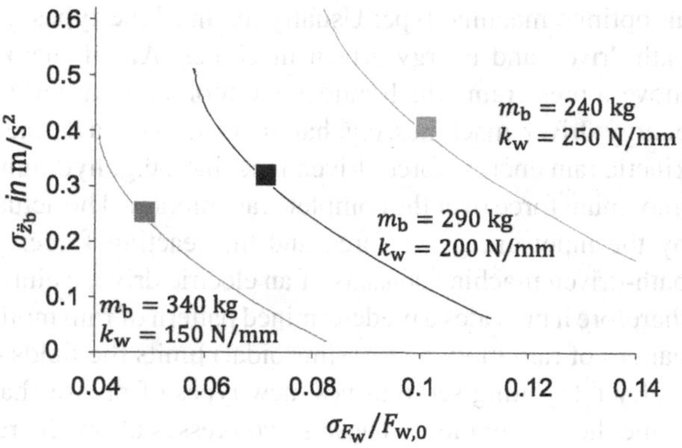

Fig. 3.33 Simulated influence of uncertain system parameters on the Active Air Spring when using the skyhook controller with preview (square) when riding on a highway with 100 km/h [20]

Table 3.3 Important characteristics of the Active Air Spring [20]

Parameter	Value
Static load F_0	2850 N
Actuator force $\Delta F_a(z = 0)$	±1180 N
Static pressure p_0	14 bar (a.)
Total load-bearing area A_0	(2195 ± 910) mm^2
e maximum deflection z_{max}	±70 mm
Air spring volume V_0	2.2 l
Upper piston	
Piston diameter $d_{p,1}$	(105 ± 6) mm
Outer diameter $d_{o,1}$	140 mm
Load-bearing area A_1	(11.740 ± 570) mm^2
Segment high $h_{seg,1}$	76 mm
Lower piston	
Piston diameter $r_{p,2}$	(94 ∓ 5) mm
Outer diameter $r_{o,2}$	127 mm
Load-bearing area A_2	(9545 ∓ 340) mm^2
Segment high $h_{seg,2}$	76 mm

Table 3.4 Implemented skyhook controller with preview

Parameter	Value
$k_{\dot{z}_b}$	−1750 Ns/m
k_{z_r}	−3860 N/m
$k_{\dot{z}_r}$	−1016 Ns/m
$k_{\ddot{z}_r}$	−18.29 Ns^2/m

an optimal machine type. Usually the machine types are divided into force-driven, path-driven and energy-driven machines. All of them have in common that they move a press ram which guides the tool along a horizontal or vertical axis. While energy-driven machines, e.g. hammers, provide a defined forming energy in terms of kinetic ram energy, force-driven machines, e.g. hydraulic presses, provide a specific maximum force over the complete ram motion. The actual ram motion is then defined by the input energy or force and the reacting forces. In contrast, a conventional path-driven machine consists of an electric drive, a clutch, flywheel and crank drive; therefore it provides a predetermined pattern of ram motion [54]. The non-modifiable pattern of ram motion (e.g. sinusoidal) limits the fields of application.

By integrating servo drives, new types of presses have been introduced to overcome these limitations. These servo presses allow the ram motion to be adjusted by means of control systems and algorithms. However, these machines so far only allow ram movements in one direction. To achieve more demanding geometries to make use of properties difficult to deform, special machines or complex tool designs are often unavoidable.

One prominent case of special processes is orbital forming. During orbital forming a tool with small contact areas rolls over the workpiece and progressively shapes it [8]. Forming forces are reduced drastically in orbital forming presses, the translatory ram motion is extended by an additional rotary motion. While the translatory ram motion can be freely programmed, the rotary motion can be adapted to the process only in restricted levels.

Classical path-driven presses with a one-dimensional motion are designed for high productivity and have proven their value over the last decades. Especially path-driven presses with flywheel are able to produce a huge number of parts per minute. But they are limited to one degree of freedom ram motion. Today's servo presses provide an increased flexibility, as described in Sect. 6.21.

But they are not able to offer motion patterns which special presses use to get the most out of the individual forming process.

As seen from the previous paragraphs, classical and special presses either offer a high productivity or specialised motions. After an investment decision for a forming machine is made, the flexibility with respect to motion pattern and productivity is very limited. But due to uncertain conditions and requirements on the upstream and downstream market, modern production systems should be able to switch over to a multitude of processes and process chains. This is only possible through the cost-intensive acquisition of different machine types.

Furthermore, disturbances that occur during production, such as varying material properties, are difficult to control. Since existing machine concepts only offer very limited possibilities to compensate for these, tight tolerance requirements are set for material and process parameters which are continuously monitored. Yet, increasing the adaptability of the process has been accompanied by productivity reductions and the installation of additional drive systems.

Our objective is therefore to combine the advantages of different press types and to enable adaptability without loss of productivity. A new approach is the independent actuation of the press ram at three points [16]. An independent motion of these points allows any desired motion of the ram in three degrees of freedom (dof). While mechanical presses highly rely on a stiff design to achieve a high accuracy, a free 3-dof ram motion allows for a compensation of mechanical inaccuracies, elasticities and disturbances by means of control systems [17]. Thus it is possible to break away from previous design constraints and to pursue new design objectives.

Scheitza [45] proved the feasibility of these theoretic considerations of a new 3D Servo Press concept with a mechanical demonstrator with 1 ton press force, Fig. 3.34. Each of the three ram points is driven by an independent press gear which consists of eccentric and modified knuckle joint kinematics. To adapt the shut height of the tools, two additional spindle kinematics modify the press gear. Several studies have shown that the new press concept allows to extend process limits and combine processes, thus increasing flexibility and productivity. The ability to react to disturbances paved the way to control product properties, see Sect. 5.3.2.

After successful validation of the 3D Servo Press concept, it was scaled up from a prototype with 1 ton press force to a 160 ton press, as shown in Fig. 3.34. While the press force is upscaled by a factor of 160, geometrical dimensions are only upscaled by a factor of approximately 4. This leads to a number of challenging size effects. The rolling bearings used in the prototype reach a technical limit in terms of load capacity and dimension at this scale. Plain bearings normally used in press design are accompanied by bearing clearance, which adds up through the kinematic chain. Investigations on this scaling effect led to the development of novel combined roller and plain bearings, as described in Sect. 5.4.2. With the integration of such a combined bearing design the bearing size has been reduced, while at the same time increasing the service life and stiffness of the bearing [47].

Scaling the ability to mount larger tools comes with a dimensional increase in the ram size. But to achieve the same tilting angle with the 1600 kN press, the ram drive points have to cover a larger distance, which is called the stroke height. Therefore, the stroke height had to be scaled up from 41 to 100 mm with an additional adaption of the shut height of 200 mm. Besides technical constraints and scalability laws, also requirements and restrictions by the infrastructure department had to be complied with when scaling up the press. Due to the associated maximum installation area and height, linear scaling of the press kinematics proved to be not practicable. In order to realise the three gears in the required dimensions, a new arrangement of individual levers was carried out, compare Fig. 3.34. The scaling of the stroke height results in a smaller width to height ratio.

During the scaling of the gear dimensions the absolute manufacturing tolerances increase, adding up in the kinematic chain. These have a significant effect on the accuracy of the ram positioning and thereby on product quality [17]. Challenges are to produce individual parts of up to 2000 mm $\pm$ 20 μm in length with homogeneous material properties.

Next to manufacturing individual parts, assembly and delivery bring up new scaling challenges. Due to infrastructural limitations, the press could not be fully assembled at the operating site. Developments of innovative products come with a significant amount of uncertainty, especially with respect to functionality. This requires strategies to reduce both uncertainty and cost. Therefore, the press had to be assem-

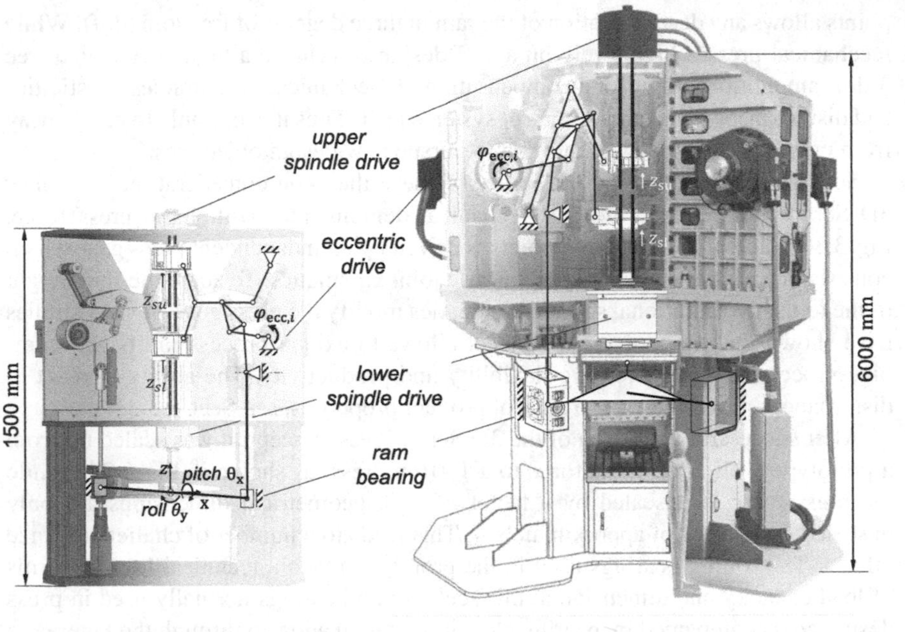

Fig. 3.34 Kinematic comparison of 1 and 160 ton 3D Servo Press

bled at an assembly location where also all functions had to be tested. To reach a cost minimum between transportation costs and assembly hours, the press was designed to be disassembled into two parts, transported to the operating site and reassembled with special equipment within a few hours.

The electromechanical devices had to be scaled to drive the scaled gear mechanisms and provide the pressing force. Presses require high torques but typical electrical machines provide high velocities at low torque. The prototype was equipped with gearboxes to increase the torque transmission, but that comes with the disadvantages of reduced efficiency, lower stiffness, large installation place, increased weight and maintenance. Hence, a gearbox cannot be upscaled for the 160 ton press; thus direct drive design is desired which requires high-torque drives. The diameter of high-torque drives is significantly larger than those of typical drives, as the torque in electrical machines scales linearly in stator length and quadratically in stator diameter. To achieve a symmetric torque on the eccentric shaft, the 160 ton press was equipped with 2 torque drives on each eccentric axis, the shafts of which being positively coupled with the eccentric.

Typical ram motions such as a press stroke or an orbital motion can be adopted from state of the art press control in which the drive motion defines the ram motion. But the 3D Servo Press is designed to control the actual product properties as described in Sect. 5.3.2. Therefore it allows for arbitrary 3D motions of the ram, which requires a paradigm shift in the control. Control methods can be adopted from robot control in which the drives are controlled in a closed-loop process and the kine-

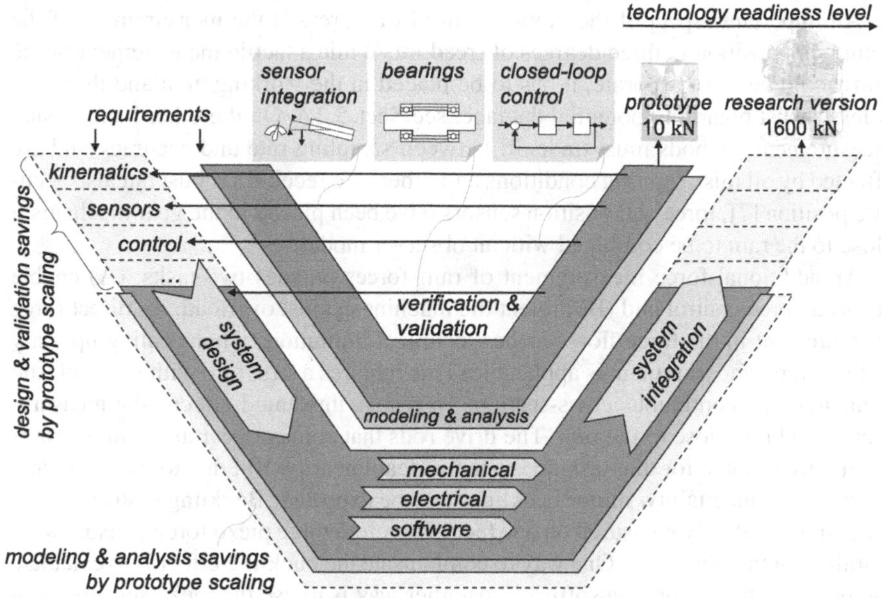

Fig. 3.35 Scaling of development method and design of the 3D Servo Press using the V-model

matic is controlled in an open-loop by means of inverse kinematics [28]. First control approaches were investigated in simulations, being implemented and validated on the prototype. This method corresponds to the V-model shown in Fig. 3.35, based on feeding back experience in the validation phase to the design phase. As lack of knowledge plays a major role in innovative research projects, early validation phases are crucial to be able to adapt to new insights. By performing time-consuming design and validation steps on the prototype, these steps can be significantly reduced for the 160 t 3D Servo Press whereby commissioning time and risk can be cut.

Early validation phases have shown that since the limited installation space did not allow a press design with maximum mechanical stiffness, the press elasticity has a greater effect on accuracy. Hence, the lower passive stiffness has to be increased by means of stiffness models and active measures, i.e. a closed-loop control of the kinematics as described in Sect. 5.4.1. A closed-loop control of the kinematics requires real-time adaption of the drive motion and hence real-time solution of inverse kinematics. As no explicit solution is available for the inverse spindle and ram kinematics, real-time optimisation routines were applied but are computationally expensive. But as higher control speed only comes with a higher sampling time [28], the optimisation steps had to be parallelised with the control. Stability analyses were carried out in experiments on the prototype and have demonstrated the effect of uncertainty that inhibits industrial closed-loop ram control today. The early analyses on the prototype also allowed to develop new robust control methods shown in Sect. 6.1.7.

An important aspect of the motion control of a press is the measurement of the actual ram position in three degrees of freedom. While a tactile measurement of the ram would be most accurate, it has to be placed in the working area and therefore is less robust regarding potential damage, see Sect. 5.2.4. On the other hand, visual measurement methods must trade off between sampling rate and accuracy and are affected by oil mist, lighting conditions and others. To receive a robust but also accurate position [7], force and position sensors have been placed in the gear mechanism close to the ram to be combined with an observer model.

An additional force measurement of ram forces pursues two tasks: (A) enable accurate force control and (B) protect the machine against overload. As direct force measurement in the force flow reaches technical limitations when scaling up, only indirect force measurement is applicable. This requires a structure with linear elastic behaviour and continuous cross-section. To reduce unwanted effects, the measurement should be close to the ram. The drive rods that connect gear and ram bearings are highly suitable for that task due to their ideal linearity. But due to its length and geometrical uncertainty, minor buckling is to be expected. Buckling distorts force-measurement if only measured on one face. Therefore three piezo force-sensors were installed on the drive bars. One way to compensate the buckling effect is to calculate the mean of the sensors via software. Another way is to use the piezo sensors as an electric circuit to calculate the mean. By connecting the piezo crystals in series their charge physically sums up; this performs the same function as the mean, however by even tripling the accuracy.

Besides the control system, the majority of the software modules were likewise developed and validated on the basis of the prototype. This includes functions, such as the user interface, sensor evaluation, logic, motion control and real-time communication. While a user-interface is run by an industrial PC communicating with a programmable logic controller (PLC) via non-real-time ethernet, a real-time communication between PLC, sensors and drives is mandatory. This involves also the synchronisation of the positively coupled torque drives. In contrast to the prototype, each shaft that drives an eccentric gear is equipped with two 3500 Nm torque drives. To add up the torques of both drives and prevent that both drives operate against each other, both drives synchronise their torque in a master-slave setup. While the master drive is controlled by the kinematic press controller, the second drive is directly linked to the measured torque of the master and thus supports its motion.

As research tasks changing over time require modifiable software, safety functions must be outsourced to non-modifiable hardware. Hence, safety functions are performed by sensors and an additional safety PLC. Although the design of the functional safety of the 3D Servo Press for compliance with the Machinery Directive 2006/42/EC [12] is based on typical presses, its machine safety has required several innovations.

While the actual press force on one hand serves as a process and control variable, the force sensors also protect against overload. When the maximum force is reached in case of a fault, the machine motion is decelerated to zero. As three degrees of freedom are involved in the motion, a simply upward motion of the ram might damage the tool. The new closed-loop force control allows to safely move the ram up to a limited force and stop once the tool is locked. In contrast to typical servo presses, the spindle kinematics of the 3D Servo Press is designed to continuously adapt the dead centres during the process. Due to the high kinetic energy in the spindles during the process, multiple adaptions of the initial prototype design have been performed. This involves a large amount of kinetic energy being converted into heat. Therefore, highly efficient spindles have been installed in the 3D Servo Press. However, the converted heat amount depends on the load and motion history, which is why both spindle nuts are additionally monitored with temperature sensors. The second spindle related safety aspect is maintaining the mechanical working area. On the one hand, the mechanical limitation of each individual spindle must be protected. In contrast to the prototype, the spindle force of the upscaled version is larger than the mechanical load capacity and therefore requires a sensor-based end stop. However, due to the high potential kinetic energy, it is impossible to ensure that the spindles decelerate in time when the end stop is reached. Therefore, a second sensor for speed reduction was integrated before the end stop. On the other hand, operating the window nonlinearly depends on both spindle positions. While the control software guarantees to keep these boundary conditions, an additional non-tactile sensor monitors compliance with the process window. To improve the control performance and accuracy, combined roller-plain bearings were integrated in each joint. Those are lubricated via multiple circuits which are being monitored in terms of flow, pressure and temperature.

Due to its 8 servo drives, the 3D Servo Press requires a maximum electric power of 1.2 MW resulting in a nominal mechanical press force of 160 tons at a maximum speed of 100 strokes/min. While eccentric drives contribute to the maximum speed with 100 kW each, the spindle drives come with 300 kW. The spindle kinematics is designed for a high force transmission which allows to freely adapt the gear during the process without being disturbed by reacting forces. But the spindles are only allowed to move in a defined process window which yields in the achievable stroke and shut height. The process window of the spindle positions (Fig. 3.36a can also be mapped to the shut and stroke height, see Fig. 3.36b). This results in an shut height range from 500 to 700 mm and a stroke height adjustment from 25 to 100 mm, see Fig. 3.36b.

The eccentric kinematics results in a position-dependent force transmission of the eccentric drives and therefore position-dependent maximum ram force. Under the assumption of an inelastic system, an infinite transmission ratio can be reached at the top and bottom dead centre. The nominal force of the 3D Servo Press is reached $\varphi_{\text{ecc,i}} = \pm 16°$ before the bottom dead centre. The maximum force reaches its minimum between the two dead centres.

Due to the 120° arrangement of the three gears, which are coupled via the central spindles, the eccentrics can be controlled independently, as shown in Fig. 3.34. As a result, both a translatory stroke and an orbital motion, as well as anything in between can be realised.

The maximum stroke height of 100 mm and the distance between the ram bearings result in a maximum pitch angle of 3.44° and maximum roll angle of 2.97°.

Starting from an initial design of the 1 ton press in 2008, the 160 ton press was developed while starting the production of first parts in 2011, see Fig. 3.37. In the

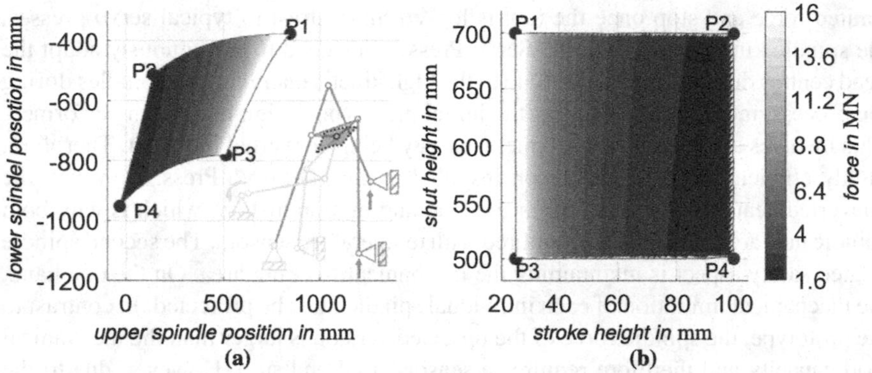

Fig. 3.36 Process window of the 3D Servo Press in combination of minimum process force at maximum engine power, **a** as a function of the spindle positions and **b** transformation to the shut and stroke height

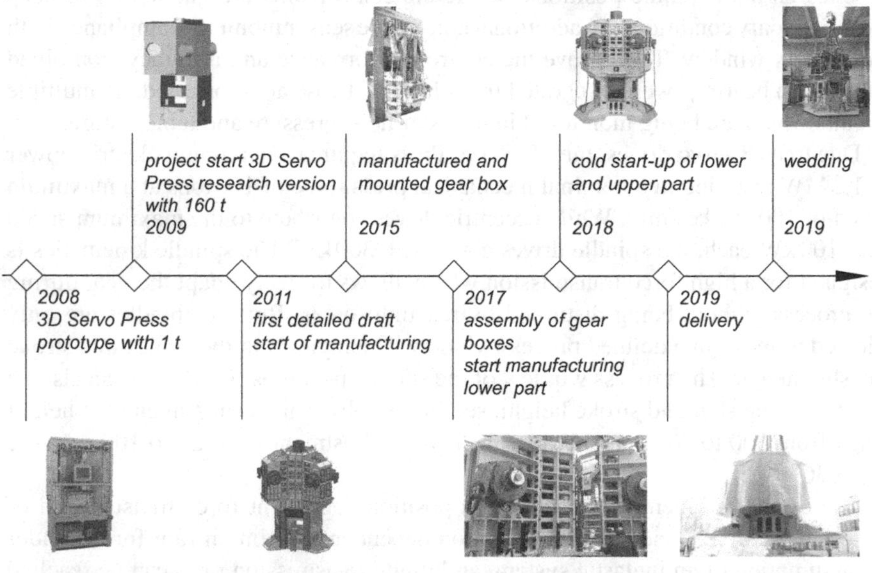

Fig. 3.37 Timeline of development 3D Servo Press from 2008 to 2019

following years, the press was further detailed as the production of individual parts progressed. In 2015, the first major milestone was reached with the assembly of the first gearbox. In the same year, the centrepiece of the 3D Servo Press was realised. The completion of the upper part began in 2017 and was completed with the assembly of the last gear box to form a fully integrated gearbox. Subsequently, the assembly of the lower part and cold commissioning of the press was carried out at the assembly site and assured the functionality of the 3D Servo Press. Following the cold commissioning the press was transported to the operating site, where the final positioning and commissioning took place.

References

1. Ackermann J (2002) Robust control: the parameter space approach, 2nd edn. Communications and control engineering. Springer, London
2. Ashby MF, Jones DRH (2005) Engineering materials. Elsevier/Butterworth-Heinemann, Amsterdam
3. Baucells M, Borgonovo E (2013) Invariant probabilistic sensitivity analysis. Manag Sci 59(11):2536–2549. https://doi.org/10.1287/mnsc.2013.1719
4. Bedarff T (2017) Grundlagen der Entwicklung und Untersuchung einer aktiven Luftfeder für Personenkraftwagen. Forschungsberichte zur Fluidsystemtechnik, vol 10. Shaker, Herzogenrath. Dissertation, TU Darmstadt
5. Bedarff T, Hedrich P, Pelz PF (2014) Design of an active air spring damper. In: Murrenhoff H (ed) 9th international fluid power conference (9th IFK). HP - Fördervereinigung Fluidtechnik, Aachen, pp 356–365
6. Bretz A, Calmano S, Gally T, Götz B, Platz R, Würtemberger J (2015) Darstellung passiver, semi-aktiver und aktiver Maßnahmen im SFB 805-Prozessmodell. Preprint
7. Calmano S, Schmitt S, Groche P (2013) Prevention of over-dimensioning in light-weight structures by control of uncertainties during production. In: International conference on "New developments in forging technology". MAT INFO Werkstoff-Informationsgesellschaft mbH, Fellbach, pp 313–317
8. Calmano S, Hesse D, Hoppe F, Traidl P, Sinz J, Groche P (2015) Orbital forming of flange parts under uncertainty. In: Pelz PF, Groche P (eds) Uncertainty in mechanical engineering II, vol 807. Applied mechanics and materials. Trans Tech Publications, pp 121–129
9. de Secondat Montesquieu C (1750) The spirit of laws (Nourse J, Vaillant P in the Strand, London). Translated from the French of M. de Secondat, Baron de Montesquieu
10. Efron B (1979) Bootstrap methods: another look at the jackknife. Ann Stat 7(1):1–26. https://doi.org/10.1214/aos/1176344552
11. Enss GC, Gehb CM, Götz B, Melz T, Ondoua S, Platz R, Schaeffner M (2016) Device for optimal load transmission and load distribution in lightweight structures (Kraftübertragungsvorrichtung). Patent DE 10 2014 106 858 A1
12. European Parliament (2006) Council of the European Union: Directive 2006/42/EC of the European Parliament and of the council of 17 may 2006 on machinery, and amending directive 95/16/EC. Official J Eur Union. https://eur-lex.europa.eu/eli/dir/2006/42/oj
13. Feldmann R, Gehb CM, Schaeffner M, Matei A, Lenz J, Kersting S, Weber M (2020) A detailed assessment of model form uncertainty in a load-carrying truss structure. In: Model validation and uncertainty quantification, vol 3. Conference proceedings of the Society for Experimental Mechanics series. Springer
14. Fowlkes WY, Creveling CM (2012) Engineering methods for robust product design: using Taguchi methods in technology and product development. Engineering process improvement series, Addison-Wesley, Reading, Mass
15. Ghanem R, Higdon D, Owhadi H (2017) Handbook of uncertainty quantification. Springer, Cham
16. Groche P, Scheitza M, Kraft M, Schmitt S (2010) Increased total flexibility by 3D servo presses. CIRP Ann 59(1):267–270
17. Groche P, Hoppe F, Sinz J (2017) Stiffness of multipoint servo presses: mechanics vs. control. CIRP Ann 66(1):373–376
18. Hagedorn P (1982) Non-linear oscillations, vol 10. The Oxford engineering science series. Clarendon Press, Oxford

19. Hanselka H, Platz R (2010) Ansätze und Maßnahmen zur Beherrschung von Unsicherheit in lasttragenden Systemen des Maschinenbaus. Konstruktion 2010(11–12):55–62
20. Hedrich P (2018) Konzeptvalidierung einer aktiven Luftfederung im Kontext autonomer Fahrzeuge. Dissertation, TU Darmstadt. https://tuprints.ulb.tu-darmstadt.de/8469/
21. Hedrich P, Pelz PF (2018) Einzelradaufhängung für ein Kraftfahrzeug. Patent DE102018127301A1, 31 Oct 2018
22. Hedrich P, Johe M, Pelz PF (2015) Aktor mit einem linear verlagerbaren Stellglied. Patent DE 102015120011 A1, 18 Nov 2015
23. Hedrich P, Lenz E, Pelz PF (2017) Modellbildung, Regelung und experimentelle Untersuchung einer aktiven Luftfederung in einer Hardware-in-the-Loop-Simulationsumgebung. VDI-Fachtagung Schwingungen 2017. VDI-Berichte Band 2295. VDI Verlag, Düsseldorf, pp 447–460
24. Hedrich P, Lenz E, Pelz PF (2018) Minimizing of kinetosis during autonomous driving. ATZ Woldwide 120(7–8):68–75
25. Hedrich P, Brötz N, Pelz PF (2018) Resilient product development – a new approach for controlling uncertainty. In: Pelz PF, Groche P (eds) Uncertainty in Mechanical Engineering III, vol 885. Applied mechanics and materials. Trans Tech Publications, pp 88–101. https://doi.org/10.4028/www.scientific.net/AMM.885.88
26. Heinrichs D (2015) Autonomes Fahren und Stadtstruktur. In: Maurer M, Gerdes JC, Lenz B, Winner H (eds) Autonomes Fahren: Technische, rechtliche und gesellschaftliche Aspekte. Springer, Berlin, pp 219–239. https://doi.org/10.1007/978-3-662-45854-9_11
27. Holl M, Rausch L, Pelz PF (2017) New methods for new systems - how to find the techno-economically optimal hydrogen conversion system. Int J Hydrog Energy 42(36):22641–22654. https://doi.org/10.1016/j.ijhydene.2017.07.061
28. Hoppe F, Pihan C, Groche P (2019) Closed-loop control of eccentric presses based on inverse kinematic models. Proc Manuf 29:240–247. https://doi.org/10.1016/j.promfg.2019.02.132
29. Ihn T (2016) Data analysis – Fehlerfortpflanzung. Lecture Notes, ETH Zürich
30. Iooss B, Saltelli A (2016) Introduction to sensitivity analysis. In: Ghanem R, Higdon D, Owhadi H (eds) Handbook of uncertainty quantification. Springer, Cham, pp 1–20. https://doi.org/10.1007/978-3-319-11259-6_31-1
31. Isermann R (2008) Mechatronische Systeme: Grundlagen, 2nd edn. Springer, Berlin
32. Lenz E, Hedrich P, Pelz PF (2018) Aktive Luftfederung – Modellierung, Regelung und Hardware-in-the-Loop-Experimente. Forschung in Ingenieurwesen 82:171–185. https://doi.org/10.1007/s10010-018-0272-2
33. Miner M (1945) Cumulative damage in fatigue. J Appl Mech 12:159–164
34. Mitschke M, Wallentowitz H (2014) Dynamik der Kraftfahrzeuge, 5th edn. Springer, Wiesbaden
35. Newton I (1726) Philosophiae naturalis principia mathematica. Tomus Primus, London
36. Pahl G, Beitz W, Feldhusen J, Grote KH (2007) Engineering design: a systematic approach, third edn. Springer, London. https://doi.org/10.1007/978-1-84628-319-2
37. Palmgren A (1924) Die Lebensdauer von Kugellagern. Verein Deutscher Ingenieure 68:339–341
38. Pelz P, Buttenbender J (2004) The dynamic stiffness of an air-spring. In: ISMA2004 international conference on noise & vibration engineering, pp 20–22
39. Pelz PF, Rösner J (2008) Schwingungsfluiddämpfung- und/oder -federung. Patent DE102008007566B4, 05 Feb 2008
40. Pelz PF, Holl M, Platzer M (2016) Analytical method towards an optimal energetic and economical wind-energy converter. Energy 94:344–351. https://doi.org/10.1016/j.energy.2015.10.128
41. Phadke MS (1989) Quality engineering using robust design. Prentice-Hall, London
42. Rams D, Klatt J (1995) Weniger, aber besser: Skizze für Phonokombination TP 1. Jo Klatt Design und Design Verlag, Hamburg
43. Rexer M, Brötz N, Pelz PF (2020) Much does not help much: 3D pareto front of safety, comfort and energy consumption for an active pneumatic suspension strut. In: 12th international fluid power conference, pp 37–42. https://doi.org/10.25368/2020.91

44. Saltelli A, Aleksankina K, Becker W, Fennell P, Ferretti F, Holst N, Li S, Wu Q (2019) Why so many published sensitivity analyses are false: a systematic review of sensitivity analysis practices. Environ Model & Softw 114:29–39. https://doi.org/10.1016/j.envsoft.2019.01.012
45. Scheitza M (2009) Konzeption eines flexiblen 3D-Servo-Pressensystems und repräsentative Basisanwendungen. Shaker
46. Sethi A, Sethi S (1990) Flexibility in manufacturing: a survey. Int J Flex Manuf Syst 2(4). https://doi.org/10.1007/BF00186471
47. Sinz J, Niessen B, Groche P (2018) Combined roller and plain bearings for forming machines: design methodology and validation. In: Congress of the German academic association for production technology. Springer, pp 126–135
48. Sivak M, Schoettle B (2015) Motion sickness in self-driving vehicles. Technical report, UMTRI-2015-12, Transportation Research Institute, University of Michigan. http://hdl.handle.net/2027.42/111747
49. Sivia DS, Skilling J (2006) Data analysis: a Bayesian tutorial, 2nd edn. Oxford University Press, Oxford
50. Stevin S (1582) Tafelen van interest. Christoffel Plantijn, Antwerpen. Neudruck
51. Sullivan LH (1922) The tall office building artistically considered. West Archit 1922(31):3–11
52. Vaughan D (2007) The Challenger launch decision: Risky technology, culture, and deviance at NASA. University of Chicago Press, Chicago. Reprint
53. Vergé A, Pöttgen P, Altherr LC, Ederer T, Pelz PF (2016) Lebensdauer als Optimierungsziel – Algorithmische Struktursynthese am Beispiel eines hydrostatischen Getriebes. O+P – Ölhydraulik und Pneumatik 60(1–2):114–121
54. Wegener K (2014) Forming presses (hydraulic, mechanical, servo). In: Laperrière L, Reinhart G (eds) CIRP Encyclopedia of production engineering. Springer, Berlin, pp 547–553. https://doi.org/10.1007/978-3-642-20617-7_16695

Chapter 4
Analysis, Quantification and Evaluation of Uncertainty

Maximilian Schaeffner, Eberhard Abele, Reiner Anderl, Christian Bölling, Johannes Brötz, Ingo Dietrich, Robert Feldmann, Christopher M. Gehb, Felix Geßner, Jakob Hartig, Philipp Hedrich, Florian Hoppe, Sebastian Kersting, Michael Kohler, Jonathan Lenz, Daniel Martin, Alexander Matei, Tobias Melz, Tuğrul Öztürk, Peter F. Pelz, Marc E. Pfetsch, Roland Platz, Manuel Rexer, Georg Staudter, Stefan Ulbrich, Moritz Weber, and Matthias Weigold

Abstract This chapter describes the various approaches to analyse, quantify and evaluate uncertainty along the phases of the product life cycle. It is based on the previous chapters that introduce a consistent classification of uncertainty and a holistic approach to master the uncertainty of technical systems in mechanical engineering. Here, the following topics are presented: the identification of uncertainty by modelling technical processes, the detection and handling of data-induced conflicts, the analysis, quantification and evaluation of model uncertainty as well as the representation and visualisation of uncertainty. The different approaches are discussed and demonstrated on exemplary technical systems.

The book at hand is devoted to portraying our holistic approach to master the uncertainty of technical systems in mechanical engineering over all the phases of the product life cycle. The conceptual basis of our specific approach, as motivated in Chap. 1 and elaborated in Chap. 3, as well as the consistent classification and definition of uncertainty in Chap. 2, form the foundation of this approach, see Fig. 1.12.

This chapter deals with the analysis, quantification and evaluation of data and model uncertainty in mechanical engineering as an essential first step to master uncertainty. This will then be extended and completed by the methods and technologies to

M. Schaeffner (✉) · E. Abele · R. Anderl · C. Bölling · J. Brötz · I. Dietrich · R. Feldmann · Christopher M. Gehb · F. Geßner · J. Hartig · P. Hedrich · F. Hoppe · J. Lenz · D. Martin · T. Melz · T. Öztürk · Peter F. Pelz · R. Platz · M. Rexer · G. Staudter · M. Weber · M. Weigold
Department of Mechanical Engineering, TU Darmstadt, Darmstadt, Germany
e-mail: maximilian.schaeffner@sam.tu-darmstadt.de

S. Kersting · M. Kohler · A. Matei · Marc E. Pfetsch · S. Ulbrich
Department of Mathematics, TU Darmstadt, Darmstadt, Germany

T. Melz
Fraunhofer Institute for Structural Durability and System Reliability LBF, Darmstadt, Germany

P. F. Pelz et al. (eds.), *Mastering Uncertainty in Mechanical Engineering*,
Springer Tracts in Mechanical Engineering,
https://doi.org/10.1007/978-3-030-78354-9_4

master uncertainty presented in Chap. 5 and the strategies to master uncertainty introduced in Chap. 6. We provide both a mathematical and an engineering perspective to the analysis, quantification and evaluation of data and model uncertainty. Examples of this interdisciplinary approach are among others presented in Sects. 4.3.1 and 4.3.2. Furthermore, the methods are illustrated and their application is demonstrated using the technical systems presented in Sect. 3.6. The examples given appear in all phases of the product life cycle: design, production and usage, see Sect. 3.1 and, thus, offer a broad overview of the activities presented within this book.

We start with the identification of uncertainty by modelling technical processes in Sect. 4.1 with the aim to gain information on data uncertainty as introduced in Sect. 2.1; we consider uncertainty in single processes and its propagation in process chains. An important aspect in this domain is the detection and handling of data-induced conflicts and thus data uncertainty, which will be covered in Sect. 4.2; here the main goals are the prevention of critical failures, finding correlations among the data, and isolating faults. Computer models based on physical or empirical knowledge of a technical system are useful tools in the design phase. Since reality is commonly complex and cannot completely or exactly be represented by mathematical models, one faces the problem that all models are imperfect, i.e. model uncertainty occurs, see Sects. 1.3 and 2.2. In Sect. 4.3, the analysis, quantification and evaluation of model uncertainty are being studied; here different methods to identify sources of model uncertainty and quantify model uncertainty are discussed with the aim to analyse the accuracy of a model belonging to a technical system. Finally, in the case uncertainty is detected, it is often unclear how to represent and visualise the information in an informative way. In Sect. 4.4, a three-layer architecture is presented to solve this issue.

4.1 Identification of Uncertainty During Modelling of Technical Processes

Maximilian Schaeffner

Uncertainty occurs if properties in the life cycle process product design, production and usage, as introduced in Sects. 1.2 and 3.1, cannot be determined completely or at all. However, these are no measurable characteristics of an individual product. Uncertainty becomes obvious in, e.g. deviations between the actual and the planned product geometry as a consequence of incompletely determined production processes or undesired behaviour during usage processes. This section covers the identification of data uncertainty during modelling of technical processes as a step towards mastering uncertainty.

Besides model uncertainty, which has been introduced in Sect. 2.2 and is covered in this chapter in Sect. 4.3, possible uncertainty in the modelling of technical systems has to be considered during the product design as data uncertainty, see Sect. 2.1, of the model parameters used for design and dimensioning.

In Sect. 4.1.1 we show how random deviations of the component properties can be taken into account probabilistically during system design for the example of passive and active vibration isolation. In Sect. 4.1.2, we present the improvement of mathematical model predictions for the simulation of systems by means of a Bayesian inference based parameter calibration.

Uncertainty propagates in process chains and can ultimately lead to undesirable behaviour in production or usage processes, see Sect. 3.2. Section 4.1.3 proceeds with the model-based description and analysis of uncertainty in consecutive machining processes, such as drilling and reaming or drilling and tapping.

4.1.1 Analysis of Data Uncertainty Using the Example of Passive and Active Vibration Isolation

Roland Platz and Jonathan Lenz

The quantification and evaluation of uncertainty in load-bearing structures is of growing importance for decision-making in the early product design phase as introduced in Sect. 1.2. This may especially become necessary due to the increasing complexity and scope of structures with multi-functional properties like mechatronic, semi-active or active systems. For example, active vibration control in mobile applications, such as an active suspension strut of a car, needs additional energy sources, sensors, actuators and a controller, see Sect. 3.4. This makes the active system more complex compared to a passive system with tailored, but only fixed inertia, damping and stiffness properties [155]. In this section investigations to numerically compare the influence of aleatoric data uncertainty in the model parameters are summarised; according to Sect. 2.1, this is on predicting the dynamic behaviour from a passive and an active technology for vibration isolation of a one mass oscillator [128–130]. The variation and uncertainty of model parameters of the passive system may lead to inadequate tuning. In addition and due to growing complexity of the active system, new uncertainty in the dynamic behaviour may arise compared to the passive system. Most importantly, the energetic effort and possible reduced availability of the active system may influence the acceptance of the active technology, see Sect. 1.6. Figure 4.1a shows the simple mechanical model of a one mass oscillator with only four model parameters mass m, damping coefficient b and stiffness k for passive vibration isolation, as well as an additional gain factor g for active vibration isolation [130].

The mass m oscillates in z-direction when excited by the harmonic base point stroke $w(t) = \widehat{w}\cos(\Omega t + \delta)$ with the excitation frequency Ω, excitation amplitude $\widehat{w}$, time t, and phase shift δ. For simplification, $\delta = 0$ throughout the analysis. We assume linear characteristics of the internal damping force F_b, stiffness force F_k, and actuator force F_a in Fig. 4.1b. With $2D\omega_0 = b/m$ and $\omega_0^2 = k/m$ referring to the damping ratio D and the angular eigenfrequency ω_0 as well as with the frequency relation $\eta = \Omega/\omega_0$ and the factor $\zeta = \Omega/(m\,\omega_0^2)$, the complex amplification function (CAF) of mass displacement in z-direction in the frequency domain is

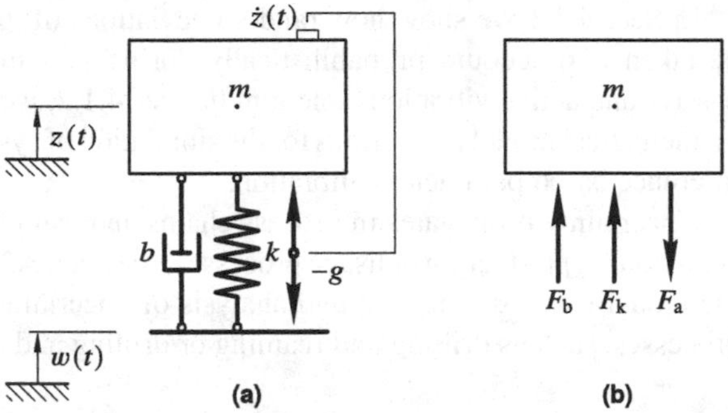

Fig. 4.1 One mass oscillator with base excitation **a** simple mechanical model and **b** internal forces [130]

$$\underline{V}(\eta) = \frac{\widehat{z}_p}{\widehat{w}} = \frac{1 + \mathrm{i}\, 2\, D\, \eta}{(1-\eta^2) + \mathrm{i}\, (2\, D\, \eta + g\, \zeta)} \tag{4.1}$$

with the amplitudes $\widehat{z}_p$ and $\widehat{w}$ from the complex particulate integral approach $\underline{z}_p(t) = \widehat{z}_p\, e^{\mathrm{i}\,\Omega\, t}$ and $\underline{w}(t) = \widehat{w}\, e^{\mathrm{i}\,\Omega\, t}$ as derived in [128]. The amplitude of (4.1) is

$$|\underline{V}(\eta)| = \sqrt{\frac{1 + (2\, D\, \eta)^2}{(1-\eta^2)^2 + (2\, D\, \eta + g\, \zeta)^2}} \tag{4.2}$$

and its phase is

$$\psi(\eta) = \arctan \frac{-2\, D\, \eta^3 - g\, \zeta}{1 - \eta^2 + (2\, D\, \eta)^2 + 2\, D\, \eta\, g\, \zeta}. \tag{4.3}$$

Deterministic case studies for different damping

Figure 4.2 shows the amplitude (4.2) and phase (4.3) of the CAF (4.1) for different *damping* cases (a)–(f) depending on different damping coefficients $b_1 < b_2 < b_3$ and feedback gains $g_1 < g_2 < g_3$.

For the passive system in Fig. 4.2, cases (a)–(c), the mass m and stiffness k are assumed constant while three different damping coefficients b_{1-3} are chosen, with gain $g = 0$. The higher the damping, the lower the maximum amplitude $V_{\max}$ at resonance frequency ω_0. However, the amplitudes beyond the isolation frequency $\Omega > \omega_{\text{iso}}$ remain higher with increased damping, which is well known. In case of active vibration isolation, cases (d)–(f), different gains g_{1-3} are chosen with assumed low passive damping b_1. A higher gain leads to a lower maximum amplitude $V_{\max}$ at resonance frequency ω_0 and keeps a low amplitude beyond the isolation frequency $\Omega < \omega_{\text{iso}}$, which is the benefit of the active approach.

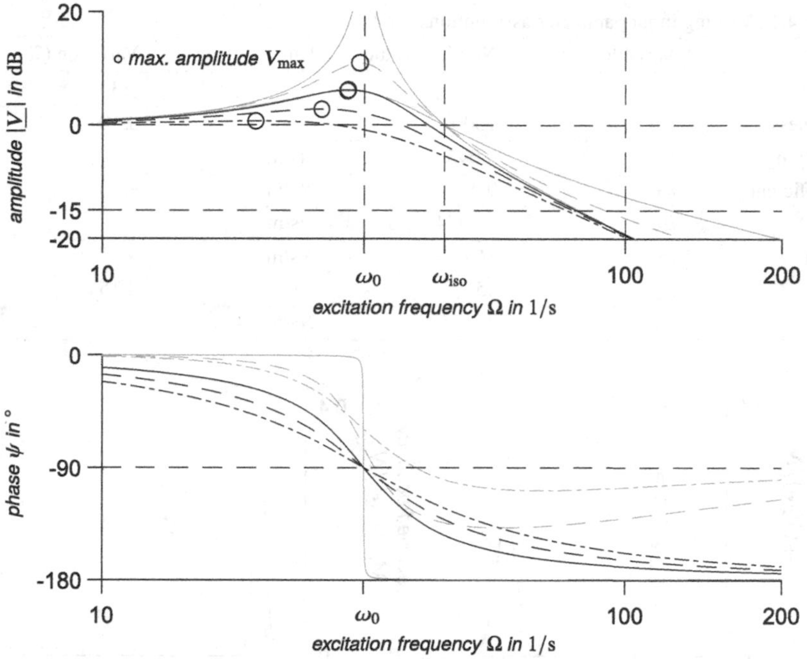

Fig. 4.2 Amplitude $|\underline{V}(\Omega)|$ and phase $\psi(\Omega)$ responses of the complex amplification function (CAF) (4.1) according to (4.2) and (4.3) for *damping* cases **a**–c with varying damping coefficients b_1 (——), b_2 (— —) and b_3 (–·–) and without gain $g = 0$ for passive, and *damping* cases d–f with low damping coefficient b_1 and varying gains g_1 (——), g_2 (— —) and g_3 (–·–) for active vibration isolation [128]

Probabilistic case studies for different CAF-points-of-interest

The influence of aleatoric data uncertainty on the numerical simulation of the dynamic behaviour of the passive and active one mass oscillator subject to vibration isolation is investigated with a Monte Carlo Simulation (MCS), see Sect. 3.3. For that, additional *CAF-point-of-interest* case studies (i)–(vi) for the *damping* cases (a)–(f) are discussed: (i) varying maximum amplitude V_{max}, (ii) varying vibration amplitudes $|\underline{V}_0|$ at the undamped resonance frequency ω_0, (iii) varying isolation frequency $\omega_{iso} = \sqrt{2}\,\omega_0$, (iv) varying amplitudes $|\underline{V}_{100}|$ at the excitation frequency beyond the passive system's fixed isolation frequency, $\Omega = 100\,1/s > \omega_{iso}$, (v) varying excitation frequency ω_{15} for -15 dB isolation attenuation, and (vi) varying decaying time $t_{0.01}$ until steady state vibration is reached or, respectively, initial transient vibrations are damped, so only 1% is left, see also [128]. The model parameters m, k, b_{1-3}, and g_{1-3} in Table 4.1 vary around an assumed nominal mean value, maximum and minimum values of the variations in % are considered as the $\pm 3\sigma$ interval per model parameter according to experience and literature [120, 128, 142]. The MCS uses 10,000 samples that meet the convergence criteria [101, 128].

Table 4.1 Varying input parameter assumptions for MCS

Property	Variable	Nominal value	Unit	Variation (%)
Mass	m	1	kg	±3
Stiffness	k	1000	N/m	±10
Damping Coefficient	b_1	0.095	Ns/m	±30
	b_2	9.487	Ns/m	
	b_3	18.974	Ns/m	
Gain	g_1	16	Ns/m	±15
	g_2	25	Ns/m	
	g_3	35	Ns/m	

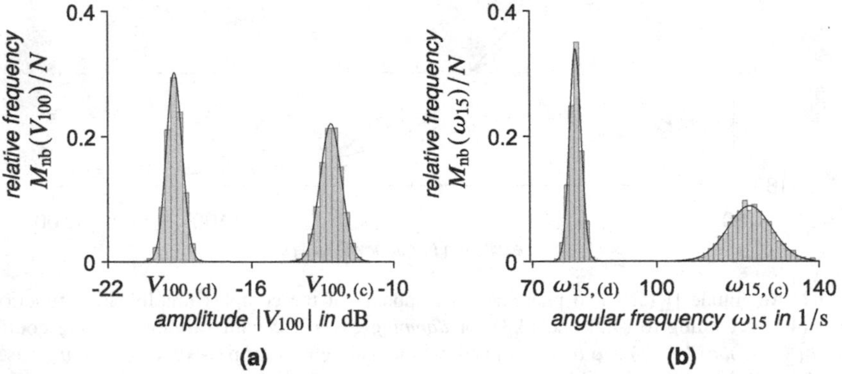

Fig. 4.3 Histograms of the relative frequency $M_{nb}(x)/N$ for constant bin-width delta and varying amounts $n_b = 1, \ldots N_b$ of bins per output **a** $x = V_{100}$, case study (iv), and **b** $x = \omega_{15}$, case study (v), for damping cases **c** and **d** for $n = 1, \ldots, 10,000$ samples [128]

As an example, Fig. 4.3 shows histograms of the relative frequency $M_{nb}(x)/N$ for varying number of bins N_b, with $n_b = 1, \ldots, N_b$ bins and with $n = 1, \ldots, N = 10{,}000$ samples, and constant bin-width Δ per varying output $x = V_{100}$ and $x = \omega_{15}$ according to the *CAF-point-of-interest* case studies (iv) and (v) for damping cases (c) and (d) [128].

In summary, Fig. 4.3a shows that for case (iv), the relative frequency $M_{nb}(V_{100})/N$ of the amplitude V_{100} becomes relatively less narrow around the empirical mean $\overline{V}_{100,(c)} = -12.66$ dB with a relatively small standard deviation $s_{V100,(c)} = 0.45$ dB due to high damping b_3 for the passive approach, damping case (c). However, for the active approach, damping case (d), the standard deviation $s_{V100,(d)} = 0.33$ dB and empirical mean $\overline{V}_{100,(d)} = -19.23$ dB are smaller although the lowest gain g_1 is used. For case (v) in Fig. 4.3b, the relative frequency $M_{nb}(\omega_{15})/N$ of the angular frequency ω_{15} at -15 dB vibration attenuation becomes relatively less narrow around the empirical mean $\overline{\omega}_{15,(c)} = 122.82$ 1/s with relatively small standard deviation $s_{\omega 15,(c)} = 5.45$ 1/s at higher passive damping b_3, damping case (c). Again, for damping case (d), the empirical mean $\overline{\omega}_{15,(c)} = 80.48$ 1/s and the standard deviation $s_{\omega 15,(d)} = 1.41$ 1/s are smaller than for the passive approach of damping case (c).

Conclusion

The observations described in this contribution show that if aleatoric data uncertainty occurs, high active damping results in less scatter at angular frequencies beyond the isolation point compared to the passive approach, see also [128]. Furthermore, the scatter of the amplitude attenuation beyond the angular isolation frequency is smaller with the active approach. Investigations are under way to validate the numerical comparison of uncertainty in passive and active vibration isolation with an experimental example.

4.1.2 Bayesian Inference Based Parameter Calibration for a Mathematical Model of a Load-Bearing Structure

Christopher M. Gehb, Tobias Melz, and Roland Platz

Load-bearing structures with kinematic functions such as the suspension of a vehicle and an aircraft landing gear enable and disable degrees of freedom and are part of many mechanical engineering applications. For an adequate numerical prediction of their load path, being e.g. necessary to develop a controller during the design phase, see Sect. 3.1, we need an adequate mathematical model with calibrated model parameters. Therefore, in this section, the adequacy of an exemplary load-bearing structure's mathematical model is evaluated with its predictability being increased by model parameter uncertainty quantification and reduction, compare Sect. 2.1. Conventionally, optimisation algorithms are used to calibrate the model parameters deterministically, as e.g. investigated in [51, 104, 161]. In contrast and as presented here, the model parameter calibration is formulated to achieve a statistically consistent model prediction with the data gained from experiments [87, 118, 144]. The most influential parameters being of interest for the model prediction, i.e. the load path through the load-bearing structure represented by the support reaction forces, are identified for calibration by a sensitivity analysis. Subsequently, the mathematical model is adjusted to the actual operating conditions of the experimental load-bearing structure via the model parameters by applying a Bayesian inference based calibration procedure. Uncertainty represented by originally large model parameter ranges is reduced and quantified to increase the model prediction accuracy.

Load-bearing structure

The investigated load-bearing structure in Fig. 4.4 is derived from the more complex load-bearing system MAFDS intended to provide the possibility of intentionally introducing uncertainty in an exemplary technical system, see Sect. 3.6.1.

The load-bearing structure consists of a translational moving mass m_A connected to a rigid beam with mass m_B and mass moment of inertia Θ_B via a spring-damper with stiffness k_S, a damping coefficient b_S, as well as two semi-active guidance

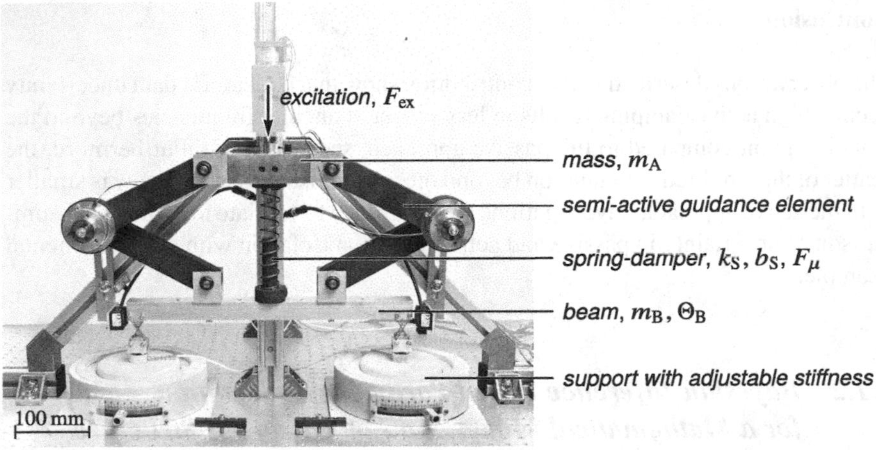

Fig. 4.4 Exemplary load-bearing structure with semi-active guidance elements for load redistribution, see Sect. 5.4.8, according to [52]

elements. The two semi-active guidance elements provide an approach to redistribute loads, e.g. in case of weakened or damaged structural components, see Sect. 5.4.8. Two supports at the ends of the beam are equipped with an adjustable stiffness to simulate weakened or damaged structural components. A weakened or damaged structural component is represented by a reduced support stiffness depicting a reduced load-bearing capacity [53–55].

Mechanical and mathematical model

Having in mind to achieve load redistribution, according to [52] the mathematical model of the load-bearing structure in Fig. 4.4 comprises parts to describe the general system dynamic, the friction and the electromagnetic actuator. The model part describing the general dynamic is chosen for model parameter calibration in this section. The friction model calibration is described in detail in [52, 56].

Figure 4.5 depicts the mechanical model and the free body diagram of the load-bearing structure. The mechanical model consists of a movable mass m_A, a rigid beam with mass m_B, length l_B and mass moment of inertia Θ_B in the x-z-plane, see Fig. 4.4. The associated independent degrees of freedom (DOF) are the vertical displacements z_A, z_B and rotation φ [55]. The linear equation of motion system of the load-bearing structure becomes

$$\boldsymbol{M}\ddot{\boldsymbol{r}} + \boldsymbol{D}\dot{\boldsymbol{r}} + \boldsymbol{K}\boldsymbol{r} = \boldsymbol{F} \tag{4.4}$$

with the $[3 \times 3]$ mass $\boldsymbol{M}$, damping $\boldsymbol{D}$ and stiffness $\boldsymbol{K}$ matrices, and the $[3 \times 1]$ acceleration $\ddot{\boldsymbol{r}} = [\ddot{z}_A, \ddot{z}_B, \ddot{\varphi}]^T$, velocity $\dot{\boldsymbol{r}} = [\dot{z}_A, \dot{z}_B, \dot{\varphi}]^T$ and displacement $\boldsymbol{r} = [z_A, z_B, \varphi]^T$ vectors. The $[3 \times 1]$ force vector $\boldsymbol{F}$ contains the excitation force F_{ex},

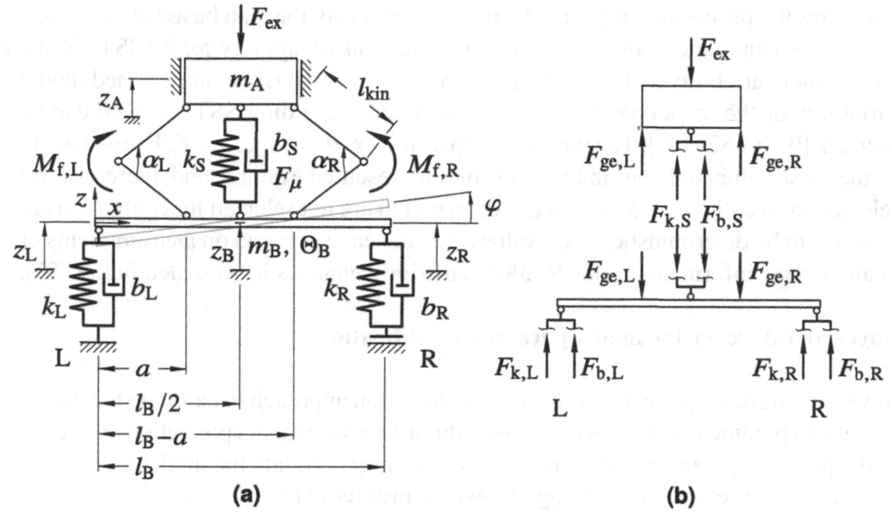

Fig. 4.5 Load-bearing structure, **a** mechanical model and **b** free body diagram with vertical forces, horizontal forces are neglected [52]

the friction induced force F_μ and the forces $F_{ge,L}$ and $F_{ge,R}$ for load redistribution provided by the semi-active guidance elements, see Sect. 5.4.8. A more detailed derivation of (4.4) is presented in [52, 55].

The mathematical model of the load-bearing structure is derived to capture the load path through the structure and to predict and evaluate the load redistribution capability in case of the semi-active structure in [52, 55, 56]. The derived mathematical model underlies model simplifications such as the assumption of lumped masses and rigid bodies. Furthermore, the spring-damper and the guidance elements are assumed to be free of mass, the model is assumed to be planar and undesired friction is summarised in a single dissipative force F_μ [52, 57]. Although these model simplifications can be attributed to model uncertainty, they may contribute to data uncertainty, as introduced in Sects. 2.1 and 2.2, and can be—at least partly—considered via parameter calibration in the following.

Sensitivity analysis

The sensitivity of the mathematical model predictions on parameter variations is assessed by calculating the statistical significance of parameter variations on the model prediction variation. Thus, the influence of the model parameters with respect to a model prediction of interest is identified. We assess the statistical significance by an analysis of variance (ANOVA) using the coefficient of determination R^2 [9, 136]. The coefficient of determination of model parameters $\boldsymbol{\theta}$

$$R^2_\theta = \left(1 - \frac{\mathrm{SSE}_\theta}{\mathrm{SST}}\right) \cdot 100\% \tag{4.5}$$

calculates the proportion of the model output variability that can be ascribed to each calibration candidate parameter variation. The *sum of squares total* (SST) is the total model variability and the *sum of squares error* (SSE) is the unexplained model variability of the model parameters $\boldsymbol{\theta}$. More details regarding SST and SSE can be found in [9, 36, 52, 136]. The three model parameters $\boldsymbol{\theta} = [m_{\mathrm{A}}, b_{\mathrm{S}}, F_{\mu}]$ turned out to be the most influential ones in the scope of the presented example and, therefore, are selected to be calibrated. Model parameters which are not selected for calibration are assumed to be deterministic. Their values are chosen, e.g. based on measurements or manufacturer information. The detailed sensitivity analysis is presented in [52, 57].

Bayesian inference for model parameter calibration

Bayesian inference is used as a statistical calibration approach to calibrate the uncertain model parameters identified as most influential on the model prediction of interest in the previous paragraph. The aim is to statistically correlate the model predictions with the measurements by solving an inverse problem [118]. The relation between measurements and simulations according to [75, 87, 144] is given by

$$Y_n^{\mathrm{E}}(t) = Y_n^{\mathrm{M}}(t, \boldsymbol{\theta}) + \varepsilon_n(t), \; n = 1, \ldots, N \tag{4.6}$$

where $Y_n^{\mathrm{E}}(t)$ represents the experimental results and N is the number of measurements. The model prediction of interest $Y_n^{\mathrm{M}}(t, \boldsymbol{\theta})$ is supplemented by the measurement error $\varepsilon_n(t) \sim \mathcal{N}(0, \sigma^2)$, that is assumed to be independent and identically distributed as well as normally distributed with zero mean and standard deviation σ [144]. Through the Bayesian inference approach, we can update current knowledge of the system and its model parameters with new information obtained from experimental tests. Thus, the parameter uncertainty is quantified and reduced by systematic inference of the posterior distribution [87, 144]. Using the Bayes' Theorem [13, 144], the posterior parameter distribution given the experimental results can be stated as

$$P(\boldsymbol{\theta}, Y^{\mathrm{M}}|Y^{\mathrm{E}}) = \frac{L(Y^{\mathrm{E}}|\boldsymbol{\theta}, Y^{\mathrm{M}}) \times P(\boldsymbol{\theta})}{P(Y^{\mathrm{E}})} \propto L(Y^{\mathrm{E}}|\boldsymbol{\theta}, Y^{\mathrm{M}}) \times P(\boldsymbol{\theta}) \tag{4.7}$$

with the likelihood function $L(Y^{\mathrm{E}}|\boldsymbol{\theta}, Y^{\mathrm{M}})$ representing the probability of experimental results Y^{E} given a set of parameters $\boldsymbol{\theta}$ for the model prediction of interest Y^{M} [52, 144]. The total probability $P(Y^{\mathrm{E}})$ is typically not computable with reasonable effort and is only normalising the result anyway [65]. It is more practical to sample from a proportional relationship of the posterior parameter distribution.

The parameter space is explored using the Marcov Chain Monte Carlo (MCMC) sampling to approximate the posterior parameter distributions $P(\boldsymbol{\theta}, Y^{\mathrm{M}}|Y^{\mathrm{E}})$ by drawing multiple samples from these posterior parameter distributions. That is, the histograms of the model parameters $\boldsymbol{\theta}$ of all random samples produce the approximated posterior parameter distributions $P(\boldsymbol{\theta}, Y^{\mathrm{M}}|Y^{\mathrm{E}})$ in (4.7) [117, 144]. Figure 4.6 depicts the model parameter calibration results obtained from 25,000 MCMC

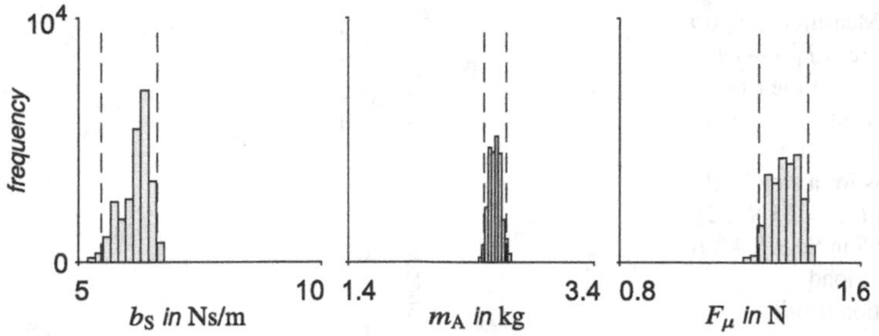

Fig. 4.6 Posterior distribution with 95% inter-percentile intervals (— —) for the viscous damping b_S, the mass m_A and the friction induced force F_μ according to [52]; x-axis limits represent the prior parameter ranges

runs. The parameter distributions are depicted as histograms representing approximations of the posterior parameter distributions for the three model parameters $\boldsymbol{\theta} = [m_A, b_S, F_\mu]$ [57]. Furthermore, the narrow histograms graphically depict the knowledge gain and the uncertainty reduction in the model parameter ranges. The model parameter ranges covering the 95% inter-percentile can be reduced by approximately 89% for the mass m_A, by approximately 82% for the viscous damping b_S and by approximately 84% for the dissipative force F_μ compared to the prior bounds represented by the limits of the x-axis in Fig. 4.6.

Comparison of the non-calibrated and calibrated model predictions

The effect of the statistical calibration procedure on the model prediction accuracy is exemplarily shown in Fig. 4.7 for a step load excitation $F_{ex} = 25$ N. The envelopes of each of the 300 Monte Carlo (MC) simulation runs for non-calibrated and calibrated model parameter ranges are conducted and compared to the related support reaction force measurements F_L and F_R averaged for 10 measurement repetitions. The support reaction force measurements F_L and F_R are quite similar as the load-bearing structure is undamaged. The non-calibrated model parameter ranges are equally distributed between the lower and upper prior bounds. The calibrated model parameter ranges are distributed according to the histograms in Fig. 4.6.

The simulations using calibrated model parameters tend to be closer to the measurement with smaller envelopes. Even though calibrated and non-calibrated envelopes widely encompass the measurements, the envelope area of the calibrated MC simulations is reduced significantly by 75% compared to the envelope area of the non-calibrated MC simulations [57], qualifying the Baysian inference as suitable calibration method for the presented example.

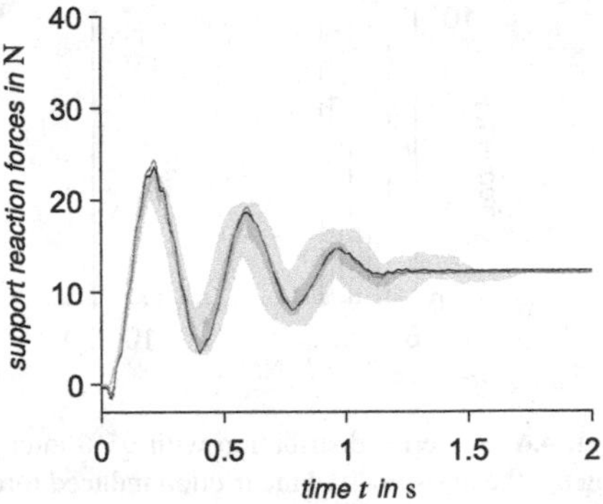

Fig. 4.7 Measured support reaction forces F_L (—) and F_R (—) versus time t and non-calibrated ( ) and calibrated ( ) model predictions for a step load excitation $F_{ex} = 25\,N$ [52], cf. Fig. 5.65 in Sect. 5.4.8 for semi-active load redistribution results

4.1.3 Model-Based Analysis of Uncertainty in Chained Machining Processes

Felix Geßner, Christian Bölling, Eberhard Abele, and Matthias Weigold

In the production of components for technical systems, as described in Sect. 3.6, several processing steps are used. To generate the final product geometry, the processes are linked to a process chain, see Sect. 3.2. A widely used process chain in machining is roughing and finishing. A process with a high material removal rate is linked to one or more processes capable of generating the required machining quality. The production of high-quality bore holes is often realised by the process chain drilling—reaming, as shown Fig. 4.8. The reaming process slightly enlarges the diameter of a bore hole in order to improve the surface quality and the circularity. Another frequently applied process chain is drilling – tapping. The desired thread geometry is created by removing material from the pre-drilled bore wall.

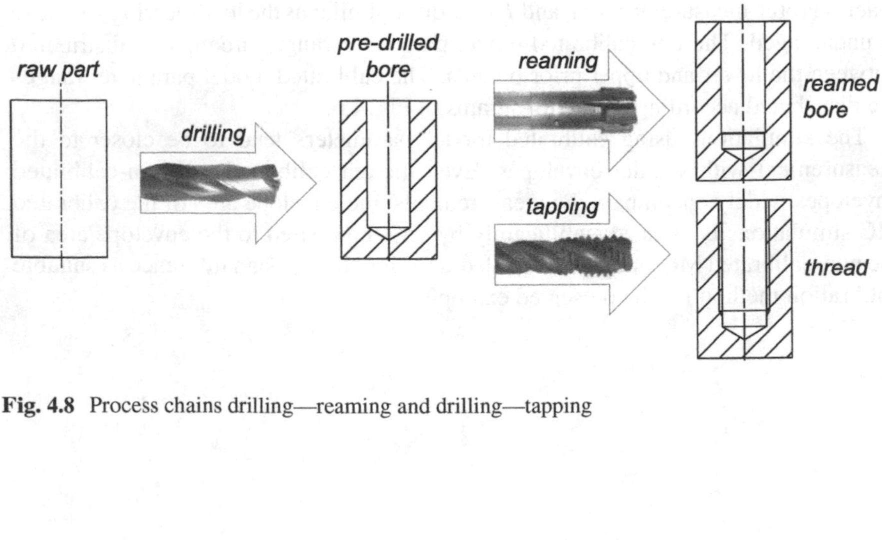

Fig. 4.8 Process chains drilling—reaming and drilling—tapping

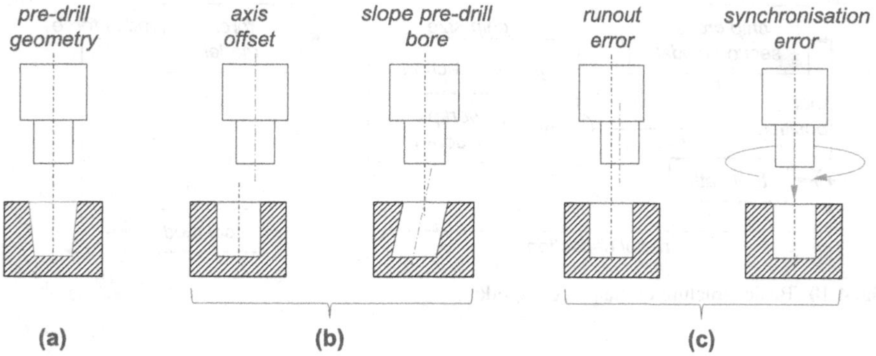

Fig. 4.9 Geometrical uncertainty in drilling-reaming and drilling-tapping process chains caused by **a** the pilot process, **b** the process chain or **c** the final process

Machining processes are generally affected by data uncertainty in form of incertitude, see Sect. 2.1.3. For this reason, functional parts of a component are always provided with interval-based tolerances, which ensure functionality in the overall system. Forms of geometrical uncertainty, which occur in drilling process chains, are shown in Fig. 4.9. The occurring uncertainty can be categorised according to their origin. One source of uncertainty are geometric deviations of the pilot hole, e.g. variations in diameter, straightness or cylindrical shape. Those deviations can be caused e.g. by hardness deviations in the workpiece material. Another source is the process chain in which positioning variation between the pilot hole and the following process step occur. Since uncertainty is accumulated, deviations can also neutralise each other. The uncertainty must therefore be evaluated in the overall context. Axes misalignment of up to 0.03 mm occur due to limited accuracy of the machine tool and re-clamping operations. In industrial applications, e.g. reaming of valve guides in a cylinder head of a combustion engine, misalignments of up to 0.1 mm occur, which are caused by the joining process of the blanks [72]. A radial deviation of the pre-drilled bores is induced due to oblique and uneven surfaces, incorrectly placed centring, cavities, transverse bores, blowholes and inclusions. The resulting radial forces lead to elastic bending of the pilot drill and thus to a slope bore [125].

Additionally, uncertainty is caused by the final processes without being influenced by pre-processing. A runout describes the radial displacement of the tool in the chuck. In industrial applications, a radial runout during reaming can be limited to 0.003 mm by adjustable adapters [72]. Earlier investigations on tapping indicate a radial runout of 0.03 mm [37]. During tapping, a synchronisation deviation between the translatory and the rotational axes often occurs. This is generated by a deviation between the axis movements when reversing the direction of rotation.

Different approaches are used to model machining processes in order to predict process variables or raise process understanding. Zabel [166] differentiates models with regard to whether they are based on the finite element method (FEM) or not. In the simulation of machining processes, which are characterised by material deforma-

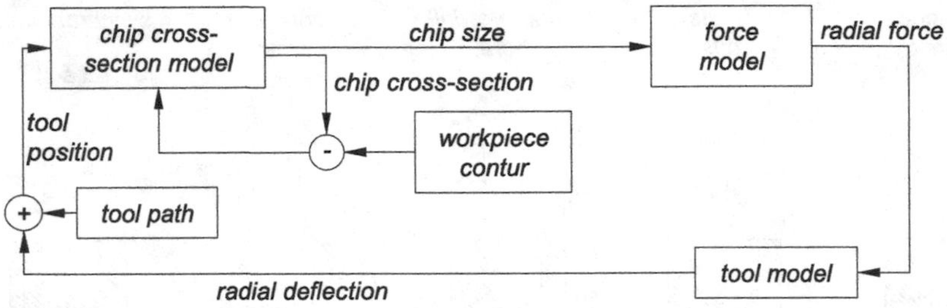

Fig. 4.10 Basic structure of the process model

tion as well as occurring dissolution of the material bond, considerable computing time is required using FEM-based models.

Non-FEM-based modelling requires more detailed knowledge of the particular process but is less demanding with regard to computational effort. This allows for the simulation of the effects of uncertainty in less time. Basically, these approaches are divided into analytical and geometrical models. Analytical models use closed mathematical expressions to describe the considered phenomena [166]. Thus, models of chip formation, shear planes, temperature and process forces can be established. The geometric approaches determine the geometric quantities of machining processes, which are often used as input for analytical models. Mechanistic modelling represents a combination of geometric and analytical models. It is fundamentally based on the assumption that process forces, occurring during the machining process, are proportional to the chip cross-section [85].

Examples of mechanistic models within the context of reaming operations and the disturbance variables occurring can be found in [21, 72]. For tapping, the first mechanistic model aiming towards the analysis of the process and the influence of uncertainty are developed by Dogra [37].

The basic structure of the mechanistic process model based on [85] is shown in Fig. 4.10. The main input of the model is the original workpiece contour. We use a chip cross-section model, as shown in [1], to determine the chip sizes, resulting from the tool geometry and position. The model is based on the intersection of 2D elements. The process forces are determined based on these chip sizes in an empirical force model. Summation of the forces caused by each of the tool's cutting edges enables calculating the resulting force that leads to a radial deflection of the tool. In the model, the deflection of the tool is e.g. calculated by a combination of dynamic and static modelling approaches [21]. When selecting suitable approaches, we consider the rotational speed of the tool and the prevailing geometric constraints. Due to the low rotational speeds in tapping, dynamic considerations can be largely neglected. However, the complex geometric boundary conditions must be mapped. Based on the tool's radial deflection and its path, the tool position is determined. This information is used to specify the geometric intersection in the next calculation step.

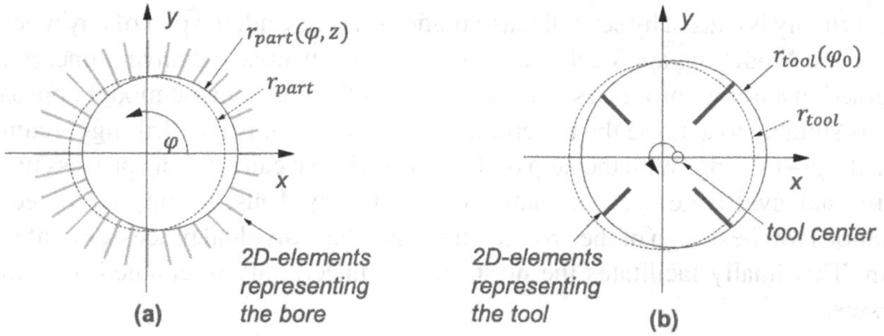

Fig. 4.11 Mapping **a** positioning errors and **b** runout errors in the tapping model

In linked machining processes, individual process steps are linked via the generated geometry of the feature created in the previous step, e.g. the pilot hole geometry. This serves as the starting point for the subsequent machining process. For simultaneous processes, however, successful model linkage requires further connection points, as each step's stability may affect the others. For the combined machining of e.g. valve guide and valve seat, the process forces of each tool step are taken into account in an overall system. Here a Jeffcott rotor with several masses is loaded with the resulting forces of each individual tool step, so that the mutual influences can be mapped [21].

In the discussed mechanistic model approaches, we represent the geometry of the pilot hole by individual plane elements arranged in a star shape around the rotational axis of the model, see Fig. 4.11. Therefore, each point on each of the planes has the same angle φ when viewed in cylindrical coordinates. In order to implement for example geometric deviations, such as slope pre-drill bore and axial offset, we modify the individual plane elements. For this purpose, we vary the radius of the pilot hole as a function of the angle φ and the cutting depth z, as shown in Fig. 4.11a. A similar procedure can be used for the tool, since it is also mapped using plane elements. By shifting the radius of these elements depending on the angle, we can also map runout errors, see Fig. 4.11b.

For mapping deviations in the synchronisation, or more general deviations in the tool path, we alter the displacement of the tool after each calculation step. In addition, we can model tooth chipping by altering the geometry or by completely removing a plane element of the tooth. Further disturbance variables can be implemented externally to the geometric intersection model. For example, hardness gradients in the component can be considered by manipulating the used force model.

Our model approaches for tapping and reaming show that the axial offset between the pilot hole and the subsequent process step is the disturbance variable of greatest influence [72]. Due to the lack of radial guidance during tool immersion, radial forces can lead to tool misalignment and subsequently to tool inclination. As a result of the radial guidance of the tool after the immersion, its inclination over the drilling depth remains almost constant [2].

Uncertainty is caused by several reasons and is an unavoidable part of any machining process. Modelling approaches can describe and evaluate occurring uncertainty in chained machining processes. One approach is the mechanistic model approach, which is suitable to analyse the uncertainty in process chains like drilling—reaming and drilling—tapping. With the help of these models, we can raise the process understanding and investigate an accumulation of uncertainty. Thus, we may derive recommendations for the design of the process chain and the individual processes contained therein. This finally facilitates the mastering of uncertainty in chained machining processes.

4.2 Data-Induced Conflicts

Florian Hoppe

Active systems, as presented in Sects. 3.4 and 5.4, have proven their effectiveness in mastering uncertainty. But in turn, they rely completely on the veracity of data. In many applications, the fusion of redundant data has therefore become common practice. However, if the confidence intervals of data from two or more sources do not overlap, this leads to so-called data-induced conflicts, which cannot be resolved with classical fusion techniques. Such data-induced conflicts reveal ignored model or data uncertainty, see Sect. 2.2. In the case of real-time controls, they require an instantaneous decision on which source to trust. Data-induced conflicts aid in uncertainty identification and are therefore a valuable tool in mastering uncertainty, see Sect. 3.3.

In the past, unresolved and ignored data-induced conflicts led to several severe incidents. According to the European Space Agency (ESA), the crash of the ExoMars Schiaparelli probe on 14 March 2017 began when calculating the altitude from a saturated inertial measurement unit (IMU) signal, which resulted in a large negative altitude. A conflict with the radar Doppler altimeter unit was detected. Since no other verification methods had been implemented at that time, the true value could not be determined. Even though the conflicting IMU had been detected at that moment, this information was not passed to other subsystems and thus caused a chain of fatal decisions during the decent resulting in a crash at 150 m/s [152]. In the following two years the Boeing 737 MAX repeatedly encountered problems with the angle of attack (AoA) sensor. One of the malfunctions manifested in an ignored conflict between left and right sensor of 20° [151]. Hence, the control system anticipated a stall and automatically pushed the nose down, which caused the fully manned plane to crash.

These incidents show that a general framework for verifying data as well as identifying and isolating a cause is required. Often a multitude of metadata, which include models, parameters and sensors, are involved in the generation of data. We refer to the general set of metadata as a data source. A data source consisting of a model and a sensor is also called a soft sensor, see Sect. 1.4. To master data-induced conflicts, the metadata involved and their uncertainty have to be taken into account. For example, statistical dependency between the metadata of different data sources invalidates majority decisions used by voting algorithms. However, the use of systematic redundancy allows the identification of the cause of conflicts.

In Sect. 4.2.1 we present and evaluate a method to establish systematic analytical redundancy and make it available for monitoring. Linking the metadata with the actual data allows us to link occurring conflicts with their cause. Two examples outline how to make use of physical models to infer specific causes in Sect. 4.2.2 and how to scale the method to systems where a multitude of conflicts might originate from a single fault Sect. 4.2.3.

4.2.1 Dealing with Data-Induced Conflicts in Technical Systems

Georg Staudter and Jakob Hartig

In the product life cycle phases of production and usage of modern technical systems, see Sect. 3.1, data is increasingly being generated redundantly. Redundancy is not necessarily physical redundancy of the sensor, but can also be established in the form of analytical redundancy, where measured values of the system are converted into the desired values via models. As introduced in Sect. 1.4, the combination of a sensor with a model to estimate target quantities using easily accessible auxiliary variables is called a soft sensor [48, 81]. An overview of the use of soft sensors can be found in a monograph by Fortuna et al. [48].

Redundancy increases the availability of information but may lead to contradictory statements and conflicts. These conflicts can be used to identify the uncertainty in the information about the system and, therefore, contribute to mastering uncertainty, see Sect. 3.3.

This section introduces the concept of data-induced conflicts, discusses the advantages and challenges, and presents a method for dealing with data-induced conflicts in technical systems. The method is a slightly extended version of [70].

Data-induced conflicts

Contradicting values of different redundant data sources are in conflict when their confidence intervals do not overlap. These so called data-induced conflicts can therefore be attributed to the model, see Sect. 2.2, to the parameters of the model, see Sect. 2.1 and to sensor errors; also they are a symptom for lack of knowledge, see Sect. 1.4. If uncertainty is not sufficiently taken into account or if too few or uncertain data sources are considered, these conflicts remain unnoticed and thus unresolved. Figure 4.12 illustrates three redundant data sources for a target quantity with their respective uncertainty characterised by the confidence intervals and a data-induced

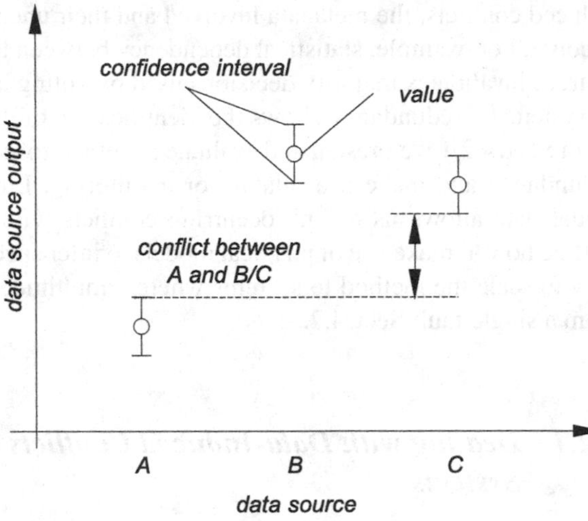

Fig. 4.12 Data-induced conflict between the sources A and B/C

conflict between source A and the consensus of sources B and C. If two sources are in consensus, their confidence intervals overlap.

Different methods have been developed to deal with conflicting data sources. On the one hand, conflicts between data sources can be seen as part of erroneous system behaviour. Thus, different methods use conflicting data for fault detection and fault isolation [76, 80]. On the other hand, conflicts can also be seen as part of the system's normal behaviour. In that case, data from multiple sources can be used to reduce uncertainty and to improve the overall level of data quality. Simple methods for data reconciliation of conflicting sensor data are voters [46]. More elaborate fusion methods are the Bayes method [27, 97], the Dempster–Shafer method [89, 165], and heuristic methods [94, 149]. In the process industry, reconciliation methods are implemented for the estimation of process state data. The goal is to fuse the conflicting data, i.e. reconcile the state of the system with the conservation laws of mass and energy. For this, the conservation laws and the measured values have to be an over-determined equation system. With a quadratic minimisation method, the system states are changed until the values satisfy the conservation laws [76].

Method for dealing with data-induced conflicts

The methods mentioned above for dealing with conflicting data sources fail to differentiate between sensor and model or do not take uncertainty into account. Therefore, we propose a methodology to support interpretation- and decision making processes in case of data-induced conflicts using the approach of redundancy via soft sensors. Through consideration of the relationships between sensor, models and information about the system, the cause of the data-induced conflicts can be isolated. For the proposed method, the following two points have to be addressed:

1. Conflicts emerge when the confidence intervals of redundant data sources do not overlap. Hence, the uncertainty in an interconnected system has to be propagated, see Sect. 3.2. How can this be done efficiently in an environment with many sensors and models?
2. The different data sources, i.e. soft sensors. How can the dependencies between different sensors and models be used in decision making processes?

The proposed approach provides a methodology to identify lack of knowledge in the interpretation of conflicting sensor data by differentiating data sources into models and sensors and spanning the investigation from the redundant observation of a single value to the interconnection between models and sensors throughout the system, see Fig. 4.13. Analytical redundancy via soft sensors is enforced by linking already existing, spatially distributed sensors with models to increase the availability of information about the desired values.

Each redundant data source Q_i, cf. Sect. 1.4, is associated with a given level of uncertainty due to precision and accuracy of the sensor, as well as model uncertainty, see Sect. 2.2, which needs to be identified and propagated in the target quantity. The first step (i) is to examine that each data source Q_i is within certain boundary conditions to ensure physical plausibility. Those limits need to be determined and individually based on the respective metadata, such as calibration data and known characteristics.

On this basis, the redundant sources are compared among themselves to detect any possible conflicts in step (ii). In case of data-induced conflicts, a method for the compact visualisation of dependencies (iii) is provided to restrict whether the conflict is caused by a sensor or due to model uncertainty. In conclusion, the provided information supports the process of interpretation of sensor data (iv) and gives evidence in which sources to trust. In the following, each step of the systematic approach is presented in detail.

(i) *Plausibility*. For checking the plausibility of data sources, sufficient metadata about the limit values derived from sensor characteristics, physical properties and technological limits are needed. With regard to the limits, there is a trade-off between sensitivity to erroneous behaviour and normal fluctuations [80]. In the case of a data source exceeding the prescribed boundary condition, the respective sensor or model can be excluded for further considerations in advance, and cross-checks with other redundant sources become unnecessary.

(ii) *Detection of Conflicts*. Data-induced conflicts can be attributed to sensor errors (technical failure or application errors) or to model uncertainty. Model errors can be caused by either insufficient simplifications or changes of the underlying physical system, e.g. due to the wear, deformations, failure of components, see Sects. 2.2 and 4.2.2. For the detection of data-induced conflicts, the uncertainty of sensors has to be considered. Especially soft sensors may have several sensor inputs to calculate the target quantity. Therefore, the sensor uncertainty has to be propagated by the model. For the propagation of uncertainty, standard methods, e.g. [83] can be used. The technical implementation of error propagation and the necessary calculation of derivatives is done with automatic differentiation

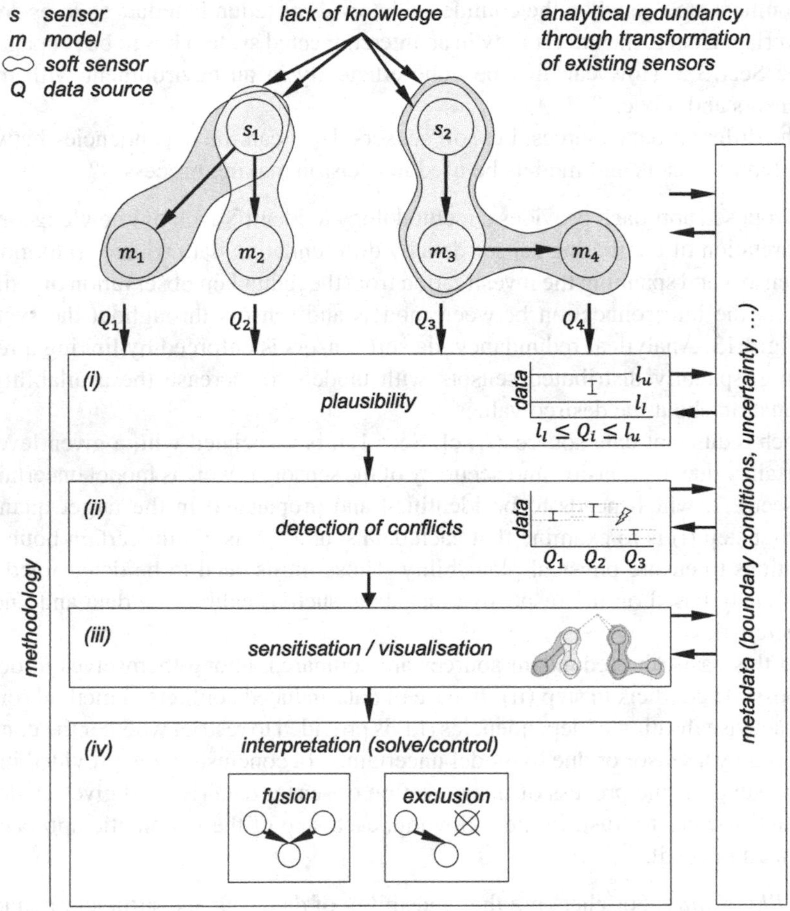

Fig. 4.13 Method for dealing with data-induced conflicts (Adapted by permission from Springer Nature Customer Service Centre GmbH: Springer, Hartig J. et al. (2020) Identification of Lack of Knowledge Using Analytical Redundancy Applied to Structural Dynamic Systems. In: Mao Z. (eds) Model Validation and Uncertainty Quantification, Volume 3. Conference Proceedings of the Society for Experimental Mechanics Series ©The Society for Experimental Mechanics, Inc. 2020)

(AD). In comparison to numerical methods AD has the benefit of calculating the exact derivative. In addition, data-driven models in the form of software code can be assigned a derivative with the help of AD [12, 109, 159].

(iii) *Visualisation*. Especially in the case of data-induced conflicts, knowledge about possible dependencies between data sources is important. Erroneous sensors or models and the consequences of the errors for other values have to be found. For downstream interpretation, it is important to depict the relationship between the soft sensors and their inputs in a human- and machine readable form. Therefore, a method for the visualisation of conflict scenarios has been developed in order to clearly depict inter-dependencies between soft sensors (sensors and models) throughout the technical system, see Sect. 4.2.3.

(iv) *Interpretation*. The interpretation of the visualised dependencies is done by reasoning. If a particular data source is only involved in other observations revealing no inconsistencies, the confidence in the respective sensor/model increases. If, on the other hand, a particular input is involved in one or many conflicting source values, it is suspected to be the cause of the conflict. For automation of the reasoning process, various classification methods can be used, e.g. pattern recognition, reasoning methods or neural networks [80]. Data sources deemed to be trustworthy can then be used in a fusion process with methods mentioned above. Data sources being suspected to cause the conflict are excluded.

Our methodology (i)–(iv) reinforces redundancy and, therefore, data-induced conflicts. Through the consideration of physical metadata and sensor data with their respective uncertainty, conflicts can be identified and the data quality increases. Furthermore, our method provides information about the relationships between sensors, models and the physical system to identify the cause of the conflict for human and machine interpretation. Section 4.2.3 shows the application of the outlined methodology on an experiment series conducted at the Modular Active Spring-Damper System introduced in Sect. 3.6 revealing data-induced conflicts.

4.2.2 Data-Induced Conflicts for Wear Detection in Hydraulic Systems

Jakob Hartig, Ingo Dietrich, and Peter F. Pelz

Due to propagation and chain reactions of contamination [99], hydraulic systems are particularly sensitive to wear and contamination. Therefore, it is of interest to detect wear in early stages during the operation of a system in order to consequently avoid high cost due to unplanned maintenance. In the context of Sect. 4.2, this chapter serves to demonstrate the identification of ignorance, as introduced in Sect. 1.3, in the form of undetected wear by means of data-induced conflicts. As shown in Sect. 4.2.1, analytical redundancy can be used to learn about sensor or model errors by observing data-induced conflicts.

Wear itself is not directly measurable during the operation but manifests itself in the changed system characteristics. Different methods exist to detect and isolate the changing system characteristics [76, 80]. In this section, we demonstrate the use of soft sensors to determine wear via data-induced conflicts between redundantly calculated flow rates as shown in [141]. For predictive maintenance, this approach is promising in terms of cost-efficiency, since existing sensors and models are used. The approach of this example is rather simple with only two data sources (soft sensors for pump and fluid system) for one calculated quantity (flow rate). A more complicated

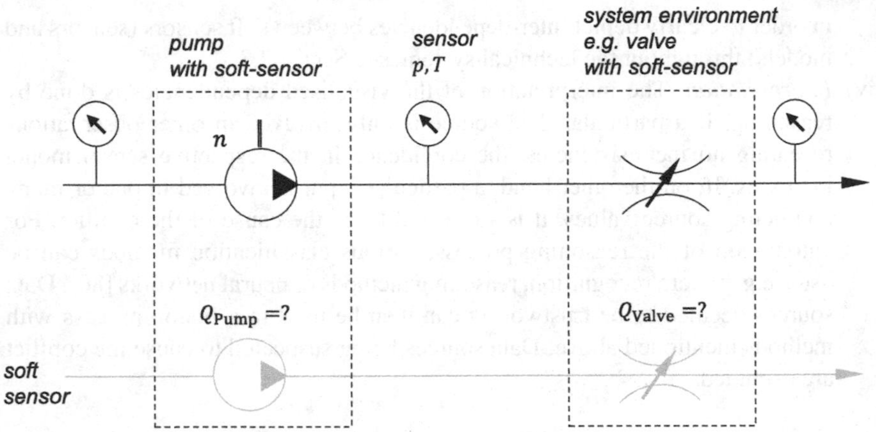

Fig. 4.14 Fluid system with measured quantities (pressures p, temperatures T, rotational pump speed n) and soft sensors (Figure adapted from [71])

system can be found in Sect. 4.2.3. In the following, first the two data sources are presented. Then wear detection via data-induced conflicts is discussed.

Analytical redundancy by means of soft sensors

To demonstrate analytical redundancy, the generic fluid system in Fig. 4.14, consisting of a positive displacement pump and a valve, is considered. The system acts as an abstraction for real fluid systems since the hydraulic resistance of a generic system is reduced to the hydraulic resistance of a valve. As indicated in Fig. 4.14, both components have an assigned soft sensor to determine the flow-rate that flows through the respective component. The purpose of both soft sensors for the pump and the valve, is to generate redundant data of the volume flow rate to make conclusions about the wear condition of the system.

For the pump, an internal leakage model is used. The ideal flow rate Q_{P}, determined by the displacement volume V and the rotational speed n, is diminished by the internal leakage Q_{L}

$$Q_{\mathrm{p}} = nV - Q_{\mathrm{L}} \tag{4.8}$$

where the gap losses Q_{L} are modelled with a semi-empirical dimensionless approach [140].

For the valve soft sensor, the definition of the K_{v} value for valves is used as a model. The valves flow-rate Q_{v} is given by

$$Q_{\mathrm{v}}(\alpha) = K_{\mathrm{v}}(\alpha)\sqrt{\frac{\Delta p_{\mathrm{v}}\varrho_0}{\Delta p_0 \varrho}} \tag{4.9}$$

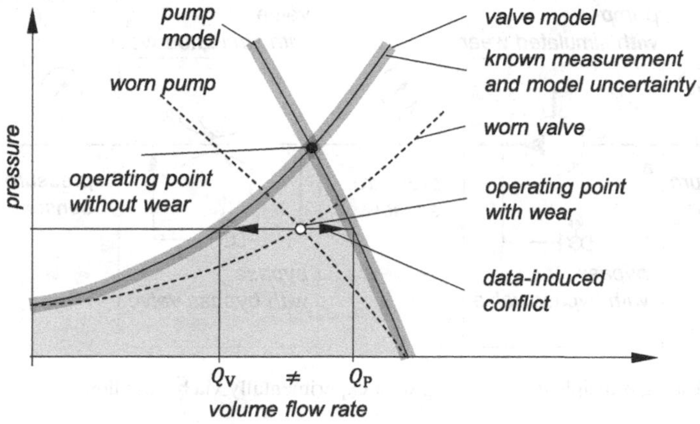

Fig. 4.15 Characteristic curves for pump and valve with and without wear; the light grey area indicates possible operating states for worn components; the arrow indicates the data-induced conflict between the two models

where $\Delta p_0 := 1$ bar and $\varrho_0 := 1000\,\mathrm{kg/m^3}$. Δp_v is the pressure difference over the valve and ϱ is the fluid density. The K_v-value is calibrated in dependence on the valve opening degree α. The uncertainty for both soft sensors is determined with error propagation. Both soft sensors depict an unworn state of the system.

Identification of wear

The fluid system in Fig. 4.14 consisting of a positive displacement pump and a valve, can be described by the flow-rate-pressure characteristics in Fig. 4.15. For the valve, the flow rate accelerates with increasing pressure, for the pump the flow rate decreases. The intersection of both curves is the operating point of the hydraulic system. Both pressure and flow rate have to be the same. When all components are new, both soft sensors are assumed to depict the relevant reality and therefore show the same flow-rate.

Now imagine a worn valve. With wear, the cross-sectional area through which the flow passes, increases. At the same pressure level more fluid can pass the valve and consequently the operating point of the system changes. Since the soft sensors depict the unworn state, they do not recognise this change and consequently deliver contradictory flow rate measurements. For a worn pump similar considerations hold.

The contradictory measurements are in conflict, if their uncertainty intervals do not overlap, see Sect. 4.2.1. The conflict can arise due to sensor break down or model error. In both cases, the data-induced conflict represents *ignorance*. A sensor breakdown can normally be excluded with limit checking, as presented in Sect. 4.2.1. We therefore concentrate on model error or, in this case, component characteristic change.

In order to review the presented soft sensor approach, the following two questions must be answered:

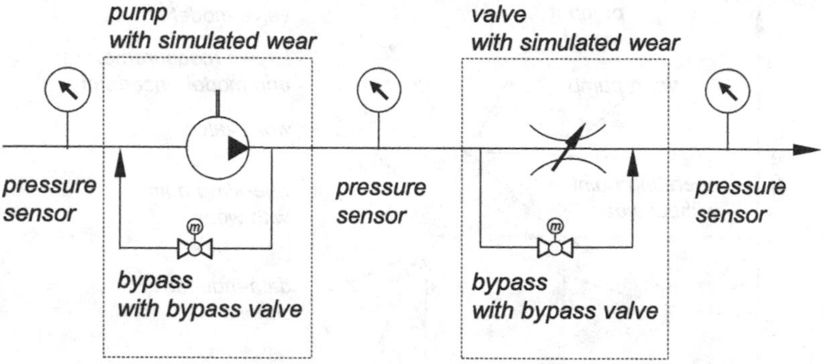

Fig. 4.16 Test-rig principle for simulating wear experimentally via bypass flows

1. Is the influence of wear greater than the soft sensor uncertainty?
2. Are two components sufficient to identify if wear occurs and where?

To answer the first question, we carried out an experimental investigation of a worn valve [127]. The study revealed that the resulting flow rate changes from wear exceed the soft sensor uncertainty. This is a data-induced conflict indicating that wear occurs.

With regard to the second question, we carried out measurements with a test rig [69, 71]. Wear changes the cross sectional area of the valve, and a worn positive displacement pump has larger gaps where fluid can flow back. Consequently, we were able to simulate wear on the test rig by installing bypass flows for the pump and the valve. This offers the benefit of easily changing wear conditions without actually destroying the components. The principle of the test rig used can be found in Fig. 4.16. The studies show that identification of wear is possible via data-induced conflicts when the flow rate outputs of the two soft sensors differ by about 6%.

From the studies, it can be concluded that the localisation of wear is not possible with only two soft sensors. According to Fig. 4.15, the possible operating range of the hydraulic system for various wear conditions is always below the characteristic curves with components in a new condition (light grey area). Therefore, the calculated volume flow rate of the valve soft sensor Q_V is always lower or equal to the calculated volume flow of the pump Q_P. This is independent from the wear condition of both components. For this reason, it is not possible to deduce the component subject to wear from these two calculated volume flows alone, the measuring system is underdetermined. However, the history of the measurement data, additional information in the two soft sensors or additional soft sensors would make it possible to deduce the worn component.

All in all, data-induced conflicts help in detecting ignored wear in hydraulic systems and can provide a measure for predictive maintenance. The localisation of the worn component is not possible with only two soft sensors. Additional system information in the form of additional soft sensors is needed for this purpose, see Sect. 4.2.3.

4.2.3 *Fault Detection in a Structural System*

Tuğrul Öztürk, Daniel Martin, and Florian Hoppe

Structural systems use mechanical elements to transfer forces along a path. The force transmission can be represented by mathematical models used to optimise the design, see Sect. 6.1.1. A combination of such models with sensors allows to determine quantities, such as forces in spatially distant elements, and can therefore be used to estimate states that are infeasible to measure.

Applications lie in the field of structural health monitoring of aeroplane to detect defects in fuselage panels [78], wing panels [167], and their connecting elements [114]. The basis for those technologies are data obtained from integrated sensors and implemented models, which are assumed to be reliable. However, unreliable data due to data uncertainty have led to numerous incidents in the past [79], as mentioned in Sect. 4.2.

Today's methods of online data validation are often based on the mere comparison of a few data sources. Information about the sensors and models involved as well as the results of these comparisons are not centrally fed back and forwarded to other subsystems. In the case of the ExoMars incident, faulty gyroscope data were detected in one subsystem but nevertheless reused at a later point in time [152].

Besides the mentioned applications, soft sensors also allow to generate redundant data throughout the system and to set up a sensor network. As a result, data sources can be continuously checked for being in conflict with each other. But when it comes to linking larger numbers of sensors and models, a single fault may cause numerous conflicts. To distinguish between possible faults, conflicts and their corresponding links in the network have to be visualised and analysed.

Both, sensor data and models, forming the soft sensor, are afflicted with uncertainty. The uncertainty of sensor data expresses itself by an unknown distribution, which is considered by means of confidence intervals. Furthermore, the models used to describe the behaviour of mechanical components often contain many assumptions and simplifications. Therefore, the uncertainty of a soft sensor can be classified as incertitude as described in Chap. 2.

In the following, the method presented in Sect. 4.2.1 is applied to a complex technical system, which is introduced briefly as a sensor network. We describe the subsequent steps of our method from conflict detection to visualisation regarding a real sensor fault case. As introduced in Sect. 3.1, uncertainty occurs over different phases of the product life cycle. The presented and evaluated method is applicable in the system's design phase to establish analytical redundancy as well as in the system's usage phase by the visualisation of data-induced conflicts to master uncertainty.

Scaling analytical redundancy to sensor networks

The method presented in Sect. 4.2.1 allows us to analyse a large number of comparative quantities from analytically redundant data sources. These data sources consist of sensors only or of sensors linked with models, thus, soft sensors. Thereby, it is taken into account that the involved data sources can be afflicted with data and model uncertainty, see Sects. 2.1 and 2.2. The method for dealing with data-induced conflicts is applied to the Modular Active Spring-Damper System (German acronym: MAFDS) presented in Sect. 3.6.1, which represents a multiple sensor-integrated structural system. The MAFDS consists of one upper and one lower truss structure, which are connected to each other via guidance elements and a spring-damper. The two truss structures in turn consist of individual beams, which are assembled to form tetrahedral elements shown in Fig. 4.17. Further, the MAFDS is equipped with several sensors, such as force transducers $F_{\mathrm{S}j}$ and $F_{\mathrm{P}j}$ as well as strain gauges in the upper, ϵ_j^{U}, and lower, ϵ_j^{L}, truss structure. More details related to the integrated sensors in the MAFDS are given in Sect. 3.6.1. To establish analytical redundancy, the measured sensor data are converted by means of models to the desired redundant quantity, denoted here as comparative quantity. As mentioned above, a linkage of measured data obtained by an arbitrary sensor with an analytical model to gain an another arbitrary quantity represents a soft sensor. In case of the MAFDS, an example for such a soft sensor is the linkage of the beam strain gauge $\epsilon_{15}^{\mathrm{U}}$ in Fig. 4.18a with a mechanical model m for the conversion of the measured beam strains to beam forces; here we assume a linear-elastic beam behaviour using Hooke's law in Fig. 4.18b. The uncertainty of parameters involved as well as the measurement uncertainty have been taken into account to estimate the confidence interval.

Based on the calculated beam forces, analytical redundancy can be established at the fixed support points, e.g. fixed support point 1 (FSP1) in Fig. 4.17. The forces measured by $F_{\mathrm{P}1}$ at FSP1 must be in equilibrium with the beam forces F_1, F_{11}, and F_{15}, which are calculated via the beam strain gauges $\varepsilon_1^{\mathrm{U}}$, $\varepsilon_{11}^{\mathrm{U}}$ and $\varepsilon_{15}^{\mathrm{U}}$ of the corresponding beams B1, B11 and B15, labelled in Fig. 4.17. It should be noted that the beam forces are converted into the components of the global coordinate system, which corresponds to the coordinate system of the piezoelectric based force transducers $F_{\mathrm{P}i}$ at the three fixed support points.

Interpretation of data-induced conflicts in sensor networks

To investigate the method presented in Sect. 4.2.1, an erroneous data set of a drop test at the MAFDS is used. In this case, the triaxial force transducer $F_{\mathrm{P}1}$ at FSP1 was connected incorrectly so that the measuring channels for the y- and z-component of the measured forces were switched, thus, resulting in conflicts among multiple data sources Q_i. In the first step, as shown in the proposed method according to Fig. 4.13, the plausibility of the sensor/transducer signals is verified by checking whether the measured data are within a reasonable range respecting the specific properties of the sensor, such as measuring ranges etc., as well as the structural system. After a successful plausibility check, the conflict detection for the data sources Q_i is continued.

Assuming the observed quantities to show a Gaussian normal distribution, the measured values are within a confidence interval around the mean value of all measurements with a certain probability. The uncertainty of sensors and, in this case, mechanical models is propagated throughout each data source Q_i with Gaussian

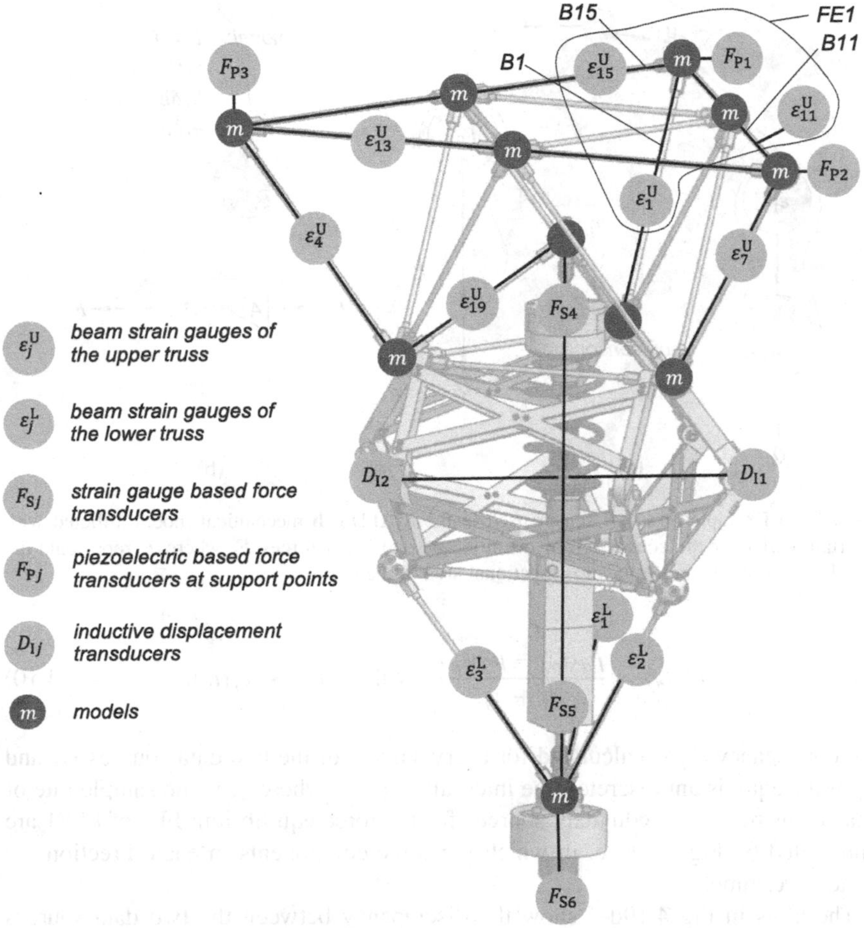

Fig. 4.17 Sensors used in the MAFDS to gain analytical redundancy and labelled beams B1, B11 and B15 for force equilibrium at fixed support point 1 with force transducer F_{P1}

error propagation, which is implemented by using automatic differentiation, see Sect. 4.2.1. The level of confidence is set to 95%. To determine data-induced conflicts, the overlap of the confidence intervals of the data sources is regarded. It is defined in terms of the set $[\mu - k\sigma, \mu + k\sigma]$ and the point where μ is the expected value and σ the standard deviation, while the coverage factor k determines the considered amount of the probability space. If the confidence intervals match completely, there is no data-induced conflict and, in turn, an absolute data-induced conflict exists if the confidence intervals do not overlap. As a measure of the severity of a data-induced conflict between two data sources Q_i and Q_j, a discrepancy d_{ij} is defined in the Eq. (4.10)

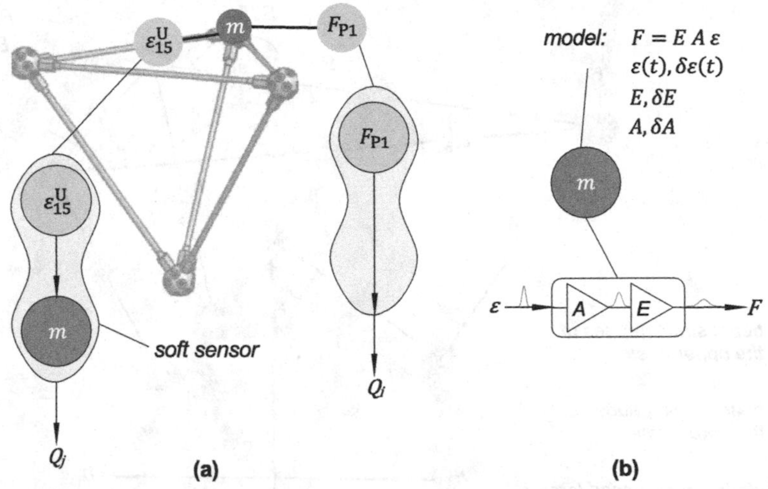

Fig. 4.18 **a** Example of a soft sensor used in the MAFDS, **b** mechanical model afflicted with uncertainty $\delta(\cdot)$ for the conversion of beam strain ε to beam forces F, where E represents the modulus of elasticity and A the cross-sectional area of the beam

$$d_{ij}(t_n) = \frac{\mu_i(t_n) - \mu_j(t_n)}{\sigma_i + \sigma_j}, \text{ with } \mu_i(t_n) > \mu_j(t_n). \tag{4.10}$$

The discrepancy d_{ij} is calculated for every sample of the two data sources Q_i and Q_j at the equidistant discrete-time intervals $t_n = \frac{n}{f_s}$, where f_s is the sample rate of data acquisition. The redundant sources for the force equilibrium FE_1 of FSP1 are represented by Fig. 4.19a–c, in which the force components in each direction are plotted over time.

The plots in Fig. 4.19d–f show the discrepancy between the two data sources over time for each force component. To define whether there is a conflict or not, we introduced a measure, denoted as degree of conflict (DOC), $c_{ij} = \overline{d_{ij}(t)}$, which is the time-averaged discrepancy over a time interval of interest $[t_0, t_1]$, which has to be defined individually for each scenario. If the calculated DOC is greater than a predefined threshold value c_{thresh}, a data-induced conflict is detected. In this case, c_{thresh} has been set to 1.96, which is equal to the coverage factor $k = 1.96$ of a 95% confidence interval. That means that the time-averaged confidence intervals of the two data sources are exactly adjoining ($\overline{\mu_i} - k\overline{\sigma_i} = \overline{\mu_j} + k\overline{\sigma_j}$).

For FE_1, a data-induced conflict emerges for the y- and z-direction. The result of the experiment shown in Fig. 4.19 is illustrated graphically by the conflict matrix shown in Fig. 4.20. The displayed vertical bars symbolise the different data sources $Q_i^{(m)}$ for the mth comparative quantity based on different sensors and models, which are listed on the left side. Two or more data sources are used to determine one comparative quantity, displayed on the bottom of Fig. 4.20. As described, these redundant values are examined for conflicts. Here, conflicts were detected in two of the evalu-

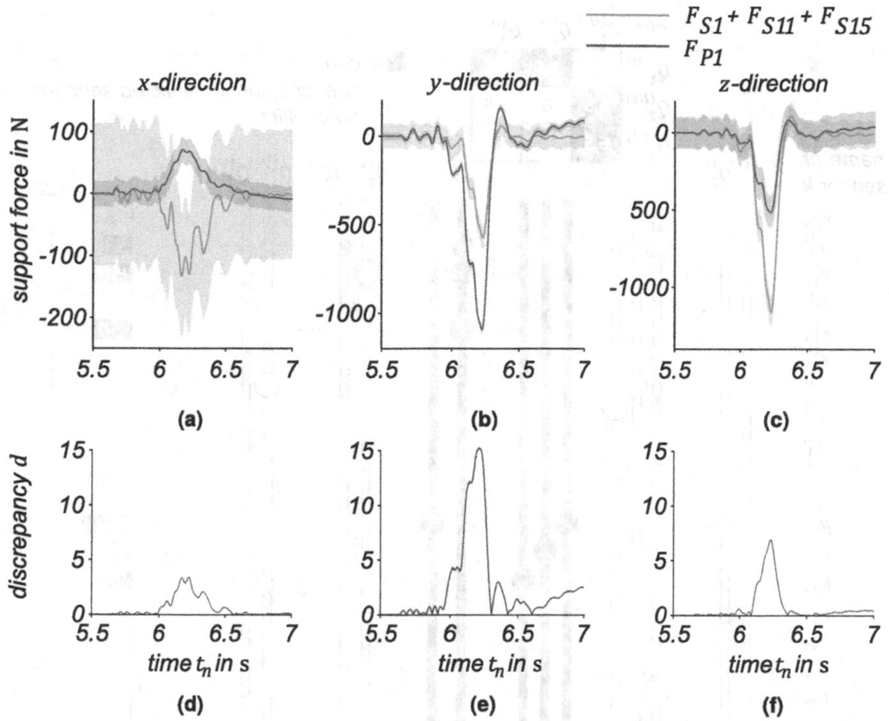

Fig. 4.19 Conflict detection for faulty force signal: **a–c** Triaxial comparison of force sensors with truss strain gauges taking the propagated measurement and data uncertainty into account, which is symbolised by the grey-shaded area, **d–f** resulting discrepancy between both data sources indicating a conflict

ated comparative quantities: 'support force 1', which is based on the equilibrium of forces FE_1, as described above, and 'force symmetry', which includes the condition that, in the case of a vertical impact, the forces in the support points 1–3 must act point-symmetrically around the centre axis due to geometric considerations. For the sake of simplicity, only sensors are shown in Fig. 4.20, but the described procedure can be extended analogously with the models contained in the applied soft sensors.

For 'support force 1', the DOC c_{ij} of the three data sources are illustrated in the conflict sub-matrix $C^{(III)}$ above. Comparisons in this sub-matrix that exhibit a data-induced conflict are marked in black as the well as the bars that represent sources involved in a conflict and the comparative quantities estimated by this sources. Other comparative quantities, which contain the same sensors, are highlighted in grey. An important benefit of the shown representation is the marking of sensors that are involved in a conflict as potentially faulty, so that this information can be considered elsewhere, which in the case of ExoMars, as mentioned above, could have supported the identification of a faulty data source.

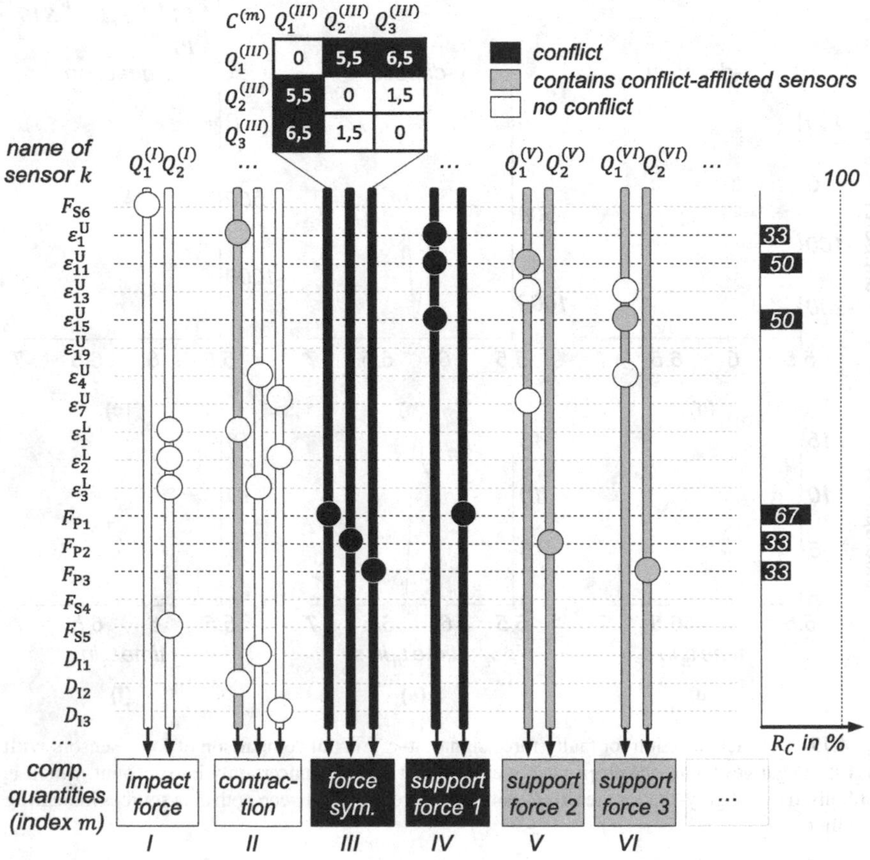

Fig. 4.20 Conflict matrix and resulting conflict rates due to sensor fault for multiple comparative quantities

Sensor F_{P1} is the common component of both conflicting scenarios, thus it is obviously recommendable to check this sensor for a variety of sensor errors. To quantify this suspicion, a conflict rate $R_{C,n}$ is introduced, which gives the operator of the system a hint on which sensor to check first. The $R_{C,n}$ of sensor n is the number of conflicts in which the sensor is involved in relation to all comparisons of sensor n with other data sources. To illustrate that, it is shown exemplary for $\varepsilon_1^{\mathrm{U}}$. This sensor is contained in one source for comparative quantity 'contraction', where it is compared with two other sources without a conflict and in one source for the comparative quantity 'support force 1' facing one other conflicting data source, so its conflict rate is 33%. If the measuring channels for the y- and z-component of the force transducer at support 1, as switched in the described case, were connected correctly, there would no longer be any conflict in the comparative quantities considered.

Conclusion

Misinterpretations can occur during the processing of sensor data due to uncertainty, especially ignorance. Data-induced conflicts occur when physical quantities used for monitoring are recorded redundantly and contradict each other. These conflicts can be used specifically to detect faults. For this purpose, we developed a method which is based on differentiating data sources into models and sensors, linking them in such a way that relevant variables are consciously recorded redundantly.

The proposed approach was applied to the MAFDS, the structural system presented in Sect. 3.6.1. An information model was built that contains all relevant metadata of the underlying sensor system, such as quantified uncertainty, as well as the used physical models. Automatic differentiation was implemented to propagate and determine the resulting uncertainty for conflict checking.

While state-of-the-art fault-detection methods only take some redundant data into account, they lack the view on the whole system. A single fault may result in a multitude of conflicts, especially in time-variant processes. To assist an operator or developer in finding the fault, the amount of data from the conflict checks must be reduced and visualised in a way that makes it easier for humans to recognise a pattern. We presented a conflict interpretation method that furthermore takes the metadata into account. Hence, the method is scalable in both, the number of soft sensors and the model depth, for example to identify faulty model parameters.

4.3 Analysis, Quantification and Evaluation of Model Uncertainty

Christopher M. Gehb, Marc E. Pfetsch, and Stefan Ulbrich

Trying to predict the future is deeply rooted in mankind. In almost every field of science and engineering, more or less sophisticated models are used to predict processes or properties and finally make decisions or draw conclusions [144]. Along these lines, models can be mathematical formulations, e.g. physical axioms and constitutive equations, or physical simplifications of reality, e.g. scaled prototypes. However, every prediction made by models comprises uncertainty, see Sect. 1.3. The uncertainty in model predictions arises essentially from the sources *data uncertainty* and *model uncertainty*, cf. Chap. 2, supplemented by numerical errors in case of mathematical models [87]. This section focuses on the analysis, quantification and evaluation of *model uncertainty*.

Reality is complex and cannot be completely represented by models, neither in mathematical formulations nor in prototype realisations, cf. Figure 1.5. Simplifications, assumptions, conceptualisations, abstractions and approximations all result in model uncertainty [144]. No matter if underlying physics is only poorly understood or linearisations need to be used to reduce computational burdens: "Essentially, all models are wrong, but some are useful." [23].

With this in mind, model uncertainty needs to be taken into account for any kind of decision making and for the evaluation of model predictions in general. The ongoing trend towards digitalisation and the related substitution of real experiments by virtual testing or the combination to Hardware-in-the-Loop (HiL) tests emphasises the necessity of a detailed analysis of model uncertainty to get reliable predictions.

This section is less understood as a textbook, but rather includes the consideration of model uncertainty for manifold examples and applications of mechanical load-bearing structures from both an engineering and a mathematical perspective. The section shows the importance of evaluating the model uncertainty in order to improve the models themselves and their predictions, and finally the conclusions to be drawn from the predictions. Additionally, mathematical approaches and algorithms are presented to analyse and quantify model uncertainty in theory and in practical examples of mechanical load-bearing structures.

4.3.1 Detection of Model Uncertainty via Parameter Estimation and Optimum Experimental Design

Alexander Matei, Marc E. Pfetsch, and Stefan Ulbrich

In this subsection we develop an algorithm to detect model uncertainty using tools from parameter estimation, optimum experimental design and statistical hypothesis testing. The mathematical models which are investigated consist of functional relations between input and output quantities, such as model parameters, state variables and boundary conditions, cf. Sect. 2.2. Within a probabilistic frequentist framework, it is assumed that the *true values* of the model parameters can be estimated by repeated calibration and validation processes with new observational data. The latter are subject to noise, and as a consequence, uncertainty propagates to those parameter estimates. In an optimally designed experiment we then find the best choice among experimental setups, so that the extent of data uncertainty upon the model parameters which we quantify by confidence regions is minimised. If the mathematical model is correct then repeated model calibration and validation with different data sets obtained from optimal experimental setups should entail almost the same parameter values within a confidence region. We interpret *inconsistencies* in the parameter estimates obtained from different measurements as an indicator for model uncertainty, i.e. the mathematical model is incapable to explain *all* the data with the *same* set of model parameters. An important feature of our algorithm is that we neither assume any prior distribution nor a specific algebraic form of the model discrepancy term in the mathematical equations. Thus, we identify the source of model uncertainty as ignorance, see Sect. 2.2. We first proposed our approach in [50].

Mathematical setting

Let $u_j \in U_{\text{ad}}$, $j = 1, \ldots, n_{\text{u}}$ be the inputs, such as boundary or initial conditions, $p \in P_{\text{ad}} \subset \mathbb{R}^{n_{\text{p}}}$ be the model parameters, such as material constants, and $y_j \in Y$ be the corresponding state variables. The first part of the mathematical model is given by an operator $e : Y \times U_{\text{ad}} \times P_{\text{ad}} \to Y$ that defines the *state equation*

$$e(y_j, u_j, p) = 0.$$

We require that for every $p \in P_{\text{ad}}$ and every $u_j \in U_{\text{ad}}$ there exists a unique solution $y_j(u_j, p)$ of the state equation. Furthermore, the solution operators

$$\mathcal{S}_j : U_{\text{ad}} \times P_{\text{ad}} \ni (u_j, p) \mapsto y_j(u_j, p) \in Y \tag{4.11}$$

are demanded to be twice continuously differentiable in both arguments.

In order to compare the state $y_j(u_j, p)$ to experimental data it is necessary to map certain components to quantities that are actually measured. This mapping forms the second part of the mathematical model. Therefore, let us define an *observation operator* by

$$h : Y \times P_{\text{ad}} \ni (y_j, p) \mapsto h(y_j, p) \in \mathbb{R}^{n_{\text{s}}},$$

where n_{s} is the number of data collecting sensors. The experimental setup is characterised by these predefined sensor types or locations and the inputs u_j. We assume h to be twice continuously differentiable in both arguments as well.

It is commonly observed that the acquisition of measurements $z \in \mathbb{R}^{n_{\text{M}} \times n_{\text{u}} \times n_{\text{s}}}$ is subject to uncertainty, where n_{M} is the number of repeated measurement series. To this end, we assume a Gaussian noise profile that is added to the true but in general unknown value $z^\star$ of the quantity of interest:

$$z_{ij} = z_{ij}^\star + \varepsilon, \quad \varepsilon \sim \mathcal{N}(0, \sigma^2), \tag{4.12}$$

for all $i = 1, \ldots, n_{\text{M}}$ and $j = 1, \ldots, n_{\text{u}}$ where $\sigma^2 \in \mathbb{R}^{n_{\text{s}} \times n_{\text{s}}}$ is the diagonal variance matrix of the employed sensors. Thus, we assume that the noise profile is independently distributed for each sensor. If the model is correct then it is a valid explanation of the data, i.e.

$$z_{ij} = h\big(\mathcal{S}_j(u_j, p^\star), p^\star\big) + \varepsilon, \tag{4.13}$$

holds for all $i = 1, \ldots, n_{\text{M}}$ and all $j = 1, \ldots, n_{\text{u}}$ with the true but in general unknown parameter values $p^\star$. Now, the following questions arise:

1. How can we estimate $p^\star$ from z and quantify the uncertainty in the estimation?
2. What are useful criteria to determine whether the Eq. (4.13) is incorrect?

The first question is extensively explored in the literature [10, 14, 92, 150] and we briefly introduce our method of choice below. The second question is strongly related to the detection and quantification of model uncertainty, cf. Sect. 2.2. This is still an active field of research. In the following we present our approach to detect model uncertainty as described in [50].

Parameter estimation

For given measurements z the following nonlinear least-square problem with state equation constraints [19] is solved to obtain an estimate of the true values of the model parameters:

$$\begin{aligned} \min_{p, y_1, \ldots, y_{n_u}} \quad & \sum_{k=1}^{n_s} \sum_{j=1}^{n_u} \sum_{i=1}^{n_M} \frac{\omega_k}{2\sigma_{kk}^2} \left[z_{ijk} - h_k(y_j, p)\right]^2 \\ \underset{p, y_1, \ldots, y_{n_u}}{\text{s.t.}} \quad & p \in P_{\text{ad}}, \quad e(y_j, u_j, p) = 0, \quad \text{for } j = 1, \ldots, n_u, \end{aligned} \tag{4.14}$$

where σ_{kk}^2 are the variances of the sensors introduced above and $\omega_k \in \{0, 1\}$ are their weights, i.e. $\omega_k = 1$ if, and only if, sensor k is used. We allow sensors to remain unused to save operational costs. Since the parameter estimate depends on the data z as well as on the weights ω, which both remain fixed, we associate a solution operator $(z, \omega) \mapsto p(z, \omega)$ with Problem (4.14).

We choose $n_z = n_s n_u n_M$ as the new dimension to rewrite problem (4.14) in vector form and further insert the solution operators $\mathcal{S}_j$ of the state equation (4.11). Let $\tilde{z} \in \mathbb{R}^{n_z}$ be the data vector obtained from rearranging z and let $\tilde{h}$ consist of $h(\mathcal{S}_j(u_j, p), p)$ for all $j = 1, \ldots, n_u$ in a row and copied n_M times. We define $\mathcal{S}(u, p) := \mathcal{S}_j(u_j, p)_{j=1,\ldots,n_u}$ for brevity. Then

$$r(z, p, \mathcal{S}(u, p)) := \tilde{z} - \tilde{h}(\mathcal{S}(u, p), p) \in \mathbb{R}^{n_z} \tag{4.15}$$

are the residuals in vector form. The diagonal weight matrix $\Omega \in \mathbb{R}^{n_z \times n_z}$ consists of copies of $\omega \in \mathbb{R}^{n_s}$ and the diagonal variance matrix $\Sigma \in \mathbb{R}^{n_z \times n_z}$ contains copies of σ^2. Then problem (4.14) can be rewritten into

$$\min_{p \in P_{\text{ad}}} \frac{1}{2} r(z, p, \mathcal{S}(u, p))^\top \Omega \Sigma^{-1} r(z, p, \mathcal{S}(u, p)). \tag{4.16}$$

Each locally optimal solution of (4.16) is a random variable. In general, its probability distribution differs from the one of the measurements z. This is due to the fact that the mapping $(z, \omega) \mapsto p(z, \omega)$ is nonlinear. The computation of confidence regions would lead to non-ellipsoidal sets which are difficult to handle. We therefore choose for a given confidence level $1 - \alpha$, where $\alpha \in (0, 1)$, a linear approximation of the confidence region K in the parameter space around $\mathbb{E}[p(z, \omega)] = p^\star$, see [92]. In fact, we approximate the distribution of $p(z, \omega)$ to be Gaussian with expected value

$p^\star$ and covariance matrix C. Then the set K is an n_p-dimensional ellipsoid determined by the covariance matrix C:

$$K(p^\star, C, \alpha) := \left\{ p \in \mathbb{R}^{n_\mathrm{p}} : \left(p - p^\star\right)^\top C^{-1} \left(p - p^\star\right) \leq \chi^2_{n_\mathrm{p}}(1-\alpha) \right\},$$

where $\chi^2_{n_\mathrm{p}}$ is the quantile function of the χ^2 probability distribution with n_p degrees of freedom. We consider the following approximations for the covariance matrix C coming from a Gauss–Newton approach [19, 38] and a sensitivity analysis [10, 38], respectively:

$$C_\mathrm{GN} = \left(J^\top \Omega \Sigma^{-1} J\right)^{-1}, \qquad C_\mathrm{S} = H^{-1} J^\top \Omega \Sigma^{-1} J H^{-\top},$$

where J is the total derivative of the residual vector r with respect to the model parameters p and

$$H := J^\top \Omega \Sigma^{-1} J + \sum_{i=1}^{n_z} r_i \Omega_{ii} \Sigma_{ii}^{-1} \frac{\mathrm{d}^2 r_i}{\mathrm{d} p^2}. \tag{4.17}$$

Our choice is determined depending on the application and the computational effort for the Hessian H, which requires the calculation of second order derivatives of the solution operator, compare Eq. (4.17) with Eq. (4.15).

Optimal design of experiments

In general, the goal in optimum experimental design is to minimise the confidence region of the parameter estimates by changing the experimental setup, namely, sensor locations and types represented by the variable ω, boundary and initial conditions described by the inputs u_j, etc. Since we employ a linear approximation of the confidence region the aim is to minimise the "size" of the covariance matrix C. There exists extensive research on different design criteria Ψ that measure the "size" of a matrix in the context of optimum experimental design [49, 143, 158]. We list a few prominent options:

$$\Psi_A(C) = \operatorname{trace}(C), \qquad \Psi_D(C) = \det(C), \qquad \Psi_E(C) = \lambda_{\max}(C).$$

Depending on the application, the computational effort, and the adaptability of the experimental setup, we formulate slightly different optimisation problems. If the calculation of the Hessian H is fast, the number of sensors is small and the experimental setup is limited to adapting sensor positions only, we consider the matrix C_S from above in the optimisation model:

$$\begin{aligned} \min_{\omega} \ & \Psi\left[C_\mathrm{S}(\omega, p(z,\omega), S(u, p(z,\omega)))\right] \\ \text{s.t.} \ & \omega \in \{0,1\}^{n_\mathrm{s}}, \quad G(\omega) \leq 0. \end{aligned} \tag{4.18}$$

Note that in an iterative solver scheme a new parameter value $p(z, \omega)$ and new solutions $\mathcal{S}_j(u_j, p(z, \omega))$ to the state equation have to be computed after each step for all $j = 1, \ldots, n_u$. The inputs u_j remain fixed here as the experimenter can only adjust sensor positions. Moreover, C_S depends on the measurements, and this would require new data, if any input values are changed. The constraints $G(\omega)$ describe user-specific restrictions on sensor combinations and on the minimal number of used sensors, see [50, 92] for more details. Problem (4.18) is a non-convex mixed-integer nonlinear program. Since we assume the number of sensors to be small, we employ heuristic methods to solve it.

If the computational effort for the Hessian H is large, the number of available sensors is high, and the experimental setup can be adapted in sensor positions and inputs, we use the covariance matrix arising from the Gauss–Newton approach in the following optimisation problem:

$$\begin{aligned} \min_{\omega,u} \quad & \Psi\left[C_{GN}(\omega, p, \mathcal{S}(u, p))\right] + \beta_1 R(u) + \beta_2 P_\varepsilon(\omega) \\ \underset{\omega,u}{\text{s.t.}} \quad & \omega \in [0, 1]^{n_s}, \quad u \in U_{ad}, \end{aligned} \tag{4.19}$$

where β_1, β_2 are positive constants. Note that the experimenter is now given the possibility to optimise both sensor weights ω and input variables $u = \left(u_j\right)_{j=1,\ldots,n_u}$. Besides, C_{GN} is independent of experimental data, and the parameter values p stay fixed in this setting. However, this approximation of the true covariance matrix may be less accurate than C_S where the parameter values are continually updated within the optimisation scheme. The function $R(\cdot)$ serves as a regulariser for the inputs to guarantee smoothness. For a fixed $\varepsilon \in (0, 1]$ the penalty term $P_\varepsilon(\cdot)$ is chosen to be a smooth approximation of the ℓ_0 "norm". We refer to [3] for a detailed mathematical description. This penalty is intended to yield sparse and $\{0, 1\}$-valued optimal sensor weights. To achieve this, we proceed in the following way. Problem (4.19) is first solved with $\varepsilon_1 = 1$ and we obtain optimal weights $\overline{\omega}_{\varepsilon_1}$ and optimal inputs $\overline{u} = \left(\overline{u}_j\right)_{j=1,\ldots,n_u}$. Then we choose another ε such that $0 < \varepsilon < \varepsilon_1$ and solve the following optimisation problem with $\overline{\omega}_{\varepsilon_1}$ as starting point and fixed inputs $\overline{u}$:

$$\begin{aligned} \min_{\omega} \quad & \Psi\left[C_{GN}(\omega, p, \mathcal{S}(\overline{u}, p))\right] + \beta_1 R(\overline{u}) + \beta_2 P_\varepsilon(\omega) \\ \text{s.t.} \quad & \omega \in [0, 1]^{n_s} . \end{aligned} \tag{4.20}$$

By successively solving (4.20) with diminishing ε_i such that $0 < \varepsilon_i < \varepsilon_{i-1}$ and with $\overline{\omega}_{\varepsilon_{i-1}}$ as starting point, the optimal sensor weights tend to become sparse and $\{0, 1\}$-valued after a few iterations $i = 1, 2, \ldots$, see [4].

Model uncertainty

We now employ the two previously introduced methods, parameter estimation and optimum experimental design, to identify model uncertainty. Hereby, it is assumed that the model $\mathcal{M}$ is valid in all sensor locations specified by ω, for all inputs $u_j \in U_{ad}$

Input: Model $\mathcal{M}$, test level TOL (e.g. 5 %), n_{tests}, computational complexity and scope.
Output: Should $\mathcal{M}$ be rejected? YES (1) or NO (0).

01: Initialise $i := 1$.
02: Generate initial data z_{ini} in all feasible sensor locations.
03: **if** computational complexity is small and scope is restricted to sensor positions **then**
04:     Solve (4.18) using z_{ini} and obtain optimal $\overline{\omega}$.
05:     Acquire data z within the optimised experimental setup $\overline{\omega}$ for the inputs u_j.
06: **else**
07:     Calculate p_{ini} from (4.16) using z_{ini}. Fix p_{ini} and solve (4.19). Obtain optimal $\overline{\omega}$ and $\overline{u}$.
08:     Acquire data z in optimal sensor locations $\overline{\omega}$ for inputs close to the optimum $\overline{u}$.
09: **end if**
10: Check whether measurement errors are Gaussian. Otherwise go to line 05/08 or exit.
11: Divide z into a calibration set z_{cal} and a validation set z_{val}.
12: Calculate $(p_{\text{cal}}, C_{\text{cal}})$ using z_{cal}. Likewise, obtain p_{val} using z_{val}.
13: Determine $\alpha_{\min} \in (0, 1)$, such that p_{val} lies on the boundary of $K(p_{\text{cal}}, C_{\text{cal}}, \alpha_{\min})$.
14: **if** $\alpha_{\min} \geq \text{TOL}/n_{\text{tests}}$ **then**
15:     **if** $i < n_{\text{tests}}$ **then**
16:         $i := i + 1$. Go to line 11.
17:     **else**
18:         **return** 0.
19:     **end if**
20: **else if** $\alpha_{\min} < \text{TOL}/n_{\text{tests}}$ **then**
21:     **return** 1.
22: **end if**

Fig. 4.21 Algorithm for the detection of model uncertainty (adapted from [50])

and for the same true model parameters $p^\star$. Since $p^\star$ are in general unknown we state the hypothesis that a particular solution of (4.16) serves as a good approximation. Then repeated solutions of (4.16) for measurement series taken at different sensor locations with possibly differing inputs should lie in the confidence region of previously estimated parameter values. But, if certain data sets lead to estimates that lie outside the confidence region of previous tests then the model is unable to predict the results of *all* experiments, i.e. the underlying model is inadequate.

Figure 4.21 summarises our algorithm to detect whether a model $\mathcal{M}$ is inadequate. In line 02 initial (or artificial) data are introduced because they appear in the covariance matrix C_S in Problem (4.18). In the alternative way (line 07), it is necessary to compute an initial parameter estimate from this data before solving Problem (4.19).

The acquisition of experimental data sets z in line 05 happens at the optimal sensor locations $\overline{\omega}$ for those inputs u_j that entered the optimisation problem (4.18). Thus, the size of the predicted confidence region for the model parameters is at its minimum provided that the measurement error has the previously stated variance σ^2, see Eq. (4.12). In line 08, experimental data is acquired at the optimal sensor locations but with inputs in the vicinity of the computed optimum $\overline{u}$. By continuity of the objective function in Problem (4.19) with respect to the inputs, the size of the confidence region for the model parameters stays close to the minimal one.

A fundamental assumption of our methodology is that the measurement errors are Gaussian. To check whether the measurement errors are normally distributed (line `10`) we refer to conventional techniques as described in [30], for example. We do not consider experiments that yield data with non-Gaussian noise since this violates our fundamental assumption in Eq. (4.12).

The choice of the calibration and the validation set in line `11` is crucial. The model $\mathcal{M}$ may or may not pass the test depending on that choice. It is possible to divide the data set randomly as in a Monte Carlo cross-validation [39]. However, there are applications where an expert judgement is necessary to perform a meaningful division. Additional help to target the worst-case split can improve the performance of our algorithm. An example for this is given in Sect. 4.3.2 where we detect uncertainty in mathematical models of the 3D Servo Press.

From lines `13` onward, a classical hypothesis test with Bonferroni correction [40] is performed. The null hypothesis and the alternative hypothesis are

$$\begin{aligned} \mathrm{HYP}_0 &: p^\star = p_{\mathrm{cal}} \text{ are the true parameter values for all } u_j \in U_{\mathrm{ad}}, \\ \mathrm{HYP}_1 &: p^\star \neq p_{\mathrm{cal}}. \end{aligned}$$

Let $\overline{\mathtt{TOL}} = \mathtt{TOL}/n_{\mathrm{tests}}$ be the corrected test level. The null hypothesis HYP_0 is rejected if $p_{\mathrm{val}} \notin K(p_{\mathrm{cal}}, C_{\mathrm{cal}}, \overline{\mathtt{TOL}})$. Recall

$$K(p_{\mathrm{cal}}, C_{\mathrm{cal}}, \overline{\mathtt{TOL}}) = \left\{ p \in \mathbb{R}^{n_{\mathrm{p}}} : (p - p_{\mathrm{cal}})^\top C_{\mathrm{cal}}^{-1} (p - p_{\mathrm{cal}}) \leq \chi^2_{n_{\mathrm{p}}} \left(1 - \overline{\mathtt{TOL}}\right) \right\}.$$

Since we are usually performing more than one hypothesis test on similar data sets, we need to account for the problem of multiple testing. The Bonferroni correction of the test level, $\overline{\mathtt{TOL}} = \mathtt{TOL}/n_{\mathrm{tests}}$, is a very conservative method to control the family-wise error rate (FWER), i.e. the probability of rejecting at least one true null hypothesis. It is reasonable to choose a small threshold for the FWER, e.g. 5% since it represents the error of the first kind which we want to be small when rejecting a model. The $\alpha_{\min}$ (line `13`) is the p-value of the statistical test, which is the smallest test level under which the null hypothesis can only just be rejected.

The greater the number of test scenarios n_{tests} the easier it becomes for a null hypothesis to pass a particular test. Since we are interested in the overall null hypothesis that the true values $p^\star$ of the parameters stay within the computed confidence regions, we interpret any rejected null hypothesis as significant, i.e. then the mathematical model itself is subject to uncertainty. In practice, it may occur that an inadequate model passes quite a lot of tests. This behaviour can be explained by the fact that even an inaccurate model may provide satisfactory results on a particular range of inputs. However, provided that enough data are available one can identify ranges of inputs for which an inadequate model fails the hypothesis test.

In summary, we proposed a new algorithm to detect model uncertainty and to quantify the quality of our decision when rejecting a mathematical model via error probabilities. We combined methods from parameter estimation, optimum experimental design and hypothesis testing to achieve this. Furthermore, our approach is suited to identify particular ranges of inputs for which the model fails to explain the data. This is especially helpful when reconsidering the system design phase in product development, see Sect. 3.1, to improve the models that have been used so far.

4.3.2 Detection of Model Uncertainty in Mathematical Models of the 3D Servo Press

Alexander Matei and Florian Hoppe

The method proposed in Sect. 4.3.1 is demonstrated here using a component of the 3D Servo Press, a multi-technology forming machine that combines spindles with multiple eccentric servo drives, see Sect. 3.6.3. Forming machines have the task of performing accurate motions of the tool centre point (TCP) under high process forces. Besides control actions, this requires the acquisition of the TCP position, see Sect. 5.4.1. Since direct measurements of the TCP are technically infeasible, elastic models shall provide the basis for the state estimation of the TCP [66]. To calibrate and validate the elastic models, measurements were taken on the small-scale prototype of the press. Furthermore, the costs for obtaining these measurements are reduced in view of future experiments on the full-scale 160t press. In this subsection, we briefly sketch mathematical models of the 3D Servo Press and present numerical results on the detection of model uncertainty from [50].

In order to model the elastic 3D Servo Press, components were classified according to their load scenario and their functional setting, respectively. The press mainly consists of coupling links and bearings, see Fig. 4.22. Additionally, friction between these components needs to be taken into account. In the following we describe the mathematical models that were employed for the different parts of the press and for the description of their behaviour:

- A *bar* model is employed for those coupling links where the stress under load is very small. Each bar is discretised by the finite element method. Each finite element stands for two masses connected by a spring. However, the actual bar elements do not have a uniform cross-sectional area. To take this into account, a mass-spring model is derived from a finite element analysis [50].
- The remaining coupling links which experience bending moments are modelled as *beams*. Each beam is again discretised by the finite element method and reduced to a mass-spring model with lumped masses, i.e. all non-diagonal elements in the mass matrix are neglected. The governing equations come from the Euler–Bernoulli beam theory [58].
- Due to their progressive stiffness characteristics, all bearings are modelled as *non-linear spring elements*, located between the joints of the coupling links.
- As expected, *friction* was observed in all bearings that move. Thus, the results of the experiments reveal a hysteresis behaviour in the load-displacement curve. Since the

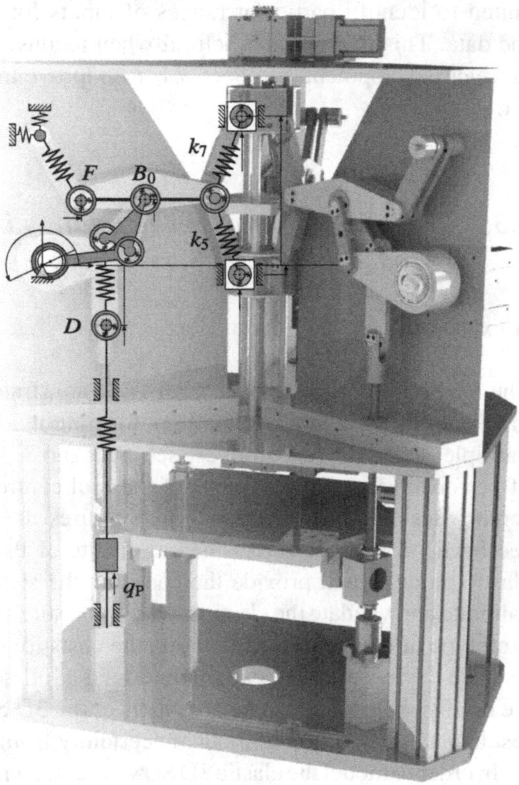

Fig. 4.22 Linkage mechanism of the 3D Servo Press [50]

complete physical modelling of friction is very challenging, we use application-specific substitute models and evaluate them experimentally. We propose three different models $\mathcal{M}_1$, $\mathcal{M}_2$ and $\mathcal{M}_3$ to deal with this phenomenon:

$\mathcal{M}_1$: linear model where friction is neglected,
$\mathcal{M}_2$: discontinuous Coulomb's friction model,
$\mathcal{M}_3$: continuous friction model with rate-independent memory.

In order to validate each of these models, several experimental data sets were collected. The measurements were conducted with $n_u = 29$ different process forces, which we call input variables. The first 15 forces were part of the loading and the last 14 were part of the unloading cycle. Our quantities of interest are the vertical displacements in point D, the horizontal displacements in point F, and vertical displacements in point B_0, when a vertical process load q_P is applied to the press, see Fig. 4.22.

There are $n_s = 3$ sensors installed at these locations which measure the displacements. Each series of measurements was repeated $n_M = 6$ times, although with slightly different process forces. To deal with this variability, we work with the

Table 4.2 Results for the optimum experimental design problem for model $\mathcal{M}_3$. A sensor combination is defined by a three-digit number, where each digit stands for the usage (1) or non-usage (0) of the corresponding sensor [50]

Sensor combination	$\Psi_A(C)$	$\Psi_D(C)$	$\Psi_E(C)$
111 (initial)	4.959×10^{09}	1.168×10^{16}	4.956×10^{09}
101	1.118×10^{29}	7.183×10^{35}	1.118×10^{29}
011	6.258×10^{09}	1.484×10^{16}	6.256×10^{09}
110	3.514×10^{09}	2.756×10^{16}	3.506×10^{09}

known setpoint values of the applied forces and linearly interpolate the data, see [50] for more details. We deviate from the algorithm presented in Sect. 4.3.1 to some extent in that we do *not* distinguish between initial data and the actual acquisition of measurements.

After the experimental data had been acquired and the measurement errors were checked to be Gaussian, our goal was to minimise costs to obtain these measurements by selecting only two out of the three sensors. The model parameters to be estimated are the stiffnesses of two bars, k_7 and k_5, see Fig. 4.22. Since there are only two parameters to be estimated, it suffices to employ two sensors and repeat the measurement process. Table 4.2 shows the results for the most important design criteria for the model $\mathcal{M}_3$, where we used the matrix C_S as covariance matrix, see Sect. 4.3.1.

From Table 4.2 we infer that the absence of the second sensor entails an increase in all design criteria by a factor of $\approx 10^{20}$ compared to the initial value where all sensors are employed. This is a strong indication that the covariance matrix became singular, i.e. it is impossible to estimate the model parameters with that sensor choice. The absence of the first sensor, though, increases the maximal expansion, which is related to the design function $\Psi_E(C)$, and the volume, which is related to the design function $\Psi_D(C)$, of the confidence ellipsoid. However, the sensor combination displayed in the last row of Table 4.2, i.e. measuring the vertical displacements in point D and the horizontal displacements in point F only, leads to the smallest expansion of the confidence ellipsoid. We choose this sensor pair. Computations for the models $\mathcal{M}_1$ and $\mathcal{M}_2$ bring us to the same conclusion.

As already mentioned, the experiments revealed a hysteresis behaviour. We want to apply the algorithm introduced in Sect. 4.3.1 to see whether model uncertainty is recognised in the friction models $\mathcal{M}_1$, $\mathcal{M}_2$ and $\mathcal{M}_3$. The output of these models together with the measurement data is shown in Fig. 4.23 for comparison.

The continuous friction model is trained by an artificial neural network using real and simulated data [16, 112]. Hence, we used four of the six data series. Thus, only $n_{\mathrm{M}} = 2$ measurement series remained for the application of our algorithm. To stay fair, we used these two measurement series for *all* models alike during the validation which we perform by hypothesis testing. The splitting of the data set into a calibration and a validation set was done in four different ways, see [50]. First, we split the test set into loading z^l and unloading z^u. For the loading case, we again split the data

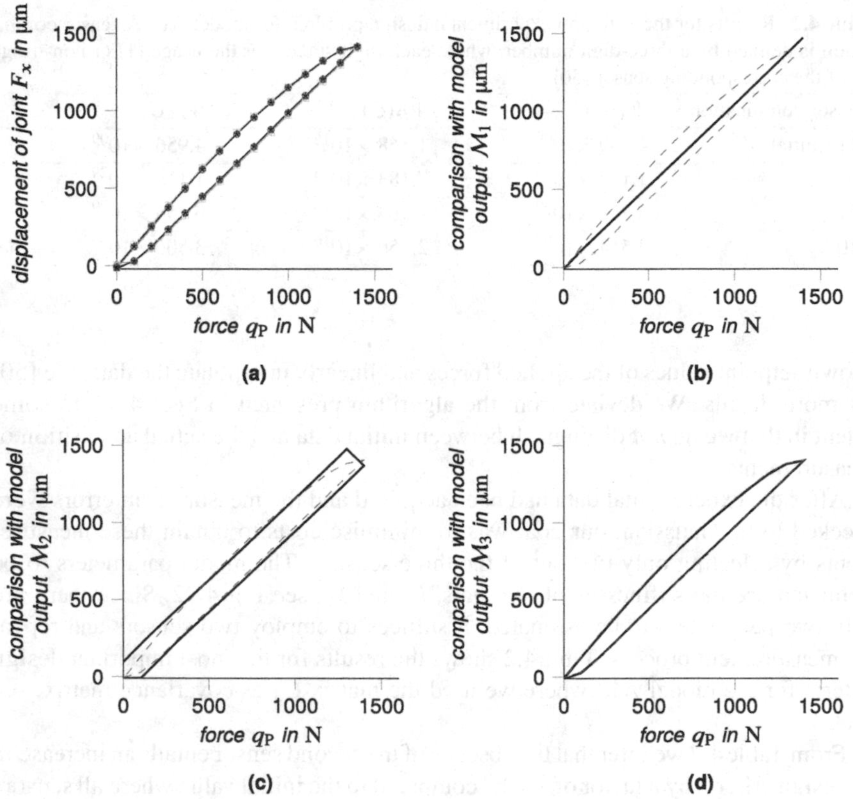

Fig. 4.23 **a** repeated measurements of the force-displacement curve and **b–d** comparison of measurements (- - -) with the model output $\mathcal{M}_1$, $\mathcal{M}_2$ and $\mathcal{M}_3$ (—), respectively [50]

homogeneously into one calibration $z_c^{l_1}$ and one validation $z_v^{l_2}$ set. The same was done for the unloading case. Next, the loading scenario was tested against unloading, such that we split the data homogeneously into one calibration z_c^l and one validation z_v^u set. Finally, data points from both loading and unloading were tested against each other, i.e. we had z_c^{lu} for calibration and z_v^{lu} for validation.

For each of these $n_{\text{tests}} = 4$ test scenarios we computed the $\alpha_{\min}$, respectively, as shown in Table 4.3. Adopting the usual threshold $\texttt{TOL} = 5\%$ for the error of the first kind and applying the conservative Bonferroni correction with $n_{\text{tests}} = 4$, see Sect. 4.3.1, the corrected test level becomes $\texttt{TOL}/n_{\text{tests}} = 1.25\%$. Then, it is clear that model $\mathcal{M}_1$ is rejected in all four test scenarios. As expected, the experimental data cannot be described by a linear model that neglects friction. A first attempt to model hysteresis which is caused by friction is given by $\mathcal{M}_2$. This model seems to be able to accurately model loading and unloading separately. However, it is insufficient in describing both phenomena with the same set of parameters. Our algorithm is able to detect this deficiency in the third and fourth test scenario. This result can be

Table 4.3 Test results for models $\mathcal{M}_1$, $\mathcal{M}_2$ and $\mathcal{M}_3$ [50]

Calibration	Validation	$\alpha_{\min}$ in %		
		for $\mathcal{M}_1$	for $\mathcal{M}_2$	for $\mathcal{M}_3$
$z_c^{l_1}$	$z_v^{l_2}$	0.02	78.78	92.99
$z_c^{u_1}$	$z_v^{u_2}$	$\ll$0.01	23.33	66.06
z_c^{l}	z_v^{u}	$\ll$0.01	$\ll$0.01	24.59
z_c^{lu}	z_v^{lu}	0.81	$\ll$0.01	93.45

explained with the fact that the Coulomb model is discontinuous whereas friction is a continuous effect. The last column of Table 4.3 shows that $\mathcal{M}_3$, which has been trained by an artificial neural network, passes all tests successfully. Thus, this model is able to explain the present type of hysteresis with the same set of parameters which are valid within their confidence region.

In conclusion, we have seen that the algorithm introduced in Sect. 4.3.1 performs well, if applied to the 3D Servo Press. The choice of the calibration and validation test sets has been done by expert judgement because of the special behaviour of the technical system, namely, the loading-unloading cycles. By splitting the data set this way, we directly target the worst-case test scenario, so that a Monte-Carlo-like splitting is not necessary. Since further development steps and online algorithms of machines rely on a valid model, a statement about model uncertainty is a valuable indication for the engineer. Furthermore, our hypothesis test could be used as a stopping criterion for the performance training of an artificial neural network. Besides the optimal placement of sensors for the model calibration, the presented method can also be used to identify uncertainty in different complex models, to be considered in the model selection.

4.3.3 Assessment of Model Uncertainty for the Modular Active Spring-Damper System

Robert Feldmann, Christopher M. Gehb, Maximilian Schaeffner, and Tobias Melz

Research on methods to quantify model uncertainty in structural engineering has intensified more and more in recent years. The information gain of such methods typically relies on using experimental data, such as structural responses or vibration analysis. Examples for methods range from the well-known Bayes-factor [137] to methods based on the Bayesian inversion [60], an error-domain model falsification approach [122], or a technique using the adjustment factor method [124]. In Sect. 2.2 a comprehensive overview over methods for quantification of model uncertainty can be found. In this section, we introduce and compare two different methods to quantify model uncertainty by applying them exemplarily to the MAFDS, as presented in

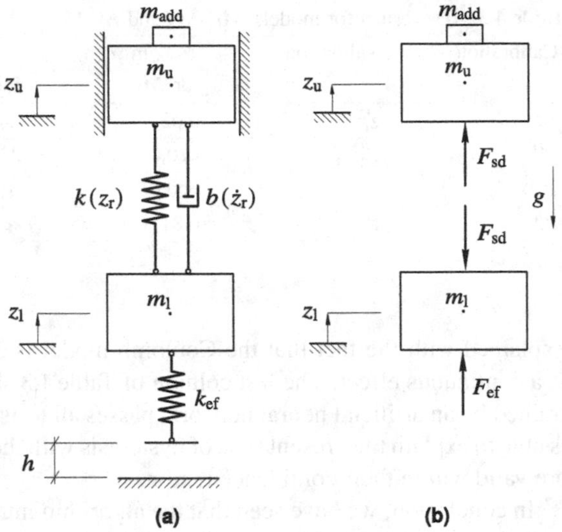

Fig. 4.24 2 DOF model and free body diagram of the MAFDS

Sect. 3.6.1. First, a method based on the direct application of the Bayes' theorem is presented and, subsequently, a method based on the modelling of a discrepancy function by means of a Gaussian process is shown.

Figure 3.23 depicts the MAFDS and Fig. 4.24 the two degrees of freedom (2 DOF) model to capture its dynamic behaviour, where the drop height is denoted by h. The position of both the upper and lower mass is determined by the coordinates z_u and z_l of the 2 DOF model, where $z_r = z_u - z_l$ denotes their relative displacement. The equations of motion are

$$\begin{pmatrix} m_u + m_{add} & 0 \\ 0 & m_l \end{pmatrix} \begin{pmatrix} \ddot{z}_u \\ \ddot{z}_l \end{pmatrix} + \begin{pmatrix} b(\dot{z}_r) & -b(\dot{z}_r) \\ -b(\dot{z}_r) & b(\dot{z}_r) \end{pmatrix} \begin{pmatrix} \dot{z}_u \\ \dot{z}_l \end{pmatrix}$$
$$+ \begin{pmatrix} k(z_r) & -k(z_r) \\ -k(z_r) & k(z_r) + k_{ef} \end{pmatrix} \begin{pmatrix} z_u \\ z_l \end{pmatrix} + \begin{pmatrix} (m_u + m_{add})g \\ m_l\, g \end{pmatrix} = \mathbf{0}, \tag{4.21}$$

where $k(z_r)$ denotes the stiffness and $b(\dot{z}_r)$ denotes the damping of the spring-damper system, as functions of the relative displacement z_r and k_{ef} denote the stiffness of the elastic foot, as can be seen in Fig. 3.23. The structure is subject to gravitation g. For details on the derivation of the equations of motion see [107]. Regression studies on the stiffness and damping properties of the spring-damper system yielded several model candidates to describe the dynamic behaviour by combinations of linear, bilinear and power functions [105]. This leads to uncertainty regarding which model candidate is most adequate to predict the dynamic behaviour of the MAFDS. This model uncertainty is assessed and analysed subsequently by comparing the different model candidates in terms of model adequacy. The main content of this section is based on [45, 108].

As outlined in Sect. 3.6.1, the inputs to the system are the drop height h and the additional weight m_{add} that can be added to the frame. The system inputs are summarised in the input vector $\boldsymbol{x} = (h, m_{\text{add}})^{\top}$. Exemplarily, the system outputs are the maximum relative compression $z_{\text{r,max}}$, the maximum force in the elastic foot $F_{\text{ef,max}}$ and the maximum force on the spring-damper system $F_{\text{sd,max}}$, as depicted in Fig. 4.24. The system outputs were chosen in such a way that they can be calculated by simulation of the model candidates, as well as measured experimentally, and thereby enable to compare different models. In Fig. 3.24 in Sect. 3.6.1, the system outputs are shown as horizontal lines in the trajectories of the experimental drop test. For simplicity of notation, the scalar simulation model outputs η and the scalar experimental outputs y are condensed in the vectors $\boldsymbol{\eta}$, $\boldsymbol{y}$, respectively:

$$\boldsymbol{\eta} = \begin{pmatrix} z_{\text{r,max,sim}} \\ F_{\text{ef,max,sim}} \\ F_{\text{sd,max,sim}} \end{pmatrix}, \qquad \boldsymbol{y} = \begin{pmatrix} z_{\text{r,max,exp}} \\ F_{\text{ef,max,exp}} \\ F_{\text{sd,max,exp}} \end{pmatrix}. \tag{4.22}$$

Application of Bayes' theorem for quantification of model uncertainty

In [108] we presented a method to compare different mathematical models based on the extent of agreement between simulation model output $\boldsymbol{\eta}$ and experimental output $\boldsymbol{y}$. Model uncertainty was quantified for $P = 4$ selected mathematical models. In a Bayesian framework, the posterior probability gives a measure of how adequate a mathematical model represents the dynamic behaviour of the MAFDS. It estimates the probability of a simulation model output $\boldsymbol{\eta}$ of each model candidate with index $q = 1, \ldots, Q$ under the condition that the experimental output $\boldsymbol{y}$ has been observed, for which measurement errors were not considered. Assuming an event-based Bayesian approach, $H_{y,q}$ denotes the statistical event describing the output of each model candidate q and A_y is the statistical event associated with the observed experimental output $\boldsymbol{y}$. Bayes' theorem [13] is then written as

$$P(H_{y,q}|A_y) = \frac{P(A_y|H_{y,q})P(H_{y,q})}{P(A_y)}, \quad q = 1, \ldots, Q, \tag{4.23}$$

where $P(H_{y,q})$ denotes the prior probability that the model $H_{y,q}$ is the true model and is assumed equal for all $Q = 4$ model candidates: $P(H_{y,q}) = 1/4$. The likelihood $P(A_y|H_{y,q})$ is the probability that experimental output $\boldsymbol{y}$ is observed when assuming a model q. Similar to a distance metric, it is estimated by the Cartesian vector distance d_p of the simulation model and experimental outputs (4.22):

$$P(A_{y_e}|H_{y,q}) = e^{-\left(\frac{d_q}{\min_{q=1,\ldots,P} d_q}\right)}, \quad \text{where} \quad d_q = \|A_{y_e} - H_{y,q}\|_2. \tag{4.24}$$

The total probability $P(A_y)$ serves as a normalisation constant and is determined analytically as the sum of the product of the likelihood $P(A_y|H_{y,q})$ and the prior probabilities $P(H_{y,q})$ for all mathematical models q. Subsequently, the posterior probability (4.23) is calculated for $K = 9$ different, independent events. Here, an

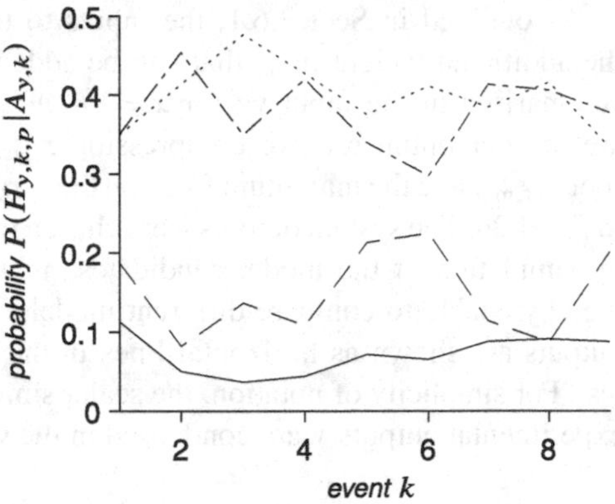

Fig. 4.25 Comparison of posterior probabilities $P(H_{y,k,q}|A_{ye,k})$ for each hypothetical event k predicted by model $q = 1$ (——), model $q = 2$ (– –), model $q = 3$ (–·–), and model $q = 4$ (······) according to [107]

event constitutes an experimental output $\boldsymbol{y}$ and simulation model output $\boldsymbol{\eta}$ for the $Q = 4$ model candidates. For each $k = 1, \ldots, K$ event, the numerical values of the system inputs $\boldsymbol{x}_k$ are unique as given in [108]. Figure4.25 depicts the posterior probabilities $P(H_{y,k,q}|A_{y,k})$ for the four model candidates and the nine events.

Models 3 and 4 exhibit a higher posterior probability for all $K = 9$ events than models 1 and 2, indicating that models 3 and 4 prove more adequate to predict the dynamic behaviour. In order to compare models over a multitude of events, the probability that one model holds true for all $K = 9$ events can be determined by

$$P(H_{y,q}|A_y) = \prod_{k=1}^{K=9} P(H_{y,k,q}|A_{y,k}) \tag{4.25}$$

and is used as a measure of adequacy. For models 1 and 2, the overall posterior probability amounts to $7.0 \cdot 10^{-8}$ and $5.6 \cdot 10^{-5}$. Models 3 and 4 exhibit significantly higher values of the posterior probability with 0.38 and 0.618, respectively. In conclusion, model 4 is most adequate to represent the dynamic behaviour of MAFDS. In summary, the presented method provides a straightforward, computationally non-intensive way to quantify model uncertainty for comparing different models.

A Gaussian process-based method for quantification of model uncertainty

Now, we apply a different method as presented in [45] to quantify model uncertainty. It is assumed that all models are wrong and incorporate a model error due to missing or incomplete physics in the mathematical model [23]. Based on this assumption, the method builds upon the pioneering work of Kennedy and O'Hagan [87], where a model discrepancy function is introduced to incorporate the model error; it thereby serves as a measure of adequacy of a mathematical model. In this framework, any experimental output of a system is represented as

$$y_n = \eta(\boldsymbol{x}_n) + \delta(\boldsymbol{x}_n) + \varepsilon_n, \tag{4.26}$$

where $y_n \in \mathbb{R}$, $(n = 1, \ldots, N)$ denotes the nth of a total of N measurements, η is the simulation model output (i.e. $F_{\text{ef,max}}$, $F_{\text{sd,max}}$, $z_{\text{r,max}}$) with not necessarily unique inputs $\boldsymbol{x}_n = [h, m_{\text{add}}]^\top$, δ is the discrepancy function and ε_n represents zero-mean normally distributed measurement noise for each measurement n. We model the discrepancy function $\delta(\boldsymbol{x})$ by a Gaussian process $\boldsymbol{\delta}(\boldsymbol{X}) \sim \mathcal{N}\big(m(\boldsymbol{X}), C(\boldsymbol{X}, \boldsymbol{X})\big)$, where $\boldsymbol{X}$ represents an input matrix $\boldsymbol{X} = [\boldsymbol{x}_1, \ldots, \boldsymbol{x}_N]$. The mean function is denoted by $m(\boldsymbol{X})$ and $C(\boldsymbol{X}, \boldsymbol{X})$ is the covariance matrix which is built up by the covariance function $c(\boldsymbol{x}_i, \boldsymbol{x}_j)$ where $\boldsymbol{x}_i, \boldsymbol{x}_j$ with $i, j = 1, \ldots, N$ denote the input vectors. The Gaussian process itself is fitted to the difference between measurement y_n and the model output $\eta(\boldsymbol{x}_n)$, using the data set

$$\Big[\big(\boldsymbol{x}_1, y_1 - \eta(\boldsymbol{x}_1)\big), \ldots, \big(\boldsymbol{x}_N, y_N - \eta(\boldsymbol{x}_N)\big)\Big]. \tag{4.27}$$

For the Gaussian process, a constant mean scale m and a squared exponential covariance function $c(\boldsymbol{x}_i, \boldsymbol{x}_j)$ are selected

$$m(\boldsymbol{X}) = \beta, \quad c(\boldsymbol{x}_i, \boldsymbol{x}_j) = \sigma_\text{f}^2 \exp\left(-\frac{1}{2}(\boldsymbol{x}_i - \boldsymbol{x}_j)^\top \boldsymbol{M} (\boldsymbol{x}_i - \boldsymbol{x}_j)\right) + \sigma_\text{n}^2 \delta_{ij}, \tag{4.28}$$

where δ_{ij} denotes the Kronecker delta in this case. The matrix $\boldsymbol{M}$ is set to $\boldsymbol{M} = \boldsymbol{I}\ell^{-2}$ with identity matrix $\boldsymbol{I} \in \mathbb{R}^{2\times 2}$ and length scale $\ell > 0$ [132]. The signal variance $\sigma_\text{f} > 0$ determines how much the discrepancy function values deviate from the mean value. Larger values of the signal variance σ_f lead to larger deviations of the discrepancy function. Measurement noise is accounted for by the noise level parameter σ_n in the covariance function (4.28). It is assumed to be an additive, independent identically distributed Gaussian noise with variance σ_n^2 [132].

The hyperparameters $(\beta, \ell, \sigma_\text{f}, \sigma_\text{n})$ inherent to the mean and covariance function (4.28) essentially govern the behaviour of the Gaussian process and are determined using a Bayesian optimisation scheme. Using the optimised set of hyperparameters, the quantiles of the 95%-confidence interval for the discrepancy functions are specified analytically by

$$C_{2.5} = \beta - 2\sigma_f, \qquad C_{97.5} = \beta + 2\sigma_f. \tag{4.29}$$

Comparing the 95%-confidence interval of the discrepancy functions by their 2.5 and 97.5% quantiles $C_{2.5}$ and $C_{97.5}$ yields a measure to select between competing models. The maximum absolute value of the two quantiles shows how much the discrepancy function deviates from zero and consequently indicates how adequate the model is.

For the example at hand, the Gaussian processes describing the discrepancy functions δ_{q,z_r}, δ_{q,F_sd} and δ_{q,F_ef} are determined for the $q = 1, \ldots, 4$ model candidates (The model candidates are not identical with the ones investigated in the previous section). The 95%-confidence intervals of the discrepancy functions are shown in Fig. 4.26. All models overestimate the three outputs of the system, which can be seen by the fact that the absolute values of the 2.5 and 97.5% quantiles of the discrepancy functions

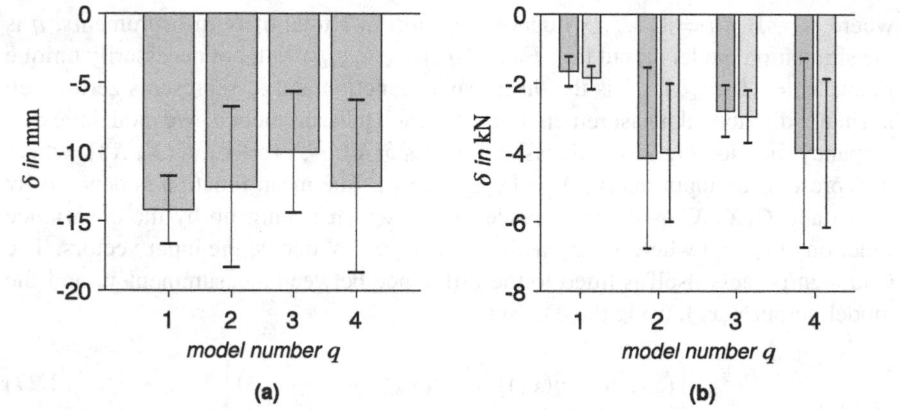

Fig. 4.26 The mean scale β (shown as bars) and the 95%-confidence intervals bordered by its quantiles calculated by (4.29) for the discrepancy function δ for the $Q = 4$ model candidates: **a** for relative displacement discrepancy function δ_{q,z_r} (▬) **b** for force discrepancy functions $\delta_{q,F_{ef}}$ (▬) and $\delta_{q,F_{sd}}$ (▬)

are consistently negative. For output $z_{r,max}$ displayed in Fig. 4.26a, the 2.5% quantile for model 3 is closest to zero, indicating a higher adequacy of the model. However for the force outputs $F_{sd,max}$ and $F_{ef,max}$ shown in Fig. 4.26b, the 2.5% quantile for model 1 is closest to zero, suggesting that model 1 is most adequate. In conclusion, no model consistently ranks best in terms of model adequacy.

Conclusion

Both presented methods provide a measure of adequacy that can be used to quantify model uncertainty. They essentially differ in their assumptions about model error. The method presented first does not differentiate between model error and measurement error. In consequence, the chosen likelihood function (4.24) does not reflect a distribution but is rather to be understood as a distance metric. In contrast, the Gaussian process based method assumes the model discrepancy as a Gaussian process and accounts for measurement error separately.

Further, the discrepancy function provides valuable information about the difference between model and measurement. For example, the mean scale exhibits if a model tends to under- or overestimate system quantities. In contrast, for the method based on the Bayes' theorem, this information is lost due to the quadratic form of the likelihood function.

For the modelling of an adequate discrepancy function, the Gaussian process based method essentially relies on assumptions on or a priori knowledge about suitable mean and covariance functions. As a priori knowledge is missing here, a rather simple choice for mean and covariance function was made. For the rare cases, in which there is a priori knowledge about the model discrepancy, the mean function could for example be polynomial, or consist of weighted basis functions.

As a concluding remark, the computing time for the first method is negligible, whereas it highly depends on the number of model inputs, outputs and measurements for the hyperparameter optimisation.

4.3.4 Model Uncertainty in Hardware-in-the-loop Tests

Manuel Rexer, Philipp Hedrich, and Peter F. Pelz

Hardware-in-the-Loop (HiL) tests investigate the behaviour of real components connected to real time simulated systems [82, 98]. As depicted in Fig. 3.20, HiL tests enable mastering uncertainty by a stepwise integration of a module into a real system by combining cyber world and real world. This section discusses the influence of the active interface between the two worlds. Therefore, compared to a simulation of the virtual component, HiL tests are in the virtual system.

The first HiL tests were used in 1936 to simulate instruments in an aircraft cockpit [82]. In the mid 1960s, electrical and hydraulic actuators were used to simulate cockpit movements [82]. Since the late 1960s, HiL tests have been used to simulate the response of structures and components to earthquakes [119]. Since the 1980s, HiL has been used at universities as well as in research and development departments for component validation [133, 139].

Formulated briefly, HiL tests are a symbiosis of an experiment and a simulation as Fig. 3.20 shows. This results in the following advantages compared to classical tests and pure simulation [82, 98, 139]:

1. Real system components can be tested in the virtual system at an early design stage. This saves costs and development time. It is a prerequisite for the agile development of physical systems.
2. Parameters of the virtual system can be changed with little effort to investigate different test configurations.
3. Components with complex non-linear behaviour can be investigated in the simulation as real components. The model uncertainty is reduced, since reality can be investigated.

Therefore, HiL tests are ideal to examine components like the Active Air Spring with the associated parts we developed, cf. Sect. 3.6.2. It is not necessary to have a two mass oscillator or a complete vehicle in hardware, since they can be virtually simulated to master the uncertainty of the component in an early design stage. HiL tests can therefore already be used in the design phase of the product life cycle introduced in Sect. 1.2. The disadvantages of HiL tests are a real time capable hardware being required with this hardware having an influence on the result; this is due to signal propagation times, measurement uncertainty, filtering, and the dynamics of the test rig, as shown in this subsection. In addition, an appropriate modelling is necessary, where a compromise between the required computing time and the complexity of the model has to be found. The relevant reality, cf. Sect. 1.3, is never represented

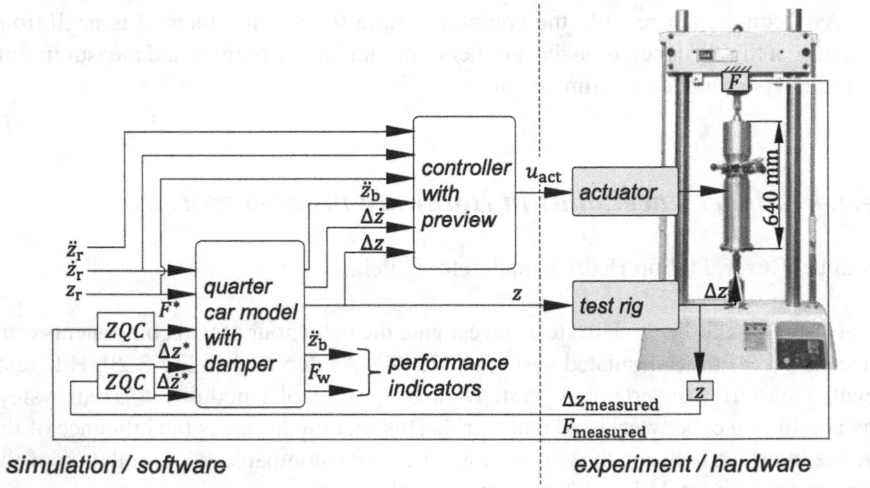

Fig. 4.27 Basic structure of the HiL experiments, consisting of simulation and hardware, as well as all signal flows [73]

completely, so there is model uncertainty. In this subsection we therefore investigate the incertitude of model uncertainty and our approach in mastering this uncertainty.

In our HiL tests, the Active Air Spring is coupled with the virtual quarter car, which is simulated in parallel in a real-time simulation environment. Figure 4.27 shows the principle structure and signal flows of these HiL tests. The air spring deflection Δz calculated in the real-time simulation is transmitted to the uniaxial servo hydraulic test rig, the active interface, which deflects the Active Air Spring. The measured axial force F is fed back into the simulation. This is therefore a "closed loop simulation". The simulated quarter car model—a foot point excited two mass oscillator—and the implemented controller are introduced in Sect. 3.6.2. The excitation is also used in the preview function of the implemented controller, which is equally integrated in the real time simulation. In order to minimise the influence of measurement noise, measured and fed back signals are filtered with a second order Butterworth filter with a cut-off frequency of 170 Hz (ZQC in Fig. 4.27). A more detailed investigation in the results of the HiL experiments can be found at Hedrich [73].

The performance indicators that are examined with these HiL tests are driving safety, e.g. wheel load fluctuation σ_{F_w}, and driving comfort, e.g. variation of body acceleration $\sigma_{\ddot{z}_b}$, being obtained from real time simulation.

The conflict diagram, Fig. 4.28, displays the measurement result of a HiL test driving on a highway with 100 km/h marked by the diamond. To this end the standard deviations of the wheel load F_w and the body acceleration $\ddot{z}_b$ are determined by the time signals measured for $T = 20$ s. Figure 4.28 shows the measured points for the designed controller with preview. The simulated result (square) is determined with a linear Active Air Spring model and the quarter car model from Sect. 3.6.2. The active Pareto line represents an ideal active system where the controller parameters have

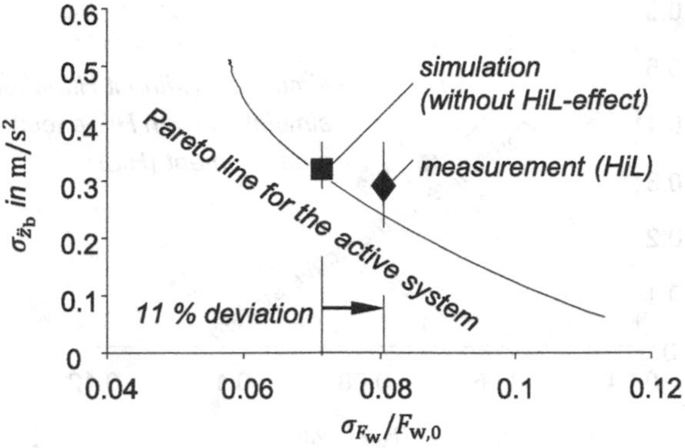

Fig. 4.28 Conflict diagram with the results for driving safety σ_{F_w} and driving comfort $\sigma_{\ddot{z}_b}$ for the HiL tests (diamond) and the basic simulations (square), as well as the simulated active Pareto line as reference driving on a highway at 100 km/h [73]

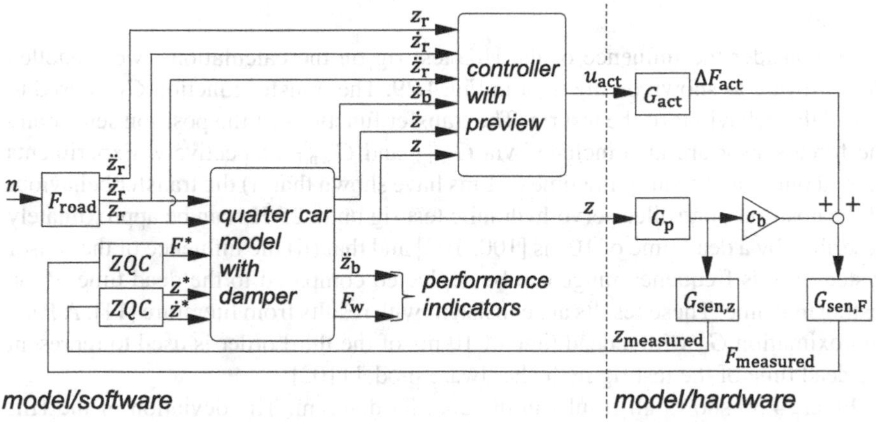

Fig. 4.29 Adapted simulation structure. By taking into account the transfer functions of the hardware, the influence of these functions can be considered in the simulations [73]

been optimised. The HiL influence appears in a deviation of the experiment from simulation in driving safety by 11%. The influence on driving comfort is negligible. Our investigations and results from literature [26] with similar HiL tests show that the test rig in particular has a mayor influence on the results of the HiL tests. The dynamics of the test rig and the sensors have not yet been taken into account in the simulation, since this has never been necessary at any open loop component measurement. Following the principle of simplicity from Heinrich Hertz introduced in Sect. 1.3, the transfer functions $G_p = G_{sen,z} = G_{sen,F} = 1$ were assumed (cf. Fig. 4.29, square). The model used for the hardware therefore does not take the reality sufficiently into account.

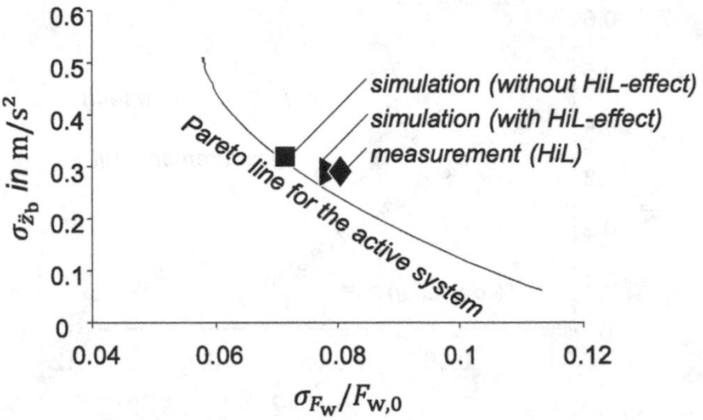

Fig. 4.30 Conflict diagram with the results for driving safety σ_{F_w} and driving comfort $\sigma_{\ddot{z}_b}$ for the HiL tests (diamond) and the basic (square) and adapted (triangle) simulations, as well as the simulated active Pareto line as reference driving on a highway at 100 km/h [73]

To consider the influence of the HiL test rig on the calculations, we modelled the hardware as shown on the right in Fig. 4.29. The transfer function G_P is used to model the behaviour of the test rig. The transfer functions of the position sensor and the force sensor are also included via $G_{\mathrm{sen,z}}$ and $G_{\mathrm{sen,F}}$ respectively. Experiments carried out with the sampling time of 1 ms have shown that (i) the transfer behaviour of the position-controlled servo-hydraulic test rig up to 25 Hz can be approximately described by a dead time of 10 ms [100, 102] and that (ii) the influence of the sensor system in this frequency range can be neglected compared to the dead time of the testing machine. These results are consistent with results from literature [11]. A Padé approximation G_{p} with a dead time of 10 ms of the third order is used to represent the dead time of the test rig in the hardware model [102].

Figure 4.30 shows all results in the conflict diagram. The deviation of the HiL test from the adapted simulation (triangle) in driving safety comes down to 2%. The following conclusions can be drawn: (i) The influence of the active interface is recognisable and it influences the results. (ii) If the influence of the test rig is considered in the calculation via a dead time element, measurement and simulation correspond quantitatively well. (iii) The remaining deviation is acceptable in the context of the linearised models used and measurement uncertainty.

The model uncertainty, i.e. the neglect of the influence of the active interface, can thus be mastered by taking the transfer function of the test rig into account. Since the influence on the driving comfort, which is the main focus of the tests, is small, the HiL influence on the experiments is tolerated. In the future we will investigate the Active Air Spring integrated in the MAFDS with foot point excitation. This enables validations of the HiL tests and the simulation of the virtual component in the virtual system with a real component in the real system.

4.3.5 *Identification of Model Uncertainty in the Development of Adsorption Based Hydraulic Accumulators*

Jakob Hartig and Peter F. Pelz

When starting a product development from scratch, not much is known about the intended system. There is general physical knowledge and experience in the form of physical axioms and constitutive equations (cf. Sect. 1.3). However, ignorance in these early stages of product development can lead to a significant model uncertainty (see Fig. 1.5 in Sect. 1.3).

In this section, we demonstrate this point in the development of innovative hydraulic accumulators. To ensure consistency with general knowledge we used axiomatic models to determine the potential of hydraulic accumulators filled with adsorptive material. In the following, we show that the omission of some of the system's numerous interconnected physical effects can lead to a large model uncertainty.

Hydraulic accumulators are used to store energy in hydraulic systems, e.g. for dynamic energy demand. The storage medium is compressed gas. Especially in mobile applications space and weight reduction by smaller and lighter components is very important. Thus the quality measured in acceptance and effort is increased, cf. 1.9. However, with hydraulic accumulators there are two opposing dependencies:

(i) The energy density of hydraulic accumulators depends on the excitation frequency due to heat transfer processes. At low frequencies hydraulic accumulators are isothermal, whereas at large frequencies the state change is adiabatic. The transition frequency between isothermal and adiabatic behaviour is inversely proportional to the accumulator volume (to be more precise, it is proportional to the specific surface) and therefore, large volume and isothermal behaviour are mutually exclusive.
(ii) The energy content of the hydraulic accumulators depends on the volume of the accumulator and thus on the size [93, 131]. Hence, energy density and energy content of conventional hydraulic accumulators cannot be maximised at the same time.

To overcome this limitation, different physical effects can be considered. One of these effects is adsorption, i.e. adherence of gas molecules to the surface of a porous material (adsorbent) which was proposed in [126, 131]. The idea behind filling hydraulic accumulators with adsorbent material like activated charcoal, is that adsorption will act as an additional gas storage capacity. In addition to that, gas molecules interact with the adsorbent and therefore lose a translatory degree of freedom during adsorption. Kinetic energy of adsorbed molecules is consequently lower than of free molecules, and energy in the form of heat has to be released during adsorption (heat of adsorption E_A). Adsorption is consequently a heat source [113]. The interdependence of these effects make it necessary to evaluate the usability of adsorption in reducing the size of hydraulic accumulators via suitable models. In the following, we show some challenges of adequately modelling hydraulic accumulators with adsorption and the potential huge impact of model uncertainty on the outcome.

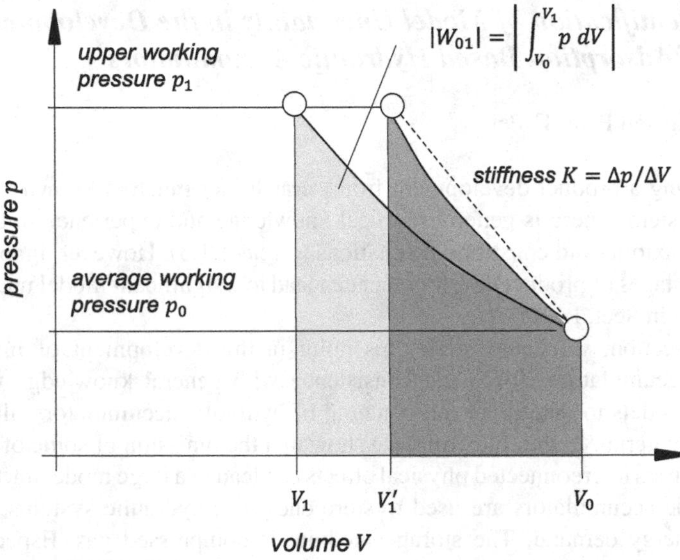

Fig. 4.31 Relationship between pressure and volume. Shaded area is the work for volume change performed on a gas volume

The increase of energy density in hydraulic accumulators is equivalent to stiffness reduction. Figure 4.31 illustrates this connection. The work for volume change performed on a gas volume corresponds to the shaded areas in Fig. 4.31. The work for the volume change performed is limited by the upper working pressure p_1 based on the mean working pressure p_0. A lower stiffness of the hydraulic accumulator is reflected by a lower gradient in the p-V diagram. With the same upper working pressure p_1, a lower stiffness leads to a higher compression from the average working volume V_0 to V_1 instead of V_1'. Thereby more volume change work can be carried out and thus more energy can be stored.

The intended function of adsorbent material for reducing the stiffness was originally thought to be the additional gas storage capacity. With this in mind an analysis of a simple model was done in the two publications [126, 131]. For completeness the model from the two publications and some results are presented below.

In comparison to the frequencies found in the application of hydraulic accumulators, the typical inherent time for adsorption is much smaller. Consequently, adsorption was modelled as an equilibrium process. The number of adsorbed molecules q in mol depends on the pressure p in the accumulator and the mass of adsorbent m_{ads}. For small deviations from equilibrium conditions, the linear Henry approximation with Henry constant H is valid

$$q = m_{ads}\, H p. \tag{4.30}$$

For the gas we assume the ideal behaviour

$$p = \varrho RT \tag{4.31}$$

to hold, where ϱ is the gas density, R is the specific gas constant and T is the absolute temperature. The equations of mass and energy conservation for the hydraulic accumulator result in

$$\varrho \frac{\mathrm{d}V}{\mathrm{d}t} + V \frac{\mathrm{d}\varrho}{\mathrm{d}t} = -M \frac{\mathrm{d}q}{\mathrm{d}t}, \tag{4.32}$$

$$c_v V \left(\frac{\mathrm{d}T}{\mathrm{d}t} \varrho + T \frac{\mathrm{d}\varrho}{\mathrm{d}t} \right) + T \varrho c_p \frac{\mathrm{d}V}{\mathrm{d}t} = -\alpha (T - T_u) A - E_\mathrm{A} \frac{\mathrm{d}q}{\mathrm{d}t}, \tag{4.33}$$

where V is the volume of the accumulator, M is the molar mass of the gas, c_v and c_p are the specific heat capacities of the gas. The heat transfer to the surrounding gas with temperature T_u is modelled with the heat transfer coefficient α and surface area of heat transfer A. All parameters for the hydraulic accumulator ($V_0, p_0, \alpha, T_u, A, \hat{V}$) were chosen to represent typical accumulators found in literature [93]. All adsorption parameters, namely isosteric heat of adsorption E_A and Henry-coefficient H were estimated for nitrogen as described in [111].

In our case the accumulator volume V is changed dynamically, denoted by

$$V = V_0 + \hat{V} \sin(2\pi f t), \tag{4.34}$$

where the index 0 denotes the pre-charged average working state of the accumulator.

Both energy density and energy content depend on the change of the pressure p with changing volume V, i.e. the stiffness

$$K = -\frac{\Delta p}{\Delta V} \tag{4.35}$$

and for comparison purposes are de-dimensioned with

$$K^+ = K \frac{V_0}{p_0}. \tag{4.36}$$

For dynamic applications the stiffness of the accumulator as a function of loading frequency $\Omega = 2\pi f$ is of interest. Therefore, the frequency response of the stiffness is shown in Fig. 4.32 (white-filled circles). Comparing the frequency response to the response of a model without adsorption (light grey curve), a stiffness reduction in the isothermal range, and a stiffness increase in the adiabatic range are visible [126, 131].

Measurements on a similar system however, showed a stiffness reduction in the whole frequency range [29]. This deviation from reality is a sign of model uncertainty (cf. Sect. 2.2. Consequently, the assumptions for the model were revisited and the assumption of temperature independence of the adsorption was given up. In the updated model, the number of adsorbed moles is a function of pressure and temper-

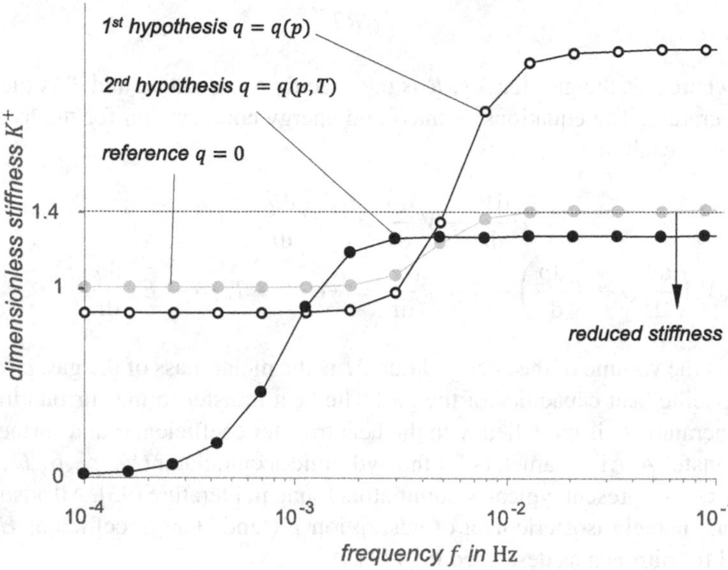

Fig. 4.32 Comparison of stiffness for model $q = q(p)$ according to 1st hypothesis as published in [126, 131], model according to 2nd hypothesis $q = q(p, T)$ and reference accumulator model without adsorption $q = 0$

ature. For the temperature dependency of the Henry coefficient $H(T)$ the following exponential Arrhenius relation can be assumed [113]

$$H(T) = H_0 \exp\left(\frac{-E_A}{MRT}\right). \tag{4.37}$$

The resulting stiffness from numerical simulation of the full nonlinear equations can be seen in Fig. 4.32 (black dots). It shows a stiffness reduction in the whole frequency range.

To find the reason for the stiffness reduction, a parameter variation of the linearised new model in the adiabatic range was carried out. The parameter variation of the new model for the parameters H_0 and E_A in the adiabatic frequency range (cf. Fig. 4.33) shows that the stiffness in the adiabatic range is mainly influenced by E_A. This indicates that the magnitude of the adsorption enthalpy is more significant for stiffness reduction than the process of adsorption itself.

To examine this issue further, a sensitivity analysis of the adsorption equilibrium was carried out. The results show that the sensitivity of the equilibrium loading with respect to temperature T is greater than on pressure, i.e.

$$\left|\frac{\partial q}{\partial T}\right| / \left|\frac{\partial q}{\partial p}\right| = \frac{-E_A}{(c_p/c_v - 1)RT^2} p \gg 1 \tag{4.38}$$

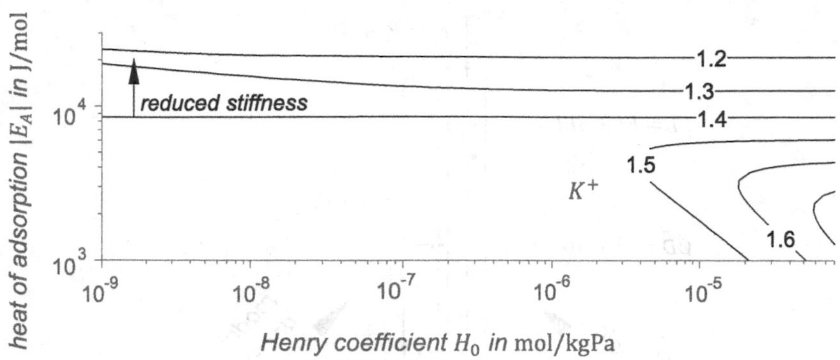

Fig. 4.33 Isolines of dimensionless stiffness K^+ in the adiabatic range for variation of H_0 and $|E_A|$. A stiffness decrease can only be seen for values of $|E_A|$ being larger than about $9 \cdot 10^3$ J/mol. In this parameter range the size of the Henry coefficient H_0 has little influence on the stiffness

Due to $\partial q/\partial T$ being larger than $\partial q/\partial p$, the number of adsorbed molecules decreases when compressing. Therefore, both results suggest that the stiffness reduction in the adiabatic range is reduced by a lower increase of temperature due to E_A being drawn from gas for desorption. The pressure and temperature rise of compression are diminished due to the heat E_A being released in the adsorption. In other words: In contrast to the original assumptions, the adsorptive material is an additional mass source and a heat sink instead of being a mass sink and heat source.

This is a totally different effect than originally intended, and therefore demonstrates a large model uncertainty due to omission of relevant effects. The discovery of this unexpected behaviour was only possible by comparing results from related areas with the model in early stages of the product development process. Inspired by these results, the model uncertainty, i.e. relevant but ignored reality was identified (cf. Sect. 1.3). It emphasises the large effect model uncertainty can have on the results, especially in systems with interdependent physical effects.

4.3.6 Uncertainty Scaling—Propagation from a Real Model to a Full-Scale System

Johannes Brötz and Peter F. Pelz

Models may be mathematical, but they may also be physical, i.e. scaled real models representing a full-scale component or system. The real model is usually scaled in size or material. Geometrically scaled models are common in architecture. In mechanical engineering, scaled models are equally gaining more and more relevance. When it comes to agile development, rapid prototyping is increasingly used resulting in scaled real models.

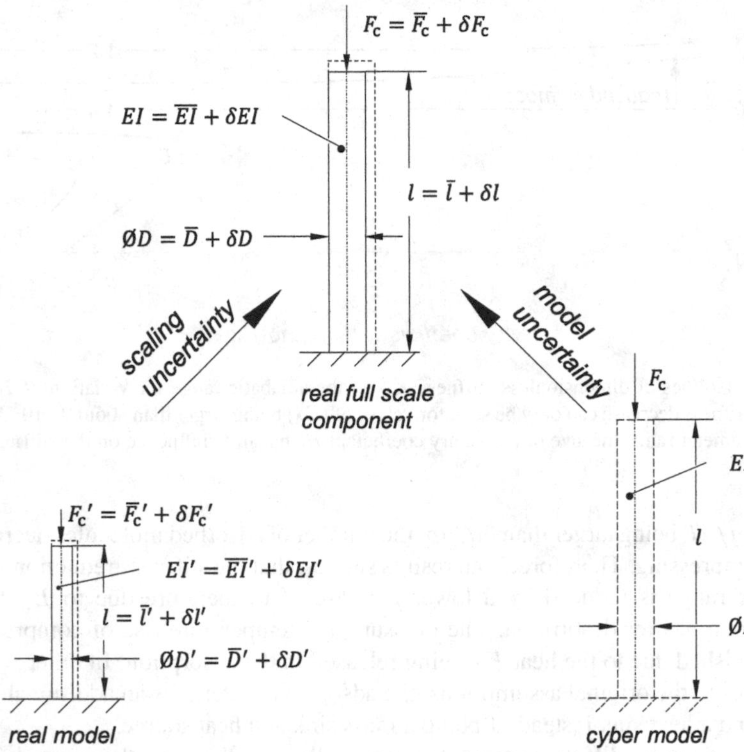

Fig. 4.34 Methods for predicting the function of a full-scale component using the example of a buckling beam

Figure 4.34 shows two methods to predict the functionality of the full-scale component taking the example of a buckling beam: firstly, scaling from real model measurements; secondly, predicting the function of the full-scale prototype by means of a cyber model. This cyber model is a mathematical model of the component, e.g. a finite element model.

Here, we focus on the first method, the scaling of the prototype's function from model measurements. Compared to the cyber model, the advantage of the methodology presented here is that there is no need to consider the uncertainty of mathematical modelling, see Sect. 2.2 So far it has remained an open question how to scale the uncertainty in shape and measurement of the physical model test to the full-scale component.

State of the art for scaling are the four steps: (i) produce a scaled physical model, (ii) measure, (iii) undimension and (iv) scale, see Fig. 4.35. When mastering uncertainty, it is no longer sufficient to only take the parameter uncertainty of a real model into account. In addition, the uncertainty must be scaled in a fifth process step, see Fig. 4.35. In this subsection we introduce a new methodology to propagate the uncertainty from a physical model to the real prototype. The beam and the related buckling

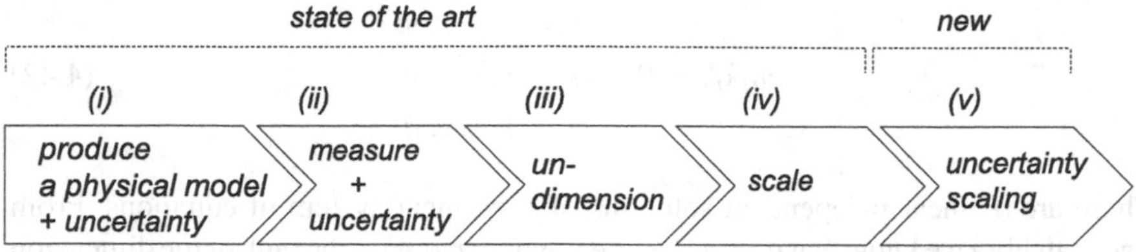

Fig. 4.35 Five steps in the process of scaling

load are used as an application-related example being an important predefined functional restriction $g \leq 0$ of a load carrying structure.

The following subsections describe the procedure of dimensional analysis, scaling and newly introducing the propagation of scaling uncertainty. We refer to the application of uncertainty scaling according to Vergé et al. [156].

Dimensional analysis

The following recap of dimensional analysis is based on Spurk [148]. A system function g and/or quality F is prescribed by n dimensional physical measures p_j, $j = 1, \ldots, n$. The unit of each physical measure p_j is given as a monomial of the $i = 1, \ldots, m$ base units P_i. The dimension of the measure is

$$[p_j] = \prod_{i=1}^{m} P_i^{a_{ij}}. \tag{4.39}$$

The matrix $A = \left(a_{ij}\right)_{n,m}$ is the dimension matrix being central in dimensional analysis. The coefficients a_{ij} are the exponents of the ith base unit for the jth physical measure.

As a consequence of the Bridgeman's postulate [24] the relation of p_j, $j = 1, \ldots, n$, is equivalent to the relation of Π_r with $r = 1, \ldots, d$ dimensionless measures. Each Π_r is a monomial of the $j = 1, \ldots, n$ physical measures.

$$\Pi_r = \prod_{j=1}^{n} p_j^{k_{rj}}, \quad r = 1, \ldots, d. \tag{4.40}$$

The demand $[\Pi_r] \overset{!}{=} 1$ for a truly relative quantity yields

$$1 \overset{!}{=} \prod_{j=1}^{n} P_i^{a_{ij} k_{rj}}, \quad r = 1, \ldots, d. \tag{4.41}$$

This is only satisfied for

$$a_{ij} k_{rj} = 0, \quad \begin{aligned} i &= 1, \ldots, m \\ j &= 1, \ldots, n \\ r &= 1, \ldots, d. \end{aligned} \tag{4.42}$$

There are d linear independent solutions of this linear system of equations. From linear algebra we know that $d = n - rg(A)$, where $rg(A)$ is the rank of the dimension matrix $A = \left(a_{ij}\right)_{n,m}$.

As an illustrative example we look for the buckling load of a beam. The analytic, i.e. mathematical model goes back to Euler [43]. This analytic model is not in focus here. The buckling beam in Fig. 4.34 with fixed-free clamping is assumed to be a cylindrical beam of circular cross-section with nominal diameter D, length l, Young's modulus E and the second moment of area I. For predicting the demanded buckling load F_c of the full-scale component we use the measured buckling load F_c' of the physical scaled model.

For the system there is only one dimensional product

$$Fn(F_c, l, EI) = 0 \quad \Leftrightarrow \quad \Pi = \frac{F_c l^2}{EI} \propto \frac{F_c l^2}{ED^4} = const. \tag{4.43}$$

Scaling

Scaling is used to predict the function g and quality F of the full-scale component. Not only geometric quantities, but other physical quantities, such as the buckling load F_c, can be scaled. The physical properties of the physical model p_j' (values of our physical model are marked by a prime) correlate with the full-scale p_j by

$$p_j := p_j' M_j, \quad j = 1, \ldots, n, \tag{4.44}$$

with the scaling factors M_j. If the dimensionless products of a real physical and a full-scale model are equal, both are said to be similar [148]:

$$\Pi_r = \Pi_r', \quad r = 1, \ldots, d. \tag{4.45}$$

If there is equality of all dimensionless products, we speak of complete similarity. With Eq. (4.44) and (4.45) we demand

$$1 \stackrel{!}{=} \prod_{j=1}^{n} M_j^{k_{rj}}, \quad r = 1, \ldots, d \tag{4.46}$$

for complete similarity.

For the beam we assume complete similarity in the dimensionless product Π given in Eq. (4.43). Hence, using Eq. (4.46) we get the scaling factor M_{F_c} of the critical buckling load F_c

$$M_{F_c} = \frac{M_E M_D^4}{M_l^2}. \tag{4.47}$$

The use of such a scaling law is straight forward: Usually the geometric scaling factors M_D and M_l are known. The same is true for the scaling factor M_E for the Young's modulus. Hence, the scaling law helps predicting the full scale function F_c from the measured model function F_c'.

Uncertainty scaling

In order to take uncertainty into account, the true value p_j is given as the combination of the nominal value $\overline{p}_j$ and the tolerance range δp_j for incertitut, cf. Sect. 2.3:

$$p_j = \overline{p}_j \pm \delta p_j, \quad j = 1, \ldots, n. \tag{4.48a}$$

With the definition of relative uncertainty $U_j := \delta p_j / \overline{p}_j$, the true value is

$$p_j = \overline{p}_j (1 \pm U_j), \quad j = 1, \ldots, n. \tag{4.48b}$$

The same can be applied to the dimensionless products

$$\Pi_r = \overline{\Pi}_r \prod_{j=1}^{n} (1 \pm U_j)^{k_{rj}}, \quad r = 1, \ldots, d. \tag{4.49}$$

When considering uncertainty, the product of the scaling factors reads

$$1 \stackrel{!}{=} \prod_{j=1}^{n} M_j^{k_{rj}} = \prod_{j=1}^{n} \overline{M}_j^{k_{rj}} \prod_{j=1}^{n} \left(\frac{1 \pm U_j}{1 \pm U_{j'}} \right)^{k_{rj}}, \quad r = 1, \ldots, d. \tag{4.50}$$

Since we assume complete similarity of the dimensionless products, $\prod_{j=1}^{n} \overline{M}_j^{k_{rj}} = 1$ applies. Hence, we obtain

$$1 = \prod_{j=1}^{n} \left(\frac{1 \pm U_j}{1 \pm U_{j'}} \right)^{k_{rj}}, \quad r = 1, \ldots, d. \tag{4.51}$$

With Eq. (4.51) the uncertainty of the function and/or quality of a physical system can be calculated. The equation only needs to be solved according to the uncertainty sought.

The manufacturing of the physical model of the full scale beam entails production tolerances. For uncertainty quantification, we refer to the ISO 2768-1 standard [33]. In ISO 2768-1 general tolerances are given for components not specified in great detail. Here the incertitude of the distribution is covered by intervals. For the physical model we choose a length $l' = 200$ mm and a diameter $D' = 10$ mm. We assume that the material used is not changed. Hence, there is a complete similarity of the material $M_E = 1$. The uncertainty of the critical buckling load of the model U_{F_c} that is gained by model measurements, is defined for our example $U_{F'_c} = 0.085$. With Eq. (4.51) the uncertainty of the critical buckling load results in:

$$\pm U_{F_c} = \left(\frac{1 \pm U_D}{1 \pm U_{D'}}\right)^4 \left(\frac{1 \pm U_{l'}}{1 \pm U_l}\right)^2 (1 \pm U_{F'_c}) - 1. \tag{4.52}$$

Equation (4.52) shows that the ratios of the uncertainty of the model and full-scale parameters have an influence on the calculation of the full-scale function uncertainty. These ratios are multiplied by the term for the measured uncertainty of the model function. The uncertainty scaling is illustrated in Fig. 4.36. A geometric scaling factor $M_D = M_l = 1$ represents our real physical model. For scaling factors greater than one (upscaling), the relative uncertainty decreases. This is due to higher precision being possible in manufacturing of large diameters and lengths. For downscaling, which is for lower geometric scaling factors, there is a strong increase in the production uncertainty. This affects the uncertainty of the critical buckling load U_{F_c} of the beam, which shows a variation of higher range. Since the tolerances are defined for specific parameter regions, we obtain a discontinuous function for the uncertainty U_D and U_l and thus for U_{F_c}.

Conclusion

The analysis has shown that there is a strong need for uncertainty scaling. In the example of the buckling load, the relative uncertainty of the predicted function, the buckling load F_c, increases when scaling down. This has to be considered in the design, as it may otherwise lead to unforeseen failure due to the great uncertainty.

4.3.7 Improvement of Surrogate Models Using Observed Data

Sebastian Kersting and Michael Kohler

Computer models of technical systems are playing a more and more important role in the design and construction of complex technical systems. Implemented as computer code, such models enable the use of so-called computer experiments, i.e. an experiment with the technical system is simulated via a computer program using the underlying mathematical model. An overview of the design and analysis of computer experiments can be found in [138] or [44]. In general, these computer models are imperfect, in the sense that they do not predict the reality perfectly, as discussed in Sect. 2.2 There are several reasons, e.g. because of missing knowledge of underlying physical dependencies, or because of an approximation of those to reduce complexity. A typical example is neglecting the friction or considering it

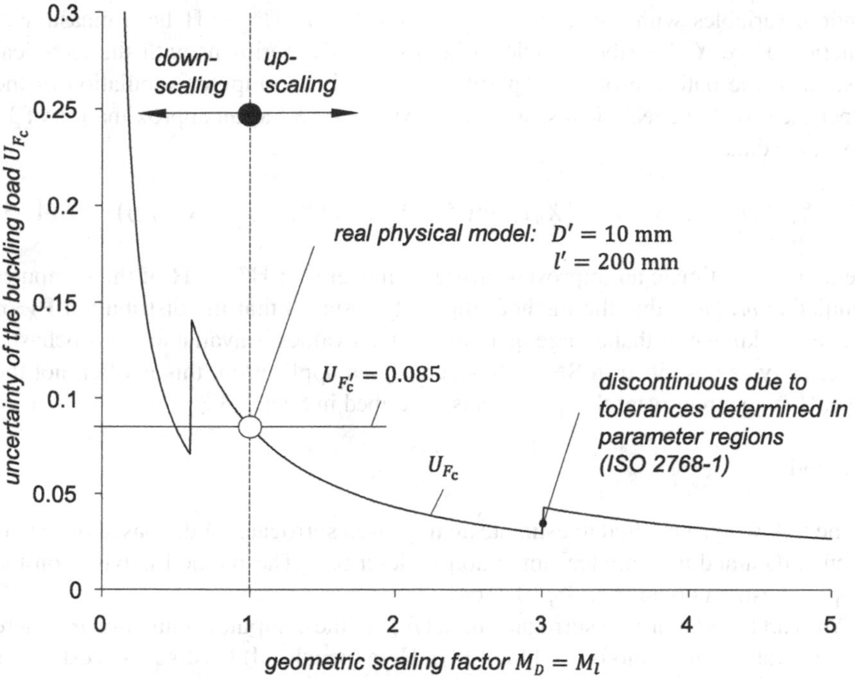

Fig. 4.36 Uncertainty of the buckling load of the full-scale beam U_{F_c}

constant. Furthermore, in uncertainty quantification it is often required to perform a large number of computer simulations of an experiment with the technical system, which can be time-consuming, since typically these computer simulations are computationally expensive. A solution to circumvent this problem is to use a so-called surrogate model. There is a vast variety of literature on methods for estimating a surrogate model. For example [25, 31, 90] used quadratic response surfaces, [22, 32, 77] investigated surrogate models in the context of support vector machines, [121] concentrated on neural networks, [17, 86] used kriging and [160] used Gaussian processes. In the following, a method is described, which is able to circumvent the challenges of the imperfectness and the computationally expensiveness of the computer model, by estimating an improved surrogate model, which has a smaller prediction error and is faster to compute than the computer model, as shown in [62, 88, 91]. Furthermore, the improved surrogate model can then be used to quantify and analyse model uncertainty as shown in [164]. According to Fig. 3.1 the method is applied in the product or system design phase (A).

Mathematical setting

The method, which will be described below is based on the following mathematical setting: Let $(X, Y), (X_1, Y_1), (X_2, Y_2), \dots$ be independent and identically distributed random variables with values in $\mathbb{R}^d \times \mathbb{R}$, and let $m : \mathbb{R}^d \to \mathbb{R}$ be a measurable function. Here X describes (random) inputs of an experiment with the technical system, Y the outcome of the experiment and m is a computer simulation of the experiment with the technical system, thus we use $m(X)$ as an approximation of Y. Given the data

$$(X_1, Y_1), \ldots, (X_n, Y_n), \big(X_{n+1}, m(X_{n+1})\big), \ldots, \big(X_{n+L_n}, m(X_{n+L_n})\big), \tag{4.53}$$

the aim is to estimate an improved surrogate model $\hat{m}_n : \mathbb{R}^d \to \mathbb{R}$ of the computer simulation m. Note that the method implicitly assumes that the distribution $\mathbf{P}_X$ of X is either known or that a large quantity of input values is available, i.e. stochastic uncertainty as described in Sect. 1.6 occurs. In an application, this is often not the case. How to circumvent this problem is described in Sect. 4.3.8.

Method

In the following, a method to estimate an improved surrogate model based on experimental data and a computer simulation is described. The method is based on the proposed estimators in [62, 88, 91, 164].

We start by estimating a surrogate model $\hat{m}_{L_n}$ of the computer simulation m. There is a vast variety of methods (cf. [44, 138]). Here (penalised) least-squares estimates are used, defined by

$$\hat{m}_{L_n}(\cdot) \in \arg\min_{f \in \mathcal{F}} \frac{1}{L_n} \sum_{i=n+1}^{n+L_n} |f(X_i) - m(X_i)|^2 + pen_n^2(f), \tag{4.54}$$

where $\mathcal{F}$ is a set of functions, $(X_1, m(X_1)), \ldots, (X_{L_n}, m(X_{L_n}))$ is the set of input values evaluated with the computer model m of size $L_n \in \mathbb{N}$ and $pen_n^2(\cdot)$ is a penalty term which usually penalises the 'roughness' of the function and which is nonnegative for each $f \in \mathcal{F}$, i.e. $pen_n^2(f) \geq 0$. If the input dimension is smaller or equal to 3 then smoothing spline estimates can be used for $\mathcal{F}$ as shown in [91]. For bigger input dimensions neural network estimates can be applied as in [62] or [88]. Of course, there exist other estimator function classes as discussed above.

As discussed in Sect. 2.2, usually every computer model has an inherent model error. To circumvent this problem, an estimator of the residuals is constructed by first calculating the residuals of the surrogate model with respect to the experimental data

$$\epsilon_i = Y_i - \hat{m}_{L_n}(X_i) \quad (i = 1, \ldots, n) \tag{4.55}$$

and then applying a (penalised) least-squares estimate on this sample, defined by

$$\hat{m}_n^{\epsilon}(\cdot) \in \arg\min_{f \in \bar{\mathcal{F}}} \frac{1}{n} \sum_{i=1}^{n} |f(X_i) - \epsilon_i|^2 + pen_n^2(f), \tag{4.56}$$

where $\bar{\mathcal{F}}$ is a set of functions. Finally, the improved surrogate model is a composition of the estimators above, defined by

$$\hat{m}_n(x) = \hat{m}_{L_n}(x) + \hat{m}_n^{\epsilon}(x) \quad (x \in \mathbb{R}). \tag{4.57}$$

In case that only a small sample of experimental data is available, the estimator of the residuals (4.55) usually does not yield satisfying results. In this case, if an additional independent sample of input values $X_{n+Ln+1}, \ldots, X_{n+Ln+N_n}$ of size $N_n \in \mathbb{N}$ is available, one can use a weighted (penalised) least-squares estimate instead of (4.56) defined by

$$\hat{m}_n^{\epsilon}(\cdot) \in \arg\min_{f \in \bar{\mathcal{F}}} \frac{w^{(n)}}{n} \sum_{i=1}^{n} |f(X_i) - \epsilon_i|^2 + \frac{(1 - w^{(n)})}{N_n} \sum_{i=1}^{N_n} |f(X_{n+L_n+i})|^2 + pen_n^2(f), \tag{4.58}$$

where $w^{(n)} \in [0, 1]$ is a weighting term, which should be chosen data-dependent. Here, adding the weighted mean square of the euclidean norm of the vector $(f(X_{n+Ln+1}), \ldots, f(X_{n+Ln+N_n}))$ of function values of the additional sample is used as a regularisation.

Application

In order to demonstrate the usefulness of the above described approach, we apply it to the drop tests with the MAFDS, which are described in Sect. 3.6.1; here we only consider the drop height as input variable and neglect the additional payload, as in [91]. The system outputs are the maximum relative compression $z_{r,max}$. For $\mathcal{F}$ and $\bar{\mathcal{F}}$ we use a smoothing spline estimator as implemented in the *MATLAB* routine *csaps()*. A smoothing spline estimator depends on an additional smoothing parameter. In the estimation of $\hat{m}_{L_n}$ this smoothing parameter is chosen by generalised cross-validation, cf. [157]. The smoothing parameter and the weighting parameter $w^{(n)}$ in the estimation of $\hat{m}_n^{\epsilon}$ are chosen by a k-fold cross-validation, cf. [68], where the smoothing parameter is chosen from the fixed set $\{2^l : l \in \{-8, \ldots, -1\}\}$ and the weighting parameter is chosen from the set $\{0, 0.1, \ldots, 1\}$.

The result is illustrated in Fig. 4.37. To conclude, we observe that model uncertainty occurs. The computer model overestimates the outcome of the experiments, whereas the improved surrogate model fits the experimental data quite accurately.

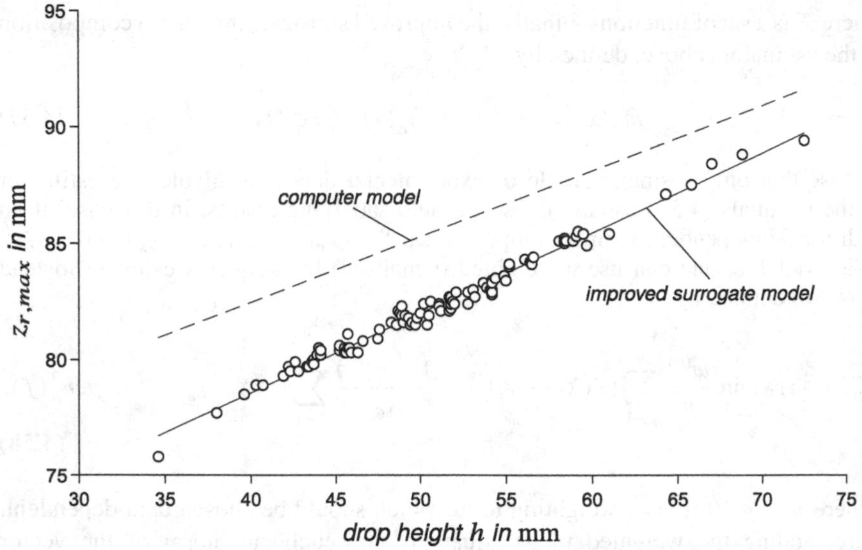

Fig. 4.37 Measured data $(X_1, Y_1), \ldots, (X_n, Y_n)$ (black circles), computer model m (dashed line) and improved surrogate model $\hat{m}_n$ (solid line)

4.3.8 Uncertainty Quantification with Estimated Distribution of Input Parameters

Sebastian Kersting and Michael Kohler

Methods of uncertainty quantification are frequently applied in an experimental setting. This serves to quantify the uncertainty in the outcome Y of an experiment with a technical system, depending on an input X. This would be easy, if a large quantity of experimental data is available, but in most cases running experiments is expensive and time consuming. In order to circumvent this problem, one can use knowledge (e.g. physical knowledge) of the experiment with the technical system to implement a computer model m and use this to generate a data set of computer experiments. In this context, the input-output tuple (X, Y) is modelled as an $\mathbb{R}^d \times \mathbb{R}$ valued random variable, i.e. the experiment depends on a d-dimensional real valued input and has a real valued output. Then, if the input distribution $\mathbf{P}_X$ is known, one can generate realisations of the input X and evaluate them with the computer model m to generate the data set

$$\big(X_1, m(X_1)\big), \ldots, \big(X_n, m(X_n)\big) \tag{4.59}$$

of computer experiments. This data set can then be used as an approximation of reality to apply a method of uncertainty quantification, for example see Sect. 5.2.6. In the case that the computer model does not fit reality and a sample of experimental data is available, one can also use the method described in Sect. 4.3.7 to construct an improved surrogate model, which then can be used instead of m.

Frequently, we see the situation that the distribution $\mathbf{P}_X$ is unknown and instead only a (rather small) data set of experimental data is available. In the following a method to estimate the probability density function $g\colon \mathbb{R} \to \mathbb{R}$ of Y based on the set of experimental data and a computer model $m\colon \mathbb{R}^d \to \mathbb{R}$ is described. Comparing the probability density function estimated by the method with an estimate of the probability density function based on the computer model enables the detection of model uncertainty. The method is according to Fig. 3.1 applied in the product or system design phase (A).

Mathematical setting

The method described in the following is based on the subsequent mathematical setting: Let $(X, Y), (X_1, Y_1), (X_2, Y_2), \ldots$ be independent and identically distributed random variables with values in $\mathbb{R}^d \times \mathbb{R}$, and let $m : \mathbb{R}^d \to \mathbb{R}$ be a measurable function, i.e. stochastic uncertainty as described in Sect. 1.6. Here Y describes the outcome of an experiment with the technical system, X the (random) inputs of the experiment and m is a computer model of the experiment with the technical system, thus we use $m(X)$ as an approximation of Y. Given the data

$$(X_1, Y_1), \ldots, (X_n, Y_n) \tag{4.60}$$

the aim is to estimate the probability density function $g\colon \mathbb{R} \to \mathbb{R}$ of Y. Note that to apply the method described below, it will be necessary that the evaluation of m on specific input values is possible.

Method

The method described in the following is based on [88], which is an extension of [62, 91]. In the following, we will assume that X is multivariate normally distributed to estimate its distribution and generate a sample based on this estimated input distribution. An overview of methods to generate a data set based on a specific class of distribution can be found in [35].

In order to estimate the parameters of the distribution $\mathbf{P}_X$ of X, a maximum likelihood estimate based on the data (4.60) defined by

$$\hat{\mu} = \frac{1}{n} \sum_{i=1}^{n} X_i \tag{4.61}$$

and

$$\hat{\Sigma} = \left(\frac{1}{n} \sum_{k=1}^{n} (X_k^{(i)} - \hat{\mu}^{(i)})(X_k^{(j)} - \hat{\mu}^{(j)}) \right)_{1 \le i, j \le d} \tag{4.62}$$

is used, where $X_k^{(i)}$ denotes the ith component of the d-dimensional random variable X_k. Alternatively, one can use the unbiased version of (4.62) defined by

$$\hat{\Sigma} = \left(\frac{1}{n-1} \sum_{k=1}^{n} (X_k^{(i)} - \hat{\mu}^{(i)})(X_k^{(j)} - \hat{\mu}^{(j)}) \right)_{1 \le i,j \le d} . \quad (4.63)$$

Given these estimators of the parameters μ and Σ of the input distribution $\mathbf{P}_X$, a sample of size $N_n = N_{n,1} + N_{n,2} \in \mathbb{N}$ can be generated which is independent and multivariate normally distributed with mean $\hat{\mu}$ and covariance $\hat{\Sigma}$. Therefore, we first generate an independent sample $Z_1, \ldots, Z_{N_n}$ of d-dimensional vectors, where for each vector the components are independent and standard normally distributed, and set for every $i = 1, \ldots, N_n$

$$\bar{X}_i = \hat{O} \hat{\Lambda}^{1/2} Z_i + \hat{\mu}, \quad (4.64)$$

where $\hat{O}$ and $\hat{\Lambda}$ are defined by the eigendecomposition

$$\hat{\Sigma} = \hat{O} \hat{\Lambda} \hat{O}^T \quad (4.65)$$

of $\hat{\Sigma}$. Here $\hat{\Lambda} = \mathrm{diag}(\hat{\lambda}_1, \ldots, \hat{\lambda}_d)$ is a diagonal matrix consisting of eigenvalues of $\hat{\Sigma}$ and $\hat{O}$ is an orthogonal matrix whose columns are eigenvectors of $\hat{\Sigma}$.

To estimate an improved surrogate model $\hat{m}_n$ of m we use the method described in Sect. 4.3.7, with a few minor changes. To estimate the surrogate model $\hat{m}_{L_n}$ of m we first generate the data set

$$U_{1,n}, \ldots, U_{L_n,n} \quad (4.66)$$

of size $L_n \in \mathbb{N}$, where the values in this set are independent and uniformly distributed on the centred cube $B_n := [-c \cdot (\log L_n), c \cdot (\log L_n)]^d$, for some suitably chosen constant $c > 0$. This set is then used to construct the surrogate model $\hat{m}_{L_n}$ of m, i.e. we define the estimator by

$$\hat{m}_{L_n}(\cdot) = \arg\min_{f \in \mathcal{F}} \frac{1}{L_n} \sum_{i=1}^{L_n} |f(U_{i,n}) - m(U_{i,n})|^2 + pen_n^2(f), \quad (4.67)$$

where $\mathcal{F}$ is a set of functions. In case the data set (4.60) is sufficiently large, the estimator of the residuals can be defined as estimate (4.56) in Sect. 4.3.7. Otherwise we make a modification of estimate (4.58) from Sect. 4.3.7, where we replace the sample of additional input data by the first $N_{n,1}$ of the generated input data, i.e. the estimator is defined by

$$\hat{m}_n^{\epsilon}(\cdot) \in \arg\min_{f \in \bar{\mathcal{F}}} \frac{w^{(n)}}{n} \sum_{i=1}^{n} |f(X_i) - \epsilon_i|^2 + \frac{(1 - w^{(n)})}{N_{n,1}} \sum_{i=1}^{N_{n,1}} |f(\bar{X}_i)|^2 + pen_n^2(f), \quad (4.68)$$

where $\bar{\mathcal{F}}$ is a set of functions.

The improved surrogate model is constructed as in Sect. 4.3.7, i.e. it is defined by

$$\hat{m}_n(x) = \hat{m}_{L_n}(x) + \hat{m}_n^{\epsilon}(x) \quad (x \in \mathbb{R}). \tag{4.69}$$

In order to estimate the probability density function g of Y, the kernel density estimator of [123, 135] is applied on the sample $\hat{m}_n(\bar{X}_{N_{n,1}+1}), \ldots, \hat{m}_n(\bar{X}_{N_{n,1}+N_{n,2}})$, i.e. it is defined by

$$\hat{g}_{N_{n,2}}(y) = \frac{1}{N_{n,2} \cdot h_{N_{n,2}}} \sum_{i=N_{n,1}+1}^{N_{n,1}+N_{n,2}} K\left(\frac{y - \hat{m}_n(\bar{X}_i)}{h_{N_{n,2}}}\right), \tag{4.70}$$

for some bandwidth $h_{N_{n,2}} > 0$ and some kernel $K\colon \mathbb{R} \to \mathbb{R}$, which is usually chosen as a symmetric and bounded density, e.g. the Gaussian kernel $K(t) = \frac{1}{\sqrt{2\pi}} \exp(-\frac{1}{2}t^2)$.

Application

As an example we consider a lateral vibration attenuation system with piezo-elastic supports. A visualisation of the technical system can be found in [103, Fig. 1].

This system consists of a beam with circular cross-section embedded in two piezo-elastic supports A and B. Support A is used for lateral beam vibration excitation and support B for lateral beam vibration attenuation, as proposed in [61]. The two piezo-elastic supports A and B are located at the beam's end; each consists of one elastic membrane-like spring element made of spring steel, two piezoelectric stack transducers arranged orthogonally to each other and mechanically prestressed with disc springs, as well as the relatively stiff axial extension made of hardened steel that connects the piezoelectric transducers with the beam. For vibration attenuation in support B, optimally tuned electrical shunt circuits are connected to the piezoelectric transducers [63].

Our aim is to quantify uncertainty, i.e. to estimate the probability density function of the maximal amplitude of the vibration occurring in an experiment with this attenuation system. Five parameters vary during the construction of the attenuation system and influence the maximal vibration amplitude: the lateral stiffness $k_{lat,y}$ and $k_{lat,z}$ in direction of y and z, the rotatory stiffness $k_{rot,y}$ and $k_{rot,z}$ in direction of y and z, and the height of the membrane h_x. In our setting these five values are the input X of the experiment with the technical system. A computer model (above denoted by m) is available with which we can compute an approximation $m(X)$ of the maximal vibration amplitude Y to a corresponding input value X. To apply the density estimator (4.70) we measured the corresponding parameters for ten real built systems. As a result we got the data in Table 4.4.

Since the parameters vary in scale, it does not make sense to estimate the surrogate model $\hat{m}_{L_n}$ on $U_{i,n} \sim U([-c \cdot \log(L_n), c \cdot \log(L_n)]^d)$. Instead we rescale the

Table 4.4 Measured data for the ten built systems. The values of $k_{rot,y}$ and $k_{rot,z}$ are given in Nm/rad, the values of $k_{lat,y}$ and $k_{lat,z}$ are given in N/m, the values of h_x are given in m and the values of the maximal vibration amplitude y are given in $\frac{\text{m}}{\text{s}^2}$/V

	1	2	3	4	5	6	7	8	9	10
$k_{rot,y} \times 10^2$	1.31	1.34	1.31	1.23	1.14	1.29	1.35	1.28	1.04	1.20
$k_{rot,z} \times 10^2$	1.31	1.28	1.43	1.25	1.30	1.34	1.22	1.16	1.18	1.11
$k_{lat,y} \times 10^7$	3.27	3.28	3.35	3.29	3.22	3.26	3.19	3.54	3.21	3.42
$k_{lat,z} \times 10^7$	3.07	3.22	3.29	3.25	3.30	3.18	3.16	3.51	3.37	3.44
$h_x \times 10^{-4}$	6.79	6.77	6.82	6.80	6.79	6.76	6.81	6.74	6.68	6.84
$y \times 10^1$	1.45	1.42	1.44	1.42	1.43	1.35	1.47	1.32	1.31	1.63

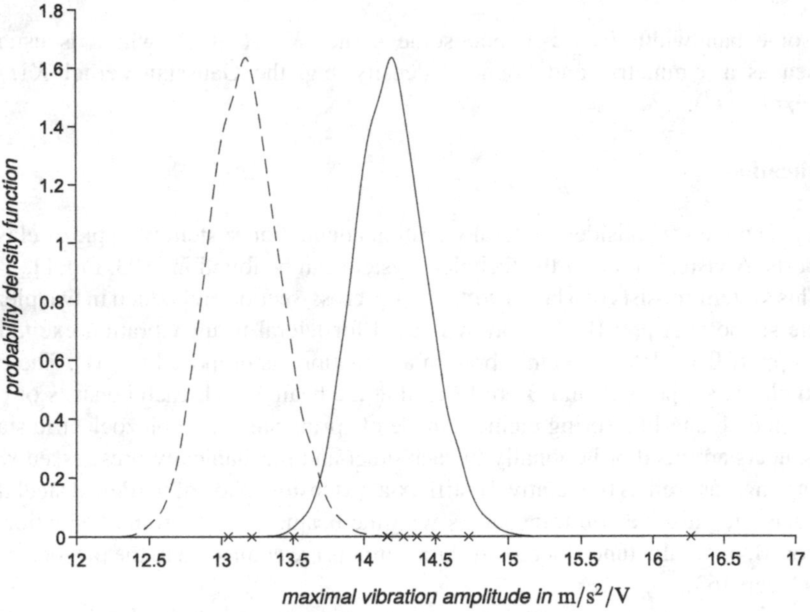

Fig. 4.38 Density estimator based on surrogate model (dashed line), density estimator based on improved surrogate model (solid line) and as reference the data $Y_1, \ldots, Y_n$ indicated on the x axis

components of $U_{i,n}$ so that for each component $U_{i,n}^{(j)} \sim U([\hat{\mu}^{(j)} - 2 \cdot \sqrt{\hat{\sigma}_{jj}}, \hat{\mu}^{(j)} + 2 \cdot \sqrt{\hat{\sigma}_{jj}}])$ holds.

We apply the density estimator (4.70) to the data and obtain as a result Fig. 4.38, where we compare it to a density estimator based on the surrogate model $\hat{m}_{L_n}$. The result shows that the estimator based on the improved surrogate model fits the data better, i.e. the improved surrogate model is able to predict the reality more accurately than the surrogate model; hence the density estimator is more accurate. Model uncertainty occurs which leads to the conclusion that the computer model does not fit reality.

4.4 Representation and Visualisation of Uncertainty

Moritz Weber, Georg Staudter, and Reiner Anderl

Product development is a knowledge-intensive process where, despite its uncertainty [42, 95, 106], designers define what the product has to achieve in the physical domain and how this has to be accomplished, potentially, according to customer specifications, cf. Sect. 1.2. As to the definitions it is determinated which tests are necessary, how and in which quantity the product must be manufactured and at what time it needs maintenance. Uncertainty may have negative effects on decisions, which can lead to over-sizing, unfulfilled customer demands and unforeseen failures [74]. As pointed out by Anderl et al. [6], software applications used in the engineering context rarely consider uncertainty quantification, which partly explains the lack of awareness about uncertainty by designers. This section introduces our approach to overcome this issue by visualising uncertainty and its consequences for developers and the required digital representation of knowledge about uncertainty. This aims to support engineers and designers to better understand uncertainty regarding product and process properties, and thus helps the engineers to recognise, evaluate and analyse uncertainty in their designs (cf. Sect. 1.7).

A three-layer architecture that includes representation, presentation and visualisation of uncertainty [6] is the basis for the approach introduced in this section. The representation layer is dedicated to the digital representation of data uncertainty with all of its subtypes (cf. Sect. 2.1). For this purpose, it uses an ontology-based information model, see Sect. 4.4.1. The presentation layer serves as an auxiliary layer for the visualisation and creates use-case defined objects, which serve as an intermediate representation for visualising uncertainty. The concept of uncertainty cloud (uCloud) enables the tangible presentation of geometric tolerance uncertainty. To this end, it creates an Euclidean space that describes the probability distribution of a body existence of a physical part [6]. The visualisation layer uses the instances of the presentation layer and maps its objects into the functionality of computer graphics, see Sect. 4.4.2. The MADFS (see Sect. 3.6.1) serves as an application example for the outlined approach and its methods, see Sect. 4.4.3.

4.4.1 Ontology-Based Information Model

Moritz Weber and Reiner Anderl

For the purpose of identifiying the uncertainty in early stages of the product development process and thus enabling uncertainty management, information about all product life cycle phases is necessary. Therefore, a suitable model is required to digitally represent information about uncertainty. The ontology-based approach offers the opportunity of an appropriate conceptual space based on scientific knowledge about uncertainty. In addition, it provides high semantic value.

Ontologies are defined as "formal models of selected aspects of the real world" [67]. They digitally represent objects or assets and their relations for the use in advanced applications of information and communication technology. Ontologies are designed for the specification of semantics of higher-order to enable knowledge representation. Ontologies use triples to formalise information. Each triple comprises subject, predicate, and object.

For authoring the ontology, we use a variant of the Web Ontology Language in version 2 (OWL 2). OWL 2 is standardised by the World Wide Web Consortium (W3C) and comprises three language variants of various expressive power. Here, we chose the variant OWL Description Logics (DL), since it comes with the greatest possible expressive power, while maintaining the computational completeness and decidability necessary for inference and validity checking. OWL 2 supports serialisation using the Extensible Markup Language (XML), which enables the easy exchange of information. A further advantage of OWL 2 for dealing with uncertainty and especially ignorance is the Open World Assumption made by OWL 2, so a statement can be true irrespective of whether it is known to be true [64].

Since ontologies are based on description logic, so-called inference machines can infer new knowledge based on already known information. In addition, they can be used to verify the integrity of the knowledge [7, 20]. An ontology comprises an Assertional Box (A-Box) and a Terminational Box (T-Box). The T-Box formalises the knowledge about the concepts—also called classes—of the described domain, whereas the A-Box contains the knowledge about the specific instances of these concepts in the domain.

This section describes the information model used for the exchange and visualisation in load-bearing systems and therefore forms the fundament for the methods described in Sect. 4.4.2 and Sect. 4.4.3 and is therefore a contribution to the modelling of uncertainty in information technology. Figure 4.39 provides an overview of the ontology-based information model named Collaborative Ontology-based Property Engineering System (COPE), which we have developed with its major components [145]. It incorporates the property-driven development approach [162] and comprises the three life cycle phases development, production, and usage (cf. Sects. 3.2 and 1.7). Data uncertainty is characterised as uncertain property value and uncertain relationship that specifies the effect of the uncertain value. The product model and a process model define the context of these two components. A major approach of the ontology is property and process classification. The following paragraphs describe four of the major components of the ontology-based information model "COPE" in more detail.

Uncertain property value

For the representation of the uncertain property values, we have developed a partial model referred to as 'Uncertainty Data Type' (UDT) [146]. Its aim is to represent digitally the uncertainty of product and process properties to improve the interchangeability of these data types. The approach is based on the digital uncertainty

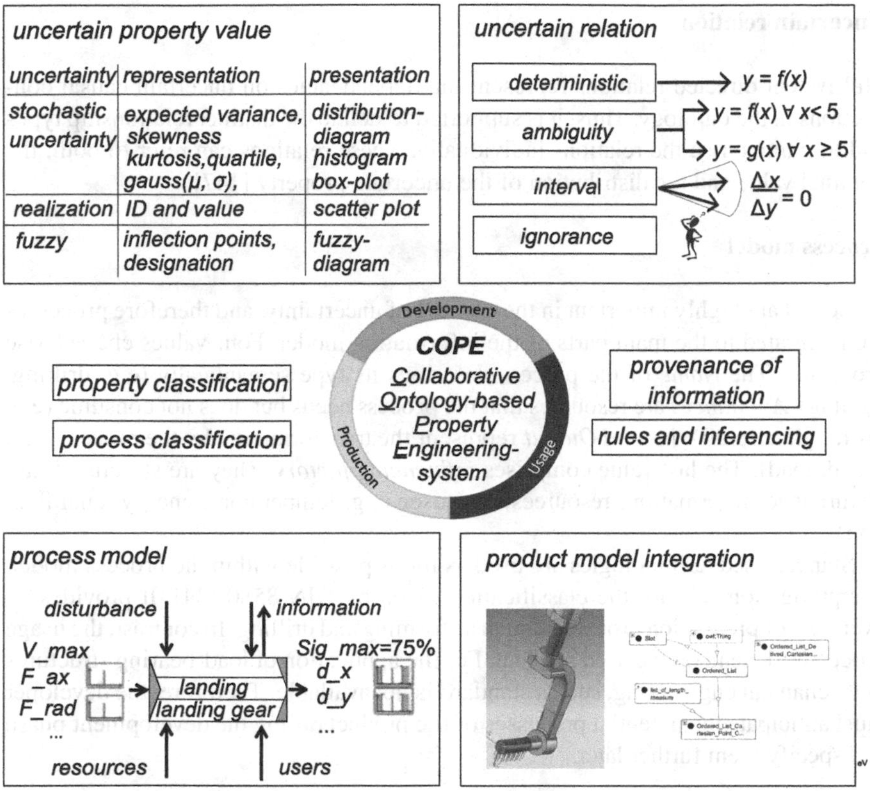

Fig. 4.39 Overview of COPE [145]

representation introduced and discussed in Sect. 2.1, and it covers all three types of data uncertainty described there.

Uncertain relation

Multivalent directed relations represent the dependencies on uncertain causal connections in the ontology. Thus, it is supported to define the distinct relationship types and to parametrise the relations individually. These relations can refer to both, the nominal value and the distribution of the uncertain property [147].

Process model

Processes are highly important in the context of uncertainty, and therefore processes are integrated to the main parts of the information model. Four values characterise processes. The *Name* of the process describes its type semantically (e.g. drilling, landing). *Appliances* are resources that the process needs but does not consume (e.g. drill, light aircraft). *In- and Output* represent the transformation by the process (e.g. speed, load). The last value comprises *influencing factors*. They are structured into disturbance, information, resources, and user (e.g. temperature, energy, qualification).

Standardised terminologies are used as far as possible within the process model. For production, we use the classification given by DIN 8580 [34]. It provides an overview of production processes, such as forming and drilling. In contrast, the usage processes depend on the used product. For the application of load-bearing structures in mechanical engineering, such a standard is not available. Therefore, the developer must anticipate the potential processes of the product during the development phase and specify them further later.

Product model integration

The information model is integrated into the product model for two reasons. It references the uncertainty in the integrated product model and is used to assign uncertain property values to parts of the product model. This approach enables unique identification throughout all life cycle phases and improves the usability of uncertainty information. Furthermore, the integration into the product model provides an appropriate basis for the visualisation of the uncertainty information (see Sect. 4.4.2) in the respective product context. The item references entities of the Boundary Representation (BRep) to localise the uncertainty information.

The ontology-based information model extends the product model based on ISO standard 10303-108 for parameterisation and geometric boundary conditions of explicit product models of parts and assemblies. In this context, the ontology constitutes the T-Box. Specific CAD models and attached data constitutes the A-Box. The definition of uncertain geometric entities results in a geometrically underdetermined state in the A-Box. Systems of equations cannot further characterise the relations between the geometric entities without reducing the degrees of freedom of the geometric entities and thereby removing the information of the geometric variation. The A-box describes geometry and topology of the geometric entities and

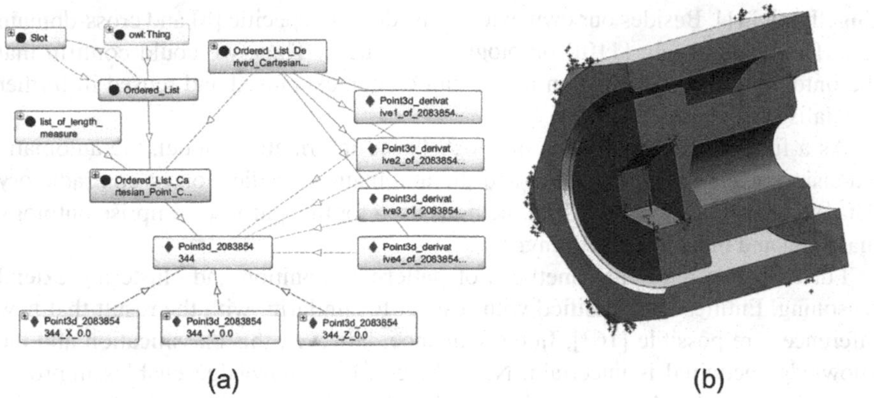

Fig. 4.40 **a** Representation of points in the information model and **b** visualisation in the CAD-model

allows the characterisation of the solutions of the system of equations, which are algebraically identical but geometrically different.

For the processing of time-variant uncertainty information (see Sect. 3.4), we have developed a concept with the corresponding implementation for the bidirectional connection of a CAD system (Siemens NX), and a numerical linear equation system solver (Matlab). The ontology serves as a mediator between the two systems so that the results of the ontology queries are applied in the CAD system, as well as in Matlab. The representation of time-variant uncertainty extends the A- and T-Box representing design variants in the parametric product model. Furthermore, time-variant changes in the geometry of assembly components are also represented [168]. Figure 4.40a shows a graph-based visualisation of a small section of the information model. It shows individual points and their connections. Circular symbols indicate concepts, and diamonds indicate specific instances. This example depicts a point with its three Cartesian coordinates and four points derived from it. The derived points represent possible corner points after production and after consideration of the uncertainty. The figure equally shows a small part of the class hierarchy. Figure 4.40b is a visualisation of the CAD-Model and a larger quantity of derived points for selected vertices. A designer interprets the selected geometry and decides whether the boundary conditions meet the requirements.

The automated generation of the T-Box is based on the software OntoStep developed by the National Institute of Standards and Technology (NIST) [96]. This software tool has been extended for the extraction of product parameters for the generation of the instances. In this way, data sets concerning uncertainty and its distributions are integrated.

The ontology-based information model was also adapted for a specific domain [170]. Here, we extended the ontology for the application scenario Uncertainty Mode and Effects Analysis (UMEA) for human effects in aerospace. UMEA is an extension of the failure mode and effects analysis (FMEA), and was proposed by

Engelhardt [41]. Besides our own extensions, domain-specific [8] and cross-domain (e.g. Dolce UltraLight [110]) ontologies were used. Thus, we could confirm that the ontology-based information model can be contextualised and reused in further specialised use-cases.

As a further extension of the ontology-based information model, the automatic extension of the knowledge base and the automatic classification of contradictory data were taken into account. The methods used for this purpose comprise ontology matching and inductive reasoning.

For inductive reasoning, methods of pattern recognition and clustering extend reasoning. Entities are classified with respect to similarity with the result that new inferences are possible [163]. In consequence, however, this classification and the knowledge acquired is uncertain. Nevertheless, this knowledge enables improvement of product development decisions. Therefore, the designer is provided with the inferences including a measure of confidence. Ontology matching is applied to integrate knowledge from heterogeneous and distributed sources automatically. Thereby, analogies between two or more ontologies need to be identified and used to join the knowledge.

Domain-specific rules for ontology matching and inductive reasoning of axioms of geometric relationships are the core of the integration of both methods in the ontology-based information model. This enables the integration of methods to detect and control data and model uncertainty into the ontology-based information model.

We presented an ontology-based information model that combines domain-specific knowledge to support product developers. In addition, it provides a basis for further analyses and the visualisation of the effects of uncertainties. We chose an ontology-based information model that is based on description logics and OWL 2. Thereby, the advantages of an expressive, descriptive language are combined with those of decisive formal semantics. In contrast to alternative forms of data representation, such as databases, ontologies not only allow data queries but automated classification, validation of the integrity of data, and extension of the knowledge base by inference. Furthermore, due to the high semantic value, knowledge interpretation improves, and the exchange of information is simplified. The ontology-based information model offers a functionality to store not only time-invariant but time-variant information about uncertainty, too. Furthermore, instances can be generated, and ontology matching and inductive reasoning can extend the knowledge automatically. The use of an ontology-based approach allows to extend the information model further. The integration into a digital twin, for example, can enlarge the knowledge base and thus increase the quality of product development decisions, see Sect. 4.4.3.

4.4.2 Visualisation of Geometric Uncertainty in CAD Systems

Georg Staudter and Reiner Anderl

The visualisation of geometric uncertainty comprises the graphical presentation of the statistical distribution of data obtained from measurements conducted during production and usage, by utilising the functionality of computer graphics [47] to generate an appropriate appearance of uncertainty [6, 84, 154]. The following section introduces our approach for the visualisation of the geometric uncertainty in CAD (computer-aided design) systems, focusing on stochastic data uncertainty associated with geometrical model parameters, see Sect. 2.1. The visualisation of uncertainty is part of the middle layer of the framework of mastering uncertainty introduced in Sect. 1.7 and thus, an important element within the analysis, quantification and evaluation of uncertainty in mechanical engineering.

Despite the fact that the consideration of uncertainty associated with geometry is crucial during the design process, today's CAD systems provide only a limited design-oriented view with functionalities to specify nominal geometry and geometric tolerances. There is still a lack of functionalities for the visualisation of geometric uncertainty [6]. The effect of the different geometric tolerances on the part (e.g. shape, dimensions, features, locations) cannot be graphically visualised either. Advanced tools, such as Computer-Aided Tolerancing (CAT), focus mainly on geometric dimensioning and some basic stack-up analysis, but do not provide harmonised solutions for the graphical visualisation of tolerance and uncertainty associated with measurement. Therefore, there is a need to integrate geometric uncertainty in the geometric product model in order to explicitly depict uncertainty [6, 145].

Our approach for the visualisation of geometric uncertainty focuses on the integration of information about uncertainty and its correlations into CAD-systems via an ontology-based information model. Therefore, the geometric product model representing the 3D-CAD model is decomposed into appropriate elements, such as features and boundary representation elements (BREP-elements), which enable the association of uncertainty. The hierarchical structure of the product model, as well as its uncertainty information, are mapped into the ontology-based information model in terms of a product and process representation, see Sect. 4.4.1. The mapping assures that the product model can be transformed into an ontology-based representation and vice versa [6, 116, 145].

When integrating information about geometric uncertainty into the product model, it is necessary to do both, specifying uncertainty explicitly and deriving a presentation appropriate for visualisation. For the presentation of geometric uncertainty associated with tolerance, we have developed the concept of the uncertainty cloud or "uCloud". The uCloud concept creates a three-dimensional space that visualises the probability distribution of the physical part surface location. The uCloud is generated by a set-theoretical operation, which compromises two volumes, each representing a maximum, respectively minimum value of a particular geometrical property [6].

$$V_{\text{maxBody}} = \text{Volume with maximum geometric property}$$
$$V_{\text{minBody}} = \text{Volume with minimum geometric property}$$
$$V_{\text{uCloud}} = V_{\text{maxBody}} \backslash V_{\text{minBody}} = \{x | (x \in V_{\text{maxBody}}) \bigwedge (x \notin V_{\text{minBody}})\}$$

The resulting uCloud element is then used to apply visualisation techniques for geometric tolerances. Conceptually, visualisation techniques are divided into the domains of (i) graphical, (ii) symbolic, (iii) structural and (iv) verbal visualisation [6].

(i) Graphical visualisation uses functionality of computer graphics, such as colour, colour intensity, transparency or coloured patterns. To attach the semantics of uncertainty to the graphical appearance of uncertainty, a cross-reference table is needed [6]. (ii) Symbolic visualisation associates predefined symbols to presentation objects, and it enables the attachment of uncertainty information. In the domain of (iii) structural visualisation presentation objects are mapped onto structures, such as lists, tables, tree- and graph-structures. (iv) Verbal visualisation expresses uncertainty lexically and creates a textual output using the ontology approach [146, 147]. Figure 4.41 shows graphical visualisation techniques for uncertain properties and their uncertain value description with respect to the Uncertainty Data Type (UDT).

With the concept of uCloud combined with visualisation techniques for uncertain geometric properties, we provide an approach for the static visualisation of time-invariant uncertainty on individual parts by creating a cloud-like space, which contains the part surface of the real product. The result is a visualisation through transparently shaded offset bodies, which are linked to the corresponding UDTs (Uncertainty Data Type). Additionally, uncertainty information related to product structures such as lists, tables or tree structures can be displayed.

Figure 4.42 shows the uCloud approach for a geometric deviation of the diameter of a shaft with different types of uncertain value descriptions resulting from manufacturing process specific tolerances. Figure 4.42a illustrates the uCloud for an interval, or more specifically for a geometric tolerance, where an upper deviation, a nominal diameter and a lower deviation are graphically visualised [145]. With respect to this information, the diameter of the manufactured surface of the shaft lies within the transparent shaded offset. Since the geometric deviations are small in relation to the dimension of the shaft, the technique of enlarged detail, known from technical drawings, is used. This approach is also available for visualising stochastic tolerance data. Therefore, a sigma level (e.g. Six Sigma) or a confidence interval of the distribution function (e.g. 99.9997%) is selected, depending on the available input data. Both sigma levels and the expected value result in three characteristic points, as indicated by an additional specific symbol [145].

Figure 4.42b shows the visualisation of the geometric deviation of the diameter of a shaft with stochastic uncertainty information regarding its geometric tolerance. With a given histogram as an input, the colour density range is mapped onto the minimum and maximum frequency, and it is visualised by elements which correspond to the classes of the histogram. The element colour density corresponds to the probability distribution given by the representation of the histogram. In the case of a given

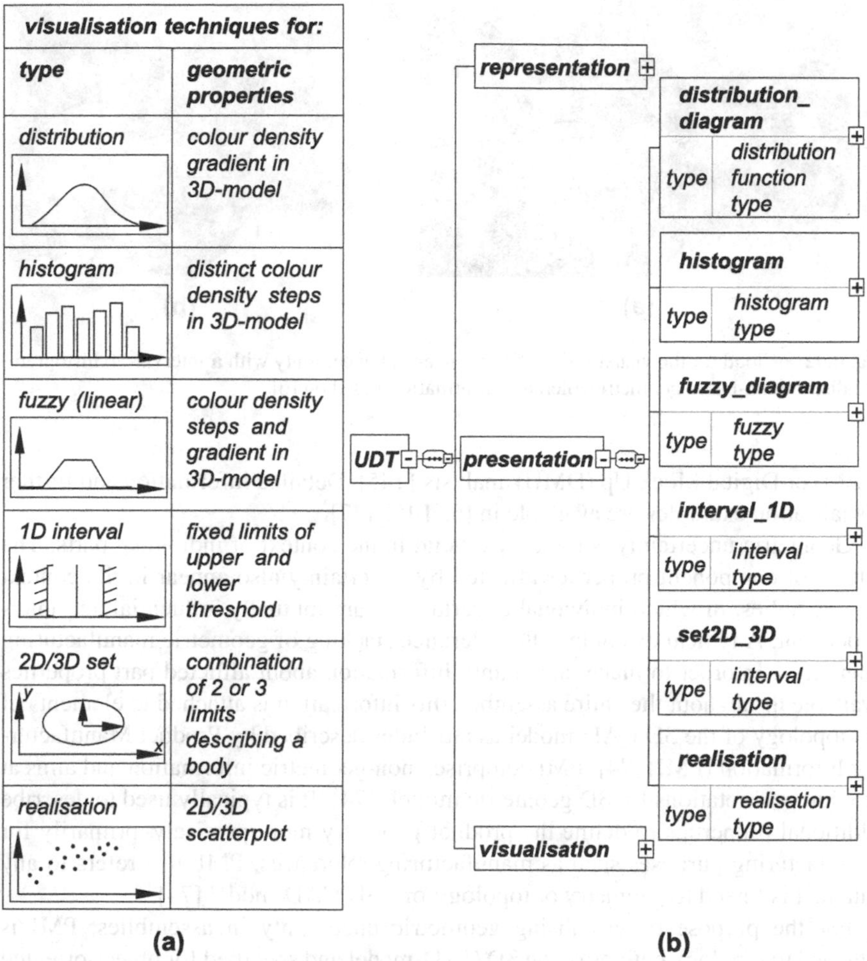

Fig. 4.41 Graphical visualisation techniques for different types of uncertainty (**a**) in relation to the UDT (**b**)

distribution function, the uCloud element colour density is mapped to the probability of the function [6].

Each uCloud element has to be generated respectively to the given type of the uncertainty value description. With the approach of a sectional view we enable engineers and designers to interpret different influences, which occur in the product life cycle, such as imperfect manufacturing, wear and corrosion. Engineers are able to interpret uncertainty occurring within a single part or an assembly. Furthermore, the uCloud concept complements the 3D data model without manipulating the idealised description of the geometry allowing its further usage for e.g. Finite Element Method

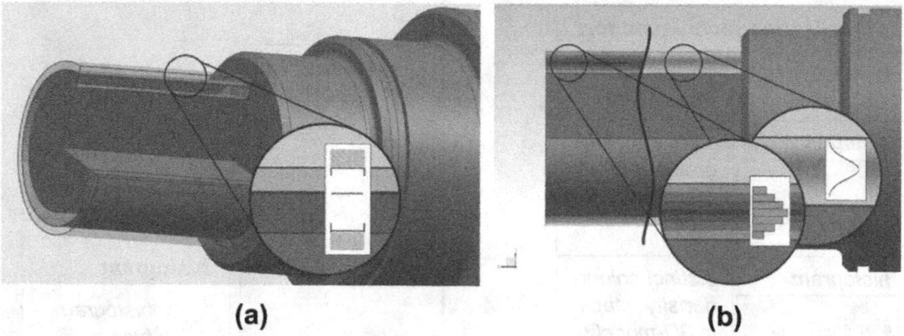

Fig. 4.42 uCloud for the visualisation of time-invariant uncertainty with **a** interval-, **b** histogram- and distribution-based geometric tolerance information based on [6]

(FEM) or Digital Mock Up (DMU) analysis [145]. Detailed information and further visualisation examples are available in [6, 145, 147].

Geometric uncertainty is not only crucial in the context of individual parts. The effects of component properties affected by uncertainty also appear in the context of assemblies, in which individual uncertainties are mutually influential and interdependent. A typical example is the tolerance stacking of geometric manufacturing tolerances. In order to make uncertainty information about afflicted part properties available throughout the entire assembly, this information is attached to elements of the topology of the 3D CAD-model as attributes described by Product Manufacturing Information (PMI) [74]. PMI comprises non-geometric information and aims at providing annotations for 3D geometric models [74]. It is typically used to describe additional properties to define the product geometry more precisely, primarily for manufacturing purposes such as manufacturing tolerances. PMI also refers to any data that is linked to geometry or topology of a 3D CAD-model [74].

For the purpose of visualising geometric uncertainty in assemblies, PMI is attached to topology entities of the 3D CAD-model and specified for object-oriented implementation. In this context, the target topological entities for referencing PMI are body-, face-, and edge-attributes, as they are important for assembly constraints. The body-attribute serves as an individual part specific information carrier and contains all information from face- and edge-attributes, that belong to an entire body. Through the configuration of individual parts via assembly constraints, the body-attributes associate corresponding parts with one another and enable bidirectional PMI exchange [74]. One individual part contains exactly one body-attribute but multiple face- and edge-attributes; these comprise uncertainty information mapped into a specific PMI being associated with the afflicted object property. The ontology-based information model provides the informational context, which is linked to the different attributes [74].

Object attributes for edges refer to the information for the mathematical description of the geometrical instance of the edge in a three-dimensional space. Object attributes for faces reference information for the corresponding surface, such as:

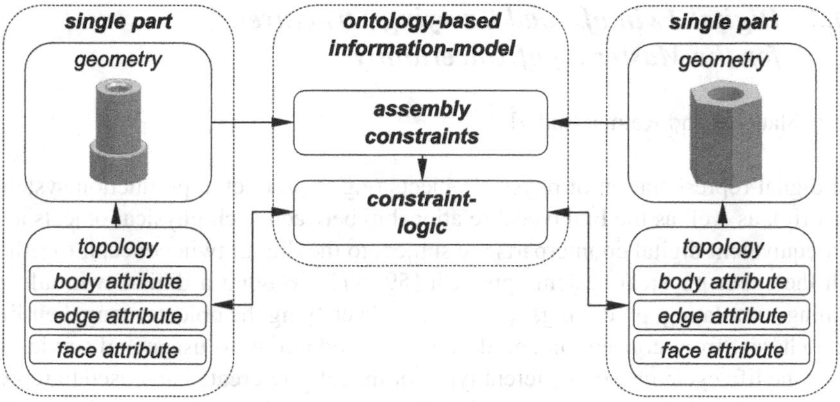

Fig. 4.43 Informational context for the connection of single parts with assembly constraints based on [74]

direction of the normal surface, radius and central axis of cylindrical surfaces, surface contents as well as the mathematical description of the surface and the uncertainty type in relation to the geometric deviation [74]. Thus, the geometric deviation in the x-, y- and z-direction in a three-dimensional space is described. Object attributes for bodies collect the information from attributes attached to a part's surfaces and edges to provide a complete attribute bundle for the neighbouring parts. In order to reference individual parts within an assembly, the designer assigns assembly constraints, referring to different reference elements. Figure 4.43 illustrates the interlinking between attributes, individual parts, assembly constrains and configuration logic.

Through the referencing of two individual parts using an assembly constraint, the body-, edge- and face-attributes of the individual parts are automatically linked bidirectionally with one another [74]. As a result, PMI containing geometric uncertainty are referenced to the neighbouring part. The configuration logic for assembly constraints defines how the geometric uncertainty is being propagated into the neighbouring parts. With the help of object attributes and their internal processing in the ontology-based information model, the concept of uCloud can be extended from single parts to assemblies. Combined uncertainty zones of individual parts within an assembly are visualised with respect to an absolutely positioned, freely selectable individual part. The validation of the concept, applied to the MAFDS (see Sect. 3.6.1), is outlined in Sect. 4.4.3, see [74].

4.4.3 Digital Twin of Load Carrying Structures for the Mastering of Uncertainty

Georg Staudter and Reiner Anderl

The digital representation of physical objects (e.g. a product, a production system, a test rig), as well as the biunivocal relationship between such physical objects and their equivalent digital counterparts are subject to the digital twin concept, together with the cyber-physical system approach [59, 115]. Having a digital twin allows defining, simulating, predicting, optimising and verifying the objects along their life cycle phases, from conception and design, via production, to usage and servicing. Along the life cycle phases, different types of models are created and used to represent physical objects, e.g. system models, functional models, 3D geometric models, multiphysics models, manufacturing models, and usage models, see Sect. 1.3. These models constitute the digital twin.

The transfer of data from the physical domain to the digital domain is a key approach to generate the digital twin. In the widest sense, a digital twin requires to implement a data flow where data, acquired from testing, production, maintenance and operation are integrated into a digital domain to support such models and assist in predictive and decision-making processes, see Sect. 1.4. This section addresses the challenges of mastering uncertainty associated with the respective data (Sect. 2.1) and models (Sect. 2.2 and introduces our approach to the visualisation of data-induced conflicts (Sect. 4.2) for uncertainty identification (Sect. 3.3) in the digital twin context.

The benefits derived from the digital twin implementation, depend on incorporating data from the physical domain into the digital domain. In the physical domain, data acquisition requires measuring physical magnitudes. The result of measurement should be a threefold structure: nominal value of the magnitude, measurement unit, and uncertainty of the measurement [18]. The most widely used data quality dimensions are: accuracy, completeness, currency, and consistency [15]. Within the accuracy dimension, the uncertainty of a measured magnitude is a significant contributor to the indicator of the data validity, see Sect. 2.1. However, literature shows that explicitness of the uncertainty of measured data is still a challenge. There is a lack of bidirectional semantic harmonisation of the uncertainty representation in the standards used to transfer data, both from the digital domain to the physical domain and vice versa [134].

Geometric data obtained from the physical domain are used to recreate 3D geometric models of the physical objects. In the literature, these models are named with the terms as-built, as-fabricated and as-manufactured [28, 59, 153]. The aim of having an as-manufactured 3D model is to represent the geometric deviations caused by the manufacturing processes and use the model representation to perform simulations that previously were executed using an as-designed 3D model. Consequently, it is necessary to explicitly represent the uncertainty of the reconstructed as-manufactured 3D model, see Sect. 2.2

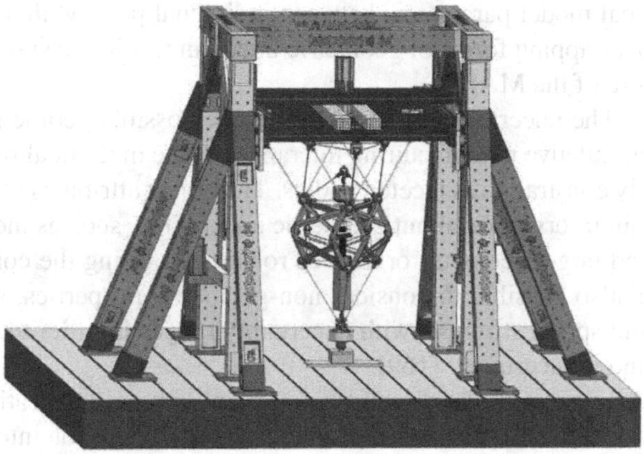

Fig. 4.44 Virtual Demonstrator of the MAFDS from Sect. 3.6 in Siemens NX

Measurements are necessary to capture the main geometrical dimensions of the physical components. We used measurement results with their corresponding uncertainty to create a 3D model of the physical test rig MAFDS (see Sect. 3.6.1) referred to as the virtual demonstrator. The virtual demonstrator includes material and physical properties in addition to part geometries and product structures. It consists of a Multi-Body-Simulation (MBS) model with a set of virtual sensors to simulate the functional behaviour while visualising the behaviour and movements of the test rig. The dynamic analysis allows the determination of velocities, accelerations and displacements of the moving components, as well as the reaction forces. Figure 4.44 shows the implementation of the virtual demonstrator in Siemens CAD-System NX12.

Internally, the moving components, joints and drivers are converted into a mathematical system of differential equations, which can then be solved to determine the desired quantities. This can be performed using different solvers which depend on the respective CAD-system and are mostly proprietary. Additionally, moving components are simplified to their mass, inertia properties and geometrical dimensions, while deformation properties are neglected. This leads to a major challenge for the quantification of the respective model uncertainty [5].

In the context of a digital twin, another effect that must be taken into account for MBS is the influence of geometric tolerances of the physical component, see Sect. 4.4.2. Since such effects are often not taken into account, it may appear, for instance, that an interference fit situation occurs in the simulation, when in reality there is a slight clearance in the joint or vice versa. Therefore, it is not only necessary to explicitly represent the uncertainty of the reconstructed as-manufactured 3D model, but to consider effects, such as the classical tolerance stacking of geometric manufacturing tolerances in assemblies. With the help of object attributes and their internal processing in the ontology-based information model, as outlined in Sect. 4.4.2, the concept of uCloud allows the visualisation of combined uncertainty zones of individual parts within an assembly.

Figure 4.45 shows the visualisation of an uncertainty zone by the bidirectional exchange of information about stochastic data uncertainty associated with geometrical model parameters between individual parts with a maximum deviation due to overlapping forms of geometric uncertainties in the subassembly of the upper structure of the MAFDS.

The uncertainty zone visualises the possible geometric deviation resulting from cumulative manufacturing tolerances of the individual parts in the context of assembly constraints as faceted bodies. The object attributes of the topology elements contain information about geometric uncertainty, such as the divergence between actual and target geometry or surface roughness. Using the concept of object attributes, it is also possible to consider non-geometric properties, such as damping properties and spring stiffness, with uncertainty ranges in order to simulate product behaviour under uncertainty [169].

In general, the digital twin concept aims at integrating measured data acquired from testing, production, maintenance and operation into the digital domain to assist in decision-making processes. These processes depend strongly on the quality of the underlying information base. The data to quantify and evaluate a system response is typically gathered by a variety of sensors, see Sect. 1.4. Because of the complexity of the context, several data streams must be integrated, and possible data-induced conflict situations must be identified, see Sect. 4.2. To identify a possible erroneous sensor behaviour, values of interest are observed redundantly in the physical domain. The objective is to avoid situations where possible errors remain unnoticed, see Sect. 4.2.1. Redundancy increases the availability of information and thus, contributes to the verification of the data. On the other hand, if several sources provide inconsistent or conflicting data, a defective interpretation may occur. Therefore, it is necessary to provide methods for explicitly representing and visualising data-induced conflicts in the digital twin context, see Sect. 4.2.3.

Section 4.2.1 presents a methodology for the identification of data-induced conflicts and the interpretation of conflicting sensor data. The approach is based on differentiating data sources, such as soft sensors, into models and sensors, by spanning the investigation from the redundant observation of a single value to the interconnec-

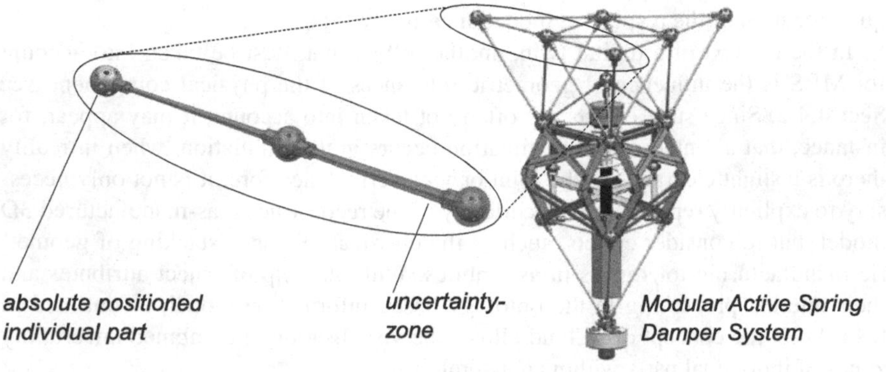

Fig. 4.45 Visualisation of geometric deviation of individual parts in assemblies based on [74]

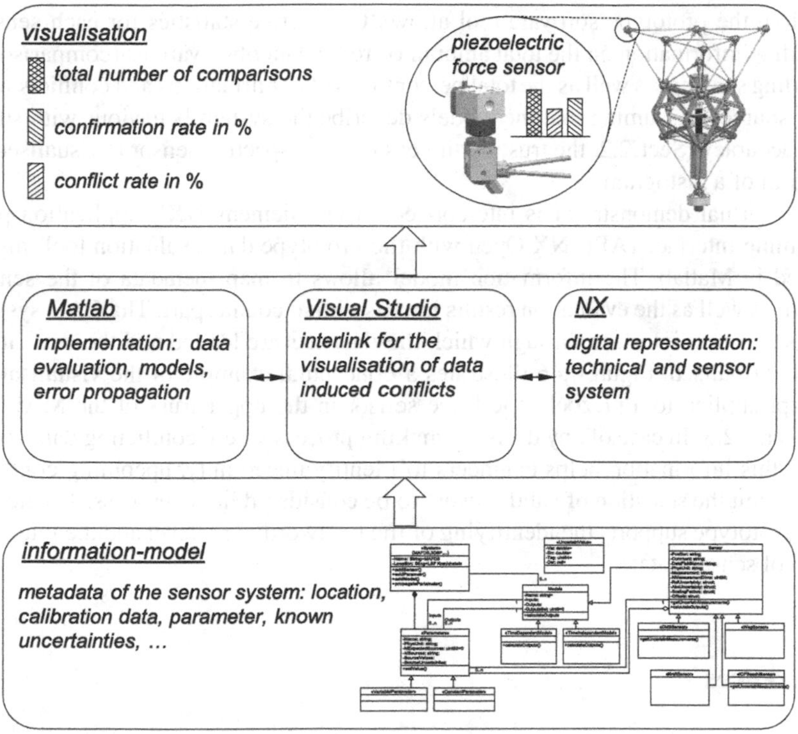

Fig. 4.46 Concept for the visualisation of data-induced conflicts in Siemens NX

tion between models and sensors throughout a technical system. Here, the proposed methodology is applied to the virtual domain for the purpose of the visualisation of data-induced conflicts in CAD systems, see Fig. 4.46.

In addition to information on components, such as dimensions and parameters with their respective uncertainty, the information model represents the underlying sensor system as well as models for generating analytical redundancy. Each sensor of a physical test rig is represented by its metadata, including identification data, calibration data, known uncertainty, as well as relative and absolute placement within the test rig. Soft sensors are used to convert the measured data values and to generate analytical redundancy, see Sect. 1.4. In order to represent information about the models, system knowledge, such as symmetry characteristics and orientation of the system components, are integrated into the information model.

The information model is the basis for the data evaluation and is as such implemented in Matlab. The prototype development comprises also methods and classes for the propagation and calculation of the resulting uncertainty. The result is a software tool for the automatic detection of data-induced conflicts, see Sect. 4.2.3. The output of the system provides detailed information on data-induced conflicts, as well as on the interconnections between models and sensors throughout the system. In addition, the prototype software tool allows to generate statistics for each sensor, providing information on the total amount of redundant observations (comparisons) including sensing, as well as the total percentages of confirmations and conflicts with other sources. Assuming that the models describe the system behaviour with sufficient accuracy (Sect. 2.2, the trustworthiness of the respective sensor is visualised in the form of a histogram.

The virtual demonstrator is interconnected via Siemens NX's application programming interface (API) NX Open with the prototype data evaluation tool implemented in Matlab. The information model allows to map metadata of the sensor system as well as the evaluation results into its virtual counterpart. The CAD system serves as a user interface through which the data sets are loaded and the evaluation results visualised. Figure 4.46 illustrates a conceptual example of the visualisation concept applied to a piezoelectric force sensor in the upper truss of the MAFDS, see Sect. 4.2.3. In case of any decision-making process where conflicting data could occur, this information helps engineers to identify uncertainty, upcoming conflicts and to limit the selection of valid sensors to be considered in the process. The developed prototype supports the identifying of the trustworthiness level and the interpretation of sensor data.

References

1. Abele E, Geßner F (2018) Spanungsquerschnittmodell zum Gewindebohren: Modellierung der Auswirkung von Unsicherheit auf den Spanungsquerschnitt beim Gewindebohren. wt Werkstattstechnik online 108(1–2):2–6
2. Abele E, Hauer T, Haydn M, Bölling C (2011) Reduzierte Unsicherheit bei der Bohrungsfeinbearbeitung - Neue Erkenntnisse zum Vorbohrungseinfluss auf den Reibprozess. Werkstattstechnik online 101(1–2):81–87
3. Alexanderian A, Petra N, Stadler G, Ghattas O (2014) A-optimal design of experiments for infinite-dimensional Bayesian linear inverse problems with regularized ℓ_0-sparsification. SIAM J Sci Comput 36(5), A2122–A2148. https://doi.org/10.1137/130933381
4. Alexanderian A, Petra N, Stadler G, Ghattas O (2016) A fast and scalable method for A-optimal design of experiments for infinite-dimensional Bayesian nonlinear inverse problems. SIAM Journal on Scientific Computing 38(1), A243–A272
5. Anderl R, Binde P (2017) Simulationen mit NX/Simcenter 3D: Kinematik, FEM, CFD, EM und Datenmanagement. Mit zahlreichen Beispielen für NX 11, 4th edn. Carl Hanser Verlag. https://books.google.de/books?id=QDqZDgAAQBAJ
6. Anderl R, Maurer M, Rollmann T, Sprenger A (2013) Representation, presentation and visualization of uncertainty. In: CIRP design 2012. Springer, pp 257–266
7. Antoniou G, Franconi E, van Harmelen FF (2005) Introduction to semantic web ontology languages. In: Eisinger N, Maluszynski J (eds) Reasoning web: first international summer school, vol 3564. Lecture notes in computer science. Springer, Berlin, pp 1–21. https://doi.org/10.1007/11526988_1
8. Ast M, Glas M, Roehm T (2013) Creating an ontology for aircraft design: an experience report about development process and the resulting ontology. In: Deutsche Gesellschaft für Luft- und Raumfahrt – Lilienthal-Oberth e.V. (ed) Publikationen zum DLRK 2013, pp 1–11
9. Atamturktur S, Hemez FM, Laman JA (2012) Uncertainty quantification in model verification and validation as applied to large scale historic masonry monuments. Engineering Structures 43:221–234. https://doi.org/10.1016/j.engstruct.2012.05.027

10. Bard Y (1974) Nonlinear Parameter Estimation. Academic Press, New York
11. Batterbee DC, Sims ND, Plummer AR (2005) Hardware-in-the-loop simulation of a vibration isolator incorporating magnetorheological fluid damping. In: ECCOMAS thematic conference on smart structures and materials
12. Baydin AG, Pearlmutter BA, Radul AA, Siskind JM (2018) Automatic differentiation in machine learning: A survey. The Journal of Machine Learning Research 18(153):1–43
13. Bayes T (1763) An essay towards solving a problem in the doctrine of chances. Philos Trans R Soc Lond 53:370–418. https://doi.org/10.1098/rstl.1763.0053
14. Beale E (1960) Confidence regions in non-linear estimation. J Roy Stat Soc: Ser B (Methodol) 22(1):41–76
15. Bertino E (2015) Data trustworthiness – approaches and research challenges. In: Garcia-Alfaro J, Herrera-Joancomartí J, Lupu E, Posegga J, Aldini A, Martinelli F, Suri N (eds) Data privacy management, autonomous spontaneous security, and security assurance. Springer, pp 17–25
16. Bertotti G, Mayergoyz ID (eds) (2006) The Science of Hysteresis. Academic, Amsterdam and Boston
17. Bichon BJ, Eldred MS, Swiler LP, Mahadevan S, McFarland JM (2008) Efficient global reliability analysis for nonlinear implicit performance functions. AIAA Journal 46(10), 2459–2468. https://doi.org/10.2514/1.34321
18. BIPM, IEC, IFCC, ILAC, IUPAC, IUPAP, ISO, OIML (2008) Evaluation of measurement data – guide to the expression of uncertainty in measurement. JCGM 100
19. Björck A (1996) Numerical Methods for Least Square Problems. SIAM, Philadelphia, Pa
20. Bock J, Haase P, Ji Q, Volz R (2008) Benchmarking owl reasoners. In: van Harmelen F, Herzig A, Hitzler P, Lin Z, Piskac R, Qi G (eds) Proceedings of the workshop on advancing reasoning on the web: scalability and commonsense, ARea 2008, at the 5th European semantic web conference, ESWC08, CEUR workshop proceedings, vol 350
21. Bölling C (2019) Simulationsbasierte Auslegung mehrstufiger Werkzeugsysteme zur Bohrungsfeinbearbeitung am Beispiel der Ventilführungs- und Ventilsitzbearbeitung. Dissertation, TU Darmstadt
22. Bourinet JM, Deheeger F, Lemaire M (2011) Assessing small failure probabilities by combined subset simulation and support vector machines. Structural Safety 33(6), 343–353. https://doi.org/10.1016/j.strusafe.2011.06.001
23. Box GEP, Draper NR (1987) Empirical Model-Building and Response Surfaces. Wiley, New York
24. Bridgman PW (1922) Dimensional Analysis. Yale University Press, New Haven
25. Bucher C, Bourgund U (1990) A fast and efficient response surface approach for structural reliability problems. Structural Safety 7(1), 57–66. https://doi.org/10.1016/0167-4730(90)90012-E
26. Burrows CR (ed) (1994) The active control of vibration. Mechanical Engineering Publ
27. Castanedo F (2013) A review of data fusion techniques. The Scientific World Journal 2013:. https://doi.org/10.1155/2013/704504
28. Cerrone A, Hochhalter J, Heber G, Ingraffea A (2014) On the effects of modeling as-manufactured geometry: toward digital twin. Int J Aerosp Eng 2014. https://doi.org/10.1155/2014/43927 Article ID 439278
29. Coakley J, Elliot AS (2012) An air spring. Patent WO002012052776A1
30. D'Agostino RB (1986) Goodness-of-fit-techniques. CRC Press
31. Das P, Zheng Y (2000) Cumulative formation of response surface and its use in reliability analysis. Probabilistic Engineering Mechanics 15(4), 309–315. https://doi.org/10.1016/S0266-8920(99)00030-2
32. Deheeger F, Lemaire M (2010) Support vector machine for efficient subset simulations: 2SMART method. In: Proceedings of the 10th international conference on applications of statistics and probability in civil engineering (ICASP10)
33. Deutsches Institut für Normung (1991) DIN ISO 2768-1:1991-06. General tolerances – tolerances for linear and angular dimensions without individual tolerance indications

34. Deutsches Institut für Normung (2003) DIN 8580. Fertigungsverfahren—Begriffe, Einteilung. https://doi.org/10.31030/9500683
35. Devroye L (1986) Non-Uniform Random Variate Generation. Springer-Verlag, New York
36. Dodge Y (ed) (2010) The concise encyclopedia of statistics. Springer, New York. https://doi.org/10.1007/978-0-387-32833-1_62
37. Dogra APS, DeVor RE, Kapoor SG (2002) Analysis of feed errors in tapping by contact stress model. Journal of Manufacturing Science and Engineering 124:248–257. https://doi.org/10.1115/1.1454107
38. Donaldson JR, Schnabel RB (1987) Computational experience with confidence regions and confidence intervals for nonlinear least squares. Technometrics 29(1), 67–82
39. Dubitzky W, Granzow M, Berrar DP (2007) Fundamentals of Data Mining in Genomics and Proteomics. Springer
40. Dunn OJ (1961) Multiple comparisons among means. Journal of the American Statistical Association 56(293), 52–64
41. Engelhardt R, Birkhofer H, Kloberdanz H, Mathias J (2009) Uncertainty-mode-and effects-analysis – an approach to analyze and estimate uncertainty in the product life cycle. In: Norell Bergendahl M (ed) DS 58-2: Proceedings of ICED 09, the 17th international conference on engineering design, vol 2. Design theory and research methodology, ICED. Design Society, Glasgow, pp 191–202
42. Engelhardt R, Koenen J, Enss G, Sichau A, Platz R, Kloberdanz H, Birkhofer H, Hanselka H (2010) A model to categorise uncertainty in load-carrying systems. In: 1st MMEP international conference on modelling and management engineering processes, pp 53–64
43. Euler L(1744) Methodus inveniendi lineas curvas: Maximi minimive properietate gaudentes, sive solutio problematis isoperimetrici latissimo sensu accepti. Marcum-Michaelem Bousquet
44. Fang KT, Li R, Sudjianto A (2006) Design and Modeling for Computer Experiments. Chapman & Hall/CRC, Boca Raton, FL
45. Feldmann R, Platz R (2019) Assessing model form uncertainty for a suspension strut using Gaussian processes. In: Proceedings of the 3rd international conference on uncertainty quantification in computational sciences and engineering (UNCECOMP 2019)
46. Fischer MJ (1983) The consensus problem in unreliable distributed systems (a brief survey). In: Karpinski M (ed) Foundations of computation theory. Lecture notes in computer science, vol 158. Springer, Berlin, pp 127–140. https://doi.org/10.1007/3-540-12689-999
47. Foley JD, Van FD, Van Dam A, Feiner SK, Hughes JF, Angel E, Hughes J (1996) Computer graphics: principles and practice, vol 12110. Addison-Wesley
48. Fortuna L, Graziani S, Rizzo A, Xibilia MG (2007) Soft sensors for monitoring and control of industrial processes. Advances in industrial control. Springer, London. https://doi.org/10.1007/978-1-84628-480-9
49. Franceschini G, Macchietto S (2008) Model-based design of experiments for parameter precision: State of the art. Chem Eng Sci 63(19):4846–4872
50. Gally T, Groche P, Hoppe F, Kuttich A, Matei A, Pfetsch ME, Rakowitsch M, Ulbrich S (2020) Identification of model uncertainty via optimal design of experiments applied to a mechanical press. Optim Eng, to appear
51. Gaul L, Albrecht H, Wirnitzer J (2004) Semi-active friction damping of large space truss structures. Shock and Vibration 11(3–4), 173–186. https://doi.org/10.1155/2004/565947
52. Gehb CM (2019) Uncertainty evaluation of semi-active load redistribution in a mechanical load-bearing structure. Dissertation, TU Darmstadt
53. Gehb CM, Platz R, Melz T (2016) Active load path adaption in a simple kinematic load-bearing structure due to stiffness change in the structure's supports. Journal of Physics: Conference Series 744(1):012168. https://doi.org/10.1088/1742-6596/744/1/012168
54. Gehb CM, Platz R, Melz T (2017) Global load path adaption in a simple kinematic load-bearing structure to compensate uncertainty of misalignment due to changing stiffness conditions of the structure's supports. In: Barthorpe RJ, Platz R, Lopez I, Moaveni B, Papadimitriou C (eds) Model validation and uncertainty quantification, vol 3. Conference proceedings of the society for experimental mechanics series. Springer, Cham, pp 133–144. https://doi.org/10.1007/978-3-319-54858-6_14

55. Gehb CM, Platz R, Melz T (2019) Two control strategies for semi-active load path redistribution in a load-bearing structure. Mechanical Systems and Signal Processing 118:195–208. https://doi.org/10.1016/j.ymssp.2018.08.044
56. Gehb CM, Atamturktur S, Platz R, Melz T (2020) Bayesian inference based parameter calibration of the LuGre-friction model. Experimental Techniques 44(3), 369–382. https://doi.org/10.1007/s40799-019-00355-7
57. Gehb CM, Platz R, Melz T (2020) Bayesian inference based parameter calibration of a mechanical load-bearing structure's mathematical model. In: IMAC – 38th international modal analysis conference
58. Gere J, Timoshenko S (1997) Mechanics of Materials, 4th edn. PWS, Boston
59. Glaessgen E, Stargel D (2012) The digital twin paradigm for future NASA and US Air Force vehicles. In: 53rd AIAA/ASME/ASCE/AHS/ASC structures, structural dynamics and materials conference 20th AIAA/ASME/AHS adaptive structures conference 14th AIAA, p 1818
60. Goller B, Schuëller GI (2011) Investigation of model uncertainties in Bayesian structural model updating. Journal of Sound and Vibration 330(25–15), 6122–6136
61. Götz B, Schaeffner M, Platz R, Melz T (2016) Lateral vibration attenuation of a beam with circular cross-section by a support with integrated piezoelectric transducers shunted to negative capacitances. Smart Materials and Structures 25(9):095045. https://doi.org/10.1088/0964-1726/25/9/095045
62. Götz B, Kersting S, Kohler M (2018) Estimation of an improved surrogate model in uncertainty quantification by neural networks. Submitted for publication
63. Götz B, Platz R, Melz T (2018) Effect of static axial loads on the lateral vibration attenuation of a beam with piezo-elastic supports. Smart Materials and Structures 27(3):035011
64. Grau BC, Horrocks I, Motik B, Parsia B, Patel-Schneider P, Sattler U (2008) OWL 2: The next step for OWL. Journal of Web Semantics 6(4):309–322
65. Green PL, Worden K (2013) Modelling friction in a nonlinear dynamic system via Bayesian inference. In: Allemang R, de Clerck J, Niezrecki C, Wicks A (eds) Special Topics in Structural Dynamics, vol 6. Springer, New York, pp 543–553
66. Groche P, Hoppe F, Sinz J (2017) Stiffness of multipoint servo presses: mechanics vs. control. CIRP Ann. 66(1):373–376. https://doi.org/10.1016/j.cirp.2017.04.053
67. Gruber TR (1993) A translation approach to portable ontology specifications. Knowledge acquisition 5(2):199–221
68. Györfi L, Kohler M, Krzy?zak A, Walk H (2002) A distribution-free theory of nonparametric regression. Springer series in statistics. Springer, New York. https://doi.org/10.1007/b97848
69. Hartig J, Schänzle C, Pelz PF (2019) Concept validation of a soft sensor network for wear detection in positive displacement pumps. In: 4th international rotating equipment conference – pumps and compressors
70. Hartig J, Hoppe F, Martin D, Staudter G, Öztürk T, Anderl R, Groche P, Pelz PF, Weigold M (2020) Identification of lack of knowledge using analytical redundancy applied to structural dynamic systems. In: Model validation and uncertainty quantification, vol 3. Springer, pp 131–138
71. Hartig J, Schänzle C, Pelz PF (2020) Validation of a soft sensor network for condition monitoring in hydraulic systems. In: 12th international fluid power conference. Technische Universität Dresden
72. Hauer T (2012) Modellierung der Werkzeugabdrängung beim Reiben – Ableitung von Empfehlungen für die Gestaltung von Mehrschneidenreibahlen. Schriftenreihe des PTW. Shaker, Aachen. Dissertation, TU Darmstadt
73. Hedrich P (2018) Konzeptvalidierung einer aktiven Luftfederung im Kontext autonomer Fahrzeuge, Forschungsberichte zur Fluidsystemtechnik, vol 20. Shaker, Aachen
74. Heimrich F, Anderl R (2016) Approach for the visualization of geometric uncertainty of assemblies in cad-systems. Journal of Computers 11(3), 247–257
75. Higdon D, Gattiker J, Williams B, Rightley M (2008) Computer model calibration using high-dimensional output. Journal of the American Statistical Association 103(482), 570–583. https://doi.org/10.1198/016214507000000888

76. Hodouin D (2010) Process observers and data reconciliation using mass and energy balance equations. In: Sbárbaro D, del Villar R (eds) Advanced Control and Supervision of Mineral Processing Plants, Advances in Industrial Control. Springer, London, pp 15–83
77. Hurtado JE (2004) Structural Reliability: Statistical Learning Perspectives, vol 17. Lecture Notes in Applied and Computational Mechanics, Springer, Berlin
78. Ihn JB, Chang FK (2008) Pitch-catch active sensing methods in structural health monitoring for aircraft structures. Structural Health Monitoring: An International Journal 7(1):5–19. https://doi.org/10.1177/1475921707081979
79. Imai S, Blasch E, Galli A, Zhu W, Lee F, Varela CA (2017) Airplane flight safety using error-tolerant data stream processing. IEEE Aerospace and Electronic Systems Magazine 32(4), 4–17. https://doi.org/10.1109/maes.2017.150242
80. Isermann R (2006) Fault-diagnosis systems: an introduction from fault detection to fault tolerance. Springer, Berlin. https://doi.org/10.1007/3-540-30368-5
81. Isermann R, Ballé P (1997) Trends in the application of model-based fault detection and diagnosis of technical processes. Control Engineering Practice 5(5), 709–719. https://doi.org/10.1016/S0967-0661(97)00053-1
82. Isermann R, Schaffnit J, Sinsel S (1999) Hardware-in-the-loop simulation for the design and testing of engine-control systems. Control Engineering Practice 7(5), 643–653. https://doi.org/10.1016/S0967-0661(98)00205-6
83. ISO (2008) Uncertainty of measurement – Part 3: Guide to the expression of uncertainty in measurement
84. Johnson CR, Sanderson AR (2003) A next step: Visualizing errors and uncertainty. IEEE Comput Graphics Appl 23(5):6–10. https://doi.org/10.1109/MCG.2003.1231171
85. Kapoor SG, DeVor RE, Zhu R, Gajjela R, Parakkal G, Smithey D (1998) Development of mechanistic models for the prediction of machining performance: Model building methodology. Mach Sci Technol 2(2):213–238
86. Kaymaz I (2005) Application of kriging method to structural reliability problems. Structural Safety 27(2), 133–151. https://doi.org/10.1016/j.strusafe.2004.09.001
87. Kennedy MC, O'Hagan A (2001) Bayesian calibration of computer models. Journal of the Royal Statistical Society: Series B (Statistical Methodology) 63(3):425–464. https://doi.org/10.1111/1467-9868.00294
88. Kersting S, Kohler M (2019) Uncertainty quantification based on (imperfect) simulation models with estimated input distributions. Submitted for publication
89. Khaleghi B, Khamis A, Karray FO, Razavi SN (2013) Multisensor data fusion: A review of the state-of-the-art. Information Fusion 14(1):28–44. https://doi.org/10.1016/j.inffus.2011.08.001
90. Kim SH, Na SW (1997) Response surface method using vector projected sampling points. Structural Safety 19(1), 3–19. https://doi.org/10.1016/S0167-4730(96)00037-9
91. Kohler M, Krzyżak A (2017) Improving a surrogate model in uncertainty quantification by real data. Submitted for publication
92. Körkel S, Kostina E, Bock HG, Schlöder JP (2004) Numerical methods for optimal control problems in design of robust optimal experiments for nonlinear dynamic processes. Optimization Methods and Software 19(3–4):327–338
93. Korkmaz F (1982) Hydrospeicher als Energiespeicher. Springer, Berlin. https://doi.org/10.1007/978-3-642-81737-3
94. Kreß R, Crepin PY, Kubbat W, Schreiber M (2000) Fault detection and diagnosis for electrohydraulic actuators. IFAC Proceedings Volumes 33(26):983–988. https://doi.org/10.1016/S1474-6670(17)39273-X
95. Kreye ME, Goh YM, Newnes LB (2011) Manifestation of uncertainty – a classification. In: DS 68-6: Proceedings of the 18th international conference on engineering design (ICED 11), impacting society through engineering design, vol 6. Design information and knowledge
96. Krima S, Barbau R, Fiorentini X, Sudarsan R, Sriram RD (2009) Ontostep: OWL-DL ontology for step. NIST Pubs

97. Kumar M, Garg DP, Zachery RA (2006) A generalized approach for inconsistency detection in data fusion from multiple sensors. In: American control conference. IEEE Operations Center, Piscataway, NJ, p 6. https://doi.org/10.1109/ACC.2006.1656526
98. Ledin JA (1999) Hardware-in-the-loop simulation. Embedded Systems Programming 12(2), 42–60
99. Lehner S, Jacobs G (1997) Contamination sensitivity of hydraulic pumps and valves. In: Totten GE (ed) Tribology of hydraulic pump testing, STP/ASTM, pp. 261–276. ASTM, Philadelphia, Pa. https://doi.org/10.1520/STP11852S
100. Lenz E (2017) Methodischer Reglerentwurf für eine aktive Luftfeder unter Unsicherheit. Internal report, TU Darmstadt
101. Lenz J, Platz R (2019) Quantification and evaluation of parameter and model uncertainty for passive and active vibration isolation. In: Barthorpe R, Platz R, Lopez I, Moaveni B, Papadimitriou C (eds) Model validation and uncertainty quantification, vol 3. Conference proceedings of the society for experimental mechanics series. Springer, Cham, pp 135–147
102. Lenz E, Hedrich P, Pelz PF (2018) Aktive Luftfederung – Modellierung, Regelung und Hardware-in-the-Loop-Experimente. Forschung in Ingenieurwesen, pp 1–15. https://doi.org/10.1007/s10010-018-0272-2
103. Li S, Götz B, Schaeffner M, Platz R (2017) Approach to prove the efficiency of the Monte Carlo method combined with the elementary effect method to quantify uncertainty of a beam structure with piezo-elastic supports. In: Proceedings of the 2nd international conference on uncertainty quantification in computational sciences and engineering (UNCECOMP 2017), pp. 441–455. https://doi.org/10.7712/120217.5382.16762
104. Liu DP (2006) Parameter identification for LuGre friction model using genetic algorithms. In: Proceedings of 2006 international conference on machine learning and cybernetics. IEEE, Piscataway NJ
105. Locke R, Kupis S, Gehb CM, Platz R, Atamturktur S (2019) Applying uncertainty quantification to structural systems: Parameter reduction for evaluating model complexity. In: Barthorpe RJ (ed) Model validation and uncertainty quantification, vol 3. Conference proceedings of the society for experimental mechanics series. Springer, Cham, pp 241–256
106. Lutters E, Van Houten FJ, Bernard A, Mermoz E, Schutte CS (2014) Tools and techniques for product design. CIRP Annals 63(2), 607–630
107. Mallapur S, Platz R (2018) Quantification of uncertainty in the mathematical modelling of a multivariable suspension strut using Bayesian interval hypothesis-based approach. In: Pelz PF, Groche P (eds) Uncertainty in mechanical engineering III, vol 885. Applied mechanics and materials. Trans Tech Publications, pp 3–17
108. Mallapur S, Platz R (2019) Uncertainty quantification in the mathematical modelling of a suspension strut using Bayesian inference. Mechanical Systems and Signal Processing 118:158–170. https://doi.org/10.1016/j.ymssp.2018.08.046
109. Margossian CC (2019) A review of automatic differentiation and its efficient implementation. Wiley Interdisciplinary Reviews: Data Mining and Knowledge Discovery 9(4):e1305. https://doi.org/10.1002/widm.1305
110. Mascardi V, Cordi V, Rosso P (2007) A comparison of upper ontologies. In: Baldoni M, Boccalatte A, de Paoli F, Martelli M, Mascardi V (eds) WOA 2007: Dagli Oggetti agli Agenti. 8th AI*IA/TABOO joint workshop "From Objects to Agents": Agents and Industry: Technological Applications of Software Agents. Seneca, Torino, Italy, pp 55–64
111. Maurer S, Markmann B, Mersmann A (1998) A priori Vorhersage von Adsorptionsgleichgewichten. Chemie Ingenieur Technik—CIT 70(9), 1104–1105. https://doi.org/10.1002/cite.330700960
112. Mayergoyz ID (2003) Mathematical models of hysteresis and their applications. Elsevier. https://doi.org/10.1016/B978-0-12-480873-7.X5000-2
113. Mersmann A, Kind M, Stichlmair J (2005) Thermische Verfahrenstechnik: Grundlagen und Methoden, second revised and enlarged. Chemische Technik Verfahrenstechnik, Springer, Berlin

114. Mickens T, Schulz M, Sundaresan M, Ghoshal A, Naser AS, Reichmeider R (2003) Structural health monitoring of an aircraft joint. Mechanical Systems and Signal Processing 17(2), 285–303. https://doi.org/10.1006/mssp.2001.1425
115. Monostori L, Kádár B, Bauernhansl T, Kondoh S, Kumara S, Reinhart G, Sauer O, Schuh G, Sihn W, Ueda K (2016) Cyber-physical systems in manufacturing. CIRP Annals 65(2), 621–641. https://doi.org/10.1016/j.cirp.2016.06.005
116. Mosch L, Sprenger A, Anderl R (2010) Approach for visualization of uncertainty in cad-systems based on ontologies. In: ASME 2010 international mechanical engineering congress and exposition. American Society of Mechanical Engineers Digital Collection, pp 243–249. https://doi.org/10.1115/IMECE2010-37651
117. Muehleisen RT, Riddle M (2014) A guide to Bayesian calibration of building energy models. In: ASHRAE/IBPSA-USA. https://doi.org/10.13140/2.1.1674.9127
118. Nagel JB (2017) Bayesian techniques for inverse uncertainty quantification. Dissertation, ETH Zürich
119. Nakashima M (2001) Development, potential, and limitations of real-time online (pseudo-dynamic) testing. Philosophical Transactions: Mathematical, Physical and Engineering Sciences 359(1786):1851–1867
120. Ondoua S (2016) Unsicherheit in der Bewertung von Struktur-Eigenschaftsbeziehungen zwischen aktiven und passiven Systemelementen in aktiven lasttragenden Systemen. Dissertation, TU Darmstadt
121. Papadrakakis M, Lagaros ND (2002) Reliability-based structural optimization using neural networks and Monte Carlo simulation. Computer Methods in Applied Mechanics and Engineering 191(32), 3491–3507. https://doi.org/10.1016/S0045-7825(02)00287-6
122. Park I, Amarchinta HK, Grandhi RV (2010) A Bayesian approach for quantification of model uncertainty. Reliability Engineering & System Safety 95(7):777–785
123. Parzen E (1962) On estimation of a probability density function and mode. Ann. Math. Statist. 33:1065–1076. https://doi.org/10.1214/aoms/1177704472
124. Pasquier R, Smith IF (2015) Robust system identification and model predictions in the presence of systematic uncertainty. Advanced Engineering Informatics 29(4), 1096–1109
125. Paucksch E, Holsten S, Linß M, Tikal F (2008) Zerspantechnik: Prozesse, Werkzeuge, Technologien, twelfth edn. Studium. Vieweg + Teubner, Wiesbaden. https://doi.org/10.1007/978-3-8348-9494-6
126. Pelz PF, Groß TF, Schänzle C (2017) Hydrospeichermit Sorbentien -Verhalten, Modellierung und Diskussion. O+P - Ölhydraulik und Pneumatik 61(1–2):42–49
127. Pelz PF, Dietrich I, Schänzle C, Preuß N (2018) Towards digitalization of hydraulic systems using soft sensor networks. In: 11th international fluid power conference 2018. RWTH Aachen, Aachen, pp 40–53
128. Platz R, Enss GC (2015) Comparison of uncertainty in passive and active vibration isolation. In: Atamturktur S, Moaveni B, Papadimitriou C, Schoenherr T (eds) Model validation and uncertainty quantification, vol 3. Conference proceedings of the society for experimental mechanics series. Springer, Cham, pp 15–25
129. Platz R, Melzer CM (2016) Uncertainty quantification for decision making in early design phase for passive and active vibration isolation. In: Proceedings of ISMA 2016 including USD 2016 international conference on uncertainty in structural dynamics, pp 4501–4513
130. Platz R, Ondua S, Enss GC, Melz T (2014) Approach to evaluate uncertainty in passive and active vibration reduction. In: Atamturktur S, Moaveni B, Papadimitriou C, Schoenherr T (eds) Model validation and uncertainty quantification, vol 3. Conference proceedings of the society for experimental mechanics series. Springer, Cham, pp 345–352
131. Preuß N, Schänzle C, Pelz PF (2018) Accumulators with sorbent material – an innovative approach towards size and weight reduction. In: 11th international fluid power conference, pp 504–517. http://wl.fst.tu-darmstadt.de/wl/publications/paper_180319_Aachen_11th_IFK_Proceedings_Hydrospeicher_Sorbentien_preuss_schaenzle_pelz.pdf
132. Rasmussen CE (2003) Gaussian processes in machine learning. In: Bousquet O, von Luxburg U, Rätsch G (eds) Advanced Lectures on Machine Learning, vol 3176. Lecture Notes in Computer Science. Springer, Berlin, Heidelberg, pp 63–71

133. Rieger KJ, Schiehlen W (1994) Active versus passive control of vehicle suspensions – hardware in the loop experiments. In: Burrows CR (ed) The active control of vibration. Mechanical Engineering Publ
134. Ríos J, Staudter G, Weber M, Anderl R (2019) A review, focused on data transfer standards, of the uncertainty representation in the digital twin context. In: IFIP international conference on product lifecycle management. Springer, pp 24–33. https://doi.org/10.1007/978-3-030-42250-9_3
135. Rosenblatt M (1956) Remarks on some nonparametric estimates of a density function. Ann. Math. Statist. 27:832–837. https://doi.org/10.1214/aoms/1177728190
136. Saltelli A (2008) Global sensitivity analysis: the primer. Wiley, Chichester. https://doi.org/10.1002/9780470725184
137. Sankararaman S, Mahadevan S (2011) Model validation under epistemic uncertainty. Reliability Engineering & System Safety 96(9):1232–1241. https://doi.org/10.1016/j.ress.2010.07.014
138. Santner TJ, Williams BJ, Notz WI (2018) The Design and Analysis of Computer Experiments. Springer Series in Statistics, Springer, New York
139. Sarhadi P, Yousefpour S (2014) State of the art: Hardware in the loop modeling and simulation with its applications in design, development and implementation of system and control software. International Journal of Dynamics and Control 3(4):470–479. https://doi.org/10.1007/s40435-014-0108-3
140. Schänzle C, Ludwig G, Pelz PF (2016) ERP positive displacement pumps – physically based approach towards an application-related efficiency guideline. In: 3rd international rotating equipment conference (IREC) 2016. Düsseldorf
141. Schänzle C, Dietrich I, Corneli T, Pelz PF (2017) Controlling uncertainty in hydraulic drive systems by means of a soft sensor network. Sensors and Instrumentation 5:1
142. Schuëller GI (2007) On the treatment of uncertainties in structural mechanics and analysis. Computers & Structures 85(5–6):235–243
143. Silvey SD (1980) Optimal Design: An Introduction to the Theory for Parameter Estimation, vol 1. Springer, Netherlands
144. Smith RC (2014) Uncertainty Quantification: Theory, Implementation, and Applications, Computational science and engineering, vol 12. SIAM, Philadelphia, PA
145. Sprenger A, Anderl R (2012) Product life cycle oriented representation of uncertainty. In: Product lifecycle management. Towards knowledge-rich enterprises. Springer, pp 277–286. https://doi.org/10.1007/978-3-642-35758-9_24
146. Sprenger A, Mosch L, Anderl R (2011) Representation of uncertainty in distributed product development. In: 18th annual European concurrent engineering conference 2011
147. Sprenger A, Haydn M, Ondoua S, Mosch L, Anderl R (2012) Ontology-based information model for the exchange of uncertainty in load carrying structures. In: Hanselka H, Groche P, Platz R (eds) Uncertainty in mechanical engineering, vol 104. Applied mechanics and materials. Trans Tech Publications, pp 55–66. https://doi.org/10.4028/www.scientific.net/AMM.104.55
148. Spurk JH (1992) Dimensionsanalyse in der Strömungslehre. Springer, Berlin, Heidelberg
149. Steinhorst W (1999) Sicherheitstechnische Systeme: Zuverlässigkeit und Sicherheit kontrollierter und unkontrollierter Systeme. Vieweg+Teubner, Wiesbaden. https://doi.org/10.1007/978-3-322-90927-5
150. Stuart AM (2010) Inverse problems: A Bayesian perspective. Acta Numer 19:451–559
151. Tjahjono S (2019) Aircraft accident investigation report Boeing 737-8 (MAX); PK-LQP
152. Tolker-Nielsen T (2017) EXOMARS 2016-Schiaparelli anomaly inquiry: DG-I/2017/546/TTN. Technical report, Agency, European Space. https://sci.esa.int/documents/33431/35950/1567260317467-ESA_ExoMars_2016_Schiaparelli_Anomaly_Inquiry.pdf
153. Tuegel EJ, Ingraffea AR, Eason TG, Spottswood SM (2011) Reengineering aircraft structural life prediction using a digital twin. Int J Aerosp Eng 1–14. https://doi.org/10.1155/2011/154798

154. Tufte ER (1983) The Visual Display of Quantitative Information, vol 2. Graphics Press Cheshire, CT USA
155. Verein Deutscher Ingenieure (2010) VDI 2064:2010–11 Aktive Schwingungsisolierung [Active vibration isolation]. Beuth, Berlin
156. Vergé A, Lotz J, Kloberdanz H, Pelz PF (2015) Uncertainty scaling – motivation, method and example application to a load carrying structure. In: Pelz PF, Groche P (eds) Uncertainty in mechanical engineering II, vol 807. Applied mechanics and materials. Trans Tech Publications, pp 99–108
157. Wahba G (1990) Spline models for observational data, vol 59. SIAM
158. Walter E, Pronzato L (1990) Qualitative and quantitative experiment design for phenomenological models - A survey. Automatica 26(2):195–213. https://doi.org/10.1016/0005-1098(90)90116-Y
159. Walther A, Griewank A (2012) Getting started with ADOL-C. In: Naumann U, Schenk O (eds) Combinatorial scientific computing. Chapman & Hall/CRC Computational Science, CRC Press, Boca Raton, pp 181–202. https://doi.org/10.1201/b11644-8
160. Wang S, Chen W, Tsui KL (2009) Bayesian validation of computer models. Technometrics 51(4), 439–451. https://doi.org/10.1198/TECH.2009.07011
161. Wang X, Lin S, Wang S (2016) Dynamic friction parameter identification method with LuGre model for direct-drive rotary torque motor. Mathematical Problems in Engineering 2016:1–8. https://doi.org/10.1155/2016/6929457
162. Weber C (2007) Looking at "DFX" and "Product maturity" from the perspective of a new approach to modelling product and product development processes. In: Krause FL (ed) The future of product development. Springer, Berlin, pp 85–104
163. Weber M, Staudter G, Anderl R (2018) Comparison of inductive inference mechanisms and their suitability for an information model for the visualization of uncertainty. In: Pelz PF, Groche P (eds) Uncertainty in mechanical engineering III, vol 885. Applied mechanics and materials. Trans Tech Publications, pp 147–155. https://doi.org/10.4028/www.scientific.net/AMM.885.147
164. Wong RKW, Storlie CB, Lee TCM (2017) A frequentist approach to computer model calibration. J. R. Stat. Soc. Ser. B. Stat. Methodol. 79(2), 635–648. https://doi.org/10.1111/rssb.12182
165. Yager RR (1987) On the Dempster-Shafer framework and new combination rules. Information Sciences 41(2), 93–137. https://doi.org/10.1016/0020-0255(87)90007-7
166. Zabel A (2010) Prozesssimulation in der Zerspanung: Modellierung von Dreh- und Fräsprozessen, *Schriftenreihe des ISF/Technische Universität Dortmund H*, vol 2. Vulkan-Verlag, Essen. TU Dortmund, Habilitation
167. Zhao, X., Gao, H., Zhang, G., Ayhan, B., Yan, F., Kwan, C., Rose, J.L.: Active health monitoring of an aircraft wing with embedded piezoelectric sensor/actuator network: I. Smart Materials and Structures **16**(4), 1208–1217 (2007). https://doi.org/10.1088/0964-1726/16/4/032
168. Zocholl M, Anderl R (2014) Ontology-based representation of time dependent uncertainty information for parametric product data models. In: Liu K, Filipe J (eds) KMIS 2014 – proceedings of the international conference on knowledge management and information sharing. Scitepress, Setúbal, Portugal, pp 400–404
169. Zocholl M, Trinkel T, Anderl R (2014) Methode zur Beherrschung von Unsicherheit in ex35 pliziten 3DCAD Geometrien. In: Rieg F, Brökel K, Feldhusen J, Grote KH, Stelzer R (eds) 12. Gemeinsames Kolloquium Konstruktionstechnik 2014: Methoden in der Produktentwicklung: Kopplung von Strategien und Werkzeugen im Produktentwicklungsprozess. Bayreuth, pp 173–182. https://epub.uni-bayreuth.de/1789/
170. Zocholl M, Heimrich F, Oberle M, Würtenberger J, Bruder R, Anderl R (2015) Representation of human behaviour for the visualization in assembly design. In: Pelz PF, Groche P (eds) Uncertainty in mechanical engineering II, vol 807. Applied mechanics and materials. Trans Tech Publications, pp 183–192

Chapter 5
Methods and Technologies for Mastering Uncertainty

Peter Groche, Eberhard Abele, Nassr Al-Baradoni, Sabine Bartsch, Christian Bölling, Nicolas Brötz, Christopher M. Gehb, Felix Geßner, Benedict Götz, Jakob Hartig, Philipp Hedrich, Daniel Hesse, Martina Heßler, Florian Hoppe, Laura Joggerst, Sebastian Kersting, Hermann Kloberdanz, Maximilian Knoll, Michael Kohler, Martin Krech, Jonathan Lenz, Michaela Leštáková, Kevin T. Logan, Daniel Martin, Tobias Melz, Tim M. Müller, Tuğrul Öztürk, Peter F. Pelz, Roland Platz, Andrea Rapp, Manuel Rexer, Maximilian Schaeffner, Fiona Schulte, Julian Sinz, Jörn Stegmeier, Matthias Weigold, and Janine Wendt

Abstract Uncertainty affects all phases of the product life cycle of technical systems, from design and production to their usage, even beyond the phase boundaries. Its identification, analysis and representation are discussed in the previous chapter. Based on the gained knowledge, our specific approach on mastering uncertainty can be applied. These approaches follow common strategies that are described in the subsequent chapter, but require individual methods and technologies. In this chapter, first legal and technical aspects for mastering uncertainty are discussed. Then, techniques

---

The original version of this chapter was revised with the missed out corrections. An erratum to this chapter can be found at https://doi.org/0.1007/978-3-030-78354-9_8

P. Groche (✉) · E. Abele · N. Al-Baradoni · C. Bölling · N. Brötz · C. M. Gehb · F. Geßner · J. Hartig · P. Hedrich · D. Hesse · F. Hoppe · H. Kloberdanz · M. Knoll · M. Krech · J. Lenz · M. Leštáková · K. T. Logan · D. Martin · T. Melz · T. M. Müller · T. Öztürk · P. F. Pelz · R. Platz · M. Rexer · M. Schaeffner · F. Schulte · J. Sinz · M. Weigold
Department of Mechanical Engineering, TU Darmstadt, Darmstadt, Germany
e-mail: groche@ptu.tu-darmstadt.de

S. Bartsch · A. Rapp · J. Stegmeier
Institute of Linguistics and Literary Studies, TU Darmstadt, Darmstadt, Germany

B. Götz · T. Melz
Fraunhofer Institute for Structural Durability and System Reliability LBF, Darmstadt, Germany

M. Heßler
Institute of History, TU Darmstadt, Darmstadt, Germany

L. Joggerst · J. Wendt
Department of Law and Economics, TU Darmstadt, Darmstadt, Germany

S. Kersting · M. Kohler
Department of Mathematics, TU Darmstadt, Darmstadt, Germany

P. Pelz et al. (eds.), *Mastering Uncertainty in Mechanical Engineering*,
Springer Tracts in Mechanical Engineering,
https://doi.org/10.1007/978-3-030-78354-9_5

for product design of technical systems under uncertainty are presented. The propagation of uncertainty is analysed for particular examples of process chains. Finally, semi-active and active technical systems and their relation to uncertainty are discussed.

The first Chapters of this book provide the conceptual basis for mastering uncertainty in design, production and usage of load-bearing structures in mechanical engineering. This forms the fundamental floor of our house, presented in Chap. 1. Besides the presentation of our motivation, Chap. 1 reflects on the source and quality of models, as well as on data and the term "structures". Chapter 2 provides a consistent classification and definition of uncertainty. This is essential for our specific approach on mastering uncertainty, which is described in Chap. 3, making use of three exemplary technical systems, which are introduced in Sect. 3.6. On the middle floor of the framework of mastering uncertainty, presented in Fig. 1.12, we discuss methods for analysis, quantification and evaluation of uncertainty in Chap. 4 and now introduce methods and technologies to apply our approach to technical systems.

Therefore, we first cover the technical and legal requirements for mastering uncertainty with a focus on product safety and liability in Sect. 5.1. Although legal uncertainty usually expresses itself during product usage, it has to be considered from the very beginning of the product life cycle during design and production. The technical specification of products, which has to define the technical requirements on the product, effects the uncertainty in an early stage of the development process. From a legal perspective, two points are important: firstly, to meet product safety requirements, sufficient knowledge of the legal framework regarding the specific application must be available; this might be challenging, especially for new developments; secondly, liability risks have to be minimised during the design and production phase.

In Sect. 5.2, we propose several methods to uncover and master uncertainty during the design phase. Therefore, we first introduce a method for the estimation of uncertainty occurring during the whole product life cycle, and a generic process model used to uncover uncertainty in production or application processes. Furthermore, uncertainty arising from models which are indispensable for product design, is discussed, and different ways to deal with it are presented. Our specific approach to master uncertainty in the product development process is presented using our our three demonstrator systems Modular Active Spring-Damper System, see Sect. 3.6.1, Active Air Spring, see Sect. 3.6.2, and 3D Servo Press, see Sect. 3.6.3.

The production phase usually takes place in the form of a process chain. Each single process, as well as the material passing through the process chain, is subject to uncertainty. The uncertainty is propagated through the process chain, see Sect. 3.2. In Sect. 5.3 we give examples on how to master this propagation of uncertainty in process chains.

Section 5.4 deals with methods and technologies to manipulate single processes and their application on both the production and usage phase. Based on the definition of semi-active and active processes, which is given in Sect. 3.3, we cover the manipulation of production processes using innovative components. Furthermore, we present several semi-active and active technologies to master uncertainty within the usage phase.

5.1 Technical and Legal Requirements for Mastering Uncertainty

Peter Groche and Laura Joggerst

Defining requirements in order to master uncertainty is the main goal of the following section. Both, technical and legal requirements may be established. Regarding the product phases design (A), production (B) and usage (C) presented in Fig. 1.3 legal uncertainty usually manifests itself during the usage phase, if the product causes damage. Ideally, legal requirements influence technical requirements and are already considered during the design and production phase of the product in order to achieve the highest possible certainty regarding legal liability of producers. The economic impact of legal requirements can be derived from the legal framework. We will focus on product safety and product liability. In the following, we understand product development as the totality of all steps leading to a marketable product, including design and production.

In Sect. 1.6 product design was discussed as a constrained optimisation problem, where the objective function has three challenging aspects, namely (i) minimal effort, (ii) maximal availability and (iii) maximal acceptability. Conformity with product liability is the formal aspect of acceptance.

The process of defining specifications in engineering design can be understood as a socio-technical system, whereby the functionality and quality of a product are the result of a complex and dynamic interaction of different stakeholder groups. In Sect. 5.1.1, we examine this process of how specifications are formulated. Furthermore, we consider the way specifications are used in product development. Therefore, a classification of specifications into objectives and constraints is introduced.

From the technical perspective, defining a precise and complete technical specification is the basis for the subsequent development process of any new product. In classical engineering design, uncertainty in the use of the product can derive from the misinterpretation of product requirements during product development. This uncertainty can potentially be addressed at a very early stage of the development process. We will discuss the general problem of specification uncertainty, as well as the impact technical specifications may have on the overall uncertainty of a complex load-bearing system in Sect. 5.1.2.

In order to clarify product requirements from a legal perspective, we need to apply the abstract knowledge of the legal framework to specific applications. Combining legal expertise with the fields of engineering and mathematics allows a more user-oriented approach to discuss the problem of legal uncertainty where it occurs.

Therefore, product safety requirements and the importance of technical standards will be addressed in order to prevent cost-intensive product recalls. For innovative products in particular, the problem arises that producers may not rely on technical standards in their development process, as such standards do not yet exist. The question, how producers should cope with this uncertainty is also part of our discussion in Sect. 5.1.3.

From a product liability perspective, many liability-causing legal risks can be avoided in the design of the product and during the organisation of the production process. We take an application-oriented approach to answer the question, which measures the producer needs to implement in order to minimise liability risks. The specific requirements for producers using algorithmic optimisation techniques in product development (Sect. 5.1.5) are of a different nature than the requirements for producers implementing an autonomous production process (Sect. 5.1.4). In both cases, the difficult question arises, how liability risks can be minimised, when using innovative development techniques or production methods.

Technical standards, although not always legally binding, are often the basis of contracts when selling products. Compliance with standards is therefore a requirement for products. However, technical standards must apply to many different occasions, and the language used is therefore ambiguous. In Sect. 5.1.6, an information system is proposed that detects semantic uncertainty in standards and provides suggestions to the users of standards to resolve the uncertainty.

Uncertainty does not only affect product development, but planning processes in general. In the scientific discipline of project management, concepts, methods and practices for dealing with uncertainty have been developed during the last decades. In Sect. 5.1.7, we reflect on these approaches from a historical perspective in order to provide a better understanding of current tools for planning processes.

In addition to the general technical methods for mastering uncertainty, we hope to provide another dimension to the tools which are discussed in Sect. 3.3.

5.1.1 'Just Good Enough' Versus 'as Good as It Gets': Negotiating Specifications in a Conflict of Interest of the Stakeholders

Peter F. Pelz, Michaela Leštáková, Kevin T. Logan, and Tim M. Müller

Technical specifications provide requirements of the product or services that are considered in the design process and are discussed, quantified and checked by the stakeholders involved. In classical engineering design, they are supposed to be defined at the beginning of the product lifecycle. This classic approach dating back to Pahl and Beitz [215] and being the fundament of the guideline VDI 2221 [271] is related to uncertainty for two reasons. Firstly, at the beginning of the engineering design process, various ways of using and misusing the product have to be anticipated. Secondly, in retrospect of every engineering design process, the product functionality and quality can be seen as a result of a complex and dynamic interaction of the three stakeholders (1) supplier, (2) customer, (3) society. For us the supplier is in most cases identical with the manufacturer. There are several strategies known for mastering this uncertainty, such as user-centred design, requirement management and agile project management. A review of these methods exceeds the scope of this section. Instead, the following pages embody a reflection on how specifications are used as a design method in the context of Sustainable Systems Design (SSD) and of mathematical optimisation. Moreover, it outlines the process of how specifications are formulated by the three stakeholders mentioned above.

Classification of specifications into objectives and constraints

In SSD, see Sect. 1.6, Figs. 1.9 and 1.10, a constrained optimisation problem in the form of "maximise quality subject to functionality" is solved in the design process. Quality corresponds to the objectives and the functionality to the constraints of the optimisation problem.

The tree diagram in Fig. 5.1 provides a classification into two branches, objectives and constraints, which have four specification types in the leaves of the tree. There is one objective (a) and three different types of constraints (b), (c), (d):

(a) *quality objective*,
(b) *quality constraints*,
(c) *functional constraints*,
(d) *restricted design space*.

The assignment to constraints and sub-objectives is done by each stakeholder individually even though the same system is addressed as we will see in the following. The term "constraints" originates from the field of mathematical optimisation and describes the restricted solution space of the optimisation problem. In the context of technical specifications, it describes possible restrictions and limitations, but also demands and requirements.

(a) *Quality objectives* are specified by weighted sub-objectives. Weights are impact-specific weighting factors often known as cost factors (e.g. pricing the environmental impact by CO_2 taxation). Also, impact-specific incentives (e.g. financial subsidies for improvement) can be used as weights. The sub-objectives are the three quality directions minimal effort, maximal availability, maximal acceptability, Sect. 1.6. The three quality directions are in agreement with Taguchi's quality engineering methodology [258]. Taguchi demands that manufacturers also consider the effort, i.e. the economic costs and the costs to society, as a quality measure. The weighted sub-objectives are approximated by agents (humans and/or machines) in an incremental Continuous Improvement Process (CIP) controlled by the target and impact-specific weighting factors. Hence, a quality objective defines a direction and not the required quality level.
(b) *Quality constraints* ensure such required quality level. The advantage of quality constraints versus quality objectives is the clear commitment. The disadvantage is that constraints are fixed and not optimised. Hence, optimal quality and hence sustainability will never be reached by quality constraints. A typical quality specification is a cost limit to be reached in a design-to-cost engineering process or the setting of a pollution limit.
(c) *Functional constraints* are specified on the basis of expressed or anticipated customer needs.

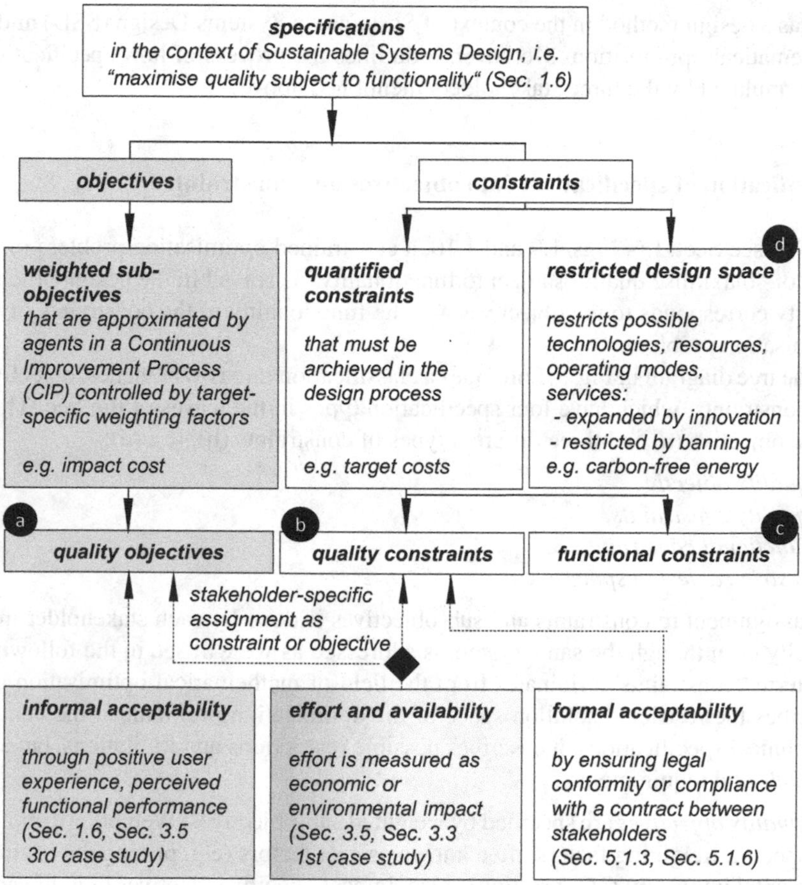

Fig. 5.1 The tree diagram (solid line) classifies system specifications in the context of SSD as presented in Sect. 1.6 into objectives and constraints. There are **a** quality objectives, **b** quality constraints and **c** functional constraints specified on the basis of named or anticipated customer needs. A restriction of the design space **d** is the third and last constraint

(d) *Restricted design space* is given by available technologies, resources, operating modes, services. The design space may be extended by innovations or restricted by banning technologies, such as demanding carbon-free energy supply.

Section 1.6 classifies acceptability into informal and formal acceptability. Formal acceptability is achieved by ensuring legal conformity or compliance with a contract between stakeholders. Hence, formal acceptability is assigned to quality specifications in the first place. Informal acceptability is fostered through positive user experience or positively perceived functional performance. The quality objective in Robust Design is a prominent example as presented in the third case study of Sect. 3.5. But how does the process of specifying objectives and constraints work? How are quality directions given and assigned to the different stakeholder groups?

Product functionality and quality is negotiated in a cybernetic system of the three external stakeholder groups (1) supplier, (2) customer, (3) society. Internal stakeholders, i.e. the employees of a company, are not the primary focus here. To understand the basic dynamics and outcome of this negotiating process, the external stakeholders can be modelled as agents in a cybernetic control system. The purpose of this section is not to provide and evaluate a detailed model, but to illustrate how the boundaries of SSD can be extended from an techno-economic system to a socio-technical system, see Fig. 5.2.

Cybernetic stakeholder system

The values of various parameters relevant in the process of negotiating and formulating the specifications (quality objectives, quality constraints, ...) in SSD are subject to processes that transcend the purely technical system. In order to understand how specifications are determined, we suggest that the socio-technical system can be modelled as a cybernetic system, see Fig. 5.2. We begin by identifying three stakeholders that are coupled in a control loop: (1) supplier, (2) customer of products and services like owners and operators and (3) society represented by one or several governments and non-governmental organisations.

This concept is now applied to analyse the dynamics of specifications using the example of the so called "energy ship", cf. 1st case study in Sect. 3.3. Here, the supplier (2) is the ship manufacturer. The customer (1) is the owner and operator of the energy ships. The customer follows the Friedman doctrine, "the social responsibility of business is to increase profit" [109]. This is reflected in the customer's *quality objective*: minimising the levelised costs of hydrogen (LCOH).

Tolerated functional and quality constraints are usually communicated to the supplier in a feedforward control. For example, *quality constraints* such as tolerated efficiency limits or service life guarantees can be required at the lowest possible price, which is a *quality objective* assigned to the supplier by the customer. However, formulating the quality constraints by the customer does not give the supplier any incentive to deliver the best possible product, but only to act according to the precept "just good enough". On the one hand, this attitude is welcome in order to avoid overachievement. On the other hand, this attitude limits sustainability. Another possibility would be to map the quality constraint to a quality objective, i.e. to accept a higher price if the component efficiency is higher. This leads to the principle "as good as it gets", which is essential for the development of sustainable products: quality objectives are followed in an incremental feed back control strategy as Fig. 5.3 implies. As discussed in Chap. 1, feed back control is robust with respect to uncertainty in the controlled systems. In contrast, feed forward control demands models of the controlled systems.

Society has many options for intervening as shown in Figs. 5.2 and 5.3. It may manipulate weighting factors in the quality objectives resulting in (i) *incentivisation*, i.e. certain technologies or solutions can become incentivised like discounted loans for fuel cell technology; (ii) *penalisation*, i.e. certain technologies or impacts are

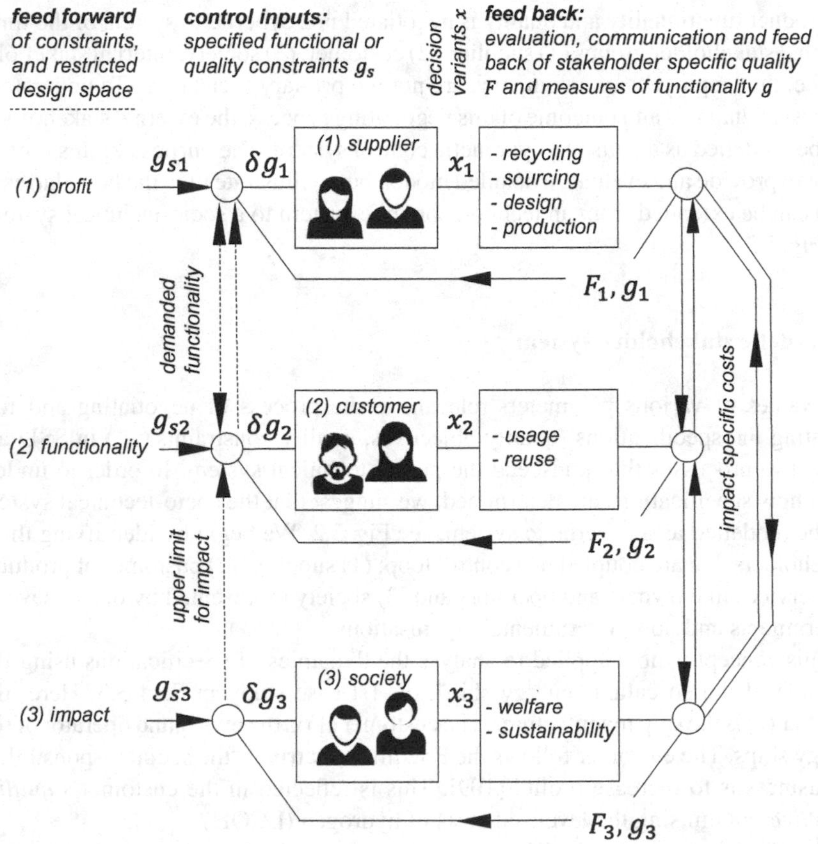

Fig. 5.2 Cybernetic model of a socio-technical system with three stakeholders: supplier, customer and society. Control inputs are the constraint specifications for the three stakeholders of the socio-technical system

penalised for example by fiscal policies like CO_2 tax; setting limits in the quality constraints by (iii) *regulation*, i.e. legal constraints are introduced to limit impact like minimum efficiency levels for fuel cells; or reduce the design space leading to (iv) *restriction*, i.e. the use of a certain technology is restricted like banning Diesel engines, which leads to an increased use of hydrogen and more profitability for the customer. This concerns both the customer and the supplier. As there is constant interaction, the respective specifications are negotiated in the market and they influence all stakeholders.

This classification is not strict and absolute. For example, companies can voluntarily overfulfil the normative requirements of the government, which is done in the context of Corporate Social Responsibility (CSR), and thus deviate from the Fried-

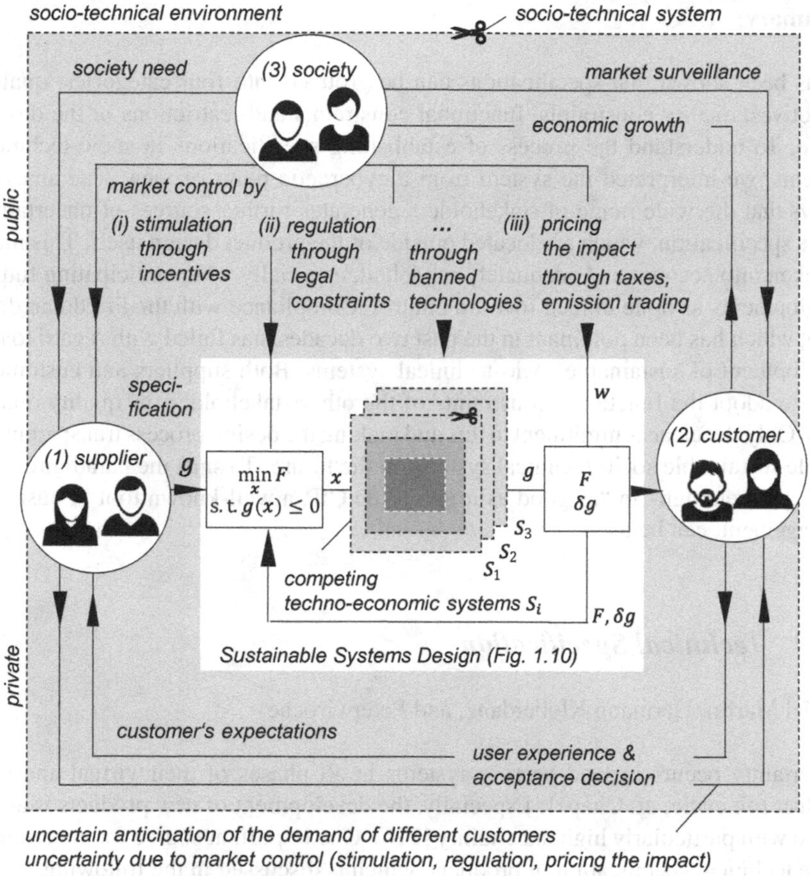

Fig. 5.3 Socio-technical system with the several competing techno-economic systems in its core and the three stakeholder groups supplier, customers, society as human controllers and sensors

man doctrine [212]. However, this unnecessary over-fulfilment of quality constraints can lead to an increased reputation in turn, increased profit and thus improve the quality objective—maximise profit – of the company [212]. Companies of the Value Balancing Alliance [267], founded by BASF SE, Deutsche Bank, SAP and other companies in 2019, go one step further by defining the impact specific cost and the positive impact of their actions themselves. The high level of uncertainty provides scope for interpretation and raises the suspicion of greenwashing [251]. In the perspective of the presented model, the responsibilities of the stakeholders are mixed up, which can cause unintended dynamics. This is why transparent and measurable Key Performance Indicators for the impact are necessary.

Summary

It has been shown that specifications can be grouped into four categories: quality objectives, quality constraints, functional constraints and restrictions of the design space. To understand the process of establishing specifications in socio-technical systems, we interpreted the system from a cybernetic point of view. The analysis shows that the wide range of stakeholders generates further sources of uncertainty in the specification, which are located outside of the product design itself. This must be taken into account and adequately modelled, especially when anticipating future developments as done in feed forward control. Compliance with the Friedman doctrine, which has been dominant in the past two decades, has failed with regard to the development of sustainable socio-technical systems. Both suppliers and customers have to adopt the functional constraints of the other stakeholders as quality objectives. Only the clear commitment to this and making the design process transparent do enable sustainable socio-technical systems in the future. To shift the paradigm from "just good enough" to "as good as it gets", the CIP, a well-known tool in business management, can be used.

5.1.2 Technical Specification

Daniel Martin, Hermann Kloberdanz, and Peter Groche

Uncertainty occurs in load-bearing systems in all phases of their virtual and real product life cycle, cf. Chap. 1. Especially, the development of new products is associated with particularly high uncertainty [216]. One important source for uncertainty is the technical specification of products, which is discussed in the following.

Uncertainty in product development can only be mastered in the long term, if it is given equal consideration to technical, economic and ecological requirements from the very first beginning. Correspondingly, reliable information from the technical specification supports decisions with far-reaching consequences in the development process.

New product ideas are often described by prospective users based on the intended application process or their contribution to the functionality of an overall system. However, the description of new product ideas is usually very imprecise. Thus, at the beginning of the development process, developers have to define the product properties in such a way that the resulting products are equipped with the desired properties in the use phase, cf. [224, 225].

The development task is analysed and detailed with regard to functionality and quality, effort, availability and acceptance, cf. Sect. 1.6. Usually, the development task is formally described in the form of requirements; and it is summarised in a document for the entire development team. The formal documentation is called technical specification. The document is frequently referred to as a requirements list, specification sheet, product concept catalogue or similar [69].

Although the technical solution is still largely unknown in the design phase, e.g. load assumptions must be made, product functions and basic dimensions have to be defined, as well as restrictions, e.g. with respect to costs and dimensions. Product development teams then use the technical specification during the entire development process as an information storage and decision basis.

The technical specification contributes to mastering of uncertainty in the development process by (i) defining requirements which are subject to uncertainty and (ii) directly specifying the requirements of the system in later use including foreseen uncertainties. Both contributions lead to a reduced modelling uncertainty.

The complete and detailed specification of the functional requirements helps product developers to reduce uncertainty during the creation of technical solutions. The completeness of the technical specification avoids that important functionalities are overlooked, whereas the sufficient level of detail prevents misinterpretations during product development. Both reduce model uncertainty, see Sect. 2.2, and structural uncertainty, see Sect. 2.3, in the development process, as well as uncertainty regarding the expectations of product use.

The models we have developed for the identification, classification and evaluation of uncertainty help designers to foresee influences relevant to uncertainty and to quantify an acceptable level of uncertainty, especially in product use. These can be formulated and documented as requirements in the technical specification. With the methods we have developed to master uncertainty in product development, uncertainty in life cycle processes can be predicted and the specified uncertainty requirements are met during product development.

Mastering uncertainty due to modelling

The development of new products is characterised by gradually more detailed models, such as functional structures, effect relationships, embodiment designs and geometry representations. The developers create these models based on assumptions, e.g. for loads, performance and available assembly space. These assumptions are based on information that is documented in the technical specification. Thus, the guidelines we developed to reduce modelling uncertainty [283] also serve as a guide to reduce uncertainty through the technical specification.

We exemplify the role of technical specifications by the development of a new bearing concept, presented in Sect. 5.4.2. The bearing is used as a key element in the challenging development of the 3D Servo Press, enabling a force of 1600 kN, see Sect. 3.6.3. It consists of a combination of roller and plain bearings. The use of bearings in such a special application leads to uncertainty in the development process, due to the complex load and installation conditions.

The choice of an appropriate and meaningful generic model for a development step is supported in particular by the most detailed possible documentation of critical system states and load cases.

For the description of the bearing function we use the model according to Holland [158], described in Sect. 5.4.2, which was adapted to the mode of operation of the combined bearing concept [247]. The model assumes a force equilibrium between the operating force F_{op}, the plain bearing force F_{pb} and the roller bearing forces F_{rb}. The plain bearing force F_{pb} is based on a pressure build-up in the lubricant, which depends on the dynamic viscosity η of the lubricant, which is supplied with pressure p_{in} and volume flow Q_{in}, as well as the speeds of the shaft, ω_{shaft}, and bearing shell, ω_{shell}, and the position of the shaft inside the bearing shell.

For common plain bearing designs, a quasi-static load condition of the rotating shaft by a constant load is assumed. In this case, the rotation speeds of shaft and bearing shell cause a non-constant pressure distribution p_{rot} in the circumferential direction due to the radial shaft displacement e. However, the detailed specification of the force curve over the angle of rotation, which occurs in mechanical presses, indicates that a significant temporal change of the shaft displacement occurs due to a force peak at the operating point. This leads to an additional pressure build-up p_{sq} in the area of the minimal bearing gap h_{min}, which is not covered by the lubrication model for steady operation. The resulting forces on the shaft can be expressed by the Sommerfeld numbers for rotation, So_{rot}, and squeezing, So_{sq},

$$\mathrm{So}_{rot} = \frac{F_{rot}\,\psi^2}{b\,d\,\eta\,\left|\omega_{shaft} + \omega_{shell} - 2\dot{\delta}\right|} \quad \text{and} \quad \mathrm{So}_{sq} = \frac{F_{sq}\,\psi^2}{b\,d\,\eta\,\dot{\varepsilon}}.$$

Here, δ describes the position of the minimal bearing gap in polar coordinates while $\dot{\varepsilon}$ is the squeezing rate. The consideration of these two forces in the model is of decisive importance for the specific application. In Sect. 4.3.5 we have seen how large the effects of model uncertainty on the results can be when a model is applied inappropriately because of ignorance. Initiated by the specification of the force curve over the shaft's angle of rotation, which varies considerably over time, the development of the combined roller and plain bearing system of the 3D Servo Press was based on the assumption that, in addition to the hydrodynamic effect, a displacement of the oil must be taken into account when designing the plain bearing. Since the relevant load conditions are thus recorded much more accurately, the uncertainty of the press development could be significantly reduced by the choice of the model.

In general, when specialising a generic model for the current development task, e.g. a functional structure, the developers define an appropriate system boundary and granularity of the model. The system boundary represents the part of reality that is considered as relevant. The granularity results from the scope and thus the number of components for modelling the system, see Sect. 1.3.

The technical specification influences uncertainty in modelling the system by information that acts as a basis for assumptions and decisions of the developers. In particular, detailed requirements regarding load assumptions, disturbance variables, resources and boundary conditions enable the developer to choose a model horizon that takes into account essential uncertainty-critical influencing parameters. The identification of critical uncertainty in the technical specification supports the definition of relevant components with adequate complexity. This ensures the conciseness of the model as a very important model feature and a prerequisite for successful product design.

We use the specification of the viscosity range of the system element lubricant which is a part of the roller and plain bearing as an example. The combined roller and plain bearing system is exposed to high loads. Due to its significant influence the temperature dependence of the dynamic lubricant viscosity η must be taken into account.

Another example concerning the adequate mapping of the combined roller plain bearing system in a model is the deviation of the assumed loading situation. A possible deviation from assumed conditions is given by an inclination of the operating load caused by uncertainty outside the system boundary of the combined bearing. The behaviour of the combined bearing under such disturbing influences is investigated in [132].

The force conduction through the roller bearings is calculated by means of $F_{\mathrm{rb}} = c_{\mathrm{rb}}\, e$ with c_{rb} representing the stiffness of the rolling elements, neglecting elastic deformations of the shaft and the plain bearing shell.

This example shows that neglecting insignificant influences contributes to simplifying models and improving their conciseness. The differentiation and detailing of critical requirements urges developers to decide explicitly on the admissibility of simplifying assumptions that affect particularly critical system properties, see Sect. 4.3.

In the example of the combined bearing system, the dependencies of viscosity on pressure and temperature were recognised by references in the technical specification, but the uncertainty by neglecting these dependencies was assumed to be acceptable.

In summary, it can be stated that the modelling of the system is based on the information of the technical specification. The technical specification supports the developers in making relevant assumptions and to work out appropriate system models by specific documentation of critical functional requirements. Thus, it can reduce the influence of uncertainty. Model uncertainty can be effectively controlled by incorporating all relevant system areas and system properties into the development process to optimally meet customer expectations, despite the simplification of reality.

Mastering uncertainty due to requirements

As shown in Sect. 1.6, informal acceptance can represent the fulfilment of customer expectations, which has a significant influence on the success of the product. As described there, a high level of acceptance is achieved by reducing effort and increasing availability. In order to ensure a high level of availability in product development, product developers detail this objective right from the start in the form of requirements that are as precise as possible or even quantified. They document these requirements in the technical specification as a basis for decision-making during product development.

However, the quantification of availability depends strongly on product usage and is not always possible, as already stated in Sect. 1.6. The models and methods we have developed for the identification, evaluation and quantification of uncertainty and uncertainty influences allow in these cases the indirect specification of availability requirements without the technical solutions being known.

The type of the product to be developed and its conditions of use determine the scope and description of the uncertainty that must be taken into account in the development. Depending on the degree of uncertainty these are documented in the technical specification in the form of requirements regarding reliability or robustness properties.

Systems and machines for stationary use in production are usually operated in factory buildings with largely determined environmental and process conditions. The reliability specification defines "the probability that the required function of a product will not fail during a defined period of time and under given working conditions" [143]. Reliability is indicated in the technical specification by performance indicators such as *Mean Time Between Failure* (MTBF), *Mean Time To Repair* (MTTR), *Uptime* or *Production Yield*. Reliability is therefore based on the assumption that the causes of uncertainty are largely known and vary only slightly. The specification must therefore ensure that disturbance influences are defined or eliminated during the operation of the systems.

When developing robust systems, information regarding uncertainty causes in the specification are a prerequisite to guarantee a high availability in use, cf. [192].

In our case of highly loaded roller bearings, lubrication must be ensured so that the calculated service life and frictional torques (and thus power loss) are achieved during operation. Regarding the combined roller and plain bearing presented in Sect. 5.4.2, the volume flow Q_{pb} of the lubricant exiting the plain bearing in the axial direction serves to lubricate the roller bearings. However, this coupling of the lubricant flows of the plain bearing and the rolling bearings has a considerable influence on the system behaviour and must therefore be taken into account in the development. However, on the one hand, the lubricant properties are predetermined by the design of the plain bearing and, on the other hand, the lubricant temperature and the lubricant flow vary depending on the operating conditions of the plain bearing. The developer can only master the undefined lubrication conditions by a robust design of the rolling bearing. A prerequisite for mastering the uncertainty in the robust design of roller bearings is therefore comprehensive information on the properties of the lubricant and the lubricant flow in the technical specification.

Conclusion

In a nutshell, during the development of new products, developers may significantly influence the uncertainty of the product to be developed with the technical specification. The specified properties of the system can have an indirect or direct influence.

Mainly the complete coverage of causes of uncertainty contributes to mastering uncertainty in the product development process. A mostly complete specification of functionality and uncertainty influences can reduce model uncertainty and indirectly enhance correct predictions and availability of the system and thus the informal acceptance.

Uncertainty can be directly reduced if uncertainty effects from and on the system environment are captured and specified comprehensively and as detailed as possible, preferably quantitatively. The checklists of the robust design methods of the early phases of product development can be particularly helpful in this respect.

Elaborating the technical specification, it is purposeful to anticipate development steps and to support the developers in making assumptions by specific information structured according to model characteristics, cf. [283].

5.1.3 Product Safety Requirements for Innovative Products

Laura Joggerst and Janine Wendt

Product recalls, as discussed in Sect. 1.1, are expensive and harmful to the producer's image and therefore should be avoided. Although product recalls become relevant in the usage phase of a product, the root cause often traces back to the design and production phase. Figure 1.3 presents an overview of the three product phases and how they are linked. But product recalls are only necessary for products that can be dangerous, i.e. they present potential risks to the body, health or life of consumers, users or third parties in general. Reasons for recalls can be manifold. For example, in certain sectors, the product quality decreases due to economic factors and strong competition [278, p. 62].

But another reason for product quality not complying with product safety requirements can be, that the requirements are becoming more and more diverse and complex, Sect. 1.1. Or even the fact that no technical standards are applicable to new and innovative products.

If a product is in use and it turns out that it poses a safety issue, it needs to be taken off the market. Legal product safety is this strict due to its preventive character. Its aim is to ensure that only safe products are in use and on the market. Together with product liability, product safety requirements provide full protection of the consumers' legally protected rights. In this subsection we aim at clarifying what is required for the producer to prove the safety of his product.

Product safety

The Product Safety Act is the main legal framework for product safety in Germany. It is accompanied by numerous technical standards, which specify requirements for the practical application [173, marg. 2]. Furthermore, it is subject to European regulation, which is why many of the technical standards are developed by European Standardisation Organisations. Basically, two groups of products need to be distinguished. Those which fall into the category of harmonised products and those which do not. In the case of harmonised products, the producer compulsorily needs to comply with the provisions, applicable for the specific case. These provisions are mostly based on European Standards. If the product complies with said provisions, it is assumed to be safe and can therefore be placed on the market [195, marg. 3]. National provisions for harmonised products often only regulate a minimum of safety which the product has to achieve. They are often accompanied by industry standards which are more detailed but not obligatory since they lack legal quality [66, p. 42]; [117, p. 1492]. In the case of non-harmonised products, the producer can exclusively follow national standards. Similarly to harmonised products, the compliance with these standards implies their safety. But, the application of these technical standards is not mandatory. Nevertheless, producers need to prove the compliance of the product with safety standards set by the national provision and technical standards, as well as state-of-the-art technology. On the contrary, the non-compliance with said standards does not automatically imply that the product is unsafe. Unsafe products are rather those which pose a potential risk to the legal rights the Product Safety Act aims to protect namely a person's health, body and life.

The Product Safety Act is the hurdle which anyone placing a non-food product on the market for the first time needs to clear. At the same time the product needs to be made available on the market for commercial use. Further, the Product Safety Act is not only applicable to products, but also to facilities which are deemed hazardous and thus require monitoring and inspection. For this discussion, we will solely focus on products.

Market surveillance

In cases of non-compliance with the safety requirements set by provisions and technical standards, Market Surveillance Authorities can direct producers to different behaviour in order to avert danger for consumers, due to the respective product. These measures may consist of requesting the producer to take action in order to end the risk accompanied with the product. They can also prohibit the further distribution of the product, or, in particularly risky cases, even demand the product to be recalled. Any of the mentioned measures are likely to have a negative impact on the producer's reputation and are certain to cost money. Therefore, the objective is to avoid authority intervention at all times.

RAPEX

The so-called RAPEX-System (Rapid exchange of Information) enables both member states and the European Commission to exchange information on dangerous products which are on the European market, but are being taken off the market, or are part of a product recall.

In addition, the RAPEX-System determines the level of risk of a product [174, p. 38]. Producers need to apply this assessment system to their product. The level of risk also determines possible measures taken by the market surveillance authorities.

Technical standards and the proof of safety

Technical standards are developed by private institutions, which collect knowledge from industry standards and compile it into one comprehensive document. These standards are more user-friendly than the requirements set by legal provisions, since they can have a very specific scope of application. Furthermore, legal provisions are vaguer in order to be applicable to a wide variety of use cases. As mentioned before, technical standards concerning non-harmonised products have no legally binding character, as they do not undergo a legislative procedure. Nevertheless, they represent the state-of-the-art technology. Therefore, when the producer designs a product following the technical standards, it implies that the product complies with the current minimum safety standard existing in the industry. As a result, the product is deemed to be safe and can be distributed on the market.

Non-compliance with existing technical standards

If producers do not comply with existing and applicable technical standards, this does not per se indicate that the product is unsafe [165, p. 722]. The reasons for this non-compliance can vary. The producer could just have overlooked the existence of the technical standard or he could have consciously neglected it. The reason behind this could be that the producer found a better, easier or cheaper option to provide the same level of safety of the product with a different construction than recommended by the technical standard. Therefore, it is necessary for the producer to prove that the minimum safety, set by the technical standards applicable, is met otherwise.

Missing technical standards

Another problem arises, if no technical standards are available to the producer when designing the product. This can be the case for innovative products, when no clear industry standard has been developed, yet. Without technical standards, there is no implication for the safety of the product in terms of the Product Safety Act [117, p. 1493]. So, how can the safety of the product be determined? This question is particularly important for the producer, as the product cannot be marketed without the positive safety implication.

In most cases, it is neither practical nor possible to apply technical standards developed for a different scope of application to another product, even if the products are similar [213].

It becomes apparent that the effort for providing proof of the safety of the product is much higher than in the case of existing technical standards. The producer must apply some kind of risk analysis to prove the safety of the product. But no official guidelines exist for how this analysis should be executed and what information needs to be taken into account. This uncertainty increases the efforts of the producer to provide the necessary proof of safety even more.

Practical guidelines would help the producer to do what is expected. This would equally make products safer compared to the case where a producer sets his or her own standards to prove the safety of any innovative product. German jurisdiction only offers guidance, as it was decided that sampling inspections are not a permissible basis for providing proof of the product series as a whole [213].

Machinery directive

The Machinery Directive [97] also demands a risk analysis and gives guidance as to which aspects the producer needs to consider during the analysis. For example, the producer needs to consider the intended use in addition to the foreseeable misuse of the product and determine the limits of the product. In the case of the Machinery Directive, the product is a machine and the Directive only applicable to such. Additionally, the producer needs to identify potential hazards and dangerous situations caused by the machine. Just by mentioning these requirements, it becomes clear that the risk analysis described by the Machinery Directive is not very specific and does not include specifications on how the information to evaluate said aspects should be gained.

Product safety and market surveillance package

The requirements stated in Article 8 of the Proposal for a Product Safety and Market Surveillance Package [96] are similarly vague. In this proposal, technical documentation for the product is required and should contain, amongst others, an analysis of the possible risks related to the product. But again, it contains no guidance as to which information is relevant for the analysis.

Non-legislative guidelines

The decision of the European Commission concerning the RAPEX-Guidelines [95] is a non-legislative and therefore non-binding document. But other than the mentioned legislative acts, this guideline provides useful information on how to perform a risk analysis for products and recommends a three-part assessment.

The foundation of this assessment is to describe the product and the inherent danger as well as identifying the group of potential users. Regarding this, intended, as well as unintended users need to be considered. Potential users should also be categorised according to their vulnerability. For instance, children and elderly people are vulnerable users in most cases. But depending on the use case or type of product usually not particularly vulnerable users might become vulnerable, for example, if warnings and instructions on the product are written in a foreign language. If different types of users can be identified, the risk assessment might have to be performed for different scenarios considering the different users in order to identify the highest risk combination. For the actual risk analysis, the producer should first anticipate a situation in which the product inherent risk manifests itself by injuring a person. This scenario will mostly revolve around the potential defect the product can have during its lifetime. This defect then causes the injury of a user or person, who comes into contact with the product.

Furthermore, it is significant for the assessment of how severe the anticipated injury would be. According to the RAPEX-Guideline, factors for the evaluation can be the duration the hazard of the product has on the potential victim, the body part which would potentially be injured and the impact on the respective body part. For example, losing a finger is a much more severe injury than lightly burning the skin on one finger. The severity of the injury can then be used as an indicator for the level of danger the product poses and vice-versa. In case different scenarios are possible, the one causing the severest injury should be considered first, as preventing this scenario provides the greatest safety for potential users. The second step would be to determine the probability of the scenario and therefore the probability of the person getting injured due to the hazardousness of the product. The RAPEX-Guideline recommends separating the anticipated scenario into smaller steps, identifying the probability of each individual step and then multiplying the individual probabilities to receive the overall probability of the scenario.

Lastly the RAPEX-Guideline recommends that the producer should combine the two steps, joining hazardousness and probability to obtain the risk of the product. Four risk categories are intended: serious, high, medium and low. Figure 5.4 demonstrates the combination of the probability of the injury, the severity of the expected injury, as well as the impact on the risk classification. Depending on the risk category, it is expected of the producer to apply a suitable effort to prevent the calculated risk. The higher the risk, the more important it is to mention that the whole process behind the described risk analysis should be documented, in particular in the case market surveillance authorities request this information. Figure 5.5 provides a simplified overview of the recommended procedure.

Finally, the recommended sources of information for all the consideration above are accident statistics, knowledge gained from previous and similar products, as well as experts' opinions. As the problem discussed regards the question of how producers of innovative products can provide proof for the safety of their product in case no technical standards exist, it is highly likely that neither knowledge nor accident statistics exist that would be applicable to the product in question. Therefore, producers should base their assessments on information gained from experts' opinions but not without checking the obtained information for plausibility.

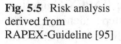

probability of damage during foreseeable lifetime of the product		severity of injury			
		1	2	3	4
high	> 50 %	H	S	S	S
	>1/10	M	S	S	S
	> 1/100	M	S	S	S
	> 1/1000	L	H	S	S
low	> 1/10000	L	M	H	S
	> 1/100000	L	L	M	H
	> 1/1000000	L	L	L	M
	< 1/1000000	L	L	L	L

S – serious risk
H – high risk
M – medium risk
L – low risk

Fig. 5.4 Risk level from the combination of the severity of injury and probability derived from RAPEX-Guideline [95], Table 4

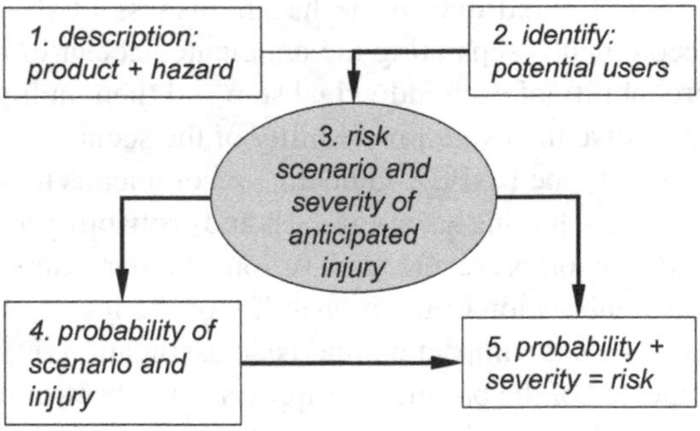

Fig. 5.5 Risk analysis derived from RAPEX-Guideline [95]

Conclusion

We conclude that technical standards are a useful tool for mastering uncertainty. Compliance with these standards implies that the minimum safety requirements are met and that the product is safe enough to be distributed on the market. In some cases of new technology, technical standards do not exist and therefore no guidance is available to the producer. As a result, the producer must provide proof of the safety of his product in other ways. As no legally binding documents exist, we found that following the risk assessment described in the RAPEX-Guideline is currently the safest way for the producer to demonstrate the safety of his product.

5.1.4 Legal Uncertainty and Autonomous Manufacturing Processes

Laura Joggerst, Janine Wendt, and Peter Groche

In the future of manufacturing more and more autonomous systems will be implemented to raise product quality and product safety. But first and foremost, they are implemented to make production processes more cost- and material-efficient. Although fully autonomously acting manufacturing processes are still to come, the discussion of relevant legal issues is very present. Without autonomous manufacturing processes on the market, there can be no jurisdiction to base legal requirements on. Uncertainty regarding legal requirements and legal risks can be one reason for the stagnation of innovation. The general issue and characterisation of uncertainty is discussed in Chap. 2. If the legal requirements that manufacturers must comply with could be specified today, innovation could be promoted and guided in the right direction. This tool for mastering legal uncertainty accompanies the methods for mastering uncertainty in a general technical sense as discussed in Sect. 3.3.

Legal literature often discusses, whether or not our existing liability regime is applicable to new and innovative technology [285]. This question arises because the current legal liability framework was established during a time way before automated processes were thinkable. Without going into more detail, it can be stated, that the current liability regime has such a broad spectrum that even autonomous systems generally fall into its scope. It goes without saying that some alterations will have to be made to the way we interpret regulations. Without these alterations, liability gaps can occur.

For the sake of a liability regime that is well suited for the innovative technology to come, and provides guidance as to what is legally expected from innovators and producers, this interpretation needs to take a practical approach. What are the specific problems that occur, when using, e.g. autonomous production processes, and how do we interpret the existing liability regime accordingly? Regarding liability risks in manufacturing processes, both conventional and autonomous, we are concerned with two main rules: Sect. 1 paragraph 1 of the Product Liability Act and Sect. 823 paragraph 1 of the German Civil Code. Contracts between the parties can also result in compensation, if they are violated. In the context of this discussion, however, we will exclusively focus on the aforementioned rules. Firstly, this is so because contracts are only binding between the parties resulting in the risk of uncertainty being much lower. Secondly, because the real legal uncertainty lies in the liability for damages inflicted on users and third parties.

Section 823 paragraph 1 of the german civil code

Section 823 paragraph 1 of the German Civil Code applies to damages done to life, body, health and property of a person. It is the key rule of German tort law. Manufacturers' liability is only one facet of this rule, for a historical account see [36]. Its broad scope applies to many different actions causing the damages mentioned above. The first requirement of this rule to be applicable is that damage is done to one of the protected rights. Secondly, the damage must be caused by an intentional or negligent action. Finally, this action needs to be unlawful, meaning, not complying with the general principles of the law. Assuming that, in most cases, the manufacturer does not intend to harm another person's rights, which is considered to be an intentional action, the large number of cases will evolve around the question of how to determine negligence. A negligent action is based on the violation of a security obligation. Generally, anyone who uses or creates a source of danger to others, needs to ensure that no harm is done [37, marg. 10]; [33, marg. 21 f.]. The security requirements that derive from this obligation depend on different factors: Potential users, whether they are consumers or professionals, intended and unintended but foreseeable use cases, which are limited by intentional misuse. All of these criteria need to be taken into consideration by the manufacturer. Users' safety expectations must also be taken into account as far as they are reasonable.

In addition, the product has to comply with the current state-of-the-art safety standards. In this context, current refers to the point in time in which the product is marketed. German legislation has developed specifications for the safety obligations, the manufacturer has to comply with. They can be categorised in the areas of design, construction, instruction and monitoring of the product during its use phase. During the design phase, the producer is expected to choose the final product design with the required safety in mind. At the same time, he needs to document his decisions in this respect. In the event of liability claims, the producer needs to prove that no other design alternative would have been better suited to prevent the occurred damage [32, marg. 16].

Product liability act

For damages done to body, health or property of a person, Sect. 1 paragraph 1 of the Product Liability Act is applicable besides the general provisions of tort law. The Product Liability Act explicitly addresses producers of defective products. A product is defective, if it does not meet the expected safety standard. Safety requirements can be established similarly to Sect. 823 paragraph 1 of the German Civil Code. The same categorisation applies, except there is no obligation to monitor the product during its use phase. This is due to the fact that the aim of the Product Liability Act is to ensure that only safe products are placed on the market. Therefore, its relevance ends with the first placement of the product onto the market. Everything past this moment can only be regulated by Sect. 823 paragraph 1 of the German Civil Code. The main difference between the two provisions is the fact that the liability stated by the Product Liability Act is not based on fault, making it more consumer-friendly.

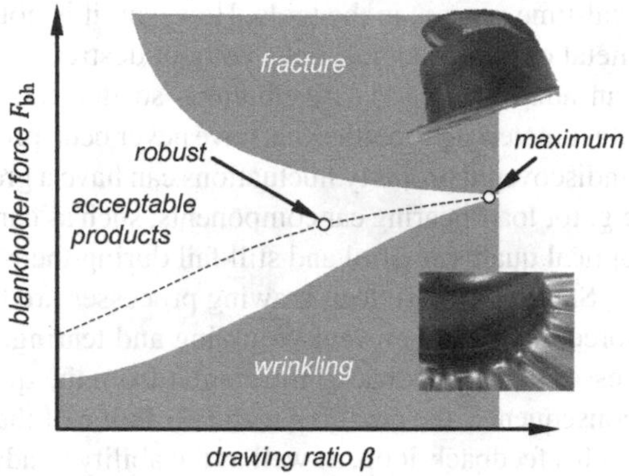

Fig. 5.6 Process window for deep drawing following [111]

Automated deep drawing

For discussion purposes, we are going to evaluate the product liability issues concerning autonomous production processes referencing an autonomous deep drawing process. In order to get a better understanding of the problem, it is necessary to evaluate current state-of-the-art closed-loop controlled processes.

The demands on geometry and ever new materials pose a challenge for conventional deep drawing processes. The wall thickness of the final part has to be further reduced in order to make better use of the material, which increases the risk of wrinkling and cracking [7]. The use of more robust and thinner materials contributes to this effect as well. In addition, there are the fluctuations inherent in the process or the fluctuations in the material. All of this narrows the process window [168]. The deep drawing process consists of different steps, starting with the production of a blank cut from supplier coil material, followed by the actual deep drawing step and the quality control. The highly automated process has the ability to take data into account, which is provided by the supplier of the coil material. Also, the process can forward information from each production step to the next. Fluctuations and uncertainty in materials and between the individual steps of the process can therefore be taken into consideration. Information on the quality of the final product can then be fed back to adapt blank holder forces in the deep drawing process. Achieving the maximum drawing ratio, as displayed in Fig. 5.6, enables better material utilisation and the deep drawing of more complex geometries. Drawing ratio in general refers to the ratio of the blank diameter to the punch diameter. In conventional deep drawing processes, the maximum drawing ratio is linked to a higher risk of wrinkling or tearing, and therefore, to potential failure of the final product. With highly automated processes, the desired maximum drawing ratio becomes more achievable without taking higher risks.

Quality control checks random samples for compliance with specified product requirements. This is achieved by both optical measurements for the geometrical properties and destructive testing for robustness. Further, it is possible to measure blank holder forces, draw-in ratio of the blank, as well as wrinkling and tearing with real-time sensors in the tools. However, it is not possible to detect thinning of sheet metal or microscopic cracks without destructive testing. Thus, although the process can adapt to material fluctuations so that the geometry produced stays constant, other material properties that have never been problematic before can fluctuate. These undiscovered property fluctuations can have a great impact on the safety of a product, e.g. for load-bearing car components, such as car doors. Therefore, products can pass optical quality control and still fail during their use.

State-of-the-art deep drawing processes are based on the control of blank holder force in order to prevent wrinkling and tearing. Still, these offline closed-loop processes do not feedback information from the quality control or crash test data. As a consequence, the next step in the evolution of the deep drawing process must include such a feedback-loop, as well as the ability to adapt to an online closed-loop process. This will also involve processes to be able to learn. But while learning from geometry, data alone will prevent wrinkling and tearing, necking for example might be left unseen. In addition, the learning process needs to consider strength parameters. Since measuring strength has not yet been feasible inline, the measuring effort would be high. For the process to learn according to given strength parameters, measurements have to be taken frequently, for example every 100 parts.

Outlook on future autonomous processes

The autonomous production process will be able to make many relevant decisions by itself without human interaction, assuming that there is enough data from the use phase and failure of the previous product available. Even decisions concerning the design will be made by the process itself, for example choosing the coil material for deep drawing. According to the desire to produce more light-weight products, the process might decide to use a material that has not been used before, by applying the knowledge it has achieved from previous iterations with other materials. Errors might occur, if the new material behaves differently, due to the fact that it is more brittle, for example. Therefore, microscopic cracks may occur in the final product that cannot be detected 100 % by optical quality control, yet. Consequently, the process could not adapt its production parameters as there is no data base to learn from. The process might have to be limited to materials that have already been tested before. In this sense, the process could not act fully autonomous. Nevertheless, the ability to learn and adapt production parameters for every product counteracts material fluctuations as well as fluctuations in tolerances from upstream processes, and therefore copes better with many of the uncertainties equally occurring in current production processes.

Discussion

If products, manufactured by an autonomous production chain, cause harm to a person's rights, the question arises who will take responsibility for the damages done. In most cases both Sect. 823 paragraph 1 of the German Civil Code and Sect. 1 paragraph 1 of the Product Liability Act apply. The Manufacturer of the product is held liable. Section 823 paragraph 1 of the German Civil Code requires a breach of duty of care, whereas liability under the Product Liability Act requires a defective product. Therefore, different categories of defects have been established: errors in the design, the fabrication, instruction or product monitoring. Errors in fabrication refer to deviations between the designed product and the final product after manufacturing. But, a 100 % error-free production is not expected of the producer. Therefore, he is not held liable for so called outliers in production [35, marg. 12]. For autonomous production processes, this category is no longer useful. Therefore, we focus on the other categories.

The interesting, and from a legal perspective, important feature of the autonomous production process is its ability to learn. During the deep drawing process, the parameters can be adapted to produce a product with a defined geometry. The process decides on the used material, the applied blank holder force and the duration of the deep drawing process. All these choices influence not only the geometry of the final product, but also its safety, for instance in the event of a crash. Conventionally, these choices would be made by the manufacturer, but in autonomous production processes, decisions are transferred to the process itself. Consequently, legal obligations concerning the safe product design can no longer be ascribed to a specific choice of the manufacturer. In order to identify legal obligations for the use of an autonomous production process, we have to consider the production process in its entirety, rather than the individual product, for a deeper discussion, also see [163].

Use of autonomous production processes

The first question to answer is whether autonomous production processes should be used at all, which is a crucial question for all new technologies that pose new risks. In order to give an answer, the interests at stake need to be considered. In an autonomous production process, parameters are set automatically during the process itself, rather than manually as in conventional production processes after a product inspection. Ideally, autonomous production processes minimise the risks of human misbehaviour and thus the risk of product failure.

In reality, there is still a lot of uncertainty, when trying to identify the actual risk of a product failure. Along the example of the automated deep drawing process, manufacturing close to the process limit also poses the risk of material thinning, which cannot be detected in a non- destructive way, yet. At the same time, completely new failures could occur, which have not been relevant for conventional production processes yet. The manufacturer needs to ensure that the products manufactured by such a process are as safe as the current state-of-the-art products [32]. The learning phase during the development of the production process needs to produce a steady product quality complying with the current state-of-the-art safety standards. Even then, a residual risk cannot be eradicated completely. Therefore, manufacturers have to find a way to quantify the risks as well as the safety-gain inherent to the autonomous production process. At which level of residual risk the autonomous production process may be used is not clear yet. This depends on the acceptance of such processes in the public and political opinion. Nevertheless, setting a standard for the permissible risk should not be left to the industry. Instead, clear guidelines should be developed by political institutions.

Obligations during the use of autonomous production processes

When applying the currently applicable legal framework to the technological innovation of the autonomous production process, two problems become apparent. Firstly, the learning character of the process is based on failure. Learning can only come from failure, whereas the legal obligation is based on avoiding such failure. Secondly, the functionality of the process is based on software. It is acknowledged that software cannot be produced 100 % free of errors. Knowing this, the manufacturer needs to take particular care in monitoring the functionality of the production process [253, p. 3147]. This obligation is conventionally not prioritised from a legal perspective, but serves as a compensation for the fact that all relevant design choices are made by the process, not the manufacturer. The producer's obligation to design a safe product is therefore less demanding.

The question then is, how this obligation to monitor the process should be implemented. It is also worth mentioning that one cannot always conclude the safety of the product from the fact that the process functions correctly. As a result, the manufacturer should also identify safety relevant properties of the product and monitor them as well. The effort, the manufacturer needs to put into monitoring the process and the product depend on how the process learns and adapts. Particularly the frequency of a learning impulse is relevant, since the process only changes and adapts the process parameters after such impulses. If the process is based on the data of every single product, parameters can change after each product. Thus, the manufacturer would also have to monitor the safety relevant properties of each product. Whereas, if the data of only every hundredth product is used as a learning impulse, a change in the process parameters only occurs at this frequency. Then, it would be sufficient to check the properties of every hundredth product, i.e. the product after the learning impulse.

Obligations concerning the products of autonomous production processes

Lastly, the manufacturer has an obligation to monitor the product during its use phase. This way, defects that occur after a certain time in combination with other products or in general "in the field" can be detected. The information derived from this monitoring process is valuable and must be used to either prevent damage to protected legal rights or simply as input for future product development processes. The measures that need to be taken to monitor the product are manifold and depend on the specific product. But in general, it can be stated that the manufacturer needs to implement a complaint management system [103, marg. 379]. In addition to managing this passively achieved information, the manufacturer needs to actively inform himself about possible defects and errors concerning his product by checking newspapers, relevant journals [34, marg. 34], test reports or internet forums [146, p. 2729]. With the advancing technological innovation, it could also be possible to monitor products in real time. This would pose new legal questions which have to be discussed in the future.

Conclusions

In summary, autonomous production processes enable different and more efficient production ways. The current legal framework adapts to the new production technology. Manufacturers need to "train" the autonomous process until reliable results are achieved. The residual risks and the safety gained are to be quantified in some way. Guidelines to what level of residual risk is permissible should be developed by political institutions. When using an autonomous production process, the focus must be on monitoring the process and the resulting products during the production process, as well as on monitoring the products during the use phase.

5.1.5 Optimisation Methods and Legal Obligations

Laura Joggerst and Janine Wendt

Computer software and hardware have improved so far that using simulation and optimisation tools in product development processes has become a common feature. These tools allow for more complex designs and larger systems, as well as improvement of development time, accuracy and safety. At the same time, optimisation methods are becoming more and more popular in developing product parts or a system topology. A new trend in mathematical optimisation is the treatment of resilient systems. Methods for developing resilient systems explicitly take failures of components into consideration, trying to guarantee at least a predefined limited function of the system itself. Although these methods are highly promising, the implementation is very complex and is usually done in the context of data, model and structural uncertainty, see Chap. 2. This uncertainty can then result in legal liability, if a person's rights are harmed. In order to minimise legal uncertainty in the development process using optimisation methods as well as resilience optimisation methods, we need to understand the way these methods work and how their implementation affects the legal product development requirements we know. The concept and use of resilience for mastering uncertainty is described in Sects. 3.5 and 6.3.1.

Legal literature usually takes a more abstract approach to discussing the impact new technologies and new algorithmic design methods have on the existing legal framework. This is why we take a rather technical approach in order to gain a more specific and concrete idea as to how our current liability regime adapts to new algorithmic design methods. We also want to discuss which legal requirements can be derived for producers to comply with, and if and how the current legal framework might have to be adapted.

Legal liability

As seen in Sect. 5.1.4, Sect. 823 paragraph 1 of the German Civil Code is the key liability rule. In addition to producer liability, the provision generally covers cases in which a person is injured, killed, or the person's property is impaired by a source of danger in circulation. The basis of this liability is the violation of a duty of care. Anyone who benefits from a dangerous product has to ensure that this product does not harm others [28]. In this sense, dangerous products are those, which are potentially dangerous for others, simply because of their nature, for example cars. The determination of necessary precautions depends on a number of factors. On the one hand, it depends on who the potential users are and therefore who is potentially at risk. On the other hand, it depends on which dangers can potentially occur. All this has to be considered by the manufacturer of a product or the operator of a system.

It should be emphasised, however, that only what is reasonable can be expected. The manufacturer or operator does not have to eliminate every residual risk at every price in order to be spared from liability. What can reasonably be expected varies from case to case. In principle, however, it can be said that the safety standard set by state-of-the-art technology has to be satisfied. On the one hand, the state-of-the-art is represented by norms and standards, but on the other hand by the solutions available on the market. Still, compliance with the current norms and standards alone does not necessarily protect against liability. In some cases, it will rather be necessary to take safety measures beyond norms and standards that are available according to the state-of-the-art technology. Norms and standards thus form a kind of minimum safety that should not be undercut. Although they are not legally binding, non-compliance with them has an indicative effect for non-compliance with safety requirements, which would have been possible according to state-of-the-art technology [31, with further references, marg. 16].

Resilience

The anticipation of several possible failures of the system or a product is a promising step into trying to make products safer. The idea is to ensure a pre-determined minimum functionality that the system or the product will still be able to achieve in case of disturbance or the malfunctioning of a specific amount of product components. Resilience can be implemented in the product development process, for instance by only considering design options that achieve the minimum functionality in the event of k component failures [9]. The product or system then has a so-called buffering capacity of k. For more than k component failures, no reliable outcome can be predicted. This is, in a way, a restriction when designing a product using resilience optimisation methods. Nevertheless, it is possible to concentrate on a defined number of component failures, rather than what causes a failure of the product or system at large. Furthermore, it is worth mentioning that the number of possible failure scenarios increases with the number of components and design options. Hence, the enumeration of all possible failure scenarios becomes practically impossible. More details on resilient system design can be found in Sect. 6.3.

Review of legal obligations

In order to identify the manufacturer's duty of care when using optimisation methods, we take a closer look at the selection of the model on which algorithmic design methods are based on and differentiate between optimisation methods in general and resilience optimisation methods.

Selection of the model

All algorithmic design methods are based on a model of the system. In order to specify the legal obligations when using algorithmic design methods, considering the selection of the model of the system should be the starting point. These models always represent reality in a simplified way, see Sect. 1.3. Therefore, one model only applies to the specific application for which it is designed. If a model is used for the development of an application for which it does not provide reliable information, considerable deviations between the modelled system behaviour and the actual behaviour can occur during the usage phase.

It is important to note that in most cases many models are available, while some represent reality better than others. This results in the important restriction of the most reliable models often being too complex and thus too time-consuming for computation. In this respect, the developer needs to find a balance between an accurate model and the appropriate computation time.

When discussing legal obligations, the level of risk to third parties' protected rights has to be determined and should be foreseen by the developer of a product [26, marg. 6]. First of all, the developer would have to ensure that the model he uses can generally provide reliable information about the product. The developer also needs to take the boundaries of the model into consideration, cf. Sect. 2.2. Furthermore, he has to have some kind of proof that the approximations made on the basis of the underlying model transfer into reality. This is usually achieved by model validation and verification. With the help of either experimental data or simulation results, it can be shown that the predicted behaviour is in accordance with the actual behaviour of the product. This procedure can be expected from the developer. Validating the model can imply that the producer of the product has fulfilled the safety obligations in the development phase. But in order to fully protect the developer from liability claims, safety aspects must be reproduced by the model as well.

Use of optimisation methods

Optimisation methods allow the development of more complex systems or products, as they simplify the design choices for the development process. These methods are based on mathematical models of the system and its environment. As already mentioned, one of the causes for uncertainty in using such methods arises from the fact that models can never be an exact representation of reality, as the latter is too complex to be computed. The outcome of the design with optimisation methods depends on the chosen model and on the chosen input parameters. These parameters are mostly economical and structural factors set by the developer. Moreover, the number of input parameters is finite. Safety of the product as such cannot be an input parameter. In fact, an optimal solution for the given optimisation problem can be structurally unsafe. This becomes clear, when considering optimisation problems that are used to minimise the material usage of a product.

As the producer cannot rely on the optimisation method to choose a particularly safe product design, he still needs to ensure that safety standards are met. So far, this process does not deviate from a development process which does not rely on optimisation methods. In a conventional development process, the developer would consider different design alternatives which solve the problem defined. These are the alternatives the developer would then have to choose from [274, marg. 972]. The decision would be made considering functionality and economic factors. Simultaneously, safety requirements can and should be considered when choosing the final product design.

In using optimisation methods for product development, the numerous possible design alternatives are not considered by the producer himself. The algorithm decides on one single product design that fits the defined problem the best, given the predefined input parameters. Output is therefore only one optimal design option. This one design alternative can of course be tested by the producer considering safety aspects. Conventionally, this does not suffice legal standards as they are developed by jurisdiction. The act of choosing the design and the consideration accompanying that process represents an important step that is ascribed to the producer. But imagining the producer would have to trace back every possible design alternative the algorithm considered during the optimisation process, all facilitation gained by using the algorithm would be lost. Apart from that, the time, cost and computational efforts would be great. Only measures are legally expectable, that are reasonable. These legal standards need to be adapted. Otherwise, innovations might be hindered. Therefore, to demonstrate the safety of a product in a way that exonerates the producer from liability claims, it needs to be possible to check only the design alternative the algorithm provides as an optimal output. Then again, it has to be taken into account that complex products or large systems, such as a bridge, cannot be tested in 'real life'. Prototypes have their limits, where the size of the system or the number of components exceeds the feasible. Consequently, producers need to check and prove the safety of a product or system by using simulation methods. Then, the problem we discussed at the beginning becomes relevant again. Simulation algorithms are based on models of the product and its environment. Hence, the producer has to be certain about the boundaries of the simulation and the underlying model.

Use of resilience optimisation methods

Resilience optimisation enables the development of systems and products that can withstand a defined number of failures of an undetermined cause. For a defined number of failures within the system, functionality can be guaranteed. It is not possible to foresee every possible damage scenario in the development. The resilience optimisation method can therefore make this uncertainty more manageable. If the identification of potential users and dangers to them and third parties has been an essential part of a developer's duty of care up to the current moment, this is no longer considered when using resilience optimisation methods. The system developed in this way guarantees a minimal functionality despite unknown failure scenarios. As long as the functionality of the system is guaranteed, no one will be harmed. At least in theory, resilience optimisation promises safer products. In reality, however, it is uncertain how these systems react to failures or damage events. Furthermore, the model and data uncertainty within the design stage might be significant. Nevertheless, resilience optimisation can be used to support engineers in the early design stage as for instance shown in Sect. 6.3.5.

As with the development of any new technology, the question is whether the product can be brought to market at all. If the risks of the product cannot be weighed against the supposed safety gain, marketing cannot be recommended [35, marg. 17]. It should also be noted, however, that currently neither standards nor practical experience can be used to develop products using resilience optimisation methods. In order to answer the question of marketing, appropriate methods for assessing the risk of the systems would have to be developed first. If the comparison with conventionally optimised systems turns out to be in favour of the safety gain, the product can be marketed.

Once this fundamental question has been clarified, the question still remains how the current legal framework can be transferred to the systems developed in this way. So far, the starting point has been, among other things, the obligation of the manufacturer to correctly assess the danger of the product for potential users, as well as third parties. This assessment should then be used as the basis for selecting an appropriate design. As already discussed regarding the conventional optimisation methods, the selection of a design alternative is now left to mathematical optimisation. Furthermore, due to resilience optimisation methods, the preceding step of assessing potential danger and user is omitted.

In the end, all that remains is the legal premise that the manufacturer must do everything reasonable to ensure the necessary safety of the product. What the manufacturer can reasonably be expected to do must be determined by weighing up the interests of the parties involved [29, marg. 10]; [27, marg. 7]. First and foremost, it will be a matter of comparing the expected safety gain from resilience optimised products and the remaining residual risk. Moreover, it must also be considered which legal interests are potentially at risk. The higher the significance of these legal interests is classified, the more resource-intensive security measures can be expected [30]; [35, marg. 8]. For example, if a person's life is at risk, the producer is expected to take more costly precautions to avoid this danger compared to a case in which a person's property is endangered. Nonetheless, the economic interests of the manufacturer can also be taken into account [35, marg. 18]. The development process becomes faster and ideally safer by implementing resilience optimisation methods. An obligation to simulate all possible failure scenarios would take a great amount of time, effort and resources. The entire reason why resilience optimisation methods are used in the development process of new products would be counteracted. One can therefore not expect the manufacturer to simulate all imaginable failure scenarios of the product for each design alternative. Otherwise, legal requirements would hinder innovation in product development which should not be the aim of technology-friendly development of the law.

The priority of all obligations concerning the development of a safe product is currently the duty of care in construction. Primarily, the producer has to ensure that the design of the product is safe [103, marg. 71 ff.]. Further obligations, such as instructing the user of the product to limit the use to a specific application or monitoring the product during the phase of usage, are only secondary. But considering the aforementioned points, it is relevant to see a shift in prioritisation of the producer's obligations towards an increased importance of monitoring the product in its usage phase. First of all, it is important to limit the risks that still accompany the use of resilience optimisation techniques, secondly, to learn from products' behaviour "in the field".

Conclusions

Considering the information gathered, we can establish the following key findings from our work on legal requirements for the use of optimisation methods. The developer must be aware of the boundaries of implemented and used models. At the same time, the developer is also expected to validate the used models. When using optimisation methods for development, the developer can check only one optimal solution provided by the algorithm for compliance with safety requirements. Checking can also take place by using simulation methods. Before using resilience optimisation methods, the developer needs to prove that the gain of safety outweighs the remaining risks that are unknown. But the necessary methods need to be developed first. If this consideration turns out in favour of the use of resilience optimisation, new priorities in legal standards still need to be set. How to set these priorities will have to depend on the methods developed in the future. The main focus of the developer is to lie on monitoring the product during the usage phase to ensure the safety of users and third parties.

5.1.6 Linguistic Analysis of Technical Standards to Identify Uncertain Language Use

Jörn Stegmeier, Jakob Hartig, Michaela Leštáková, Kevin T. Logan, Sabine Bartsch, Andrea Rapp, and Peter F. Pelz

Technical standards play a major role in the entire product life cycle by standardising rules for the exchange of information, ensuring compatibility and reducing the variety of products, services, interfaces and terms. Technical standards are developed by international, national, and regional associations and are not to be confused with legal standards [102]. As we get to know in Sect. 5.1.3, the application of technical standards is voluntary in principle, but may as well be legally binding by law or contract. In both cases, technical standards should reduce the uncertainty of the product under development by defining formal quality requirements for product development (cf. Sect. 5.1.1 and 1.6). Consequently, unclear language use in technical standards leads to uncertainty during the product development process. In this section, we therefore focus on the linguistic analysis of technical standards to identify uncertain language in technical standards.

Since we do not know the impact of "semantic uncertainty" in standards on the product, its use and on its acceptance, we have to speak of "ignorance" (see Chap. 2). The aim of this study is an annotation method to avoid this ignorance with the help of an information system. The user of the information system can master uncertain language with the help of annotations, that give him or her hints that there is uncertainty and how to proceed.

A general requirement for technical standards, apart from the use of plain language, is the use of clear and concise expressions [76]. In contrast to that, Drechsler found that among users of technical standards, there is a considerable lack of knowledge of how technical standards are to be interpreted [82]. This discrepancy is not a coincidence but rather a consequence of the need to generalise technical standards in certain parts to make them applicable in a variety of contexts. However, underspecification inevitably leads to uncertainty as to how to correctly interpret the technical standard's intent. Two conditions must be met to bridge the gap between deliberate and necessary underspecification and all the additional information needed for correct interpretation and, hence, full compliance with the intent of the technical standard. Firstly, the user of the technical standard must be aware of this gap and, secondly, the user needs to know where to find the missing information.

If we consider sentences with "should" for example: While "should" conveys a certain intent, it can also be interpreted as a mere suggestion. The uncertain parts of technical standards are not as easy to spot as the occurrence or the meaning of "should" illustrates in the following examples: Example 1: "Dies gilt insbesondere für Apparate, die einer regelmäßigen Inspektion und Wartung bedürfen". [*In particular, this applies to devices that are in need of scheduled inspection and maintenance.*] [71, p. 38]. Example 2: "Die mit Trinkwasser in Kontakt kommenden Werkstoffe [...] dürfen Stoffe nicht in solchen Konzentrationen an das Trinkwasser abgeben, die höher sind als nach den allgemein anerkannten Regeln der Technik unvermeidbar [...]". [*The materials that come into contact with drinking water are not to release substances into the drinking water in concentrations higher than unavoidable according to the generally recognised rules of technology.*] [71, p. 15].

In Example 1, there is no list of concrete devices, which makes this passage uncertain for as long as the maintenance needs of all devices used in a project are not checked and documented. While the solution—looking up technical specifications provided by the manufacturer—to the uncertainty in Example 1 is straightforward, Example 2 presents the user with a less clear-cut issue. There is no single manual or document that records the generally recognised rules of technology. To obtain the necessary information, engineers have to either rely on experience by utilising solutions from earlier projects, or they must find examples of solutions which clearly follow the generally recognised rules of technology.

The practice of how technical standards enter the process of product development also comes into play. There are various possibilities, all of which to be found on a scale between the following scenarios. Scenario 1: The users study all standards pertinent to a specific project. Scenario 2: The users study a subset of the pertinent standards whenever the need arises. Scenario 1 is unlikely due to the sheer number of pertinent standards; Scenario 2 inevitably leads to a number of uncertain statements in pertinent standards being overlooked. In any case, due to the uncertainty inherent in standards, none of the scenarios addressed leads necessarily to a standard-compliant product development without further steps.

In each of these cases, an information system which shows all (potentially) uncertain parts of pertinent norms can assist the user in dealing with uncertainty not only by alerting him to its presence but by informing him of its nature, and by providing hints, as to where further pertinent information can be found. The primary goals of this study are therefore twofold: (i) creating a taxonomy of uncertainty in technical standards, and (ii) developing an information system which displays uncertain parts and their classification to the end user. For a detailed account of the latter, see [254].

The technical standards chosen for analysis in this study are from the DIN 1988 series "Codes of practice for drinking water" due to their relevance for Sect. 6.3.5.

Basic notions and methodology

Standard-compliant product development requires knowledge of the pertinent technical standards and of anything needed to resolve any uncertainty inherent in the standards. Uncertainty can arise from ambiguous language as well as from lack of (specialised) knowledge. From a linguistic perspective, the meaning of a word is a cognitive entity (= concept) evoked by perceivable linguistic entities, such as strings of characters or sounds (= word forms). The concept is both an abstract representation of an entity in the world (= object, phenomenon) and a reference to this entity.

The relationship between words, concepts, and objects in the world can be characterised as triangular, and is commonly described as the Semiotic Triangle [211, p. 11]. Special knowledge often needs specialised language, which can be accomplished in several ways: New words like "Covid-19" may be created and spread with the express purpose of evoking a specific concept referring to a concrete object or phenomenon in the world. Also, existing words can be re-purposed by expressly restricting the concepts they evoke (= defining their meaning) or by using them in a specific way or in a specific context, such as "mouse" in the sense of "peripheral computer device".

We understand uncertainty to mean a condition in which further steps of knowledge acquisition need to be taken to ensure standard compliance. Standards are not data in the sense of numbers, but more or less vague instructions for action that must be interpreted and supplemented by knowledge acquisition. Thus semantic uncertainty does not fit directly into the uncertainty classification from Chap. 2. Knowledge acquisition, in turn, is a process of understanding by interpreting information according to previous knowledge and applicable rules. We draw on the DIKW model (data, information, knowledge, wisdom) [4, 24, 230] in order to differentiate between knowledge, information, and data: data is a record of knowledge disassociated from the human mind, information is the result of cognitively processed data, and knowledge is the result of understanding the information by way of interpretation.

Annotating has a long-standing tradition in humanities and can be regarded as both, a part of knowledge acquisition and a scholarly primitive approach [193, 265]. We used the web service Weblicht [157], provided by CLARIAH DE [169], to automatically assign the base form (= lemma) to each word. This facilitates searches, for instances of word types, even if those word types have differing surface forms. For the manual annotation of uncertain words and phrases, we used the software INCEpTION [172] which can access multiple annotation layers (like part-of-speech and lemma) and allows for bottom-up development of annotation schemes.

Taxonomy design needs to incorporate a clear "differentiation between instances and classes" [287, p. 377], which in our case has been accomplished by a simple rule: Any concrete segment of text that demands further steps of knowledge acquisition is an instance of the super-class uncertainty and needs to be classified further. We employed a mixed development process, combining top-down and bottom-up approaches [208]: Previous knowledge of technical standards led us to distinguish evident from hidden uncertainty (top-down approach). Evident uncertainty means any uncertainty that remains after studying the norm text while hidden uncertainty arises from a conflict between individual knowledge and the specifications presented in standards. Prototypically, hidden uncertainty is individual and cannot be annotated which is why we concentrated on the branch of evident uncertainty. The subcategories of hidden uncertainty were developed in a bottom-up process.

Results

Evident uncertainty, a result of linguistic ambiguity, is divided into the semantic uncertainty and referential uncertainty, see Fig. 5.7. With regard to knowledge acquisition, semantic uncertainty is situated between information and knowledge and hinders the process of understanding while referential uncertainty is situated between

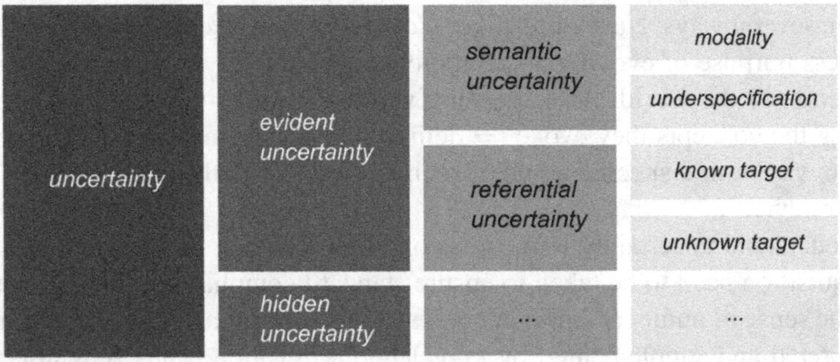

Fig. 5.7 The classes shown are a result of the manual annotation of norm texts on the level of single and multi-word expressions (phrases)

data and information, and expands the pool of information, which needs to be part of the knowledge acquisition. The sub-classes of semantic uncertainty account for different forms of meaning related uncertainty. Modality covers any instances where uncertainty arises from modal expressions, such as "can" or "should", while underspecification covers anything else. The sub-classes of referential uncertainty cover all instances of references in the technical standards. Known targets are all instances of named sources like "Diese Norm gilt in Verbindung mit DIN EN 806-2 [...]" [This standard applies in conjunction with DIN EN 806-2] [71, p. 6] while unknown targets are references to unspecified sources like "Es sind die technischen Anweisungen der Hersteller zu beachten." [The technical specifications of the manufacturer are to be observed.] [71, p. 19].

Towards automated annotation using machine-learning

The next steps focus on automation with regard to both, document import and annotation. The first steps towards a cooperation with Beuth to provide XML files instead of PDFs have been taken. The annotation process is improved by making use of Inception's in-built machine-learning (ML) and string matching capabilities. For the sake of proper ML work, a larger number of standards are annotated. For the string matching, the instances of the class *semantic uncertainty* and, to a lesser extent, of the class *unknown target* will be used to automatically retrieve synonyms from resources made available by lexicographic projects. Possible resources are the Digitales Wörterbuch der Deutschen Sprache [Digital Dictionary of German] [148], the project openthesaurus.de [204], which is also the basis for the information provided in DWDS. Further services are DeReKoVecs [100], which analyses the relationships between words occurring in the Deutsches Referenzkorpus (DeReKo) [German reference corpus] and GermaNet [140, 154], a lexical-semantic net [153]. These measures are going to enable us to annotate both, technical standards which are directly pertinent to a given project, and technical standards which are referenced by these. Thus, users of the information system are going to profit two-fold, namely gain insight into uncertain parts and their classifications to then follow and master uncertainty spanning multiple documents.

5.1.7 From Risk to Uncertainty–New Logics of Project Management

Martina Heßler

This book analyses how "to ensure product safety in a world of products with ever increasing complexity". Thereby "mastering uncertainty is central", and, as emphasised in the introduction, it "requires contributions from engineering, mathematics and law", cf. Chap. 1.

This section broadens the view in two respects. First, it examines how uncertainty is conceptualised and handled in the scientific discipline of project management. Second, it does so from a historical perspective by describing a historical shift in concepts in terms of dealing with uncertainty in the 1990s. The historical analysis of the changing term of uncertainty and the changing concepts developed by the discipline of project management leads to a better understanding and classification of current tools for planning processes in general.

In the second half of the 20th century, project management developed into a scientific discipline in order to ensure that projects are managed, controlled, and secured to run according to plan. Uncertainty is regarded as an "obstacle and a threat" [42, p. 5] to project success. Uncertainty should be minimised and made controllable. That does not come as a surprise, since challenges and problems that arise from unknown, unplanned or unexpected events and developments are obvious. In a recent study, Heidling stated that, of 318 projects "in the area of industrial, global large-scale projects 64 % (failed) to achieve the planned project goals" [150, p. 17].

The capacity of managing uncertainty is of great importance for the success of projects, especially in the context of increasing complexity and rapid change. In the relevant publications and journals of professional project management, this has been a topic of debate for many decades. The question of how uncertainty can be mastered led to many publications that questioned existing concepts and discussed new concepts for dealing with uncertainty. It is striking that the problems of dealing with uncertainty are addressed far more frequently than success factors that make uncertainty manageable.

This section asks (i) how uncertainty has been conceptualised in the discipline of project management over the last decades, and (ii) which methods have been developed for mastering uncertainty. In particular, the goal is to reveal the theoretical foundations and premises of project management, thus contributing to a fundamental understanding of the changing logic of project management. This section is based on programmatic publications within project management as a discipline. A study by Heidling, summarising and evaluating journal articles from the 2010s, should be highlighted, since it provides important insights for what follows in this section [150].

It is important to note that project management as a scientific discipline develops rather general and abstract models that do not focus on the specifics of various technological disciplines. Although project management emerged in the context of large technological projects, such as the development of the atomic bomb and space travel, models of project planning are being developed for various areas, from large technical projects to the planning of technological innovations to business projects and events. Therefore, this section does not deal with uncertainty in the process of product design, nor primarily with the control of technically induced uncertainty, but with uncertainty during the entire process of planning a project in general.

The thesis that will be developed is that, in the second half of the 20th century, we can observe a change in the concepts of uncertainty within project management, which led to the recognition of ignorance as a fundamental premise of all projects. In the theoretical foundations of project management, a paradigm shift happened: Uncertainty was redefined. The primary goal here is to analyse this change in concepts within the discipline of project management in the 1990s.

Before we can elaborate on that, it is necessary to take a closer look at the terminology used in project management and relate it to the terms used in this book. The terms of project management are strongly influenced by societal and sociological discourses. To make this clear: That which Chap. 2 of this book labels *stochastic uncertainty*, project management calls risk. In the models of project management, risk refers to possible, but not foreseeable events in the future that are known in principle and whose probability of occurrence can be calculated. However, further questions are asked, particularly in sociological discourse. It is asked how risks are produced by whom, how they are perceived by different actors and what significance is ascribed to them in each case. Thus, risk is a quantitative and qualitative category.

The technical term *incertitude*, cf. Chap. 2, is not used in the debates and models of project management at all. And while this volume differentiates further by distinguishing *ignorance* from *incertitude*, as defined in the introduction, within project management the term *uncertainty* is much less differentiated. While a paradigm shift in project management is clearly evident, the terms remain more vague than in the engineering disciplines. Models in project management only distinguish between risk and uncertainty. The interdisciplinary research field of Science Studies, which situates scientific knowledge in historical, philosophical and social context, in turn speak of the latter as *unknown-unknowns*, while this book uses the term *ignorance*.

Even before that distinction between risk and uncertainty has been drawn and a redefinition of the term uncertainty was discussed in the discipline of project management in the 1980s and 1990s, uncertainty has already been a category of project management, for example as technical uncertainty.

While in the first decades of project management the focus actually was on technical uncertainty, it has become obvious, from the 1970s and 1980s onwards, that not only the complexity of projects but also the complexity of uncertainty has increased. Societal processes became factors of uncertainty, such as civic protests against large-scale projects or the emergence of societal resistance to new technologies, for example. The pluralisation of the stakeholders involved, accelerated change processes, and changing market conditions, these developments all contributed to redefining the concept of uncertainty.

Thus, in a first step, the focus shifted from technical uncertainty to risks, thereby including technical, social and economic factors. In project management the concept of risk became central, e.g. [203]. At that very time the concept of risk also shaped social debates and the social science literature, e.g. [19, 227].

As early as the 1990s, however, the concept of risk was heavily criticised since it still aimed at controllability and mastering. In contrast, a redefined concept of uncertainty (here, exclusively meaning "ignorance"), which stressed the basically incalculable part of planning, was brought to the fore. This was accompanied by demands for new ways of dealing with such fundamental uncertainty in project planning. It is remarkable that this development can also be observed simultaneously in Science Studies and decision theory.

Similar strategies for dealing with uncertainty have been addressed within scientific project management as a discipline, as they are presented in Chap. 6 of this book, as will be shown in the following.

Risk versus uncertainty

The range of possible, unforeseen events that get in the way of project planning is wide. It ranges from technical problems, changes in the market, political or legal framework conditions to changes in the composition of the team, to name just a few factors. In an essay from 2013, Thamhain lists additional, concrete examples of uncertainty: "the failure of a certified component with proven liability in similar applications, a sudden bankruptcy of a customer organisation, or a competitor's breakthrough invention/innovation that undermines the value of your current project or threats a major line of business. By definition, unknown-unknowns are not foreseeable and therefore cannot be dealt with proactively" [260, p. 23]. If the types of uncertainty enumerated here are neither new nor surprising, the decisive point is the reference to the unknown-unknowns (ignorance), which began to shape the discourse within project management starting in the 1990s. Uncertainty, in contrast to risk, thus refers to something that is incalculable; thus it evades planning.

Since the 1990s, this has been emphasised as a new challenge in project management and can be observed analogously in Science Studies, which have dealt with the concept of the unknown-unknown. As Peter Wehling summarised, conflicts about the risks of genetic engineering and nuclear technology in the 1970s and 1980s had already made it clear that societies were confronted with a new form of the unknown, namely "the unknown as an unknown threat" [276, p. 485]. The discovery of a fundamental, unpredictable ignorance went hand in hand not only with the new technologies already mentioned, such as genetic engineering and nuclear energy, but also with the perception of living in times of accelerated scientific and social change [10, 276].

However, while this was initially addressed under the term risk, this description proved to be inadequate, since risk was associated with the possibility of calculating and thus mastering it. As already indicated, the newly perceived challenge lay precisely in the fact that phenomena exist, which are "events that are known and predictable in principle", as the term risk suggested, but that "the consequences of scientific and technological innovations [...] are simply unknown and incalculable" [277, p. 100].

Since the 1990s, project management has been asked to accept the unpredictable as an unavoidable dimension of every project and to find a different way of dealing with it. Exactly as was the case in the Science Studies, in project management, too, this was formulated as a criticism of the risk concept. Within project management, the concept of risk had also been used to refer to "events that are not completely predictable and controllable". However, the assumption still was that "the probability of their occurrence could be calculated". According to this logic, risks were transferred to the mode of predictability – thus becoming controllable. "Despite the recognition of uncertainty, certainty is nevertheless sought", summarised Böhle [42, p. 5].

However, in contrast to assuming that uncertainty is a calculable unknown, beginning in the 1990s, the concept of uncertainty has been redefined in a fundamental way within the field of project management. A newly emerging debate within project management suggested abandoning the ideas that certainty is possible at all and that uncertainty can be calculated, which were associated with detailed planning.

This was a provocation at odds with many modern ways of thinking taken for granted. For example, in the mid-1990s, J. Davidson Davidson Frame, had tied the concept of risk to the question of available information. He held the view—also common in decision theory—that little or incomplete information itself was a risk [105]. Therefore, the solution was to have as much information available as possible. This corresponded to a modern view of science in general: the assumption was that the accumulation of knowledge would eliminate ignorance [276, p. 485].

However, by the 1970s and 1980s at the latest, it became clear that as projects became more complex, ignorance in the sense of the unavailable and uncontrollable also increased, see [226, p. 423]; [189, p. 1106]. Unknown-unknowns—which, in terms of project management, constituted uncertainty—are therefore not the result of incomplete information or of something that is not yet known, but rather something that at a certain point in time is, as a matter of principle, not knowable, and thus not calculable.

According to this logic, a distinction between risk and uncertainty (as in ignorance) has been introduced to project management [203] in order to make uncertainty—understood as ignorance or unknown-unknowns—a topic of project management and to develop new methods for dealing with and ultimately mastering it. In a nutshell: In the 1990s, uncertainty was discovered as something unpredictable and recognised as an unavoidable problem. This conclusion was accompanied by the demand to accept it and therefore to "develop capacity to act *with* uncertainty" in the sense of ignorance [42, p. 7].

But what did this mean for dealing with uncertainty in project management? What does it mean for the methods developed within the discipline of project management?

Criticism of quantitative and standardised methods

The conceptual shifts from the concept of risk to a stronger emphasis on incalculable uncertainty were, unsurprisingly, accompanied by criticism of quantitative methods. Starting in the 1970s, methods were being developed to quantify risks and to make probabilities of occurrence controllable by using mathematical-statistical methods [203]; [150, p. 19]. As Heidling emphasises, the concept of risk was inseparably linked to predictability and calculability [150, p. 18]; [279], see also DIN EC 62198 [73]. But doing so in this context was considered inappropriate for the novel focus of challenges posed by uncertainty in the sense of ignorance.

In the field of Science Studies, quantitative, probabilistic methods of risk analysis have been criticised and described as the expression of a "reductionist, analytical worldview" and as such regarded as necessary to overcome [110, p. 739].

The aforementioned assumption that more information and more knowledge lead to mastering planning processes, was also questioned. For example, the pursuit of completeness of information as the basis for project management was criticised as not leading to the expected results. This was accompanied by the critical reflection of very detailed, sequential planning processes [260, p. 20f]. Archibald pointed out that the traditional approach of using large amounts of information as basis for detailed planning would not lead to solutions, especially when dealing with uncertainty [10]. It was important to not consider as many factors as possible, but rather the "most important" ones. However, these are "often those with the greatest uncertainty, and the new logic requires giving them priority over the easy, material, more certain ones" [10].

Furthermore, using concepts like controlling and monitoring to predict future events correctly and in a systematic way was not considered helpful in meeting the challenges of uncertainty [150, p. 17]).

Closely connected with such criticism of reductionist, analytical methods were the questioning of project management methods which dealt primarily with unambiguousness, which by now was regarded as inappropriate. In the 1990s, an interesting attempt was made within project management to use fuzzy logic to break up an orientation towards unambiguity in order to *calculate* uncertainty [10].

Summarising the criticism of dealing with ignorance, it can be said, in a nutshell, that the previous methods, based on calculation, quantification and detailed, sequential planning, were considered obsolete. They were criticised as ineffective when the problem of uncertainty was addressed in terms of dealing with the unplannable.

In essence, this constitutes a fundamental criticism that these methods were characterised precisely by the fact that they were a means of planning the plannable, i.e., calculating the calculable and the knowable, and therefore did not even consider the unplannable and ambiguous. That which could not be planned was, according to the criticism, simply ignored. However, it was precisely this systematic omission that led to problems during the execution of projects. Thus, a fundamental critique was formulated, which posited that the limits of the methods had not been reflected upon. Thus, the model entailed only that which could be controlled within the confines of the model (quantifiability), see similarly [42, p. 8].

This criticism, often voiced within the project management community, resulted in the demand that dealing with uncertainty required a new logic and new procedures, especially in times of "accelerating rate of changes" [10].

Paradigm shift: a new logic in project management?

If we now look at what was supposed to characterise this new logic and new methods, it quickly becomes clear that the criticism of the so-called traditional concepts was much more incisive than what was asked of the new models and procedures. Defining the problem proved to be easier (until today) than finding a solution. In the same way, this applies to Science Studies or decision theories. Here, the answers to the question of how to deal with uncertainty looked strikingly similar.

Firstly, ways of thinking and approaches based on competences considered innately human were said to be, by their "nature", counter to quantitative, standardised and rationalised procedures. It is especially the unpredictable that requires human experiential knowledge, intuition, heuristics and fuzzy thinking by real humans, as was often emphasised. As the horizon was to be extended beyond the measurable and quantifiable and diagnosed reductionisms were to be overcome, other heuristic, intuitive methods were now seen as appropriate solutions for dealing with uncertainty [10].

This was accompanied by an emphasis on the role of humans, who were no longer perceived as disruptive factors or in need of containment due to their subjectivity and assumed irrationality, but were discovered as potential. Using human skills was now considered the best way to react quickly and flexibly, which is very important when dealing with uncertainty.

Furthermore, demands were made for a holistic and synthesising approach, which would make interrelatedness and processual thinking the basis of planning, and eventually replace linear-causal thinking [10].

The calls for a new way of dealing with ignorance culminated in the demand to not only to accept uncertainty as inevitable but to see it as an opportunity: "Instead of considering uncertainty as a necessary evil, it should be considered as an extremely important, inspiring and useful factor given its inherent opportunities for making improvements and taking measures against risk. In the author's opinion, uncertainty is likely to hold some of the greatest potential for improving management skills and efficiency today" [186, p. 21].

Uncertainty was thus even interpreted as an opportunity to improve things in ongoing projects, to react flexibly with the goal of ultimately becoming more efficient. Contrary to the idea of creating robust, nearly rigid project plans, a clear plea was made for flexible, incremental, almost improvising action.

If one compares the strategies for handling ignorance, which were discussed in project management in the 1990s, to the strategies for mastering uncertainty described in this book, fundamental similarities can be stated. The strategies to manage uncertainty in technical systems, described in Chap. 6, are robustness, flexibility and resilience. Even if the term resilience did not play a role in the models of project management, the efforts to permanently adapt and flexibly ensure the success of the planning process when faced with unforeseen challenges clearly correspond to the concepts described in Sect. 6.3. That holds particularly true for the role of humans in mastering uncertainty, which has proved to be the core concept of the new logic of project management since the 1990s.

However, a remarkable difference also needs to be noted. Most of the concepts presented in this book refer to technical systems, thus having to ensure that technology itself proves robust, flexible and resilient. Therefore, both quantitative-mathematical models as well the interplay of humans and technology provide a solution for mastering uncertainty.

In contrast, project management focuses on the planning process of projects of various kinds, which inevitably implies ignorance in terms of social, economic and political developments. Thus, since the 1990s, concepts have been developed that emphasise human capabilities, thereby constructing a strong contrast between humans acting with uncertainty (ignorance) on the one hand and quantitative-mathematical models to master ignorance on the other hand.

This constitutes a move away from a paradigm of steering, controlling, measuring, calculating and detailed and sequential planning of projects; it can be summarised with a picture used by Perminova et al.: Projects should be conducted as an exploration or an open-ended journey rather, and not as a precisely planned sequence of steps [221, p. 74]; [150, p. 24]. However, this almost poetic description often clashes with the efficient culture of controlling in many organisations as well as with the design of techno-economic systems.

Summary and outlook

When looking at the premises and theoretical foundations of project management with regard to dealing with uncertainty, there is no doubt that a fundamental paradigm shift has taken place since the 1990s. The description of uncertainty as a risk, which implies the assumption that planning processes are manageable and controllable by quantification and probabilistic methods has been questioned fundamentally. This has been accompanied by a shift in terminology from risk to uncertainty in the sense of ignorance. Dealing with uncertainty, i.e., with phenomena that one does not even know one does not know, required new methods, which, in simple terms, focused on intrinsically human characteristics, such as experiential knowledge, intuition, as well as heuristics and holistic thinking

The debate within project management, which boasts many analogies to debates within science studies and decision theory in the 1990s, reflects social and technological challenges. These include not only genetic engineering and nuclear energy, which are always mentioned, especially in science studies, but also disillusionment in the context of artificial intelligence (AI) research. It was precisely here that the limits of quantitative, so-called "brute force" methods and the "Good Old-Fashioned AI" methods became very clear in the 1980s.

Putting the demand for a new logic of project management into historical context, the turn towards so-called human competencies becomes apparent, as a result of the limitations of technological methods, as well as of challenges arising from new technologies. This brought about a new dimension of ignorance. Disillusionment and the experience of the limits of mathematical-statistical-technical planning methods led to an emphasis on human-based planning procedures, and it also brought previously underestimated and unnoticed factors into focus.

The question of how to deal with uncertainty and, above all, unforeseen events in projects of all kinds remains a core question for their success. There is a lack of systematic and empirical, and above all historically comparative, studies focusing on this question, which, on the one hand, analyse concrete planning practices, and on the other hand, describe the empirical level chronologically up to the present. After all, concepts, methods and practices of project management are always time-bound. They do not only react to changing challenges by developing new approaches. The answers to new challenges also reflect contemporarily dominant approaches, as the strong focus in the 1990s on experiential knowledge, intuition and heuristics shows. This does not mean that the answers are not accurate. However, looking at things from a historical perspective makes it clear that one mainstream might chase the next. That means that we also have to critically reflect current trends: Construing a contrast between human behaviour, capabilities and technical (quantitative-mathematical) solutions, which was typical for the 1990s, is now being replaced by a stronger emphasis on the interaction and close cooperation between humans and machines.

However, no universal solution for this unpredictable uncertainty has been developed, nor are project management practices actually being driven by this insight (yet).

5.2 Product Design Under Uncertainty

Peter Groche and Hermann Kloberdanz

The entire life cycle of a product is predetermined by the product design. Since all three phases of the product life cycle design, production and usage are accompanied by uncertainty, see Sect. 1.2, an adequate anticipation of this uncertainty is the key for a successful product design. This anticipation is very demanding, because normally neither the involved stakeholders nor the events that occur in a product life are known a priori. We propose to tackle this challenge with several methods that uncover and help to reduce uncertainty in all stages of the design process in a systematic way.

The newly introduced "Uncertainty Mode and Effects Analysis" (UMEA), described in Sect. 5.2.1, allows for analysis and evaluation of occurring uncertainty in all phases of the product life cycle. One approach to reduce uncertainty in the design phase is the systematic involvement of stakeholders. This involvement can be used to reduce the fluctuation of requirements during the development and increase of the knowledge about the product to be developed by extending the designers' model horizon, see Sect. 1.3. A systematic approach for this involvement is given in Sect. 5.2.2.

Uncertainty influences the execution of all kinds of technical processes along the product life cycle, e.g. production processes or application processes. One possibility to uncover uncertainty is therefore to analyse processes with the aid of a generalised process model. Possible deviations can be attributed to system components in a standardised way and by this, uncertainty is disclosed reliably to a large extent. Such a generic process model and its application is depicted in Sect. 5.2.3.

As discussed in Sect. 1.3, models can be useful, but at the same time have their limits. Reliable information sources can help to overcome these limits. But in many cases, relevant information cannot be gained directly in experiments. A conflict between accessibility and reliability has to be solved. In Sect. 5.2.4 this challenge and possible solutions are presented for the domain metal forming processes.

Mathematical models can also be used to uncover uncertainty during the design phase. The models can be based purely on physical laws (white box models) or on data (black and grey box models), cf. Sect. 1.3. Often, a large number of uncertain design variables has to be considered. In the case of stochastic uncertainty or incertitude, information about the relevant domain of uncertainty is given, cf. Chap. 2. Due to time consuming calculations, white box models are difficult to compute for a representative number of possible combinations. Instead, fast models modelling based on surrogate models in Sect. 5.2.5 and demonstrate their usefulness for density and quantile estimation in Sect. 5.2.6.

Product design does not only comprise the initial design but also design changes that become necessary later. This applies in particular to special purpose machines, which are built to fulfil customer specific requirements. With regard to these, it is indispensable to master uncertainty, if ignorance, see Chap. 2, ultimately leads to clearly visible deviations from customer requirements. We propose a change procedure in Sect. 5.2.7, which helps to master uncertainty in these situations without jeopardising customer satisfaction.

5.2.1 The Method of Uncertainty Analysis and Evaluation: UMEA

Hermann Kloberdanz and Fiona Schulte

In this section, the Uncertainty Mode and Effects Analysis methodology UMEA is presented, which we have developed for a comprehensive, consistent and uniform uncertainty analysis in the entire development process, cf. [86, 89]. Uncertainty occurs in all phases of the virtual and real product life cycle. In the virtual product life cycle mainly uncertainty in the specification as shown in Sect. 5.1.2, and model uncertainty, discussed in Sect. 2.2, cause deviations of the planned product properties from expected properties. Products then do not fulfil the customer expectations regarding functionality, effort and availability. In addition, real products sometimes do not meet the planned properties due to uncertainty that occurs in all phases of the real product life cycle and is propagated in the process chains. In both cases, the acceptance of the products decreases due to uncertainty and endangers the market success of the products, cf. [90, 142, 143].

Especially in the case of new developments, the virtual product life cycle is characterised by a lack of secure information. Especially the early phases therefore represent a typical situation of high uncertainty. The complexity, importance and criticality of the situation is explained in Sect. 5.2.2. There, in addition to the solution of the technical task, the innovation and time pressure as well as the interconnectedness are emphasised. Misjudgements by developers regarding available resources, material properties or disturbance parameters contribute to uncertainty and can lead to critical system states in the later use phase, cf. [89].

The detailed analysis of uncertainty that can occur in the entire product life cycle is therefore a prerequisite for mastering uncertainty in the development of new systems. The uncertainty analysis has to be carried out from the beginning and throughout the entire development process. The results of the analysis act as a basis for system synthesis, e.g. by the robust design method [90]. However, previous methods of uncertainty analysis only consider a narrow range of the life cycle of a product [142]. The diverse influences in the individual life cycle processes and the propagation of uncertainty in the process chains as shown in Sect. 3.2 make it necessary to include identification and evaluation of uncertainty over the entire life cycle into the uncertainty analysis.

Analysing uncertainty in development processes

The term UMEA is an extension of 'Failure Mode and Effects Analysis—FMEA' [74]. However, the scope, consideration horizon and application range go far beyond FMEA. The UMEA is based on a procedure consisting of four phases. In addition to the identification of uncertainty and sources of uncertainty, e.g. by the method of technical process analysis that will be described in Sect. 5.2.3 in detail, the stepwise procedure for UMEA is oriented towards the main challenges of the mastering of uncertainty during the entire product development process: (i) uniform procedure for uncertainty analysis across all concretisation stages in the product development process, (ii) focus on critical product properties and uncertainty influences, (iii) consideration of the propagation of uncertainty in the product life cycle with regard to its effects in the usage phase.

The stepwise procedure at UMEA comprises four phases shown in Fig. 5.8: (i) environmental and goal analysis, (ii) identification of uncertainty and its causes, (iii) determination of uncertainty effects, (iv) evaluation and decision.

Product properties are characterised by an extreme amount, a high versatility and cross-linked uncertainty. The 'environmental and goal analysis' contributes to a purposeful focus on properties which are particularly critical with regard to uncertainty. The system is delimited from its environment and a system boundary is defined as depicted in Sect. 1.3. This is the prerequisite for distinguishing between external and

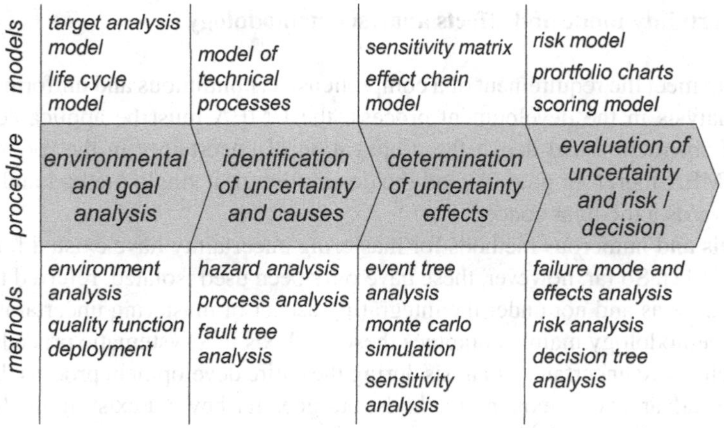

Fig. 5.8 Procedure at UMEA; process steps with associated models and methods following [90]

internal influences and supports the determination of the causes of uncertainty. Furthermore, the evaluating parties (e.g. users, stakeholders, requirement groups) are identified and their value concepts are specified. This can be used to describe the expectations of the basic system properties (e.g. maximising the benefits and quality of the system usage, minimal life cycle costs). The tolerated uncertainty can also be documented in the technical specification as explained in Sect. 5.1.2.

In the second phase, an identification of uncertainty and its sources will be carried out. For this purpose, the processes of the life cycle identified as relevant are analysed in detail and significant influences to uncertainty are identified, e.g. by a technical process analysis, as already mentioned above.

In the third phase, uncertainty effects will be analysed. The existing relationships between external influences, the process parameters and the properties of the result of the process are specified and accumulated over the process chains of the entire product life cycle as comprehensively as possible. The deviations in the result of the process are mainly assessed based on the sensitivity of the individual processes with regard to the identified causes of uncertainty, see below in this section.

In the fourth phase, the existing uncertainty is mostly assessed with regard to the usage processes. Depending on the level and probability of the accumulated uncertainty, a risk can be determined taking into account possible deviations from the expected system properties and the resulting consequences in the specific context of use. Based on this evaluation, a decision is made on adequate measures to master uncertainty in the system adaptation. The steps ‘evaluation of uncertainty and risk’ and ‘decision making’ can be combined into one phase for simplification as shown in Fig. 5.8.

The uncertainty mode and effects analysis methodology

In order to meet the requirement of a comprehensive, continuous and uniform uncertainty analysis in the development process, the UMEA must be applicable at all levels of abstraction and depict the propagation of uncertainty in the product life cycle. UMEA therefore goes beyond the description of a single method and is oriented towards a modular concept.

Models and numerous methods for mastering uncertainty have existed for some time, cf. [142]. So far, however, these have only been used isolated, referred to individual problems and not under the integrating aspect of mastering uncertainty. The UMEA methodology mainly combines these methods in a systematic procedure for a comprehensive uncertainty analysis during the entire development process. UMEA thus takes advantage of existing methods, but goes far beyond existing methods in terms of uncertainty analysis.

UMEA's analysis activities are supported by a collection of methods that can be combined according to the underlying procedure. The systematic mastering of uncertainty includes methods for identifying and describing uncertainty, for evaluating the criticality of uncertainty, and approaches to reduce uncertainty in the system design, cf. [89].

Of particular note is the combination of process models and matrix-based analyses to identify the sources of uncertainty and determine the consequences of uncertainty, cf. [86]. The detailed modelling of the single processes provides qualitative information about the relevant properties of the process's initial states and the uncertainty-relevant influencing parameters during the process execution. Based on a sensitivity matrix, the interrelationship between uncertainty in the initial state and the uncertainty-relevant influences of the process parameters can often be described quantitatively, cf. [89].

For this purpose, the expected level and deviation of the influencing parameters on the process (e.g. disturbance parameters) and the operator (e.g. work equipment in a manufacturing process) are estimated. Depending on the information available, this is made quantitatively or qualitatively according to the uncertainty model. Furthermore, the deviations in the properties of the operand (e.g. partially manufactured component, initial state) caused by previous processes are estimated. The effect of the individual influencing parameters and the uncertainty of the process execution on the properties of the operand in the final state (e.g. dimensions and shape of the manufactured component, process result) are determined by means of sensitivities. The consideration of interdependencies between the influencing parameters and effects requires a complex structure of the sensitivity matrices. The uncertainty of the process result are calculated quantitatively or at least qualitatively by adding up the partial deviations, cf. [86].

The modelling of process chains consisting of multiple processes forms the basis for quantifying the propagation and accumulation of uncertainty over the entire product life cycle. The properties including their uncertainty in the final state of a process form the initial state of the subsequent process. The uncertainty of the follow-up process can be calculated or estimated in the same way using a further sensitivity matrix. Thus, the uncertainty of production processes is accumulated up to the usage processes. The accumulation of uncertainty does not necessarily lead to an increase of uncertainty. By defining the processes and their sequence, uncertainty can also be reduced by assigning optimised processes regarding uncertainty. These processes often show a low sensitivity to critical product properties. Uncertainty from previous processes can also be mitigated or even compensated for. Examples are the process sequences of drilling and reaming discussed in detail in Sect. 4.1.3.

Applying the UMEA methodology

For evaluation we have applied UMEA exemplary to design optimisation of a very simple tripod demonstrator representing a system for uniform load distribution. In this tripod demonstrator, three legs are symmetrically and parallel attached to a load distribution ring. The device is loaded by a mass. The load is to be evenly distributed on the three legs of the demonstrator, cf. [223]. The application of the UMEA is described in detail and discussed in depth in [87] and in [86].

In the first phase 'environmental and goal analysis' of the UMEA the even load distribution on the legs and the analysis of the influence of the tripod production is recognised as a goal. In order to exclude influences by eccentric tripod load and uneven or inclined ground contact area, these are included within the system boundary.

In the second and third phase the 'identification of uncertainty and their causes' as well as the 'determination of uncertainty effects' are performed. The uneven load distribution in the legs is identified as the uncertainty to be optimised during use. In the example, different vertical distances between the load distribution ring and the contact area at the three legs, here referred to as 'total leg length', can be identified as the source of uncertainty. The further analysis shows that the 'total leg length' depends on the length dimensions of the leg components, which are influenced by the manufacturing process. Furthermore, a non-parallel and non-positional mounting of the legs during assembly influences the 'total leg length'.

In the 'evaluation and decision' of the fourth phase it is a recognised calculation that in the present system the length and the axial position of the leg components influence the uncertainty more than the angular deviation from the vertical attachment. Therefore, it was decided to develop different fastening designs. In the end, the design was selected that offers the highest manufacturing and assembly accuracy with respect to the leg length and position.

By applying UMEA to a current conducting plug connector in the engine compartment of an automobile, we have demonstrated that UMEA is in principle also applicable to more complex systems, cf. [86]. However, the disturbance influences by dirt, corrosion, vibration and temperature are a great challenge.The UMEA has contributed to mastering the uncertainty, especially by systematically incorporating the experience of the specialists.

Conclusion

The UMEA methodology developed represents a cross-process uncertainty analysis that covers the entire life cycle of the physical product with a focus on production and usage. It enables uncertainty of the entire product life cycle to be comprehensively and specifically analysed and evaluated in the entire system development process, cf. [89]:

- Methods are provided that can be used from the beginning in all phases of the development process with increasing degree of more detailed description.
- It is intended for use at a high level of abstraction as well as for very specific design phases.
- In the analysis, the planned processes of the entire physical product life cycle are included and the propagation of uncertainty in the process chains is also considered.
- A focus on relevant influences and deviations is made several times in order to master the extend and variety of uncertainty in the system development.
- Finally, UMEA goes beyond an uncertainty analysis and provides prioritised information for mastering uncertainty in system design.

In the modular concept of UMEA, well-known methods were adapted and complemented by special methods. UMEA links the different models and methods, each adapted to increasing degrees of concretisation, in a uniform procedure for the uncertainty analysis.

A fully quantified analysis of uncertainty in all steps of the UMEA is usually neither possible nor reasonable. For a methodical analysis of an entire process chain during development, qualitative approaches need to be complemented by suitable quantitative analysis methods to an appropriate extent. If, for example, significant uncertainty in the production processes are quantitatively known, the uncertainty calculation can be done without a detailed analysis.

5.2.2 Mastering Uncertainty in Product Development

Peter Groche and Maximilian Knoll

For the sake of improving product development, we present a method that takes user requirements into account while being independent of the product or personal experience. The methodology extends and improves existing approaches by taking comprehensively into account the roles for the design unit and the design process.

Design engineers have to master numerous and distinct challenges in the modern industrial environment. For this purpose, engineers work in close cooperation with specialists from various disciplines and with different skills [219]. In addition to conventional development methods, design engineers are increasingly working with agile development methods. In comparison to classical development methods, the agile methods rely on a reduced set of planning and product specifications, while at the same time strongly integrating customer feedback. Thereby, the skills and approaches of the designers must also be suitable for working in international teams. Due to increasing international division of labour, a reduction of in-house developments and a high effort for project management can be observed. In addition to these tasks, design engineers interact with customers, suppliers and various departments of the company and have to accomplish communication and strategy tasks [81]. Furthermore, they bear a huge responsibility for society, customers and the company [21]. To fulfil these responsibilities, designers are supported by a wide range of methodologies. Despite the evolution of responsibilities, the primary task for designers remains to design functional, innovative, safe and sustainable products, cf. Chap. 2. However, professional experience so far has been an essential factor in managing the daily tasks [219]. The aim of the design tasks is to ensure that the products generate economic added value for the company.

In addition to the expanded roles of responsibility, the product life cycle has changed due to the "frontloading" [164] and the shortened development times [250]. The frontloading approach shifts development decisions and activities to early stages of development, allowing challenges to be identified and saving development time. The frontloading and the shortened development times can partly lead to premature products and impact the customer satisfaction in an unfortunate way. This form of acceptance (cf. Sect. 1.7) can result in high losses and cause damages which are extended especially by merging mechanical, electrical and data processing solutions to the brand [13, 139]. Uncertainty is even more pronounced due to the simultaneously increasing complexity of products.

Thus, the application of suitable design methods is becoming increasingly important. However, current design methods focus on the design process and neglect the working environment. The working environment is determined by standardised CAD and PDM systems. Standardised CAD and PDM lead to a higher amount of available information. The use and usability of the available information depends strongly on the experience of the design engineer [47].

The economic and social changes introduced by industrial rationalisation and automation have led to the development of various methods and approaches which help to structure and simplify the design process. The step-by-step approach according to VDI guideline 2221 [269] is a universal but somewhat unpractical approach for the design of technical systems. In addition to VDI 2221, Birkhofer [40] gives a detailed overview of the design development that has taken place so far. However, the existing methods with their sequential structure do not represent real design processes [242], because they do not explicitly capture the iterative steps of product development.

The method developed by Schmitt [241] is based on known problem-solving strategies and expands these by including the working environment. The primary objectives of product development remain the realisation of customer demands and the compliance with market requirements. In order to achieve these goals, the task of product development is to generate suitable product data [217]. For this purpose, three axioms are defined as basis for the following product development method [241]. The axioms are to be regarded as equally important for successful product development and thus for the success of a company [241]. The three axioms are as follows:

(i) Products are always developed under sustainable aspects.
(ii) Decision-making and its comprehensible documentation in product development require the distinct definition of evaluation criteria.
(iii) The development process is an iterative procedure that must be individually adapted to each task and is highly dependent on the involved stakeholders.

A design unit as an organisational unit closely collaborates with marketing, sales, purchasing and production [85]. This strong integration into the corporate environment results in complex influences on the product development process. In order to take these influences into account, the involved roles of operation are considered. According to [106], stakeholder groups represent groups or individuals who influence corporate goals or may be affected by them. A differentiation is made between external and internal stakeholders. The specific roles of internal stakeholders, who are located within the company boundaries together with the design unit, are presented in more detail below.

With regard to their contributions to the overall product development process, the internal stakeholder can be divided into two groups: (i) the first internal stakeholder group is responsible for the product definition. This includes employees from sales, distribution and marketing. They have in common that they develop and define the desired properties of a product in customer contact or on the basis of market data [41]. The second internal stakeholder group is involved in the product realisation department and in the product development process. This group includes all persons in the company who are involved in the implementation of the real product. This includes production, testing and purchasing. The management can be found in both internal stakeholder groups described above. This means that there is no separation with respect to technological and economic influences.

In the general overview of the method by Schmitt [241] shown in Fig. 5.9, the core tasks of the two stakeholder groups are divided into three roles: classification, decision and design. All roles benefit from the generated product data.

As described in [241], the aim of the shown roles is to define all relevant goals resulting from the complex interaction between product components, stakeholders, expectation and the environment before the development process begins. Typical goals for the product development are, for example, functionality, time and cost requirements, cf. Sect. 1.1. The task of the internal stakeholder group "Product Definition" is to transfer the requirements and desires of the customer into dependent product properties and to communicate the goals during the process. The customer only recognises the dependent properties, whereas the design unit defines the independent properties and on the basis of this definition influences the dependent properties.

The design unit is responsible for the actual product development (Fig. 5.9 "create"). The design unit has to identify the independent properties relevant for achieving the objectives. In consultation with the internal stakeholder group product realisation, the properties must be compared with customer requirements and wishes.

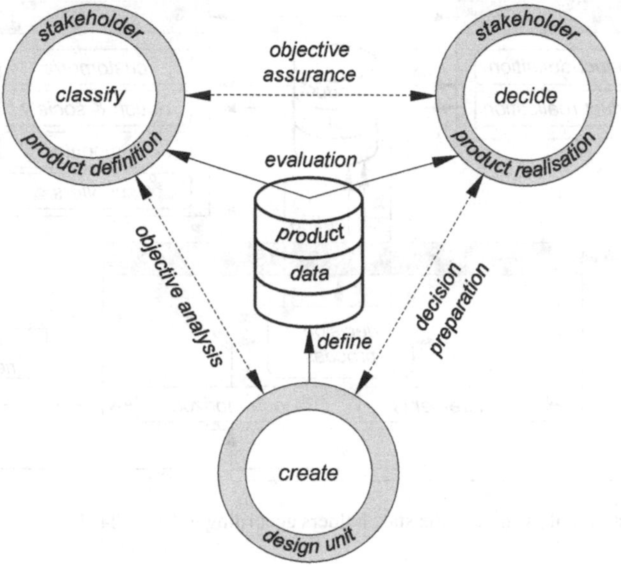

Fig. 5.9 Overview of interactions between operation roles [241, 242]

Based on the solutions developed in the Product Realisation department, the design unit prepares the decision-making process for the Product Realisation stakeholder group [241].

Decisions with great consequences for product development are decided, implemented and documented in the role "decide". It should be noted that a distinction is made between the decisions of the design unit and those of the internal Product Realisation stakeholder group. The comparison between the dependent properties with regard to the fulfilment of customer wishes and requirements is carried out during development activities. The product data compiled by the development unit represents the result of the method and serves as the basis for evaluation in the respective iterations [241].

Depending on the results of the described comparison, a further iteration of the process might be necessary. If conflicting goals cannot be resolved the internal project group, the stakeholder group responsible for the product definition contacts the customer again. However, further iterations may result from a seemingly successful completion of product development. This is the case, for example, if the customer's requirements are formulated in a misleading way or if they have been transferred to incorrect dependent properties. In this case, a closed loop is created which can be run through in an iterative manner during product development [241].

Through the iterative process and under consideration of the customer influence, the previous method can be extended to Fig. 5.10. This results in a closed product development cycle, which represents the interactions between the individual stake-

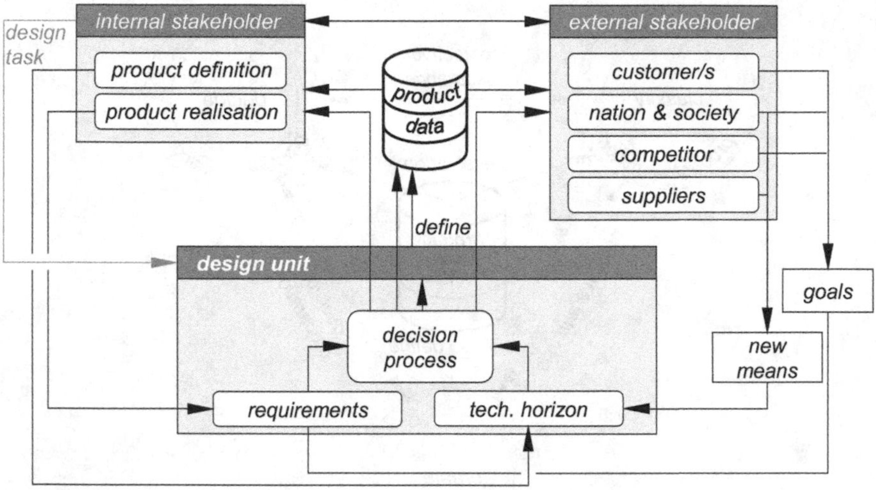

Fig. 5.10 Interaction flowchart of the stakeholders according to [241, 242]

holders. Stakeholders outside the company's boundaries can open up new opportunities for the development unit.

Potential examples are new or adapted solutions from suppliers. The government and the society, as well as competitors and external stakeholders are special because of their influences. The influence of all external stakeholders can lead to new goals as well as new opportunities for the design unit. On the one hand, the government certification of operating materials or the rejection of patent applications by competitors can create new opportunities for the design unit. On the other hand, new goals and requirements can be generated for the design unit due to laws, public interest in technologies, as well as pioneering products of the competitors. The depth of the technological horizon describes the technical and methodological knowledge of the development unit acquired over the years of professional activities. Examples are design specifications for components, detailed knowledge of machine elements and their possible applications. In order to obtain or generate the information required for the design, the development unit has two communication partners at its disposal. On the one hand, the internal and on the other hand the external stakeholders can be consulted for the absolutely necessary exchange of information. In case of high uncertainty due to limited information, it may be necessary to define new goals by communicating with the internal stakeholder group Product Definition. This influences the requirements and thus reduces the needed effort [241].

Research on design methodology has been conducted for decades, resulting in different approaches and methods to support designers. Due to the changing conditions and requirements of products and markets, new approaches have been developed and existing ones adapted. Nevertheless, there is a gap between the promises of academic methodological approaches and industrial reality. One reason for this is that methodological approaches in industry are so far perceived as inefficient.

Developing a method that extends the system boundaries of the design process to internal and external influences seems promising to update established approaches and move towards a dynamic, closed-loop system [242]. Therefore, the discussed method is based on specific participants in the product development process and their activities and responsibilities. The approach extends and improves existing ones by extending the system boundary to take into account influences on the design unit and the design process. Thus, the approach represents a possible solution to transfer industrial reality into a universal methodological approach. Preliminary analyses of the industrial application show that the method simplifies the task of the designers [241]. This shows how important it is to create a basis for a new generation of design methods [241]. By involving stakeholders in the development process, product expectations can be improved.

5.2.3 Methodical Uncertainty Consideration in Technical Process Modelling

Hermann Kloberdanz and Fiona Schulte

In this subsection, the generic model of technical processes is presented, which we developed for the comprehensive identification of uncertainty in all phases of the product life cycle. A special feature of this universal model is the purposeful localisation of uncertainty and the sources of uncertainty. Thus, the model has the character of a reasonably structured checklist and provides basic advice on effective approaches to reduce uncertainty.

As presented in Chap. 1, acceptance describes the fulfilment of expectations of a product or component in terms of functionality, availability and effort. However, acceptance is not a characteristic of the product in the narrow sense, but matters only in connection with the process steps. This applies to the process steps of all life cycle phases, e.g. in product development with regard to the planned properties, in production with regard to manufacturing tolerances, or in usage with regard to performance and efficiency.

Additionally, the term uncertainty covers both, the measurable or perceptible uncertainty of a product, and the causes why uncertainty arises in the entire life cycle of a product, cf. [143]. Consequently, the analysis of process steps has to identify and locate the sources of uncertainty, the emergence of uncertainty, and the propagation of uncertainty. Therefore, modelling the technical process steps in the product life cycle in detail has a key role in mastering uncertainty.

Mastering uncertainty in life cycle processes

A process step, here called process in simplified terms, can be considered in the context of technical systems in principle as a time dependent transformation of an operand from an initial state to a usually changed final state, cf. [72, 272]. The model of technical processes developed by Heidemann has proven convincingly to be the most suitable model for describing processes from a large number of different models, cf. [149]. In the following, the generic model and the specified models of technical processes derived from it will be referred to as process model for simplicity reasons.

It can be stated as common sense that the properties of products cannot or cannot completely be determined in both, the virtual and the real product life cycle, cf. [143]. The sources of uncertainty can be located in all life cycle phases. However, previous approaches to manage uncertainty are mostly based on an isolated consideration of uncertainty. The robust design approach of Taguchi, see Sect. 3.5, for example, primarily refers to the production of products, cf. [257, 258]. Taguchi implicitly assumes that the properties of the manufactured product, which have been planned during the development phase, correspond to the current expectations of the customer and that the product is used as intended. Further, he concentrates on the quantification of deviations in the final state of the process steps without analysing the causal influences in a well-founded structured model of the system.

In contrast, our comprehensive approach to the analysis of uncertainty is based on the modelling of processes that can be linked to process chains (or networks). In addition to the analysis of the result of a process step, the overall mastering of uncertainty in product development requires the identification of all possible sources of uncertainty in this process step. Therefore, in addition to the uncertainty of the system elements themselves, various external influences on the system that can cause uncertainty must be considered. Only two typical examples out of a huge number are thermal deformation of machine tools during production and pollution during the system's usage.

In order to be able to consider the sources of uncertainty, we have further developed the Heidemann process model for a comprehensive uncertainty analysis. Therefore, we have integrated all elements and the linking influences into the process model, which are relevant for the analysis of uncertainty in processes, cf. [175]. The generic process model is shown in Fig. 5.11.

In this process model systems are basically represented by the process itself, the initial and final state of the operand, the work equipment and the relationships between them. The modelling is particularly illustrative in production processes in which, for example, a more or less finished part is produced from a semi-finished material (operand, initial state) by a forming process. If, for example, a servo press is used for forming, it is modelled as a work equipment that starts the process by means of working factors e.g. the ram force and ram motion, cf. Sect. 3.6.3. Additionally, the system environment (e.g. buildings, adjoining systems, nature), necessary resources (e.g. power supply, coolant) and user (e.g. operator) as well as their relations to the system are represented. Uncertainty regarding the result of a process step is modelled by the deviations of the properties and state variables of the operand in the final state.

In the example of a forming process these are e.g. dimensional and form deviations of the manufactured part. Uncertainty of the process result can be caused by influences of all other elements of the process model and their interdependence. To support the

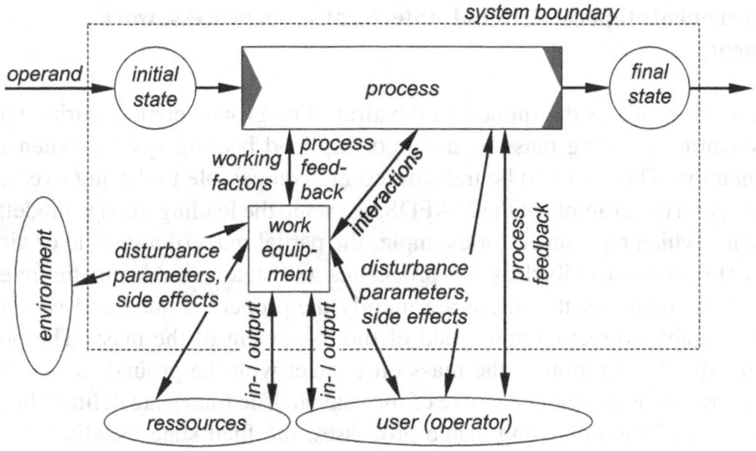

Fig. 5.11 Process model for the detailed analysis of uncertainty in processes following [175]

mostly complete identification of the sources of uncertainty, the process model can be used as a kind of checklist. The process model is mainly used in the early phases of system development. It initially serves developers to identify essential design parameters and to describe uncertainty qualitatively, when assessing robustness of working concepts based on physical effects. This is discussed later in Sect. 6.1.5.

Identification and location of uncertainty using the process model

The main purpose of process modelling for the mastering of uncertainty is (i) the largely complete identification of uncertainty and sources of uncertainty and (ii) the reasonable and sense-attributing location of the sources of uncertainty. According to their localisation in the process model, groups of typical sources of uncertainty can be identified. We recognised three groups having characteristic properties with respect to mastering of uncertainty: (i) all elements inside and outside the system boundary, (ii) planned functional interaction inside and outside the system boundary, (iii) unintended influences between all elements inside and outside the system boundary. Additionally, the effective direction of the interactions and influences have to be considered, e.g. disturbance parameters of the environment on the work equipment and emissions of the work equipment on the environment.

In the following, an overview of the utilisation of the process model is given. Taking the example of the Modular Active Spring-Damper System (MAFDS) described in Sect. 3.6.1 the identification of potential uncertainty is illustrated only very briefly and exemplarily. If necessary, reference is made to the aircraft landing gear or vehicle running gear approximated by the demonstrator. First, the usage process is considered. The special features regarding other life cycle processes are added afterwards. The description is focused on the groups of typical sources of uncertainty.

Process elements (operand, initial state, final state, process, work equipment)

The MAFDS simulates the support of the aircraft on ground contact during landing by decelerating a falling mass by a supporting load bearing system, when it hits the foundation. The forces to be transmitted are comparable to driving over a road unevenness. In the example of the MAFDS test setup the loading mass is modelled as an operand, which represents, for example, the partial mass of a vehicle or aircraft. The initial state is described by the properties and state variables of the operand. In case of the demonstrator, these are mainly the properties mass and weight and the state variables direction and speed of the movement of the mass. The process represents the deceleration of the mass on contact with the ground, which can be described as a time-dependent course of movement. The final state defines the result of the process. When planning usage processes, the final state usually represents the expected result of the process. In the demonstrator, the final state is essentially described by the position of the mass. In case of a running gear, the deceleration values of the operand are mainly decisive for the user. Acceleration values that are not or not fully accepted by the user, e.g. for reasons of comfort or safety, represent the uncertainty of the process result.

This differentiated consideration highlights the fact that, in contrast to the work equipment, the developer can only influence the properties of the operand, the process, the initial state, and as a consequence, the final state, all of which depend on the usage of the system to a limited extent. Since they cannot be determined, they represent a high degree of uncertainty. In the case of usage processes, the corresponding uncertainty can be mastered more or less only in the development process by a carefully and detailed technical specification, cf. Sect. 5.1.2. Since the initial state is significantly influenced by the preceding process step, cross-process analysis is very important in production processes. In Sect. 4.1.3 and later also in Sect. 6.1.8, this is illustrated by the example of drilling and reaming, where the uncertainty of the initial state during reaming can be influenced to a certain extent by the preceding drilling process.

Resources and inputs of work equipment

After ground contact, the MAFDS is supported on the foundation of the test setup. The supporting forces, or in a broader sense the foundation including the impact surface, are modelled as resource in the process model. Thus, it represents, for example, the condition of the road surface when looking at a chassis. Time-dependent deviations of the contact area in displacement, position, and especially in direction represent the uncertainty of the resource. For the simulation of uncertainty in the test setup, impact surfaces with different angles are provided. This allows analysing effects of an uneven road surface on running gears.

In many cases, as for example here with the running gear, the resources can practically not be influenced by the developer. The corresponding uncertainty can only be mastered by elaborating the technical specification and designing the work equipment accordingly, see Sect. 5.1.2. In the case of stationary production processes, there is usually a much smaller uncertainty regarding the supply of energy. It can be mastered either by choosing an energy source with low uncertainty or by additional measures, such as buffer storage.

Operator inputs mainly represent the operation of the work equipment. Operating errors often represent a considerable uncertainty, which is sometimes extremely difficult to master, especially in active systems. In a similar way to other resources, input uncertainty can be managed by reducing the volume and complexity of the inputs required. For example, a high degree of automation of work equipment usually reduces the uncertainty of operating errors. In case of the MAFDS, operator influences are limited to the triggering of the drop test. Thus, uncertainty of the operator is negligible here.

Unintended influences between all elements of the process model

The work equipment unintendedly interacts with the system environment, the operator and the resources. The modelling of the corresponding relationships represents on the one hand disturbance parameters on the work equipment, on the other hand side effects, which are caused by the work equipment, cf. [192]. For the MAFDS, disturbances and side effects are of minor importance, since it is shielded and operated in a low disturbance laboratory environment. In the case of represented landing gears or vehicle chassis, temperature and radiation influences, chemical and mechanical influences, as well as pollution can occur from the environment as disturbance parameters that can impair the function of the work equipment.

Conversely, the work equipment also generates side effects, which usually have a negative impact on other system elements in form of emissions, e.g. neighbouring systems or components. In the case of represented landing gears or vehicle chassis, for example, abrasion, noise and vibrations can have an environmental impact, a loss of comfort for the passengers (users) and a friction value-reducing effect on the road surface (resource). The side effects are often not acceptable and are perceived as uncertainty, cf. [175]. The process model thus indicates that disturbance parameters can be mastered by eliminating them or taking measures to reduce their influence, cf. [192]. Corresponding design strategies are discussed in detail later in Sect. 6.1.5.

Interactions between work equipment and process are, in principle, comparable to disturbances and side effects, and can be identified in a similar way; but they have a different quality. Corresponding sources of uncertainty are usually more difficult to master due to the close interlinkage of process and work equipment within the system boundary. Typical and critical cases are, e.g. soiling or contamination of food processing systems, cf. [22].

Conclusion

The process model is apparently a very simple generic model. The described analysis with the help of the process model impressively shows the complexity in scope and diversity of the resources of uncertainty. By classification and meaningful localisation in the process model, useful advice for the identification and design of robust and resilient solutions can be derived.

The findings and results of a process model analysis can be documented in the technical specification that serves as a basis for the subsequent system synthesis, cf. Sect. 5.1.2. Thus, the process model analysis is indirectly a purposeful basis for the later mathematical modelling and simulation of the systems.

When looking at production processes, the process model assist engineers to identify and locate uncertainty equally. It is very important to realise that in contradiction to usage processes in production processes the parts and components of the load-bearing systems are modelled as operands and the machine tools as work equipment.

5.2.4 *Conflicting Objectives in the Determination of Process and Component Control*

Peter Groche and Maximilian Knoll

The quality of products is one of the most important requirements in production technology. Fluctuations in semi-finished products and the forming process can influence the quality of products in a wide range, cf. Sect. 1.7. In addition to known and unknown uncertainties (Sect. 2.2) in production processes, the influence of ecological and economic effects on production technology is increasing. Furthermore, the forming processes are influenced by known uncertainties, such as temperature, tools, machines and lubricants. Especially in the last three phases (cf. Sect. 6.1.5) of the design process (cf. Sect. 1.2), sensors can be integrated cost-effectively at optimal measuring positions in tools as well as machines. The subsequent integration of sensors proves to be cost-intensive and difficult, depending on the required measurement quality. In order to consider the correlation between sensor position, costs, measuring uncertainty and product quality in the phase of the design process, a model for evaluation is presented and discussed using the example of the 3D Servo Press (Sect. 3.6.3) and conventional forming processes.

The most important semi-finished product properties in sheet metal forming are usually sheet thickness, yield strength, anisotropy and surface properties, as these influence the process and recovery behaviour [191]. In order to reduce product uncertainty, narrow tolerances for semi-finished product properties are therefore required. As a result, the testing effort, energy consumption and manufacturing costs increase during the production of semi-finished products. Another approach is the closed-loop control of product properties to make the process insensitive to process fluc-

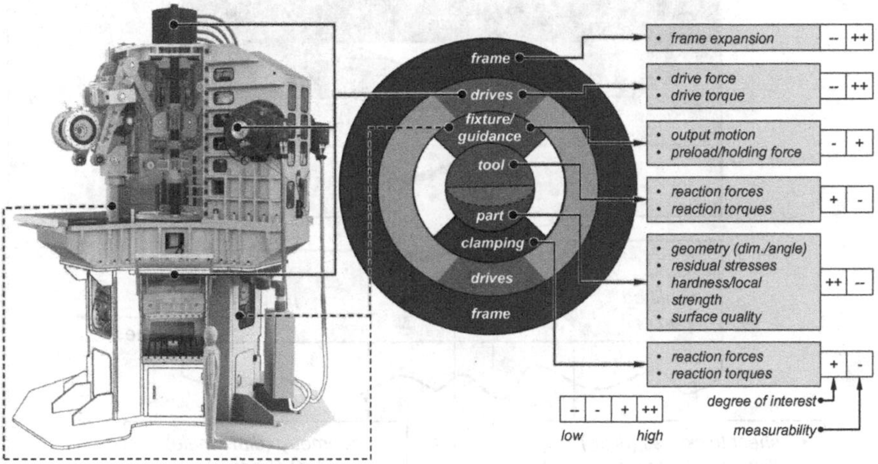

Fig. 5.12 The target diagram, given by [56], using the example of the measurement positions of the 3D Servo Press

tuations (Sect. 1.3). This requires all the components of a closed-loop control (see also Sect. 5.3.2), which consist of sensors, state observers, controllers and actuators [243]. Thereby the position of the sensors is important for the closed-loop quality control [126]. We refer to [56] for a literature review on special measurement and control approaches which are relevant for this topic. Due to the high relevance of the measuring position, it is important to consider the sensor position in the product development process to quantify uncertainty, as it has been done with the example of the 3D Servo Press (Sect. 3.6.3) in the change process (Sect. 5.2.7). Therefore, a method for the representation of the spatially opposite degree of interest in knowledge of a measurand and its measurability in the form of the target diagram in Fig. 5.12 was presented [56, 59, 122].

The target diagram of Calmano [56] shows possible measuring positions for the determination of machine, process and component parameters. For the example of a forming machine, the system is subdivided into the frame, drives, fixture, guides and clamps. A further subdivision is made for the tool, which is not directly a part of the forming machine. The innermost circle is further divided into the areas tool, product and the forming zone. By a symbolic evaluation from low (–) to high (++) (compare Fig. 5.12), an estimation of the measurability and the possible interest of the respective measuring position is possible. In the direction of the centre of the target diagram (Fig. 5.12), the knowledge gain of the measured technical/physical quantity increases, which reduces sensor uncertainty. The measurability, however, decreases for most physical quantities towards the centre of the target and proves to be most technologically challenging in the component. Compared to the forming zone, measuring positions, such as the frame and drives, are already equipped with

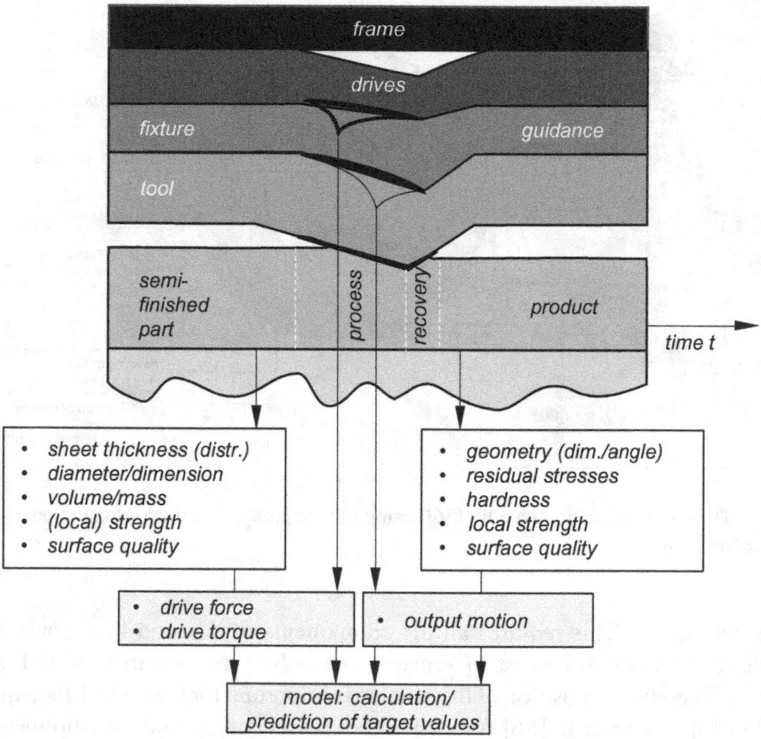

Fig. 5.13 Challenges of sensor integration with time consideration [59]

conventional measuring equipment by default or can be realised with only minor technical and financial hurdles [59].

With regard to the time sequence of the forming process it can be seen that many product properties are still changing during the recovery phase after the forming process and therefore cannot be measured directly in the process. These relations increase the uncertainty for control systems. For this reason, Fig. 5.12 can be extended by the time-dependency (Fig. 5.13).

An example of the time dependence of product properties is the resulting angle in a bending process. As long as the forming tool is in workpiece contact, the final bending angle cannot be measured. Only after the tool is no longer in contact with the workpiece, the stored elastic energy is released until a stress equilibrium is reached. Afterwards, the final bending angle can be measured. The recovery phase leads to the problem that even with successful sensor integration in the forming zone, the final product properties cannot be measured during the process. For this reason, control engineering uses condition monitors which, for example, instead of the bending angle provide indirect information about machine and tool variables, such as positions, speeds and reaction forces [59, 122].

In conventional press development, it is common practice to determine the press force by the frame elongation. In the target diagram (Fig. 5.12), the measuring position is at the outer ring. The geometry of the frame has a significant influence on the elongation, so that a calibration of the sensors is necessary. Calibration is carried out by applying known forces to the ram. Subsequently, mathematical models can be used to derive the desired measured variable (Sect. 1.3). During process operation it is possible that eccentric forces act on the ram by combining different forming processes, which were unknown in product development. Uneven process forces lead to unwanted ram tilting due to machine stiffness. This is accompanied by the possibility of a secondary force flow. This means that during process operation, there is a possibility that the occurring forces are not measured with sufficient accuracy and therefore the process has a high unknown uncertainty.

For this reason, the force measurement on the 3D Servo Press was shifted towards the centre of the target diagram (Fig. 5.12). In addition, the 3D Servo Press has a large number of sensors compared to conventional presses. The measurement of the force is realised by strain sensors similar to conventional presses. To reduce the unknown uncertainty, the strain is measured in the direct force flow of the kinematics at the drive rods (see Sect. 5.2.7). Due to the rotationally symmetrical drive rods, three strain sensors are installed per drive rod. This ensures that the most accurate force measurement possible is obtained even if the actuator stem is deflected. In order to realise a displacement control in addition to the force monitoring or force control of the press, displacement sensors with a minimum resolution of 2 μm are also integrated in the press, making it possible to reduce the uncertainty. These are located on the so-called "bearing frame", which is responsible for the force guidance and displacement of the slide in the frame. The selected positioning reduces the influence of bearing play and friction losses caused by the main gears. In addition to the sensors close to the process, the sensors of the eight servo motors are available for control. The motor torque and the angle of rotation can be determined in the motors. Due to the different measuring positions, it is possible to draw conclusions on the machine status, for example, in addition to control engineering applications, Sect. 4.2. Due to the additional degrees of freedom and the integrated sensors of the 3D Servo Press, the number of additional actuators and sensors required for flexible forming tools can be minimised, which is associated with a reduction in tool costs. The integration of the sensors and the selected measuring position affect the control quality [56] and thus the uncertainty. Therefore, the selection of a suitable control concept is important. Calmano [56] has developed a method for this purpose using different forming processes, which can be used to select suitable control concepts under consideration of economic aspects.

There is a conflict of aims between quality and productivity in a production system of forming technology under the premise of improving the economic efficiency through cost reduction and revenue increase. The methodology shown above reduces the uncertainty for the selection of the control system. The approach is a control of component properties in the process, whereby the expenditure for their implementation is set in relation to the contribution margin increase by means of investment cost calculation. The concept variants of the control approaches are created by synthesis of different measuring, determining and influencing the defined target variables. The analysis of the concept variants with regard to quality is done by an uncertainty analysis of the designed overall system. Process sequences and reject rates are taken into consideration for the analysis of productivity. The expenditure for the implementation of the concept variants is estimated on the basis of development and failure costs as well as purchased components. The economic evaluation correlates these variables by means of investment cost accounting and identifies the most economically advantageous concept variant [56].

Especially with small quantities and frequent product changes, the cost of tool-integrated actuators and sensors is a significant factor in tool costs, which is why they must be taken into account in product development. This results in a conflict of objectives between accurate measurement of the target value with a low uncertainty for closed-loop control and economic efficiency. When integrating sensors, the use of the shown methods plays an important role in the economic implementation of new measuring systems. The example of the 3D Servo Press shows that the transfer of the tool actuators and sensors to the forming machines enables an improved measuring strategy with simultaneous economic advantages. As shown, the selection of suitable measured variables, influencing variables and control concepts for the control of product properties, however, is still dependent on the process used, taking into account the economic and technological advantages and disadvantages. Therefore, these process-specific parameters have to be analysed depending on the process and evaluated under the mentioned influencing variables. The shown methods open up new possibilities for digitalised production.

5.2.5 Estimation of Surrogate Models

Sebastian Kersting and Michael Kohler

Introduction to surrogate models

Experiments with technical systems are often described by mathematical models. Such models, implemented as computer code, enable the use of so-called computer experiments, i.e. the experiment is simulated via a computer program using the underlying mathematical model. Thus, the technical system can be analysed by using computer experiments instead of real experiments. The mathematical model is given by a function $m\colon \mathbb{R}^d \to \mathbb{R}$, which models the functional relation between the d-dimensional input of the experiment and its real valued output. An overview on the design and analysis of such computer experiments can be found in Santner et al. [232] and Fang et al. [99]. Often these computer programs are complex and thus computational expensive to evaluate, but to properly analyse the experiment it is necessary to generate a large sample of computer experiments. To circumvent this problem, surrogate models can be used, cf. e.g. [15, 119, 166, 281]. A surrogate model is an estimator of the computer experiment m, which usually is much faster to evaluate. Thus, they can generate a large sample of computer experiments. A rather general method to estimate a surrogate model based on a max-min approach is described in this section. Finally, we illustrate its usefulness on the digital twin of the MAFDS, Sect. 3.6.1, as described in [15].

Mathematical setting

The method described in the following is based on the following mathematical setting:

Let $X, X_1, X_2, \ldots$ be independent and identically distributed random variables with values in $\mathbb{R}^d$, and let $m : \mathbb{R}^d \to \mathbb{R}$ be a measurable function. Here, X is the input variable of an experiment with the technical system and m is a computer experiment associated with the experiment with the technical system, where $m(X)$ describes the outcome of the computer experiment. Given the data set

$$\big(X_1, m(X_1)\big), \ldots, \big(X_n, m(X_n)\big) \tag{5.1}$$

of size $n \in \mathbb{N}$, the aim is construct an estimator $\hat{m}_n$, the surrogate model of m. Typically the data uncertainty in this case can be classified as stochastic data uncertainty as described in Sect. 2.1.1, but the method can also be applied in the other classes of data uncertainty described in Sect. 2.1.1.

Estimate a surrogate model based on a max-min approach

In most cases computer experiments are deterministic, i.e. for a given input X the result is always the same. Thus, instead of using the typical least squares approach, the surrogate model is defined as the minimiser of the maximal absolute deviation, i.e. it is defined by

$$m_n(\cdot) = \operatorname*{argmin}_{f \in \mathcal{F}_n} \max_{i \in \{1,\ldots,n\}} \big| f(X_i) - m(X_i) \big|, \tag{5.2}$$

where $\mathcal{F}_n$ is a set of functions. In other words, to construct the surrogate model a class of functions $\mathcal{F}_n$ is chosen and then a function $f : \mathbb{R}^d \to \mathbb{R}$ is selected from this set, such that the empirical maximal deviation is minimised. A possibility to compute this estimate is to use nonlinear programming, e.g. gradient descent [231] or Levenberg–Marquard [201]. Here, one has to neglect the fact that the gradient of (5.2) does not exist and use a numerical approximation of the gradient instead.

Which class of functions to use depends on the particular setting and the dimension d. Usually, if the dimension is not too big, B-splines can be applied as shown in [14]. If furthermore the input X is deterministic and the input dimension is small, a quasi-spline approximation should be used as discussed in [92].

In case of a large input dimension one possibility is to use neural networks, since they are usually able to achieve good results by circumventing the so-called curse of dimensionality as discussed in [176]. Below we will introduce a special class of sparsely connected neural networks, especially designed for computer simulations built in a modular way. We use them to estimate a surrogate model of the MAFDS, Sect. 3.6.1 and finally discuss the usefulness of this approach.

In a first step the so-called spaces of hierarchical neural networks with parameters K, M, d^*, d and level l are defined as follows, see [14]. Let $\sigma : \mathbb{R} \to \mathbb{R}$ be the so-called logistic squasher $\sigma(x) = 1/(1 + \exp(-x))$ $(x \in \mathbb{R})$.

For $M \in \mathbb{N}$, $d \in \mathbb{N}$ and $d^* \in \{0, \ldots, d\}$, we denote the set of all functions $f : \mathbb{R}^d \to \mathbb{R}$ that satisfy

$$f(x) = \sum_{i=1}^{(M+1)^{d^*}} \mu_i \cdot \sigma\left(\sum_{j=1}^{d^*} \lambda_{i,j} \cdot \sigma\left(\sum_{v=1}^{d} \theta_{i,j,v} \cdot x^{(v)} + \theta_{i,j,0}\right) + \lambda_{i,0}\right) + \mu_0$$

for some $\mu_i, \lambda_{i,j}, \theta_{i,j,v} \in \mathbb{R}$, by $\mathcal{F}^{(\mathrm{NN})}_{M,d,d^*}$, where $x^{(v)}$ denotes the v-th component of the vector x.

For $\ell = 0$, the space of hierarchical neural networks is defined by

$$\mathcal{H}^{(0)}_{K,M,d,d^*} = \mathcal{F}^{(\mathrm{NN})}_{M,d,d^*}.$$

For $\ell > 0$, define recursively

$$\mathcal{H}^{(\ell)}_{K,M,d,d^*} = \left\{ h : \mathbb{R}^d \to \mathbb{R},\ h(x) = \sum_{k=1}^{K} g_k\big(f_{1,k}(x), \ldots, f_{d^*,k}(x)\big) \right. \\ \left. \text{for some } g_k \in \mathcal{F}^{(\mathrm{NN})}_{M,d^*,d^*} \text{ and } f_{j,k} \in \mathcal{H}^{(\ell-1)}_{K,M,d,d^*} \right\}. \tag{5.3}$$

Kohler and Krzyżak [176] introduced these classes of hierarchical neural networks as an approximation for so-called hierarchical interaction models. We do not explain the advantages of hierarchical interaction models at this point, but we want to remark that this class of functions is a realistic assumption on technical systems built in a modular way.

Application to data of the MAFDS

We apply the surrogate estimator based on neural networks to a computer simulation of the MAFDS (cf. Sect. 3.6.1). In the computer experiment a Modular Active Spring-Damper System is guided on a frame and falls down on the base of the frame. Virtual sensors allow the measurement of different parameters, such as acceleration, absolute position of the Modular Active Spring-Damper System, or the forces at the point

Table 5.1 Parameters of the $N(\mu, \sigma)$-distributions of the input variables of the MAFDS in this computer experiment

System property	μ	σ
Spring stiffness N/m	27000	1200
Damping constant N s/m	140	7
Mass of spring support kg	20.35	0.25
Mass of sphere in lower truss structure kg	0.76	0.03
Mass of sphere in upper truss structure kg	0.76	0.03
Mass of crosslink in upper truss structure kg	13.74	0.5
Mass of threaded rod in truss structure kg	0.363	0.015
Mass of joint middle part kg	0.9236	0.05
Mass of joint arm kg	1.46	0.075

of impact. In the simulation, the correlation of the nine normally distributed input variables presented in Table 5.1 on the computed output variable, the maximum force at the point of impact, is analysed.

The computation of a single output value, during which a differential-algebraic equation system must be solved by the procedure *RecurDyn* of the software *Siemens NX*, takes about three minutes in this setup. Based on $n = 250$ generated realisations of the nine-dimensional input distribution and the corresponding observed outputs, we estimate the neural network surrogate model defined by (5.3).

The estimate is computed by using nonlinear programming, to be exact the Levenberg-Marquard method, where the parameters K, M, d^* and ℓ will be selected data dependent from fixed sets. The neural network parameters are chosen by a splitting of the sample, where we use $n_t = \lceil 2/3 \cdot n \rceil$ training data and the remaining $n - n_t$ data points as testing data. We use the fixed sets $K \in \{1, \ldots, 5\}$, $M \in \{0, \ldots, 12\}$, $d^* \in \{1, \ldots, d\}$ and $\ell \in \{0, 1, 2\}$. Since (5.2) is not differentiable, we generate a starting point by the approximation of a least squares approach first, and then use the Levenberg-Marquard method with a numerical approximations of the Jacobian matrix.

This surrogate model is able to compute 10000 output values in less than one second. Using this surrogate model it hence becomes possible to generate a large sample of computer simulated values, which will be used in the application of the methods in Sect. 5.2.6.

5.2.6 Density and Quantile Estimation in Simulation Models

Sebastian Kersting and Michael Kohler

In an early stage of the development, usually no prototype of the technical system is available, and thus computer simulations of the experiment with the technical system are used to analyse properties of this technical system. Often quantifying the distribution of the outcome of an experiment is the focus of the method, since the outcome often determines properties of the system, which can be used in the design. To quantify the distribution of the outcome of the experiment we describe a method to estimate its density, and a method to estimate quantiles of the distribution. In Fig. 3.2 of Sect. 3.1 the method is assigned to the product or system design phase (A).

Introduction

Constructing prototypes of technical systems is often expensive and time consuming. Thus to reduce the effort, one can use physical knowledge of the system to implement a computer simulation $m\colon \mathbb{R}^d \rightarrow \mathbb{R}$ based on a mathematical model. This model describes the functional relationship between the d-dimensional input X of an experiment and the real valued outcome of an experiment Y. For example, if the technical system is a spring damper, then the input could be the drop height, payload and the spring stiffness and the outcome could be the maximal compression of the spring damper. Usually these computer models do not perfectly describe reality, but in an early stage, if no or only a small sample of real experiments are on hand, it is usually beneficial to assume that the underlying computer model fits reality perfectly, i.e.

$$m(X) = Y \tag{5.4}$$

holds. To construct an estimator of the above mentioned characteristics of the distribution of Y, a large quantity of computer simulations is needed. In this context, surrogate models as described in Sect. 5.2.5 are playing a crucial role to generate these samples. In the following, we will present two methods to quantify uncertainty, more precisely, we will describe a method to estimate the density of Y and a method to estimate quantiles of Y.

Both methods assume that either the underlying distribution of X is known, or that a large sample of input values is available. In reality this is often not the case. But often the distribution of the input variable is easily estimated since the randomness in the inputs is often induced by measurement errors or production tolerances. To apply the methods below it becomes then necessary to estimate the underlying distribution and generate a sample of input values based on this estimated distribution. For a multivariate normal distributed input variable, one can use the method described as in Sect. 4.3.8. An overview of methods to generate a data set based on a specific class of distribution can be found in Devroye [77].

Mathematical setting

All of the methods discussed below use the following setting: Let $m\colon \mathbb{R}^d \to \mathbb{R}$ be a measurable function with $m(X) = Y$ and assume that the data set

$$(X_1, Y_1), \ldots, (X_n, Y_n), X_{n+1}, \ldots, X_{n+N_n} \tag{5.5}$$

where $n, N_n \in \mathbb{N}$ is available. According to Sect. 2.1 the uncertainty in the data can be classified as stochastic data uncertainty. Let $g : \mathbb{R} \to \mathbb{R}$ be the unknown density of Y w.r.t. the Lebesgue measure which we assume to exist.

Adaptive estimation of a density based on surrogate models

In this and the next section, two methods of uncertainty quantification based on surrogate models are presented. A detailed overview of estimating surrogate models can be found in Sect. 5.2.5, hence we will neglect how to estimate surrogate models in the presentation below.

We assume that a surrogate model m_n of m estimated on $(X_1, m(X_1)),\ldots,$ $(X_n, m(X_n))$ is already constructed. Based on this surrogate estimate we will present a method proposed in [101]. Firstly, choose $N_l \in \{1, \ldots, N_n - 1\}$, and denote $N_t = N_n - N_l$. Next, define the density estimator based on the learning data N_l and depending on a bandwidth $h > 0$ and a kernel function $K : \mathbb{R} \to \mathbb{R}$ by

$$\hat{g}_{N_l,h}(y) = \frac{1}{N_l \cdot h} \sum_{i=1}^{N_l} K\left(\frac{y - m_n(X_{n+i})}{h}\right) \tag{5.6}$$

and define the corresponding empirical measure on the test data by

$$\hat{\mu}_{N_t}(A) = \frac{1}{N_t} \sum_{i=1}^{N_t} I_A(m_n(X_{n+N_l+i})) \quad (A \subseteq \mathbb{R}), \tag{5.7}$$

where $I_A : \mathbb{R} \to \{0, 1\}$ is the indicator function which takes value 1 on A and is 0 elsewhere.

Finally choose a finite bandwidth set $\mathcal{P}_{N_n} \subseteq (0, \infty)$ and define the data-driven bandwidth choice by

$$\hat{h}_{N_n} = \operatorname*{argmin}_{h \in \mathcal{P}_{N_n}} \sup_{A \in \mathcal{A}_{N_n}} \left| \int_A \hat{g}_{N_l,h}(y)dy - \hat{\mu}_{N_t}(A) \right|, \tag{5.8}$$

where

$$\mathcal{A}_{N_n} = \left\{ \{ y \in \mathbb{R} : \hat{g}_{N_l,h_1}(y) > \hat{g}_{N_l,h_2}(y) \} : h_1, h_2 \in \mathcal{P}_{N_n} \right\} \tag{5.9}$$

and define the estimator of the density g by $\hat{g}_{N_l,\hat{h}_{N_n}}$.

Monte Carlo quantile estimator based on surrogate models

As before it is assumed that a surrogate model m_n of m estimated on $(X_1, m(X_1))$, ...,$(X_n, m(X_n))$ is already constructed. Then the most simple way to construct an estimator $\hat{q}_{m_n(X),N_n,\alpha}$ of the $\alpha \in (0, 1)$ quantile $q_{Y,\alpha}$ of Y defined by

$$q_{Y,\alpha} = \inf \{ y \in \mathbb{R} : G(y) \geq \alpha \} \tag{5.10}$$

where G is the cumulative distribution function of Y given by

$$G(y) = \mathbf{P}\{Y \leq y\}, \tag{5.11}$$

is a Monte Carlo estimator. As in Enss et al. [92] or Kohler and Krzyżak [177] the estimator is defined by

$$\hat{q}_{m_n(X),N_n,\alpha} = \inf \left\{ y \in \mathbb{R} : \hat{G}_{m_n(X),N_n}(y) \geq \alpha \right\}, \tag{5.12}$$

where

$$\hat{G}_{m_n(X),N_n}(y) = \frac{1}{N_n} \sum_{i=1}^{N_n} 1_{\{m_n(X_{n+i}) \leq y\}}. \tag{5.13}$$

Note that this is equivalent to choosing the $\lceil N_n \cdot \alpha \rceil$ biggest values of $m_n(X_{n+1})$, ..., $m_n(X_{n+N_n})$.

This method is a straight-forward Monte Carlo approach. A more complicated and thus in many cases more exact method can be found in Kohler and Tent [178].

Application to data of the MAFDS

As an application to illustrate the two methods described above we use the computer simulation of the MAFDS as described in Sect. 5.2.5 and use the same method to estimate a surrogate model as described in this section. To estimate the density by the above described adaptive approach we use $N_n = 10^4$ realisations evaluated with the surrogate model. The finite parameter set $\mathcal{P}_{N_n}$ is set as $\{2^l : l \in \{-5, -4, \ldots, 10\}\}$

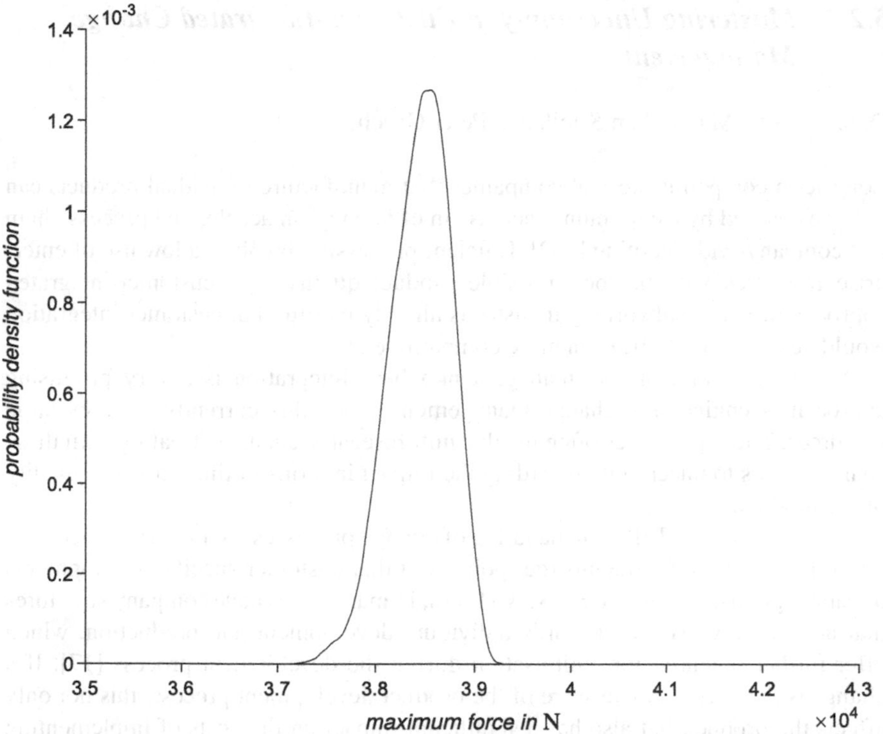

Fig. 5.14 Adaptive surrogate density estimator, defined as in (5.6), based on a sample of 10^4 realisations

Table 5.2 Estimated quantiles $\hat{q}_{m_n(X),N_n,\alpha}$, defined as in (5.12), for the maximal force

α	0.5	0.9	0.95	0.99
Quantile estimate in N	38148.75	38507.09	38600.71	38765.67

and as kernel we use the Epanechnikov kernel defined by $K(y) = (3/4) \cdot \max\{1 - y^2, 0\}$, where $(x)_+$ equals x for $x > 0$ and 0 otherwise. The result is displayed in Fig. 5.14.

The estimated density and quantiles can now be used for further development and to analyse properties of the technical system.

Since the calculation of the quantile estimate is much faster we use $N_n = 10^6$ realisations in this case. We use these realisations to calculate the 0.5, 0.95, 0.98 and 0.99-quantiles as described in Sect. 5.2.6. The result is summarised in Table 5.2.

5.2.7 Mastering Uncertainty in Customer-Integrated Change Management

Daniel Hesse, Maximilian Knoll, and Peter Groche

Long-term competitiveness of companies that manufacture individual products can only be ensured by a continuing increase in efficiency, an accelerated process chain and company-wide learning [252]. Efficient processing involves a low use of enterprise resources with the best possible product quality. The customer-integrated approach of the metalworking industry is already helpful, but customer integration would be necessary to be even more competitive [171].

A section within project management where integration is a very promising approach is engineering change management, since this currently requires large resources. These processes often involve unforeseeable changes. Dealing with these changes leads to uncertainty regarding the impact in terms of time, cost and quality of the product.

The avoidance and efficient handling of change processes are important factors to mastering uncertainty. Baumberger points out that customer-specific developments of capital goods, such as the 3D Servo Press, in many cases have company structures that are strongly oriented towards individual development and production, which offer further potential for optimisation during the development process [17]. If a change is detected at a late stage of the product development process, this not only affects the product, but also has a significant impact on the costs of implementing the change. In this context the Rule-of-Ten approach is used in numerous studies (see [63, 141, 229]). The Rule-of-Ten describes the disproportionate increase of the change effort depending on the change time [16, 64, 107]. Here, the change-costs increase exponentially with each phase of the life cycle by a "factor ten" [108].

The temporal distribution of the changes that actually occurred additionally increases the cost share of the changes in relation to the total project costs. According to a literature-based analysis by Bauer [16], three quarters of all changes occur at a time when a simple change to the product data is no longer sufficient. If changes in the planning and development phase can often be implemented by short iterations and adaptation of the product data, the lack of manufacturability and assemblability in the production phase, in particular, cause a large part of the change expenditure [84, 116].

Therefore, one of the most important tasks of change management is to avoid late changes and to shift changes into early phases by determining and recording product requirement as completely as possible [18, 62, 190]. In addition to DIN 199-4 [67], other change management approaches such as the "Generic Engineering Change Process" according to Jarratt [162], the "Strategic Automotive Product Data Standards Industry Group" (SASIG) [233], the "Implementation of a milestone-supported Customer Change Management" according to Sauer [234], the "Decentralised Change Management" according to Kleedoerfer [170], the "Design Change" process according to Yu [284], the "advanced CMII-based ECM framework" according to Wu [282], "engineering change process framework" according to Stekolschik [255] and the "Generic Engineering Change Process" according to Riviere [229] exist.

A large number of scientific approaches include a structured representation of the change process, which implies the need for clear, formalised process steps. Currently, none of the scientific publications fully meets the requirements described for the integration of a cause and effect analysis in the early phases of process flow.

Strategies for the avoidance and control of technical changes are frequent approaches of the scientific literature to make change processes more efficient, to increase quality and to reduce costs. However, the applicability of the strategies to change processes in individual production must be examined.

Many of the models presented were derived from series production processes. Since none of the work adapts the models to individual production, it can be assumed that direct transferability is not possible. Furthermore, the deficit of missing integration of customers into the change process becomes apparent. The method's orientation towards customer-related changes is only postulated in the work of Sauer, which focuses on customer changes during the change process execution. In the following, the structure of the adapted change process (see Fig. 5.15) is described.

The model aims to support companies in the implementation and execution of customer-integrated change management. In addition, the procedure model serves as a planning instrument to make the phase-specific activities transparently available to the involved stakeholders. The relevance of the customer is taken into account to the extent that the customer is the core element at the centre of the model. The change procedure itself is represented in nine steps. In addition to the customer and the phases of the change procedure, the outer layer, consisting of the three areas "Management & Organisation", "Human Resources" and "Methods & Aids", completes the basic structure.

In the following, the individual phases of the change procedure are described. The first phase of the procedure is defined as the phase of deviation analysis. This phase includes an examination of the target and actual values and can therefore be seen as a trigger for a change process. It is of high relevance that the underlying causes are identified in the case of a detected deviation. The deviations are the basis for a change request. In addition to a deviation analysis, further preliminary calculations are necessary. In the following phase of the procedure "write request", the information for the rate request is summarised. The preceding phase of write request, rate request and application decision excludes changes at an early stage which do not promise success. If the change request is approved, from a macroscopic point of view, the planning phase begins, which consists of the phases, finding solutions and detailing.

In the phase of finding solutions, classical methods support the operator in the generation of possible solutions as well as in the subsequent phase of solution evaluation, which includes a risk and impact analysis. The involvement of the customer is an important component to benefit from his extended application experience. In addition, active involvement takes place in the change decision phase in order to record all customer requirements. For this purpose, communication channels, such as meetings or e-mail exchanges, or software-assisted tools can be used. In a transparent and comprehensible decision-making process misunderstandings due to the "communication

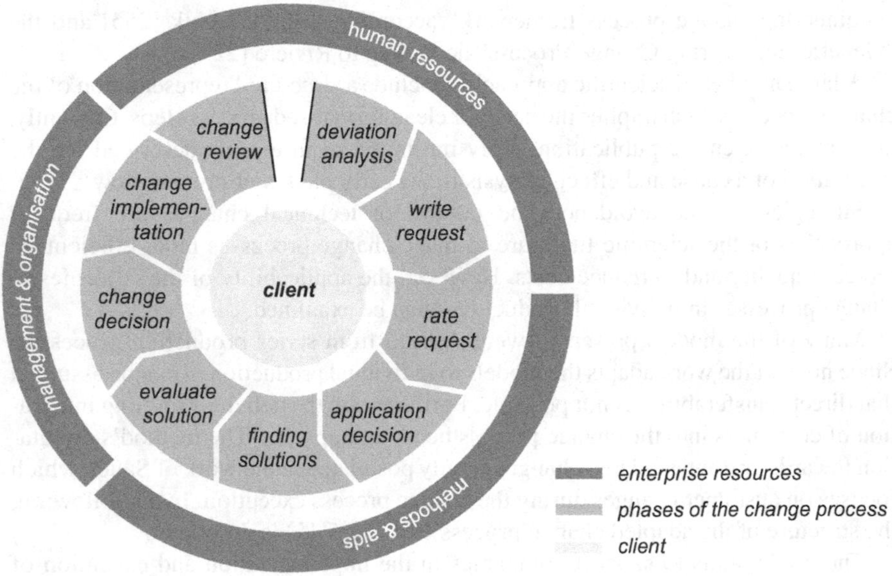

Fig. 5.15 Overview of the procedure model

barrier" between companies and customers but also between departments can be eliminated. Once a satisfactory solution has been found, evaluated and approved by the change control board taking into account all economic and technological criteria, the change implementation phase begins.

Depending on the scope of the change and the point in time when the need for change is discovered, the implementation phase is in the simplest case a simple document change; however, for complex processes the change process can extend over different departments and over a longer period of time. As in general project management, it is important that the change status, cost and schedule are continuously monitored. In addition to the technological implementation, the change procedure is concluded with a conscientious and detailed change review. In addition to the documentation of all changed data, this also includes a preparation of the data in the sense of a "change review" for individual production (Fig. 5.15). The individual phases differ in the scope and depth of the description. This is accompanied by the fact that methodological aspects are of particular importance in the analysis, evaluation and decision phases as well as in the change review phase. Special expertise and procedures predominate in the phases close to implementation. This expertise is not generally valid and can therefore only be described insufficiently [156].

Procedure model on the example of the 3D servo press

The presented modification procedure is applied to the case of sensor integration for the control optimisation of the 3D Servo Press, Sect. 3.6.3. The change was initiated by the requirement to integrate an additional monitoring and closed-loop control of the process force by means of integrated force sensors. This change was motivated by a technical innovation, with the aim of implementing the changed requirements with regard to the closed-loop control of the machine. Accordingly, an adjustment of the target property is carried out.The idea for the modification came from the Research and Development Department (R&D) at a time when large parts of the gear unit (Sect. 3.6.3) had already been manufactured but not yet assembled. Similarly, production of the lower part has not yet begun. Since the internal customer, in this case the R&D department, has initiated the change, the customer is fully involved in all processes of the change process. The change request is justified by the fact, which Groche et al. [126] have determined, that a closed-loop control of presses via the engine parameters and kinematics model is not sufficient due to friction. A first solution suggested by the customer is the integration of the sensors in the direct force flow, which is carried out for each individual gear. Thus, the integration of the sensor technology focuses on the components located in the direct force flow between bearing frame and gearbox. The drive rod, which is located in the direct force flow between the bearing frame and the gearbox, turns out to be the most promising component to be changed. The cost of the change can be identified as the procurement of the sensors and the integration into the production process. It should be noted, however, that the drive rod was already finished but not yet assembled. Furthermore, the need for action, effort and costs for the modification were estimated. To find a solution, a conventional procedure is used whereby creative methods such as brainstorming were used to develop the design solutions shown in Fig. 5.16.

The technical feasibility, costs and effort of the respective solutions were evaluated by the core team and in coordination with sensor vendors and manufacturers. The uncertainty of the respective solution regarding robustness in operation, sensor failure, overload and reduction of gear stiffness was to be evaluated. At the same time, production costs and effort must be included in the analysis. The first solution with direct force measurement using a piezoelectric-force measuring ring between the drive rod and the bearing frame proved to be the simplest solution in terms of feasibility and costs. However, this solution has the greatest influence on the strength of the gear. The second variant can be described by indirect force measurement using a force measuring ring in the force shunt. A part of the force is transmitted directly via the structure into the bearing frame, another part is transmitted via an adapter plate and the force washer in order to measure the process force. The advantages are higher rigidity and low stress on the sensor. Production, on the other hand, is more complex, because it is necessary to achieve a uniform height of the bearing surface and the drive rod on the bearing frame. The third variant involves the integration of a surface strain sensor, which also operates according to the piezoelectric effect. This approach is already used in the prototype of the 3D Servo Press and proved successful. However, three sensors each are required on the drive rod in order to

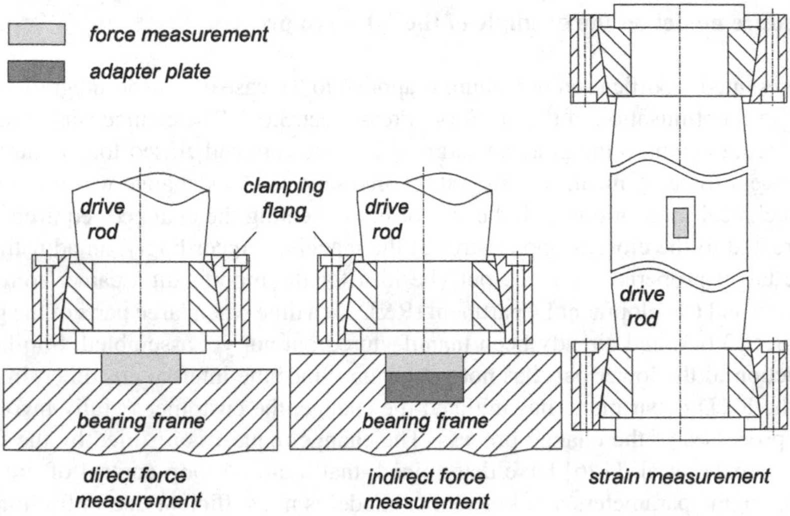

Fig. 5.16 Example of a change to the drive rod of the 3D Servo Press

obtain high-quality force measurements in the event of a possible deflection of the drive rod. A calibration of the installed sensors is necessary for all three variants and does not represent an advantage or disadvantage for any variant.

All three variants are subsequently evaluated by the internal customer and the project core team. In the questioning of the customer after the fulfilment degree of the respective variant, based on the classification of the change request the third change variant is evaluated most positively. The decision is then documented and made available transparently to all parties involved. The change variant is released and orders are issued to carry out all necessary change tasks. The responsible change manager is commissioned to take over change controlling and to regularly inform the customer and the Change Control Board about the status of implementation. The last step of the change review completes the change procedure. Further change processes follow the same procedure, as shown in the example of the 3D Servo Press.

The procedure model presented shows that the formulation of the question has a considerable influence on the classification of the customer needs. With regard to the requirements, there is a clear difference in the product to be developed. The transferability to serial production is possible, however, associated with clearly larger expenditure in the organisation.

5.3 Mastering Propagated Uncertainty in Process Chains

Peter Groche and Florian Hoppe

The production phase in the life cycle of products or systems, linking the design phase with the usage phase, see Sect. 1.2, is usually structured as a process chain. This section describes methods to master uncertainty along process chains. Following the representation shown in Fig. 3.3 in Sect. 3.2, a process chain consists of subsequent single processes, each of which is subject to disturbances. Furthermore, the material fed into the process to produce the product contributes to uncertainty, e.g. by semi-finished product variations.

Uncertainty during the production process propagates into the product and can make it infeasible to carry out subsequent steps in a production chain or to guarantee product quality. To manage this challenge, standardisation methods have been used for several hundred years. If the effect of uncertainty is particularly known, i.e. incertitude, it can be mastered by the definition of tolerances, cf. Chap. 2. However, since the influence of the uncertainty along the process chain is often unknown, it is possible that relevant product properties being subject to uncertainty and being important for quality are not considered; or the definition of tolerances for the considered properties is not appropriate. To cope with uncertainty in process chains, two processes have emerged in the last decades: firstly, robust single processes make the process chain less sensitive to quantified uncertainty, as they fulfil their function, even if the system is disturbed in a limited and predictable way, cf. Sect. 3.5; secondly, modern resilient processes, as introduced in Sect. 3.5, do not only allow to react to disturbances but also to learn from them in order to master unknown effects of uncertainty, i.e. ignorance (see Chap. 2). In the following subsections, examples are given on how to react to disturbances in order to ensure function and quality (see Chap. 1) of the system, which is in this case represented by the process chain itself. Furthermore, we illustrate how product property defects can be fed back in such a way that the process chain acquires the resilience functions of learning and anticipating, which are described in detail in Sect. 6.3. Using the example of a combined process chain which consists of forming and cutting processes, it is shown how uncertainty can be reduced (Sect. 5.3.1). In Sect. 5.3.2, the regulation of product specific requirements and the effects on the uncertainty are discussed. With this knowledge, it is shown how the propagation of uncertainty can be quantified and how the process chain can be adapted at an early stage of the production process. Section 5.3.3 provides an example for a flexible process chain using the 3D Servo Press in order to adapt product properties during the production process. One approach to master uncertainty by reducing process-related fluctuations in product quality are smart structures. Integrated in load-bearing structures and machine elements, as shown in Sect. 5.3.4, they can be used to monitor and control the production process (Sect. 5.3.5). Section 5.3.6 shows a new process-integrated calibration method to increase efficiency in the manufacture of smart structures.

5.3.1 Uncertainty Propagation in a Forming and Machining Process Chain

Felix Geßner, Maximilian Knoll, Christian Bölling, Florian Hoppe, Eberhard Abele, Matthias Weigold, and Peter Groche

In production, the interlinking of forming and machining processes represents a classic value chain. Process chains, as described in Sect. 3.2, are characterised by uncertainty, such as batch-related uncertainty in the material properties or positioning uncertainty of the machine tool. Furthermore, the open-loop or closed-loop control of the machine tool affects uncertainty (Sect. 5.3.2).

In the context of process chains the concept of the Austauschbau, as described in Chap. 2, plays a decisive role. One way to master uncertainty in form of incertitude in the Austauschbau is by defining tolerance limits, within which the component has to be positioned during the transition between the process steps. The definition of tolerance limits is usually done opposite to the direction of the value chain starting with the finished part. The required tolerances of the preceding process steps are usually determined according to the experience of the employees. As a result, the accuracy requirements of early process steps are often based on a subjective view. In particular tolerances that are too tightly selected allow little room to compensate fluctuations in raw material quality, for example, although the continuous optimisation of production steps enables the development of controlled processes or robust tools that are able to master larger uncertainty in component specifications.

Since process simulation and optimisation are often focused on one single process, predictions of component quality over the entire process chain are currently difficult to assess and heavily depend on experience and knowledge [46]. However, little attention is paid to the interfaces between the processes that are critical for linking the different process steps. One reason for this lack is that there are hardly any compatible interfaces between different simulation tools [46]. With the current production approaches, variable products with application-specific requirements, such as required by the MAFDS (Sect. 3.6.1), cannot be realised taking uncertainty into account. Therefore, we investigate a possible combined process chain consisting of cutting, forming, heat treatment and machining (see Fig. 5.17), with the main focus on the forming and cutting processes. We chose orbital forming, an incremental forming process, in which only a partial contact surface is created between the punch and the workpiece, as an exemplary forming process. Relative to a point on the workpiece-tool contact surface, the stress builds up accumulatively due to the intermittent tool contact. At the same time, the lower rolling friction between the workpiece and the punch enables a radial material flow, also known as the mushroom effect [187]. This effect can be controlled by varying the tool angle, the feed rate and the rotational speed in order to counteract fluctuations in the semi-finished product [58]. The forming machine we use is a 3D Servo Press (as described in Sect. 3.6.3) with three independently controllable degrees of freedom in the slide [126]. Since we control the material flow by adjusting the process parameters due to previous semi-finished

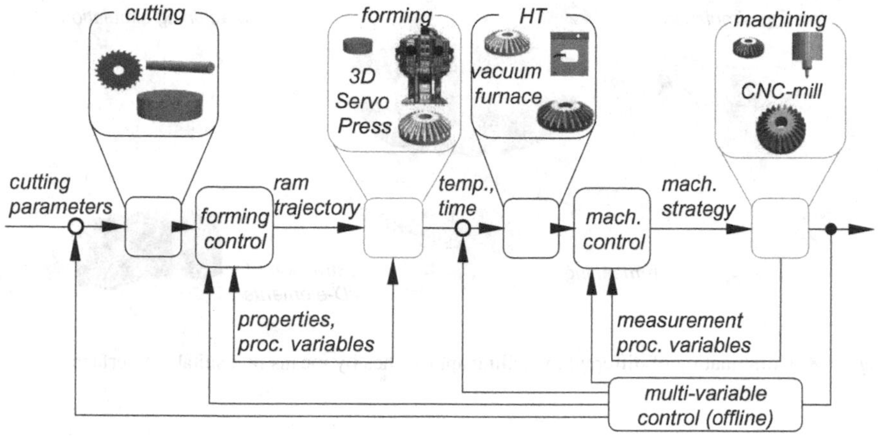

Fig. 5.17 Process chain consisting of forming and machining as well as feedback of the final component properties [46]

product uncertainties, it is necessary to predict the effects on the subsequent steps, such as reaming. By coupling the forming prediction with the reaming prediction, a closed-loop process control can be achieved, which reduces the uncertainty due to semi-finished product variations.

As a first step we implement the forming process in a 3D finite element simulation using the simulation tool Simufact Forming to map the relevant forming step [46]. In the simulation a tube with an nominal outside diameter of 48 mm and a nominal initial height of 34 mm is formed to a bevel gear with a nominal height of 17.7 mm. During the forming process a perforation with a nominal diameter of 7.5 mm is added to the centre of the bevel gear. The influence of the parameters tool angle, infeed speed and form of motion on the resulting material flow is investigated and variables for process control are identified [58]. At the same time, variations in the angle at a constant tool angle lead to a deviation of the height profile. This requires compensation by calibrating the workpiece towards the end of the orbital forming process by levelling the surface with a constant angle.

While we use the material flow as a variable in further work to control the component geometry with regard to mould filling and height, quantitative evaluation of the effect on perforation requires coupling to a reaming simulation. Furthermore, this enables the optimisation of the reaming strategy based on formed components.

Reaming is often used towards the end of the value chain. It serves the purpose of producing functional bores with shapes and of positioning within a required tolerance range. For productivity reasons, multi-bladed reaming tools that combine the functions "cutting" and "guiding" in one tool element are often used. These tools have been the subject of scientific investigations. The influence of disturbance variables on the quality of the reamed bore is investigated with regard to diameter, circularity and cylindrical shape on the one hand [38, 179] and the deflection of the tool on the other [45, 145]. Tool deflection leads to an increased diameter of the envelope

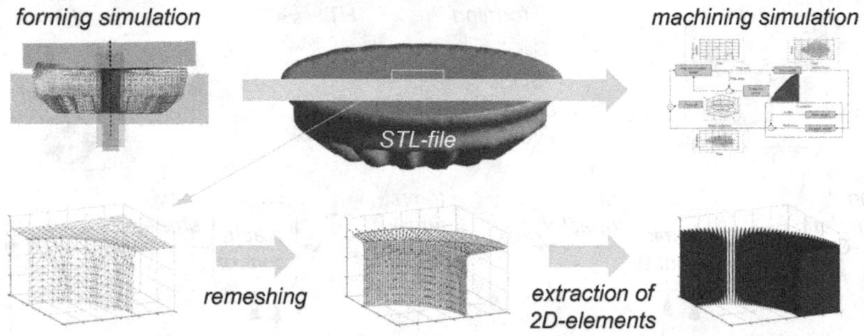

Fig. 5.18 Combination of different modelling approaches by means of a suitable interface

cylinder of all bore centres of the reamed bore. Adapting the cutting edge geometry by increasing the setting angle results in a reduction of tool deflection in the transient entry phase into the workpiece. An upstream pilot process achieves a significant reduction of the deflection, especially with long cantilevered tools with a length-to-diameter ratio of $L/D \approx 10$. The model we used to simulate the reaming process is the mechanistic model described in Sect. 4.1.3. It is based on a geometric intersection model to calculate the undeformed chip cross-section [1], an empirical cutting force model [3] and a multi-body model to represent the dynamic tool behaviour based on the Jefcott rotor theory [44].

A central point for the representation of the process chain is the linking of the individual models of the forming and machining processes. Figure 5.18 illustrates the interface to transfer the geometry between simulations. One output variable of the forming simulation is the surface of the part after forming. The component surface is represented in standard tessellation language file format (STL) by numerous triangular faces and their respective three vertex points. We then generated a new cylindrical grid using the vertices to remesh the wall of the punch hole in steps of 1°. The polygons of the 2D element model, which represent the workpiece in the reaming simulation, are then generated from the nodes of the newly meshed grid.

The perforation in the centre of the component, which is produced during the forming process, is machined using reamers in order to generate a high quality cylindrical surface. To investigate the influence of the process and geometry parameters on the displacement of the reaming tool, a tool with a diameter of $D = 8.3$ mm and a cantilever length of $L = 30$ mm is simulated. We vary the process parameters with regard to achieving a constant feed rate of $f = 0.6$ mm/rev during the entire machining process. In addition, we determine a suitable running-in strategy, using a reduced feed rate of 0.1 mm/rev until a reaming depth of 4 mm, followed by an abrupt increase to achieve the final feed rate. In the case of machining valve guides in the cylinder head of a combustion engine, we demonstrate the positive influence of an upstream pilot process using a short cantilevered reaming tool [45]. The diameter and cantilever length of the pilot reamer are reduced by 0.1 mm and 10 mm respectively.

The depth of the pilot bore is 2 mm. For all simulations we chose a cutting speed of 100 m/min.

Typically, reaming tools with a setting angle of $\kappa = 45°$ are used in industrial practice for standard applications. In theory, increasing the setting angle to $\kappa = 75°$ leads to a reduction in the passive force F_p at the individual cutting edges, reducing tool deflection especially in the face of disturbance variables. In the case of upsetting, using a reaming tool with a setting angle of $\kappa = 75°$ can reduce the tool deflection by 25% compared to a tool with a setting angle of $\kappa = 45°$. Considering orbital forming, the use of a reaming tool with a setting angle of $\kappa = 75°$ can reduce the tool deflection by 45%. In both cases, however, the tool deflection, even with adapted cutting geometry, is still about 20 µm, based on a machining depth of 17 mm (Fig. 5.19a). For functional bores, which serve to guide other components, it is often necessary to minimise the tool deflection and the associated reduction of the radial deviation of the reamed bore even further in order to fulfil the functions and reduce the wear symptoms.

Therefore, in a second step, we investigate various machining strategies for the reaming process of the formed components (Fig. 5.19b). Based on the previous results, we select $\kappa = 75°$ as the setting angle of the tool. The results of the simulation indicate the positive effect of an adapted running-in strategy (Fig. 5.19b). By reducing the feed rate during the entry phase into the workpiece, we could reduce the process forces significantly and thus minimise tool deflection. In the case of upsetting, we can reduce the tool deflection by 65% from 20 µm to 7 µm. With orbital forming, a 45% reduction in deflection from 21 µm to 11.5 µm can be achieved. Thus an improvement of the reaming quality can be obtained without the employment of further tools and with an increase of the main time from 0.5 s to 1 s. The introduction of a pilot process upstream of the actual reaming process enables a further significant reduction in tool deflection. Due to the increased stiffness of the pilot reamer, it is hardly deflected despite the occurrence of disturbance variables. In theory the following reaming process has a reduced radial allowance of $a_\mathrm{p} = 0.05$ mm up to a depth of 2 mm due to the preceding process in the transient entry phase. In addition, the influence of disturbance variables within these 2 mm can be largely eliminated by the pilot process. Therefore, the deflection of the tool can be limited to values around 1 µm.

The presented results illustrate the possibilities of a coherent consideration of forming and machining process step within the process chain. By using the simulation tools, we could show that different process parameters during the forming of the workpiece have a significant influence on the component geometry. An unfavourable selection of forming parameters leads to imperfections in the perforation after forming. Without an adapted reaming process this would cause an interrupted cut during the reaming operation and impaired bore quality. By using the knowledge from the combined models along the process chain, however, the uncertainty in the reaming process can be mastered. With the selection of a suitable cutting geometry and an adapted machining strategy, a significant reduction in tool deflection is possible. Furthermore, the results from the analysis of the process chain can be fed back into the

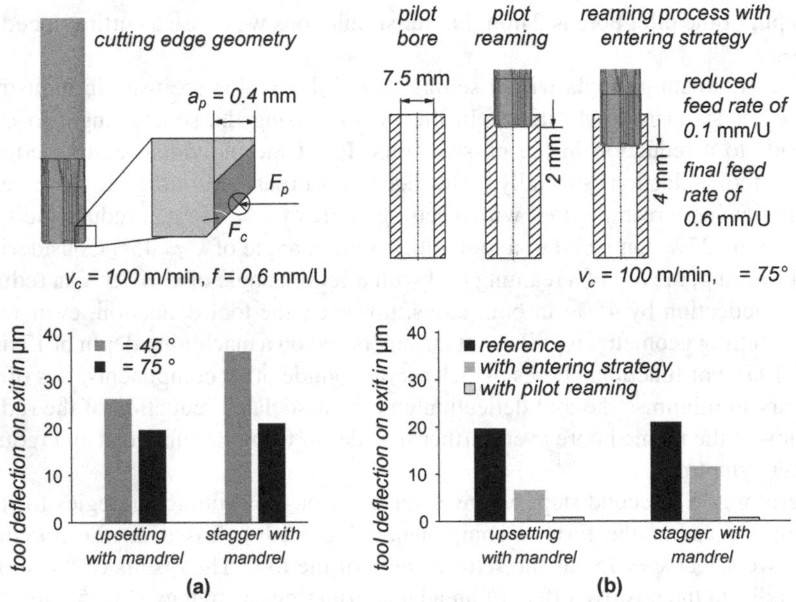

Fig. 5.19 Deflection of the reaming tool as a function of **a** the setting angle and **b** the machining strategy

forming process in order to establish a closed loop for the control of uncertainty in the process chain and thus master the chained uncertainty.

5.3.2 Closed-Loop Control of Product Stiffness and Geometry

Florian Hoppe, Peter Groche, and Maximilian Knoll

Uncertainty in manufacturing processes affects the properties of a product and as a consequence affects the product's usage behaviour. Currently, product properties are ensured in an open-loop approach by skilled workers and process planners. Product quality therefore depends mainly on the experience and qualification of these people. At the same time, changing and increasingly complex process chains present a growing challenge. Changes in the process chain and uncertainty in ambient conditions as well as properties of the raw materials affect both, product and process. Therefore, methods are needed to master uncertainty in the production phase (Sect. 1.2) and along the complete process chain.

The automatic closed-loop control of forming processes experiences a constantly growing attention, focussing on the control of individual geometric product properties [8]. The automation of production processes, however, is accompanied by the fact that several product properties need to be controlled simultaneously and boundary

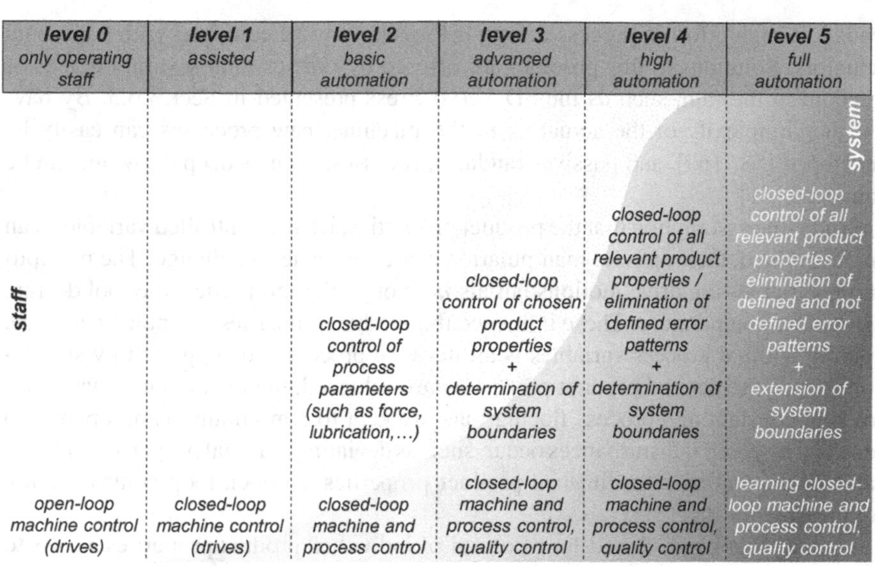

level 0
only operating staff
level 1
assisted
level 2
basic automation
level 3
advanced automation
level 4
high automation
level 5
full automation
system
staff
closed-loop control of all relevant product properties / elimination of defined and not defined error patterns + extension of system boundaries
closed-loop control of all relevant product properties / elimination of defined error patterns + determination of system boundaries
closed-loop control of chosen product properties + determination of system boundaries
closed-loop control of process parameters (such as force, lubrication,…)
open-loop machine control (drives)
closed-loop machine control (drives)
closed-loop machine and process control
closed-loop machine and process control, quality control
closed-loop machine and process control, quality control
learning closed-loop machine and process control, quality control

Fig. 5.20 Levels of automation in manufacturing processes [125] following the VDA [268]

conditions have to be met. Sometimes, the properties to be controlled are conflicting with each other. The requirements for automated driving proposed by the German Association of the Automotive Industry (German abbreviation: VDA) can serve as an analogy where a fully automated car is in charge of all the driving and does not require any human intervention [268]. When transferred to manufacturing, a fully automated process requires that all relevant product properties are controlled and that boundary conditions are not only met but also extended if necessary [125].

To differentiate the degree of automation of process chains, the VDA introduced different levels as shown in Fig. 5.20, which are determined by the autonomous abilities of the system, in the style of the levelling scheme used for autonomous driving cars.

The objective in increasing the automation level is the reduction and compensation of disturbances, i.e. to increase the robustness of the process, compare Chaps. 3 and 6. By breaking this objective down into sub-objectives, subtasks can be defined and automated piece by piece. For example, tasks that are carried out by the staff at lower levels can be replaced by automated quality inspection systems and closed-loop control at higher levels.

Both, control methods and equipment for measuring and manipulating product properties, are required to climb the automation levels. But we already face a conflict of objectives when measuring product properties, which is between good measurability and information content (Sect. 5.2.4). With increasing spatial distance between the machine drives and the product being manufactured, the manipulability of the product properties decreases. Furthermore, the uncertainty increases with increasing measuring distance to the product [56]. Moreover, the use of adjustment mechanisms leads to a high effort in process design if tools have to be equipped with additional actuators. Solutions to this problem are offered by presses with several degrees of freedom in the ram, such as the 3D Servo Press presented in Sect. 3.6.3. By having the complexity of the actuators in the machine, new processes can easily be developed [58, 160], and passive standard processes, such as deep drawing, can be enhanced [51].

Under the assumption that the product properties, i.e. the controlled variables, can be determined, their optimal manipulation represents a new challenge. The manipulating variables are drive motions but also, among other properties, the tool design, lubrication, temperature. These influence the product properties in a non-linear way. Finding optimal process variables is an iterative process and supported by simulations. Once the desired product properties are achieved, the manipulating variables are kept constant and process fluctuations are kept to a minimum using open-loop control. However, if disturbances occur, such as deviating material properties in a new batch or fluctuating semi-finished product properties, an open-loop control cannot compensate for this [57].

On the other hand, closed-loop control of individual product properties allows to adjust the process to deviant product properties by feeding back the actual product property. Incremental forming processes are particularly suitable, since they can be controlled from one increment to the next. In free bending, for example, the springback of the material is a decisive influence of the final component geometry. The latter depends strongly on the material properties of the current workpiece. By measuring the geometry after springback while still clamped, the bending operation can be adjusted and iterated to the desired geometry, as we demonstrated in [58].

Although the product's geometry is its most evident property, it is only one of many relevant properties. This is demonstrated in the Modular Active Spring-Damper System (German acronym: MAFDS), see Sect. 3.6.1, which contains, among other components, active support with spring membranes. In addition to the geometry, these must rely, above all, on the stiffness of the spring membranes. While the geometry ensures the mountability, the stiffness is decisive for the function of the shunt-damping, as well as the active buckling stabilisation. The stiffness of the membranes is mainly determined by the material properties and their geometry. Single-Point-Incremental-Forming (SPIF), a process in which a sheet metal component is incrementally formed with a tool tip, is therefore particularly suitable for production. The process makes it possible to produce almost any geometry, as long as the load-bearing capacity of the material is maintained. If variations in the material occur, the stiffness can be corrected by making minimal adjustments to the geometry within the tolerable limits [155].

Whereas open-loop control acts, closed-loop control reacts, i.e. it only takes action when a deviation is detected. This is acceptable for disturbances with slow dynamics. However, if the disturbances occur with high dynamics from part to part, as is usual in production, the controller is unable to handle them well. Examples for such disturbances are fluctuations in the lubrication and the semi-finished product properties. Nevertheless, sometimes information about the material properties of the semi-finished product can be aggregated from process data of upstream processes.

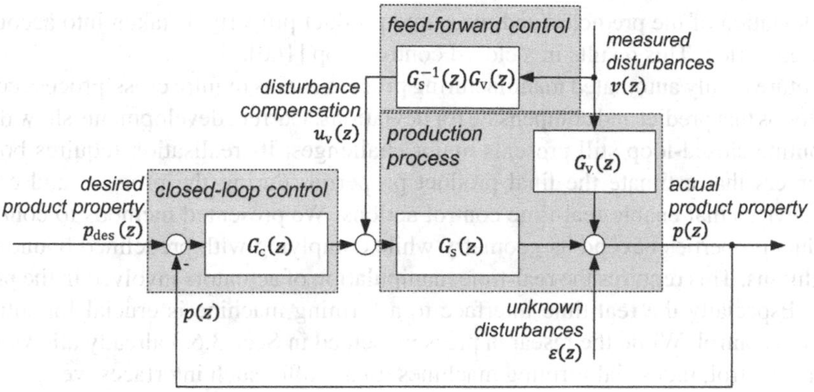

Fig. 5.21 Resilient process chain by means of closed-loop process control with disturbance feed-forward [124]

With the aid of disturbance models $G_v(z)$ we anticipate the effect of measurable disturbances v on the final product properties and compensate it using a disturbance feedforward control [124]. For example, shear-cutting processes are typically carried out prior to the actual forming operation. Disturbances are already visible in the force measurements of the cutting data. However, it remains challenging to select data features containing enough information to predict the product properties. Methods from the field of machine learning offer the possibility to identify relevant features from a data set and to create prediction models [159].

The effect of such measurable disturbances v can then be compensated for before they become noticeable in the product. However, we typically face disturbances ε that cannot be measured, or whose effects on the product properties are not known. Based on a known process transfer function $G_s(z)$, we combine the disturbance feedforward $G_s^{-1}G_v(z)$ with a closed-loop control $G_c(z)$ to a resilient process chain shown in Fig. 5.21. Such a control structure compensates for abrupt fluctuations caused by batch changes, as well as long-term disturbances, such as wear and temperature drifts. While the tool can also be equipped with additional temperature sensors, measuring the wear online is still a research issue.

The coupling of different product properties challenges their simultaneous control. When correcting one product property, additional control measures are necessary to keep the other product properties constant. Boundary conditions further intensify this problem. To decouple product properties when implementing a closed-loop control, models are required that describe the effect of the manipulated variables and the coupling of the product properties. Since these models usually can only be evaluated by simulations, extended control approaches are necessary. The model-predictive control is a suitable approach, which is based on the recurrent solution of an optimal control problem. By means of an optimisation algorithm new optimal control variables are obtained after each forming step. Uncertainty, which becomes visible in the deviation of the predicted and measured product property, is taken into account after each step. This results in a closed control loop [160].

Future highly automated manufacturing process chains require cross-process control loops that predict and compensate for deviations. Current developments show that an online closed-loop still presents major challenges. Its realisation requires both, observers that estimate the final product properties during the process, and control devices that enable real-time control actions. We presented methods to control product properties beyond its geometry while complying with predefined boundary conditions. This requires the real-time manipulation of actuators involved in the process. Especially the real-time interface to a forming machine is crucial for future process control. While the research press presented in Sect. 3.6.3 already allows for online control, industrial forming machines do not offer such interfaces, yet.

5.3.3 Controlled Partial Post-compaction of Sintered Bevel Gears

Peter Groche, Julian Sinz, and Daniel Martin

Powder metallurgy serves as a manufacturing technology for a wide range of applications, one example of which is the manufacture of gear wheels. Powder metallurgical production offers economical and ecological advantages, such as profitable cycle times, good automation possibilities [238] and an optimal use of raw materials, due to the possibility of producing near-net-shape components [78]. Furthermore, the sintering of components from powder materials offers the possibility to realise different material combinations which cannot be produced by melting metallurgy at all, or only with increased effort [238]. However, the use of gears for power transmission in safety-relevant systems places high demands on process reliability and the final component quality [83].

Sintered components are porous bodies whose mechanical properties depend on porosity and pore density [238]. The porosity P describes the volume fraction of pores present in the component, while the pore density is the ratio of the number of pores to the length of the body. Pores lead to material weakening due to notch effects under load and are therefore a possible source of component failure, especially in areas of stress concentrations [55, 137]. A local reduction of the porosity can increase the fatigue strength of sintered components [209, 259]. For this reason, a targeted post-compaction of the most heavily loaded component areas is used to minimise the local pore content. In this way the quality of the product can be improved in terms of an increased availability, cf. Sect. 1.7.

Since both process stages, i.e. sintering and post-compaction, place different demands on the functionality of the machines used, conventional post-compaction takes place in a separate process step downstream of the actual sintering process. This means that the process is associated with additional costs for machines and tools, as well as a higher logistical effort [23, 197]. In order to better exploit the potential of powder metallurgical production for safety-relevant components of load-bearing structures, we developed a process concept that enables the combination of both process steps, the sintering process and the post-compaction, in one process sequence. For this purpose, the additional degrees of freedom in the ram motion of the 3D Servo Press presented in Sect. 3.6.3 are used to realise an integrated partial post-compaction process at the example of bevel gear manufacturing.

A measure for the quality of the post-compaction process is the relative density D, defined as

$$D = 1 - P = 1 - \frac{\text{pore volume}}{\text{total volume of the porous body}}. \tag{5.14}$$

In the following we present a simulation model which serves to predict the relative density distribution in sintered components. This model is validated by experiments for an uniaxial load case and then extended for three-dimensional load conditions. The resulting model will be used to develop a process strategy for the partial post-compaction of sintered bevel gears using the 3D Servo Press.

Development of a simulation model for the uniaxial load case

A continuum-mechanical simulation model was developed. The governing equations were solved by a Finite Element (FE) solver, which was extended by an adequate material model for the process design. This model was adapted to the requirements of the specific application and validated by means of real tests to ensure that the model predicts the process behaviour in the best possible way and thus minimises model uncertainty as described in Sect. 2.2.

According to Parteder [220], the relative density D is used to assess the residual porosity at the end of the manufacturing process. The used material model must therefore primarily ensure the mapping of density development during the process. The Gurson model [136], based on the material behaviour according to Levy-Mises and described by a yield criterion, is able to predict the behaviour of porous media [220].

The development of the relative density can be calculated from the flow potential Φ as described by Parteder [220]. In order to take individual material properties into account, this model has been extended by further model parameters q_i $(i = 1, 2, 3)$ according to Tvergaard [264], which includes the dependence of the yield stress on the density [98].

The change in pore shape during the process has a significant influence on the density development [54, 182, 220]. Therefore, instead of constant Tvergaard parameters, condition-dependent parameters are used, which are adapted to the temporal deviatoric deformation ε_e during the simulation on the basis of experimentally determined material parameters. The necessary material data were generated in uniaxial compression tests with cylindrical specimens made of porous pure iron.

Results from the real uniaxial tests are compared with the simulation results. The evaluation of the model uncertainty is based on the components' densities in different stages of the compression tests. Experimental data of the porosity are derived from

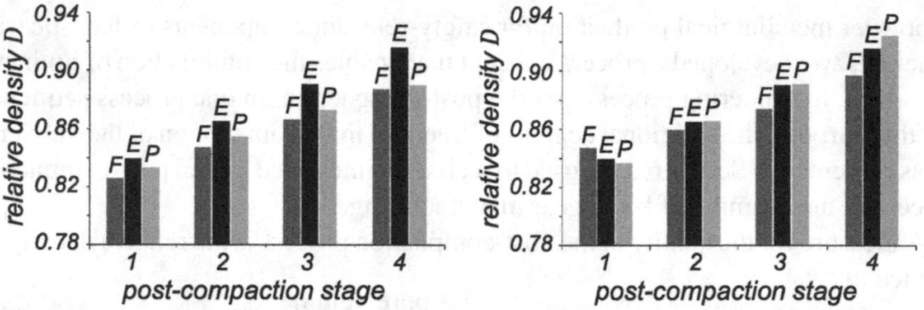

compaction stage	1	2	3	4
Δ in % between E and F	1.39	1.82	2.43	2.90
Δ in % between E and P	0.65	1.05	1.78	2.60

(a)

compaction stage	1	2	3	4
Δ in % between E and F	0.72	0.55	2.48	1.87
Δ in % between E and P	0.29	0.06	0.08	0.89

(b)

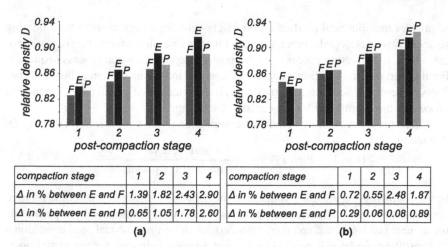

Fig. 5.22 Comparison between **a** constant Tvergaard parameters and **b** state dependent parameters after 4 consecutive post-compaction stages for an initial absolute density $\rho = 6.4\,\text{g/cm}^3$ according to Brenner [50]; the current absolute densities after each stage are $6.6\,\text{g/cm}^3$ (stage 1), $6.8\,\text{g/cm}^3$ (stage 2), $7.0\,\text{g/cm}^3$ (stage 3) and $7.2\,\text{g/cm}^3$ (stage 4); the values in the tables show the percentage deviation between experiment (E) and simulations with defined process forces (F) and defined punch displacements (P)

optically recorded microsection samples of the test components. A comparison of results from simulation and experiment is shown in Fig. 5.22. The data correspond to the relative density D after one to four post-compaction stages. The different stages represent specific values of the absolute density ρ reached in the experiment after a defined punch travel or process force. The density in the initial state is set to $6.4\ \text{g/cm}^3$. The input parameters for the simulation are either defined process forces (F) or punch displacements (P). Values of both types of parameters correspond to the measured ones in test stages with defined values of density. The black bars in the middle represent the experimentally determined values (E).

The displayed results exhibit an overall good agreement between the simulation with state-dependent model parameters (Fig. 5.22b) and the experimental results for this uniaxial load case. The deviations between simulation and experiment are small. This is especially true for the simulation with predefined punch displacements. In comparison, a larger deviation can be determined using constant Tvergaard parameters as displayed in Fig. 5.22a. This shows that the model uncertainty can be significantly reduced by using state-dependent model parameters.

Adaptation of the model for multiaxial loading

The prevailing stress conditions in forming processes are usually multiaxial. The compaction of an isotropic material is based on the mechanism of pore closure within the material, which is decisively influenced by the prevailing stress state [220]. For

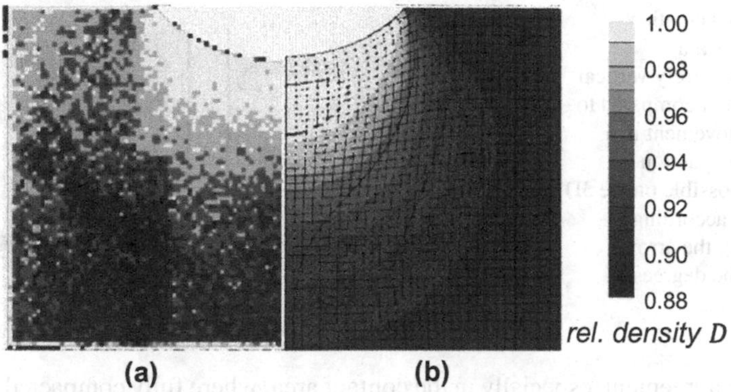

Fig. 5.23 Comparison of the spatial distribution of relative density D between **a** experiment and **b** FE simulation for a ball indentation test ($r_s = 10\,\text{mm}$) according to Strauß [256]

this reason, a suitable comparison variable is required to investigate the density development as a function of different stress states. Therefore, a modification of the Gurson-Tvergaard model including the multiaxiality X in the determination of the Tvergaard parameters q_i was performed. The multiaxiality X is defined as the ratio of the hydrostatic stress component σ_h to the deviatoric stress σ_e [220]. These can be calculated from the stress components σ_i in the three principal directions:

$$X = \frac{\sigma_h}{\sigma_e} = \frac{\sqrt{2}}{3} \frac{\sigma_1 + \sigma_2 + \sigma_3}{\sqrt{(\sigma_1 - \sigma_2)^2 + (\sigma_2 - \sigma_3)^2 + (\sigma_3 - \sigma_1)^2}}. \tag{5.15}$$

In order to improve the prediction accuracy of the simulation model for the multiaxial load case, the parameters $q_i(\varepsilon_e, X)$ depending on the deviatoric deformation ε_e and the multiaxiality X can be determined with the help of a cell model calculation according to Needleman [205] and Tvergaard [264]. The cell model calculation establishes a relationship between the microscopic properties of the material and its macroscopic damage behaviour [220].

After the successful implementation of the cell model calculation into the FE simulation using subroutines to determine the state-dependent model parameters, we performed a validation of the advanced model using ball indentation tests. A hardened (55–65 HRC) spherical tool component, which is assumed as rigid in the simulation model, with a ball radius r_s is pressed into a sample body with a press force increasing linearly over the penetration depth up to a maximum of 70 kN. The investigations were performed with sphere radii of 5 and 10 mm. The load cases of the simulation are given by the course of the press force.

The spatial distribution of the relative density at the end of the indentation process serves as a comparative quantity. The penetration depth achieved is 4.75 mm. In Fig. 5.23 the optically evaluated relative density distribution of the experiment (left) for $r_s = 10\,\text{mm}$ is compared to the simulation result. The comparison shows a good

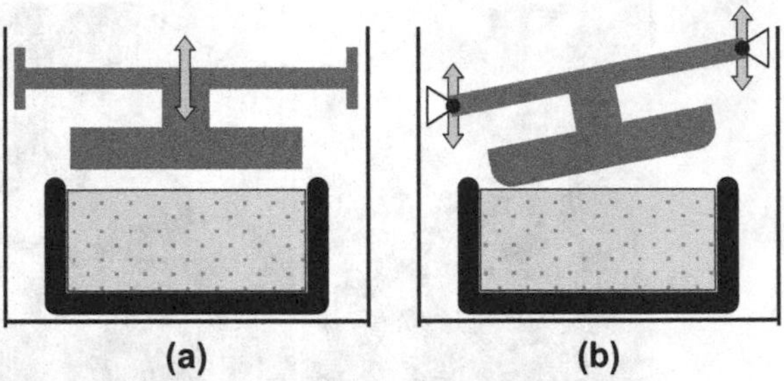

Fig. 5.24 Schematic illustration of **a** a conventional purely vertical ram movement compared to **b** the ram movement in a tumbling compaction process as possible on the 3D Servo Press according to Strauß [256]: the arrows symbolise the degrees of freedom

qualitative agreement, especially in the contact area, where fully compacted material is present. The largest deviations occur in the upper edge area, since the ejection process necessary in the experiment leads to an increase in density of the surface area due to the ejection force. In addition, the course of the density in the centre deviates noticeably in the axial direction. This is probably due to uncertain test conditions that cannot be taken into account in the simulation, such as the existence of backlash between the test specimen and the die, which leads to radial displacements of the test specimen. As a result of the validation, however, a sufficient accuracy of the simulation model for the application in multiaxial load cases can be determined.

Process strategy for partial post-compaction

In addition to the vertical degree of freedom in conventional presses, the 3D Servo Press described in Sect. 3.6.3 has the additional option of tilting the ram. This allows the press ram to be deflected by an angle of up to approximately 3°. In this way, the vertical ram movement can be superimposed by, for example, tumbling or swivelling movements in two planes [128]. Based on the developed simulation model, an FE-supported design of optimised post-compaction processes was carried out. The aim is to investigate the potential of process integration of post-compaction processes for the property adjustment of porous sintered components. The multiaxis post-compaction process takes place by utilising the additional ram degrees of freedom of the 3D Servo Press in the form of a tumbling tool movement. Figure 5.24 schematically sketches the process concept for the tumbling post-compaction compared to a conventional unidirectional compaction. The three independently controllable and adjustable eccentric drives enable the realisation of a tumbling die movement by phase shifted translational movements of the three ram drives, see Sect. 3.6.3.

In this way, the degree of compaction can be specifically influenced in the post-compaction process. The aim is to partially increase the relative density D of the end product in the critical stress zones in order to avoid component failure caused by residual porosity. In the case of gear manufacturing, these are located in the marginal zones, such as the tooth flanks and the tooth root [137]. The edge zones of the tooth tip are also the goal of further optimisation with regard to the possible degree of compaction.

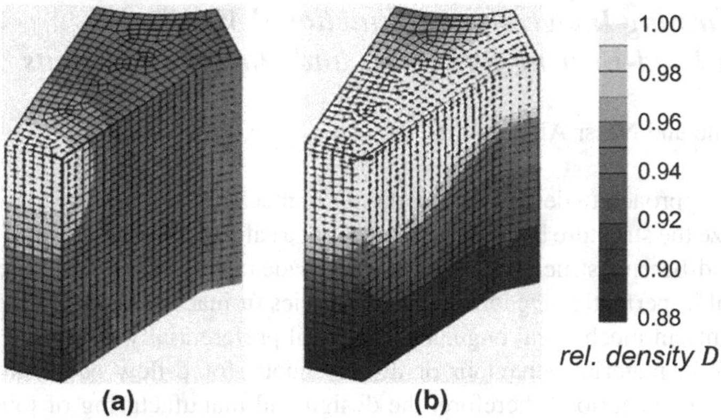

Fig. 5.25 Simulation results for the density distribution **a** after the conventional uniaxial process compared to **b** the optimised compaction process according to Strauß [256]

To estimate the achievable local degree of compaction by means of a specific ram inclination, the post-compaction of a single tooth is simulated. The modified tool paths alter the material flow, the influence of which on the result of the compaction process is investigated. The sample to be compacted consists of a simplified tooth geometry, which is derived from an involute gearing. Exploiting the symmetry only half of the geometry is modelled.

In Fig. 5.25, the spatial distribution of the relative density after the post-compaction process is compared to the density distribution of the uniaxially compacted tooth. Here, the density values in the area of the tooth flank and the tooth root are significantly lower than in the tumbling process. It is only in the area of the tooth tip where a fully compacted material area is visible. The density distribution in the remaining areas, however, is more homogeneous than in the tumbling process.

On the basis of the simulations carried out, a clear influence of the friction in the contact area between workpiece and die can be observed. Furthermore, large deformations are visible at the edge areas close to the contact zone between workpiece and die. It can be expected that the higher density in highly loaded product areas will lead to an improvement of the product's durability and strength.

The results show that extended tool kinematics can be used to positively influence compaction processes. Especially if locally high densities are required, tumbling motions can have a positive influence on the density development in critical component areas.

With the presented application, we have optimised the manufacturing process of a machine element in such a way that the properties of the end product can be specifically influenced without a subsequent separate processing step. The use of the 3D Servo Press enables a flexibilisation of the process chain. Thus, a consistent product quality can be achieved by adapting the manufacturing process to product requirements that arise in the usage phase and are not yet known during the design phase.

5.3.4 Forming Integration of Functional Materials in Load-Bearing Structures and Machine Elements

Peter Groche and Nassr Al-Baradoni

The classic approach to deal with uncertainty in mechanical load-bearing structures is to oversize the structure by taking into account a safety factor, see Sect. 1.2. Uncertainty in load-bearing structures can occur for a wide range of reasons. Typical causes are material imperfections, geometric inaccuracies or inaccurate assembly. Sources of uncertainty in mechanical engineering prevail preferentially at joints. Here, discontinuities in material behaviour or discontinuous force flow occur and lead to uncertain load scenarios. Therefore, the design and manufacturing of joints has to be carried out carefully. Bolted connections require special attention since the loads acting on the fastener can be multidimensional [121]. Smart load-bearing structures and machine elements contribute significantly to reducing the existing uncertainty. With the possibility of monitoring the actual load conditions and reacting accordingly, see Sect. 6.3 , oversizing can be avoided, resources be conserved and economic efficiency be increased [263]. Furthermore, the acquisition of process variables, e.g. process forces and torques, at the machine structure enables to verify the models of technical processes.

Smart structures comprise structures with built-in sensors and/or actuators linked via a controller [43]. So far, load-bearing structures and sensors or actuators are separately manufactured and assembled in a further process [48]. In this section, we present a novel approach for the production of smart tubular metal structures through joining by forming. The approach allows for the production of different designs of smart structures and machine elements. The sensory bending dumbbells, drive shaft and fastening element in Fig. 5.26 are some examples of possible smart structures created by this approach.

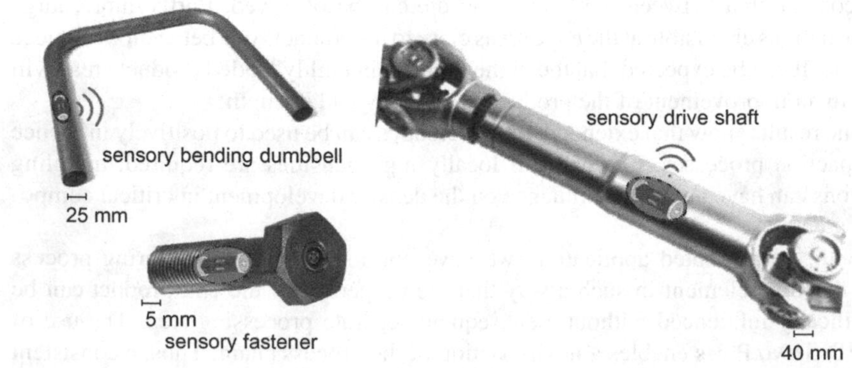

Fig. 5.26 Examples of sensory structures produced by integrating transducers into metallic load-bearing structures

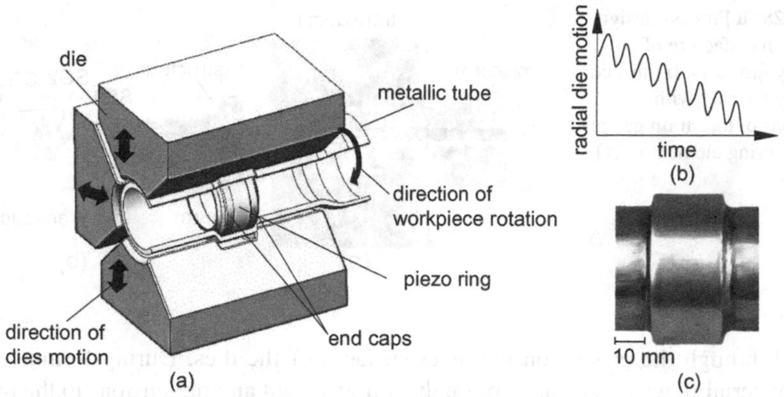

Fig. 5.27 **a** Swaging unit [49], **b** motion sequence of the dies and **c** aluminium tube with integrated piezo ring

These structures can be used in a growing range of applications due to ongoing digitalisation. Exemplary applications such as structure condition monitoring, load path analysis or lifetime estimation can be realised. Furthermore, smart structures can be used for process monitoring, allowing control approaches which boost efficiency and improve product quality. This approach offers many advantages. On the one hand, the manufacturing process is highly economical, since the functional material is integrated into the load-bearing structure during its shaping. Thus no further process steps are required. On the other hand, this integral joining ensures that the sensitive functional materials are protected from external influences, thus enhancing the lifespan of these materials.

The joining process

The novel manufacturing process for the creation of smart structures is based on joining by forming. The load-bearing structure is plastically deformed, resulting in a form and force-fit joint between the structure and the integrated functional materials [135]. For this joining operation, rotary swaging appears to be particularly suitable, as it is an established process for joining metal wires and bushings [263]. In this process, the cross-section of the workpiece is incrementally reduced by the forming dies, allowing the functional element to be integrated into the metallic structure. In order to prevent any damage of the functional element during integration, metallic end caps are used [134]. The integration of a piezo ring in a metallic tube by the recess swaging process is shown in Fig. 5.27.

Figure 5.27 shows three of the four dies of the rotary swaging unit, which oscillate simultaneously at 30 Hz in the radial direction. While the workpiece rotates, the dies move closer to each other in the radial direction, see Fig. 5.27b, leading to an incremental reduction in the tube cross-section [123]. The piezo ring is guided and

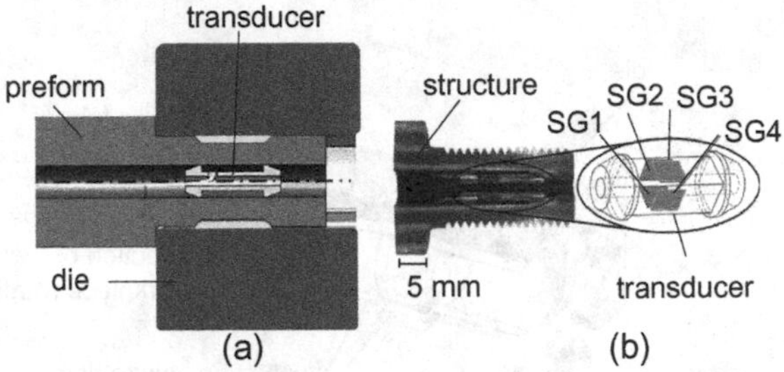

Fig. 5.28 **a** Process design for the manufacture of sensory fasteners, **b** finished sensory fastener with positions of the strain gauges on the spring element [121]

placed through the groove on the inner surfaces of the dies. During forming, the tube material flows to the end caps on the left and right and pretensions to the piezo ring [134]. This pretension defines the working range of sensory structures, as it will be reduced under tensile loads on the structure. It also ensures the functionality of piezoelectric actuators since they require a certain preload for the operation.

An adaptation of the process for the production of smart structures with homogeneous outer diameters is presented in [121]. Figure 5.28 shows a spring element with four strain gauges SG1–4 applied to measure temperature, force and bending moments integrated into a screw-shaped structure. Due to the preform of the structure and the geometrical design of the spring element, the radial motion of the dies leads to the integration of the transducer in pretension, see Fig. 5.27. In a downstream process, the fastening head and the threads can be machined [121].

Unlike the process presented in Fig. 5.27, the final structure here has a uniform radius over the whole length after the integration process. For more flexibility in adjusting the pretension on the integrated functional materials, the integration process was extended to be based on in-feed rotary swaging. Through this extension, a better control of the pretension at the integrated functional material is achieved and an elongation of the joining zone is avoided [135].

At the first step of this integration process, the inner and outer preform is created in the structure using mandrels as well as the die geometry. In the second step, the functional material is inserted into the tube and pressed by the mandrel. As soon as the dies leave the integration zone, the mandrel is removed and the rest of the tube is further formed, as shown in Fig. 5.29.

Summary

It can be summarised that the introduced approach for the creation of smart structures and smart machine elements has many advantages, e.g. economic efficiency in production and the protection of integrated functional materials against environmental influences. The main characteristic of the presented integration process is the pretension of integrated functional materials. This pretension ensures the functionality of both integrated sensors and actuators. While integrated piezo actuators, e.g. need this pretension for the operation, the force measurement range of the tensile load of the sensor structures depends directly on the value of the pretension. As has been shown in the integration process based on in-feed rotary swaging, smart structures can be produced in various designs and with a flexibly adjustable pretension. They can also be produced as semi-finished products for further processing or individualisation stages [180]. In this way, sensory structures can be used for strategies to reduce uncertainty in many applications in mechanical engineering, e.g. structural condition monitoring.

5.3.5 Process Controlling During the Production of Smart Structures

Peter Groche, Nassr Al-Baradoni, and Martin Krech

The approach presented in Sect. 5.3.4 for the production of smart structures by means of forming technology allows the simultaneous joining of functional materials, as well as the forming of load-bearing structures and machine elements by rotary swaging [121, 134]. The functional elements are integrated into metallic hollow tubes by preload, which is created by a form- and force-fit joint. As a characteristic of the applied process, the functionality of the integrated functional elements, e.g. a force transducer, depends directly on the level of preload. A too high preload can damage the functional element, too low ones can prevent reliable operation. Considering the process design in Fig. 5.29, deviations in the integration process, as generally discussed in Sect. 3.2 for process chains, can be expected. Stochastic data uncertainty in the properties of semi-finished metallic tubes as well as in the process model (see Sect. 1.6) lead to a deviation in the adjusted preload, as this preload largely depends on the material flow at the end caps of the integrated functional element [180]. In order to master the described uncertainty in the production of smart structures, an approach for process monitoring and process control is necessary and is described

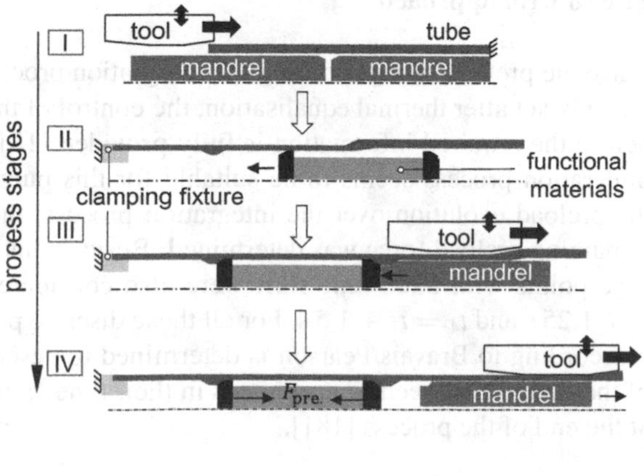

Fig. 5.29 Extended integration process for the production of smart structures based on in-feed rotary swaging [127]

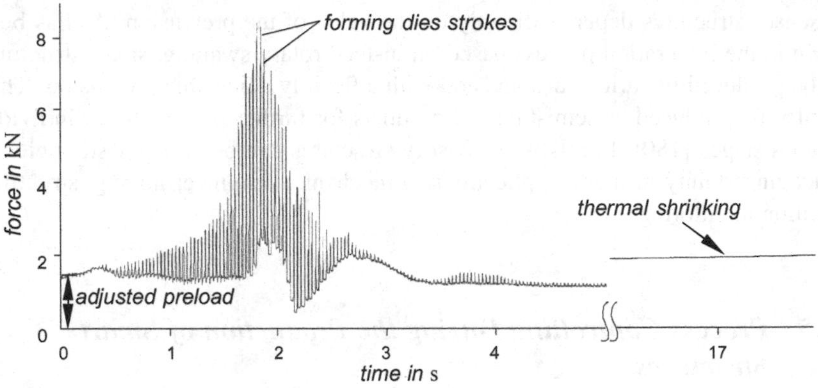

Fig. 5.30 Measured force during the integration process, adjusted preload, forming strokes and resulting preload after thermal shrinking

in the following. For this purpose, the force transducer to be integrated is calibrated and then integrated into the tube. In this way, the inline measurement of the actual preload can be achieved.

Figure 5.30 shows the force signal curve measured by the integrated axial force transducer during the integration process

As can be seen in Fig. 5.30, a certain preload of about 1.5 kN is adjusted at the beginning of the integration process. Subsequently, the dies start the forming process. With each stroke of the oscillating dies of the rotary swaging machine, the tensile force in the tube and the integrated transducer increases. Due to uncertainty in the integration process caused by thermal shrinking, fluctuation in the coefficient of friction, misalignment of components as well as geometrical deviations of the tube, a control approach to adjust a specific preload based on the measured signals of the integrated force transducer must be implemented [180].

The control approach

Since the preload changes during the integration process and the remaining preload are only set after thermal equalisation, the control of the preload must be performed before the required information is fully provided [181]. A prediction model of the integration process seems to be suitable for this purpose. For the investigation of the preload evolution over the integration process, the time t_1 with the maximum remaining preload force was determined. Besides this reference point, three further time points with fixed time shift were also considered: $t_2 = t_1 + 0.75$ s, and $t_3 = t_1 + 1.25$ s and $t_4 = t_1 + 1.5$ s. For all these discrete points, a correlation coefficient r according to Bravais/Pearson is determined to describe the degree of correlation of the preload between 12 specimens in their time values and the remaining preload at the end of the process [181].

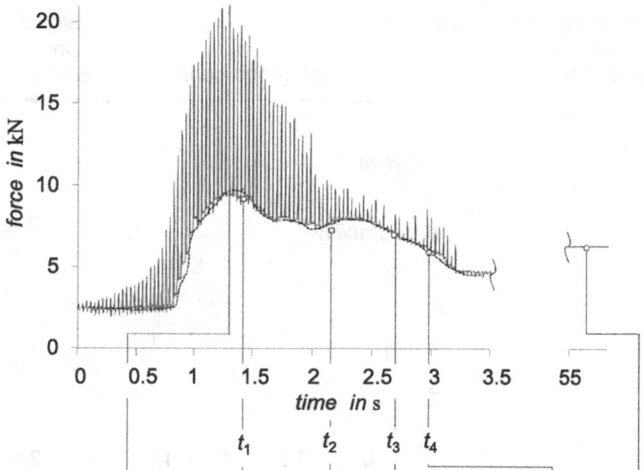

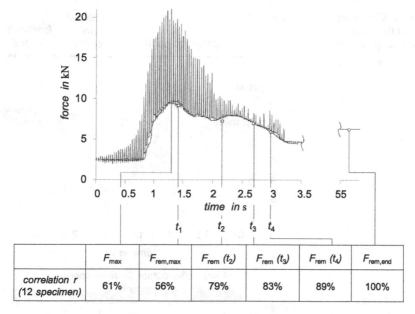

	F_{max}	$F_{rem,max}$	F_{rem} (t_2)	F_{rem} (t_3)	F_{rem} (t_4)	$F_{rem,end}$
correlation r (12 specimen)	61%	56%	79%	83%	89%	100%

Fig. 5.31 Measured force during the integration process and correlation analyses of preload evolution of 12 specimens [181]

It can be seen in Fig. 5.31 that the correlation increases towards the end of the process. The later the preload force is evaluated, the better the quality of the predicted remaining preload $F_{\text{rem,end}}$. From time t_2 on, the correlation factor r is at least 79%. For the three discrete points during the forming process F_{rem} (t_2), $F_{\text{rem}}(t_3)$, F_{rem} (t_4), a continuously improved correlation with the remaining preload $F_{\text{rem,end}}$ is observed [181].

Two process parameters exist on the experimental machine (rotary swaging), the process parameters of which can be manipulated during the forming process, and they also have a major influence on the preload force: the mandrel force F_{mand} and the in-feed speed v. On the one hand, it has been shown that the short contact effect of the mandrel, with the progress of the process, reduces the ability to affect the resulting preload. On the other hand, the in-feed speed has a better effect on the resulting preload, since it directly affects the forming forces and therefore the material flow at the end caps of the integrated force transducer [181]. For this reason, the in-feed speed is used as a manipulation parameter for the controller approach.

Based on the trials conducted in [181], the following conclusion can be drawn: in order to obtain the widest possible range of achievable preload forces, the in-feed speed must already be adjusted at $t = 2$ s. At this time, the process can be influenced most effectively (Fig. 5.32). However, the highest possible accuracy of the predicted preload is only given in the time range between $t = 2.5$ s and $t = 3$ s (see the correlation value in Fig. 5.31). In order to meet both, the requirement for high accuracy and the requirement for a wide control range, the selected control approach

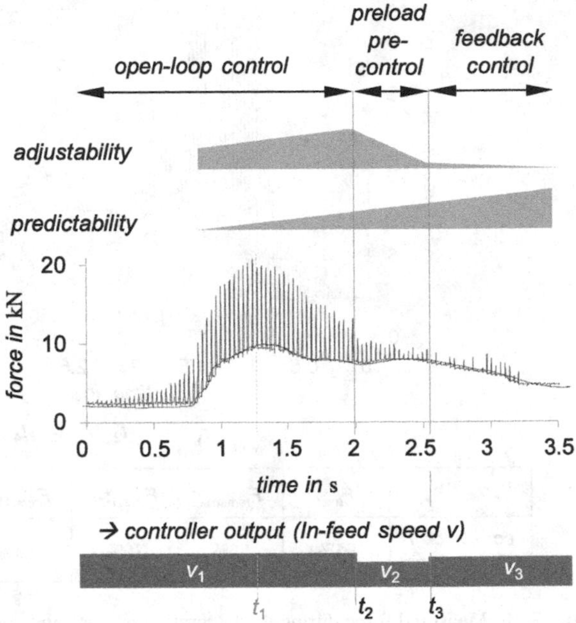

Fig. 5.32 Control approach to adjust the preload using the in-feed speed [181]

consists of three stages: An open-loop control, an early preload pre-control between $t = 2$ s and $t = 2.5$ s with the aid of the prediction model and a subsequent adaptive control for fine adjustment starting from $t = 2.5$ s. The in-feed speed is manipulated with the aid of a feed-forward parameter from $t = 2$ s according to the target preload value. The adjustability of the preload at this point is the highest (see Fig. 5.31). At $t = 2.5$ s the signal can be fedback to the controller. The predictability at this point is high enough to fine-tune the process while still providing good adjustability. The control approach therefore consists of the three phases: Open-loop control, preload pre-control and feedback control (see Fig. 5.32).

After designing the controller and the prediction model based on the correlation analysis in Fig. 5.31, sensory structures with controlled preload can be produced. Once the integration process has begun, the prediction model starts analysing the force peaks and valleys of each die stroke, with referencing the measured data to the reference point t_1. The controller works in open-loop mode until time t_2. The in-feed speed in this mode has a constant value v_1. From time t_2, the in-feed speed is adjusted to v_2 according to the preload control parameter. In the last step, at point t_3, the feedback control is activated and the in-feed speed is set to v_3. The results of this control strategy were evaluated at two different force levels to investigate the flexibility of this approach. A significant increase can be observed in the achievement of the target preload force (Fig. 5.33). The repeatability is also increased in comparison to the uncontrolled processes. The accuracy and repeatability of the controlled process are increased at higher preload forces. These results prove that the inline

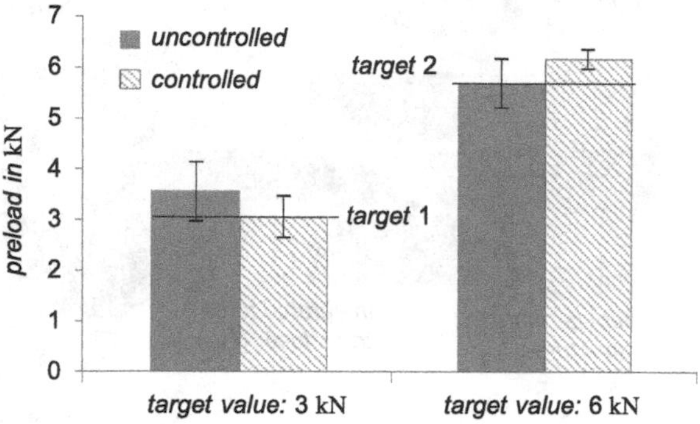

Fig. 5.33 Result of the controller approach compared to the uncontrolled process [181]

measured signals can be used to control the preload of the integrated transducer of smart structures, resulting in a higher accuracy and a smaller deviation.

Summary

Utilising the force signal of the transducer to be integrated for process monitoring and process control in the production of smart structures can significantly reduce the process-related fluctuations in the produced parts. Through correlation analysis, the resulting preload, which is directly measurable only after the process, could already be predicted in the second part of the forming process. In addition, it could be shown that it is possible to change the preload without affecting the resulting form of the smart load-bearing structure. Using the developed control approach, the fluctuation of the achieved preload force could be reduced by up to 60%.

5.3.6 Process-Integrated Calibration of Smart Structures

Nassr Al-Baradoni and Peter Groche

Smart structure and machine elements contribute to reduce uncertainties in mechanical engineering. With their abilities to monitor and react to the current load state, approaches for active process manipulation can be implemented. In Sect. 5.3.4, we discussed the efficiency in the production of smart structures by means of forming technologies. Process design for the integration of functional material into metallic tubular structure by rotary swaging is presented in [121, 127, 133]. An approach to deal with stochastic data uncertainty in the properties of semi-finished metallic tubes by controlling the joining process is presented in Sect. 5.3.5. Despite the efficiency achieved in the manufacturing process, the required time consuming downstream calibration process is still a limiting factor for a broad industrial implementation of

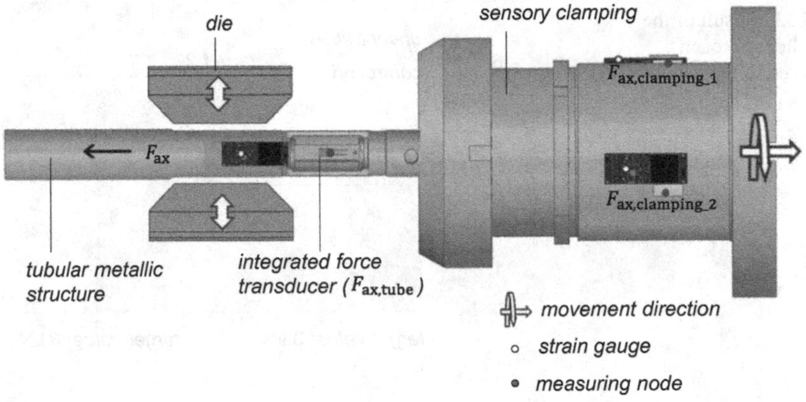

Fig. 5.34 Measurement setup for the process-integrated calibration of smart structure produced by rotary swaging [6]

sensory structures. Defined test sequences are carried out on a testing machine, for example, according to the standard DIN EN ISO 376 for force sensors in [70].

In this section, we present an approach to increase the efficiency in the production of sensory structures. By means of a suitable setup and process-adapted signal processing, it is possible to replace standard time-consuming downstream calibration processes with process-integrated calibration [6]. Once the sensory functional element is firmly integrated into the structure and the forming dies leave the joining zone (see Sect. 5.3.4), the process forces required for the subsequent forming of the structure can be used to implement a dynamic calibration of the sensory structure. For this purpose, a special sensory clamping was designed as a machine-side reference point. By measuring the forming forces in both modes, namely in the integrated sensor element and at the machine-side reference point, a calibration is realised as shown in [6]. For the sensory clamping, strain gauge sensors are applied to measure the axial force on the clamping surface. Due to the rotation of the clamping, signal amplifiers with wireless signal transmission are used (measuring nodes). The measurement setup in Fig. 5.34 contains the integrated strain gauge based axial force sensor within the tube and the two other machine-side axial force sensors on the clamping rotated 90° to each other for higher accuracy.

Signal processing

By looking at the force signals $F_{ax,tube}$ and $F_{ax,clamping}$ in Fig. 5.35, the manufacturing process can be divided into three phases. In the first phase, a certain compressive force is applied to the integrated force transducer by the mandrel to adjust the required pretension. The mandrel force remains until the forming dies leave the joining zone and the transducer is firmly joined into the structure. Once the transducer is fully integrated into the structure, the mandrel is removed and the rest of the tube is

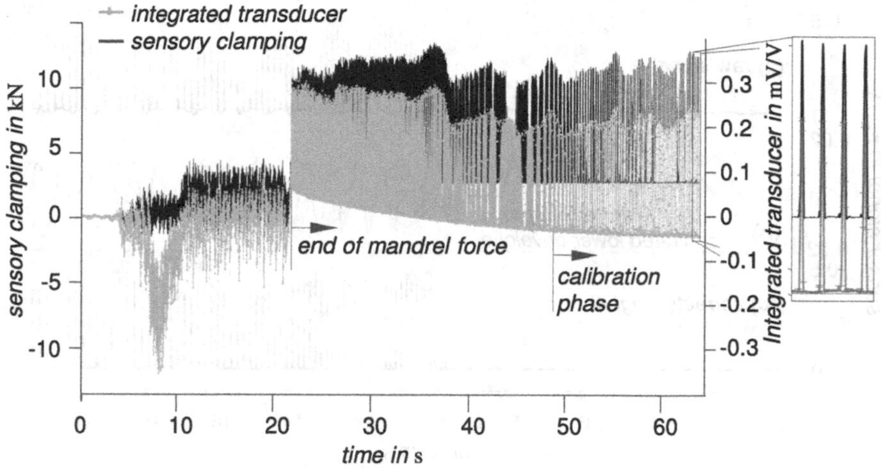

Fig. 5.35 Progression of the axial forces both in the integrated force transducer and in the sensory clamping during the manufacturing process (normed to the max values) and the beginning of the calibration phase [6]

pulled out of the swaging unit and formed by in-feed rotary swaging, see Sect. 5.3.4. As a result, the load direction changes and a high tensile force peak is induced with every stroke of the oscillating dies. At the beginning of the forming process ($18\,\mathrm{s} \leqslant t \leqslant 40\,\mathrm{s}$), a high forming frequency (approx. 30 Hz) leads to a high density of these force peaks. After the cooling-down of the joining zone in the tube, the adjusted pretension almost reaches a constant value (see also Fig. 5.34) and the dynamics of the forming dies are reduced to avoid resonance effects of the clamping [6]. The calibration phase can begin accordingly.

Even after the identified time to start the calibration, both force signals $F_{\mathrm{ax,tube}}$ and $F_{\mathrm{ax,clamping}}$ are affected by two significant interferences: On the one hand, the high temperature at the integrated transducer caused by the plastic deformation of the tube leads to a drift in the force signal $F_{\mathrm{ax,tube}}$. On the other hand, the eccentricity of the clamping during rotation causes a sinusoidal zero-point drift in both signals. Once forming starts, the sinusoidal zero-point drift becomes greater as a result of bending the tube by the oscillating dies, and it becomes less uniform as a result of the high forming moments causing the workpiece to slip, see also [6, 180]. Since neither the amplitude nor the angular frequency of the sinusoidal zero-point drift can be modulated, as the deformation behaviour is not precisely predictable, it is necessary to elaborate suitable approaches for the signal processing prior to calibration [6]. Based on the main characteristics of the incremental forming processes, i.e. that a release state occurs after each load state, an approach to correct the signals is introduced in [6]. The actual position of the zero-point in each signal is determined and then forced to the nominal position [6, 180]. Firstly, the lower envelope in each signal is determined. Secondly, the lower envelope is filtered by using a Gaussian

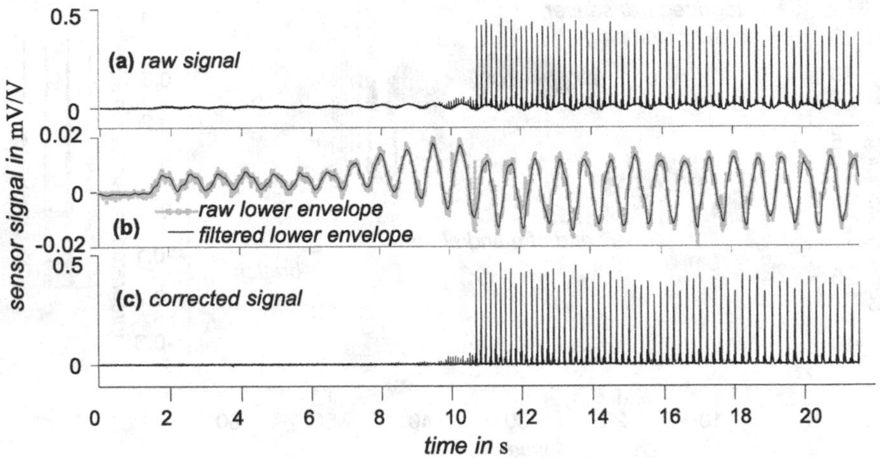

Fig. 5.36 Signal processing and correction: **a** raw signal, **b** the lower envelope (grey) and the filtered lower envelope (black), **c** force signal after correction the zero-point drift [6]

smoothing filter to get a clear zero-point drift curve. Finally, the raw signal is corrected with the determined zero-point drift curve. Figure 5.36 illustrates this procedure.

Figure 5.36a shows that the initial signal has a constant zero offset, as long as the clamping fixture does not rotate ($t < 2$ s). Once the rotation starts ($t > 2$ s), small sinusoidal zero shifts with relative constant amplitude and angular frequency occur ($t < 10$ s). This drift is caused by the gravitational force of the mass of the structure. When the deformation begins (at $t = 10$ s), the amplitude of the sinusoidal drift becomes higher with non-uniform angular frequency, which can be observed clearly in Fig. 5.36b. Once the signals are corrected, an accurate zero-point position can be observed, see Fig. 5.36c and the calibration can be carried out by correlating both signals from the integrated force sensor and the sensory clamping.

Results

To determine the achievable accuracy of the in-process calibration, the generated sensory structures were re-calibrated according to the standard calibration (DIN EN ISO 376) on a tensile testing machine (Zwick Roell 100). A comparison between the two types of the calibration shows a good correlation. In Fig. 5.37, both calibration methods are shown for an exemplary sensory structure. While the nominal force rises gradually in a standard calibration, it usually reaches the nominal value in less than one second with one stroke of the forming dies. Both calibrations, however, show good linearity, allowing the calibration coefficient to be calculated as the slope of the linear fit of the measured points. By evaluating the deviation of the process-integrated calibrations from the reference in several samples, a maximum nominal value-related deviation of 2% was observed.

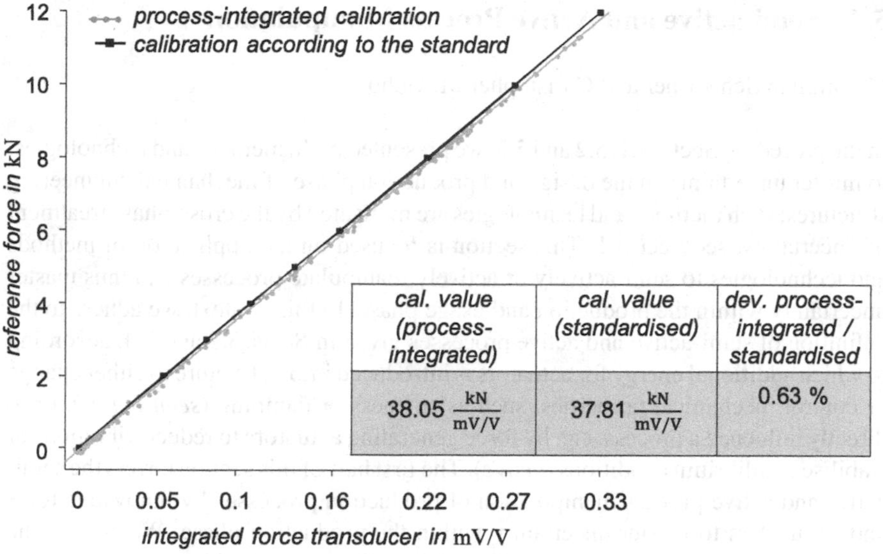

Fig. 5.37 Comparison of process-integrated and standardised calibration of a sensory structure with integrated axial force sensor and the calibration value-related deviation [6]

Summary

The novel approach, introduced in Sect. 5.3.4, for the creation of smart structures through integration of functional materials in load-bearing structures using incremental forming processes provides numerous economic and functional advantages. By means of the possibility of process monitoring (see Sect. 5.3.5) during the integration of the functional materials and, therefore, the realisation of control approaches, this manufacturing process gains the flexibility to adjust the pretension of integrated functional materials as a further advantage.

A further benefit of incremental forming for the production of smart structures could thus be demonstrated. Since the incremental process implies that every loading state is followed by a release, the acquired data can be easily processed and process-related interference effects can be corrected.

In a nutshell, smart structures with integrated functional elements help to master uncertainty in mechanical structures in their usage phase. By inline process monitoring and process-integrated calibration of smart structures, uncertainty in their manufacturing is reduced. Faulty integration processes can be detected, e.g. by evaluating the standard deviation of the process-integrated calibration [6].

5.4 Semi-active and Active Process Manipulation

Maximilian Schaeffner and Christopher M. Gehb

In the preceding Sects. 5.1, 5.2 and 5.3, we presented both, methods and technologies, to master uncertainty in the design and production phase of mechanical engineering structures; such methods and technologies are motivated by the cross-phase treatment of uncertainty, see Sect. 3.1. This section is focused on the application of methods and technologies to semi-actively or actively manipulate processes and thus master uncertainty within the production and usage phase. In this context, we adhere to the definition of semi-active and active processes given in Sects. 3.2 and 3.4, according to which additional energy for actuators is introduced into a structure to either change or control mechanical properties, such as stiffness or damping (*semi-active*), or to directly influence a process, e.g. by force generating actuators to reduce vibrations or stabilise equilibrium conditions (*active*). The first half of this section covers the semi-active and active process manipulation of production processes by innovative tools and controllers to master uncertainty within the production phase. We present the active control of press stiffness, see Sect. 5.4.1, state control with semi-active roller and plain bearings, see Sect. 5.4.2, and a sensor-integrated compensation chuck control for tapping, see Sect. 5.4.3. In the second half of this section, uncertainty within the usage phase of mechanical load-bearing structures is mastered by semi-active and active technologies. We present technologies for vibration attenuation, such as a shock absorber with integrated hydraulic vibration absorber, see Sect. 5.4.4, an Active Air Spring, see Sect. 5.4.5, and piezo-elastic supports in beam truss structures, see Sect. 5.4.6. Furthermore, we introduce approaches for active buckling control, see Sect. 5.4.7, and semi-active load redistribution, see Sect. 5.4.8.

5.4.1 Control of Press Stiffness

Florian Hoppe and Peter Groche

The functionally quality, cf. Chap. 1, of metal-formed products highly depends on the ability of the production machine to guide the tool accurately along a path. Since presses are subject to high loads, the exact positioning of the tool is challenging. Due to the elastic response of press components, variable loads lead to deviations in the tool path and thus in the product properties. Therefore, the press stiffness is one of the most important design parameters [79].

The press stiffness is defined as the ratio between a static load applied on the press ram and its deflection [68]. A common approach to modify the press stiffness is to adjust its design [80]. Numerous frame and gear designs that enhance stiffness have been established [79]. However, these also limit the accessibility of the working area [129]. Their focus is mainly on the stiffness in stroke direction, the torsional stiffness [68]. However, presses provide a degree of freedom exactly in stroke direction by means of their main drives. Especially multi-point servo presses as the 3D Servo Press presented in Sect. 3.6.3 come with the advantage of providing multiple main drives. We present an alternative method of adjusting the stiffness by means of active process compensation. This method requires additional sensors and novel observer models to detect the ram deflection and makes use of the machine's servo drives.

Furthermore, methods to identify the stiffness have to be extended. Although the complete press stiffness matrix in all six spatial directions of the ram has already been investigated in science for hydraulic [12] as well as for mechanical [61] presses, so far the stiffness has only been considered in one operating point. This is based on DIN 55189-1 [68], which prescribes that the stiffness must be determined in the bottom dead centre. Numerical press models, e.g. finite element models, are often used in press design to evaluate the press behaviour, but can only be used offline due to their high calculation effort. However, elastic models are required for the exact determination of the ram position and its control. The extent to which reduced-order models are applicable for this task still needs to be investigated. This is addressed in the following.

We evaluate the method at hand of the 10 kN Version of the 3D Servo Press, a servo press that is able to move the three eccentric drives independently. The transmission ratio of all three eccentric drives can additionally be adjusted via the common spindle drives [239]. This also affects the force transmission, which is why the gear position must be taken into account when modelling the press stiffness [126]. Due to the variable force transmission, the load on the elastic machine elements changes. An additional shift in the joint positions due to the deflection and a progressive bearing stiffness result in a nonlinear stiffness model as described in Sect. 4.3.2.

The compliance of the ram as represented by Chodnikiewicz [61] and Arentoft [11] can be described as a 6×6 press compliance matrix Λ. The deflection of the ram position at its tool centre point (TCP) under a given load vector l can be described by

$$\underbrace{\begin{bmatrix} \Delta x_{\text{tcp}} \\ \Delta y_{\text{tcp}} \\ \Delta z_{\text{tcp}} \\ \Delta\theta_x \\ \Delta\theta_y \\ \Delta\theta_z \end{bmatrix}}_{\Delta p} = \underbrace{\begin{bmatrix} \lambda_{11} & \lambda_{12} & \lambda_{13} & \lambda_{14} & \lambda_{15} & \lambda_{16} \\ \lambda_{21} & \lambda_{22} & \lambda_{23} & \lambda_{24} & \lambda_{25} & \lambda_{26} \\ \lambda_{31} & \lambda_{32} & \lambda_{33} & \lambda_{34} & \lambda_{35} & \lambda_{36} \\ \lambda_{41} & \lambda_{42} & \lambda_{43} & \lambda_{44} & \lambda_{45} & \lambda_{46} \\ \lambda_{51} & \lambda_{52} & \lambda_{53} & \lambda_{54} & \lambda_{55} & \lambda_{56} \\ \lambda_{61} & \lambda_{62} & \lambda_{63} & \lambda_{64} & \lambda_{65} & \lambda_{66} \end{bmatrix}}_{\Lambda(q)} \cdot l \tag{5.16}$$

with the translatory and angular displacements of the TCP Δx_{tcp}, Δy_{tcp}, Δz_{tcp} and $\Delta\theta_x$, $\Delta\theta_y$, $\Delta\theta_z$, respectively. Since process forces occur mainly in the vertical z-direction, i.e. the stroke direction, the λ_{i3} compliance must be reduced. Off-centre loads result in additional tilting moments that affect the ram tilting by means of the compliance parameters λ_{i4} and λ_{i5}.

The press compliance can be modelled by considering relevant machine components as trusses and beams [126]. Calculating $\Lambda(q)$ at the current drive positions q allows to estimate the machine behaviour for control algorithms. The position vector

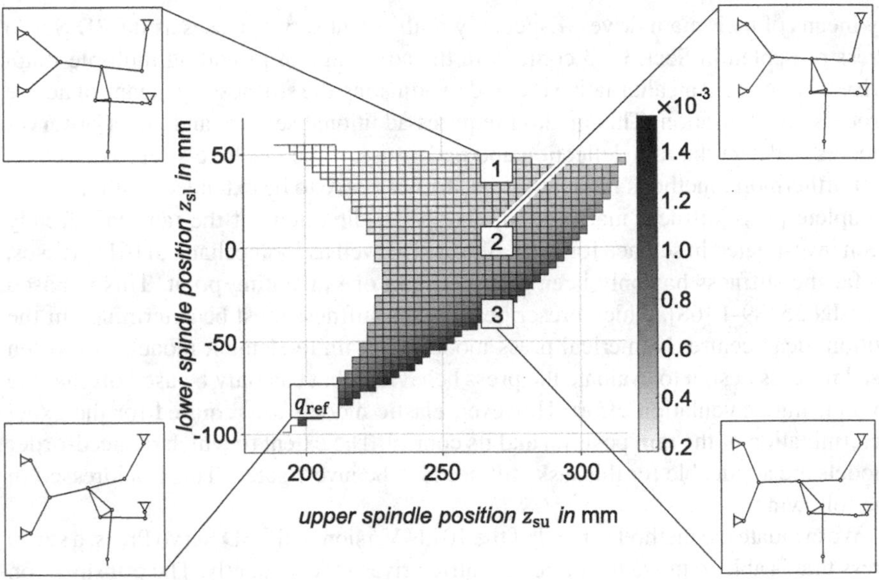

Fig. 5.38 Simulated compliance map of the 3D Servo Press prototype depending on the spindle configuration z_{su}, z_{sl} at constant $\varphi_{ecc,i} = 270°$ [126]

q consists of the three eccentric drive angles $\varphi_{ecc,i}$, $i \in \{1, 2, 3\}$ and the upper and lower vertical spindle positions z_{su}, z_{sl}. Since the compliance model is highly nonlinear, we evaluated the model for different q. The compliance of λ_{33} as a function of the upper and lower spindle positions z_{su}, z_{sl} is shown in Fig. 5.38 for constant eccentric angles $\varphi_{ecc,i} = 270°$. The operating area of the spindles is limited by the installation space and the singularities of the spindle kinematics. Figure 5.38 shows that the compliance strongly depends on the position of the spindles. The respective position of the spindles is visualised exemplarily by the corresponding kinematics. To investigate the dependence of press compliance λ_{33} on the eccentric position, three spindle positions 1, 2, 3 with significantly different compliance were chosen.

Figure 5.39 shows a comparison between simulated and experimentally determined compliance λ_{33} as a function of the eccentric angles $\varphi_{ecc,i}$ for the spindle configurations 1, 2, 3. Apparently, the compliance also strongly depends on the eccentric position and therefore cannot be assumed to be constant. The deviation between simulation and experiment indicates uncertainty. Stiffness parameters are based on the nominal values of geometric parameters, such as cross-sections and lengths, as well as material parameters. These parameters are subject to manufacturing tolerances which are propagated into the model as data uncertainty. Furthermore, the model order reduction simplifies the model and hence increases model uncertainty. How this uncertainty can be evaluated and reduced is described in Sect. 4.3.2. Nevertheless, the reduced-order model is already able to represent the variability of the compliance and is valuable for the online control.

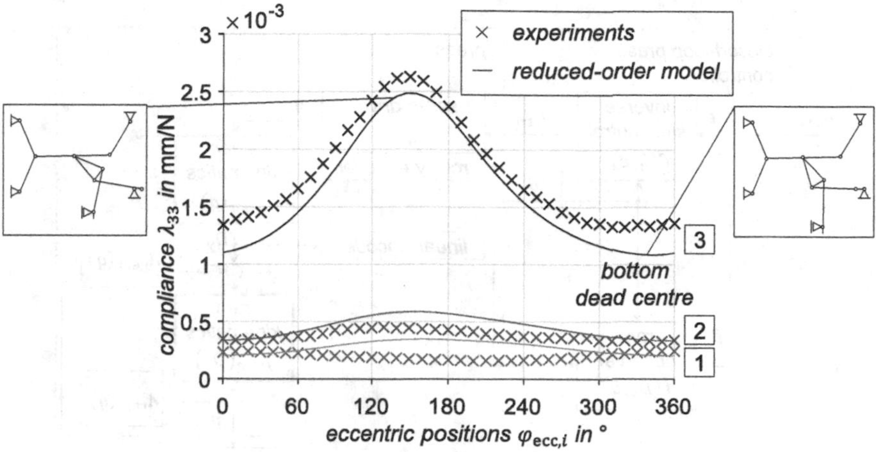

Fig. 5.39 Comparison of simulated (lines) and experimentally (cross markers) determined compliance of λ_{33}, depending on the eccentric positions $\varphi_{\mathrm{ecc},i}$ [126]

In [126] we designed an experimental setup to identify compliance matrices of both, the passive and the active system, Λ_p, Λ_a respectively. The test setup allows a variable load vector l to be applied at the ram TCP and the ram displacement Δp to be determined at the operating point q_ref (Fig. 5.38). In $n = 18$ measurements, we applied 30% of the nominal force with a pneumatic cylinder and measured the displacements and rotations of the ram TCP with dial gauges.

In comparison to the model-based estimated compliance matrix $\hat{\Lambda}_\mathrm{p}$, the measured passive compliance

$$\Lambda_\mathrm{p} = \begin{bmatrix} 0.9 & 0 & 0 & 0 & 0 & -0.7 \\ -0.1 & 1.6 & 0 & -0.2 & 0 & -0.4 \\ 0 & 0 & \mathbf{1.6} & 0 & 0.2 & 0.1 \\ 0 & 0 & 0 & \mathbf{1.0} & 0 & 0 \\ 0 & 0 & 0 & 0 & \mathbf{1.0} & 0 \\ 0.1 & 0 & 0 & 0 & 0 & 0.2 \end{bmatrix} \frac{\mathrm{mm}}{\mathrm{N}}, \quad \hat{\Lambda}_\mathrm{p} = \begin{bmatrix} 0.6 & 0 & 0 & 0 & 0 & -0.8 \\ 0 & 1.5 & 0 & 0 & 0 & 0 \\ 0 & 0 & \mathbf{1.5} & 0 & 0 & 0 \\ 0 & 0 & 0 & \mathbf{1.0} & 0 & 0 \\ 0 & 0 & 0 & 0 & \mathbf{1.0} & 0 \\ 0 & 0 & 0 & 0 & 0 & 0.3 \end{bmatrix} \frac{\mathrm{mm}}{\mathrm{N}} \tag{5.17}$$

results. The bold highlighted entries correspond to the degrees of freedom of the control z_tcp, θ_x, θ_y.

The control of the machine requires, on the one hand, a kinematic model $f(q)$ for the control law [161] and, on the other hand, a model of the ram $f_\mathrm{ram}(D_y)$ in order to estimate the ram position by means of the integrated displacement sensors D_y [126]. By extending the rigid-body model of the press to the elastic model, the estimation accuracy of the ram position p can be enhanced. As a position feedback in the press is used, it is helpful to split the press compliance models at the measuring position into the gear and the ram

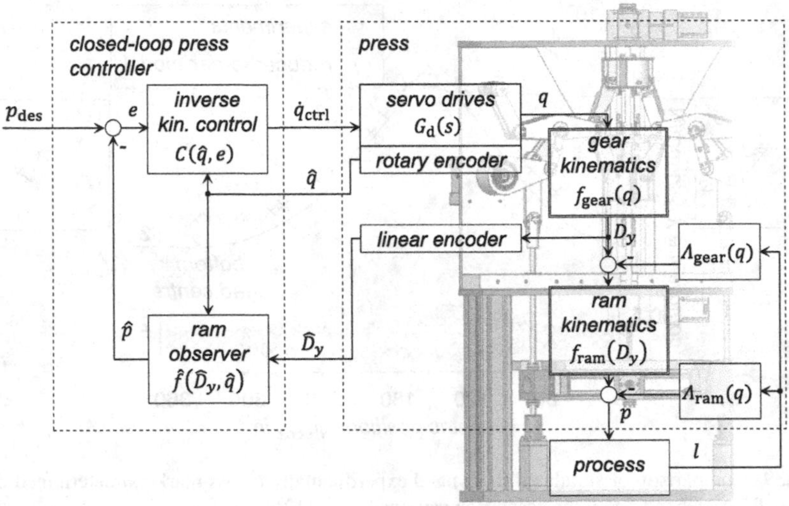

Fig. 5.40 Closed-loop motion control of the 3D Servo Press

$$\Lambda_p = \Lambda_{gear} + \Lambda_{ram} \tag{5.18}$$

$$f(q) = f_{ram}\big(f_{gear}(q)\big). \tag{5.19}$$

Thus, the control structure shown in Fig. 5.40 results. Using the extended ram observer

$$\hat{p} = \underbrace{\hat{f}(\hat{q})}_{\text{rigid-body kinematics}} \underbrace{- \hat{\Lambda}_p(\hat{p})\hat{\Lambda}_{gear}^{-1}(\hat{p}) \left[\hat{f}_{gear}(\hat{q}) - \hat{D}_y\right]}_{\text{elastic model}} \tag{5.20}$$

the measured deflection in the press gear $\hat{D}_y$ is used to estimate the actual ram position $\hat{p}$. All model assumptions and measured variables are indicated by $\hat{(\cdot)}$, as they are subject to uncertainty. Section 6.1.7 investigates how data and measurement uncertainty affects this closed-loop control and how to design a robust control.

The active press system compensates deviations detected in D_y, which are caused, among other things, by the gear compliance Λ_{gear}. In addition, the observer also takes into account the ram compliance Λ_{ram}, which cannot be measured in the process.

For the active system, we measured the press compliance matrix as

$$\Lambda_a = \begin{bmatrix} 0.7 & 0 & 0 & 0 & 0 & -0.7 \\ -0.1 & 1.7 & 0 & 0 & 0 & -0.5 \\ 0 & 0 & \mathbf{-0.024} & 0 & 0.1 & 0 \\ 0 & 0 & 0 & \mathbf{0.069} & 0 & 0 \\ 0 & 0 & 0 & 0 & \mathbf{0.071} & 0 \\ -0.1 & 0 & 0 & 0 & 0 & 0.2 \end{bmatrix} \frac{\text{mm}}{\text{N}}. \tag{5.21}$$

Since only the degrees of freedom z_{tcp}, θ_x, θ_y are controllable, only the corresponding lines $\lambda_{3,i}$, $\lambda_{4,i}$, $\lambda_{5,i}$ can be modified. While especially λ_{33}, λ_{44}, λ_{55} have been reduced significantly compared to Eq. (5.17), other entries remain almost unchanged. However, it is noteworthy that λ_{33} has a negative value, which means a displacement against the load direction. A negative stiffness is physically uncommon at first, but it is a characteristic of active systems, such as active suspensions (Sect. 3.6.2). In this case, the compliance model overestimates the actual compliance at this operating point and thus overcompensates it.

In order to increase the mechanical stiffness of presses, so far the press design has been adjusted. In contrast, an active process manipulation method has been presented to increase the stiffness by means of closed-loop control. This method requires control capabilities which multi-point presses already provide by using their main drives. Therefore, no additional force generating actuators are required. The precise measurement of the ram position is the main challenge. As direct tactile measurement is infeasible, a robust sensor positioning must be carried out at a distant position. This results in the need for observers that can determine the ram position from spatially distant measurements. For this purpose, we developed a reduced-order elastic model and included it in an observer. Comparing active and passive systems we have shown that the use of control loops and reduced compliance models allows an exact ram positioning and thus a significant reduction of the compliance.

5.4.2 *State Control of Combined Roller and Plain Bearings*

Daniel Martin, Julian Sinz, and Peter Groche

The development of industrial production plants is characterised by the demand for high productivity and optimal material utilisation. Especially in forming technology, which is characterised by high investment requirements for machines, the trend therefore goes towards flexible machine technologies with a wide range of applications and high adaptability to multiple production conditions, which makes it possible to master uncertainty occurring in the production process (see Sect. 1.2) and the resulting fluctuations in product quality.

Pioneers of such a machine technology are servo presses with an almost freely adjustable motion kinematics of the ram, thus rendering optimised motion and speed sequences of the tools possible. Due to the almost speed-independent nominal torque of servo motors, servo presses can be used to achieve high accelerations, as well as standstills under high loads [245]. This allows both, the application limits of presses to be significantly extended [214], and the cycle time to be reduced [130]. The 3D Servo Press presented in Sect. 3.6.3 is a prototype of this press technology, which is able to combine the flexibility of hydraulic presses and the economics of mechanical presses [202]. However, the increase in machine flexibility and the extension of the application limits also result in new, challenging load scenarios for the machine elements. Since these requirements can no longer be met by pure plain or roller bearings, motor acceleration in servo presses is usually limited and, therefore the potentials of servo technology in presses cannot be fully exploited [245].

An existing problem in industrial practice is the lack of knowledge of the current condition of bearings and machines. Without this knowledge, the optimum time for maintenance or servicing cannot be determined exactly, which is why fixed maintenance intervals are common in practise. This can lead to two different worst case scenarios: In the first case, maintenance is carried out at a time when it is not yet necessary, thus leading to rising costs. In the other extreme case, damage to a machine element occurs shortly after maintenance, whereby it is not immediately noticed and can therefore lead to consequential damage. Both cases can be avoided by continuous condition monitoring, which enables early damage detection. Sensor-based condition monitoring also provides the basis for preventing possible bearing or machine damage by implementing semi-active or active components. Thus, the operating limits of the machine can be extended and an extension of the service life may be achieved.

Conventional, pure roller or plain bearings are usually only suitable for a part of the application spectrum of servo presses. Roller bearings offer a lower starting torque, but have poorer damping properties, and are generally not economically feasible for high operating loads [131]. Plain bearings, on the other hand, are subject to play, which increases the cost of control [131]. Furthermore, at low speeds, mixed friction occurs in plain bearings due to local deficient lubrication, which causes increased wear, especially in the oscillating stroke due to the constant changes in direction of motion. In order to meet the increased requirements for the bearings of the 3D Servo Press, we designed a combined roller and plain bearing combining the advantages of different conventional bearing types. Monitoring of the bearings' behaviour is used to determine not only the bearings' condition but also to draw conclusions on the current machine condition. Based on these findings, it is possible to actively control relevant bearing characteristics.

The approach of combining both types of bearings in order to unite their specific advantages has been principally known for a long time [167, 207, 210, 222]. In the approach we present here, a plain bearing is complemented by two roller bearings arranged on both sides, which can be designed as cylindrical roller bearings or angular contact ball bearings. These hold the bearing shaft centrally in the plain bearing shell and reduce wear in the plain bearing during start-up [131]. This allows high loads to be transmitted and at the same time prevents backlash during start-up. Therefore, combined roller bearings offer great potential to meet the increasing requirements in the field of bearing supports for servo presses [245].

The structure of the combined bearing is shown in Fig. 5.41. The bore diameter of the roller bearings is smaller than the one of the plain bearing. This results in a shaft shoulder on which the inner rings of the roller bearings abut axially. Radial play in the cylindrical roller bearing is eliminated by an interference fit between its inner ring and the shaft shoulder. If angular contact ball bearings are used, a variable preload of the roller bearings can be achieved by means of additional bearing caps. The plain bearing, which is supplied with lubricant via a bore in the bearing shell, can be preloaded by increasing the supply pressure [245].

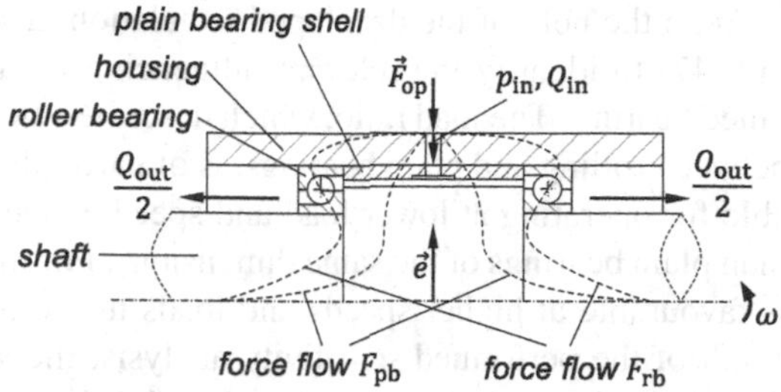

Fig. 5.41 Distribution of force flow in combined roller and plain bearing, according to Sinz [247]

The combination of the three bearings creates a statically over-determined system, which makes a robust design necessary for industrial use. After the basic feasibility of combined roller and plain bearing arrangements was demonstrated in [240], we developed a design methodology that takes into account the distribution of the operating force F_{op} between the roller bearing force F_{rb} and the plain bearing force F_{pb} as a function of the operating conditions.

Simulation model and design methodology

For the design and dimensioning of the presented bearing combination and for the estimate of operational safety, the force-path curve of the shaft is used to describe the dynamical behaviour of the bearing, resulting in the shaft displacement $e(t)$. To determine $e(t)$, we developed a simulation model according to [5, 270], which is grounded on the theory of hydrodynamic lubrication. The validity of this model, based on Reynolds [228], which is discussed in Sect. 1.3, is given for the investigated speed range of up to 400 rpm because of the small relative bearing clearance ψ of the developed bearing configuration in combination with high-viscosity oils. The shaft displacement e can be calculated by solving a differential equation based on the equilibrium of forces at the shaft according to Holland [158], which is extended by the effect of the rolling bearing forces [247] and allows to determine the load distribution on the various bearing components [130]. The force that a roller bearing exerts on the shaft is determined by means of the roller bearing stiffness c_{rb} via $F_{rb} = c_{rb}e$. The overall stiffness c_{cb} of the combined bearing serves as an indicator to assess the functional capability of the bearing combination.

The resulting simulation model enables not only the calculation of the shaft displacement path $e(t)$ but also the determination of other relevant parameters, such as dynamic viscosity μ, relative bearing clearance ψ, eccentricity ε, lubricant temperature ϑ and volume flow Q_{in}. The results are validated using an example plain bearing from [75] and experimentally determined force- and speed-dependent displacement of the bearing shaft in the plain bearing shell of the combined bearing. The validated model forms the basis of the design methodology, which is presented in the following.

With the help of the developed simulation, a sensitivity analysis was performed in [247] to identify the relevant influencing parameters in the design of the combined bearing. The load ratio, which describes the distribution of the operating force between rolling and plain bearings, is of particular interest. Roller bearings are suitable for operating at lower load and speed, since they have smaller load capacities than plain bearings of the same dimensions. For this reason, a load ratio $F_{\mathrm{pb}}/F_{\mathrm{rb}} > 1$ is favourable at higher speeds and loads in the interest of operational safety. As a result of the performed sensitivity analysis, the stiffness of the roller bearings c_{rb}, the relative bearing clearance ψ and the lubricant viscosity η can be identified as the main influencing variables. The total stiffness c_{cb} of the bearings increases linearly with increasing speed. Based on these findings, a design methodology for combined bearings can be developed. This is divided into the three steps pre-dimensioning, calculation and verification, which are carried out in an iterative process [247].

Sensor-based condition monitoring

In order to quantify the industrial benefit of the presented bearing combination, we developed a test rig in [247], which enables the condition monitoring of a sensor-equipped bearing during operation. This allows continuous testing and evaluation with regard to defined technological and economic criteria and comparison with conventional bearings. The application and load limits of the combined bearings and their emergency running characteristics are determined.

Figure 5.42 shows the developed sensor-monitored bearing. To detect the shaft displacement path $e(t)$ in the plain bearing, the test bearing is equipped with two eddy current sensors arranged at 90° to each other, which serve to detect the position of the shaft. Three temperature sensors measure the temperature on the outside of the roller bearing outer rings and the outside temperature of the plain bearing shell. In order not to influence the lubricant film in the plain bearing, the lubricant film temperature is not measured directly in the plain bearing, but only the temperature of the lubricant escaping from the bearing in the oil tank of the test rig. The operating force, which is applied by a hydraulic actuator via the housing on the opposite of the pressure inlet, is recorded by means of a piezoelectric force transducer. The oil supply pressure is recorded via a pressure gauge attached to the inlet. The accelerations that occur during operation are measured by a triaxial acceleration sensor mounted on the outside of the bearing bracket.

The radial displacement of the shaft in the plain bearing is assumed to be equal to the displacement in the roller bearings, neglecting any elastic deformation of the bearing housing or shaft in contact with the rolling elements. The stiffness of the rolling elements c_{rb} is assumed to be constant in the circumferential direction. The force F_{pb}, which is discharged via the plain bearing, results from the difference between the measured operating force F_{op} and the roller bearing forces. With knowledge of the individual bearing forces, the operating behaviour and service life of the bearings can be estimated [247].

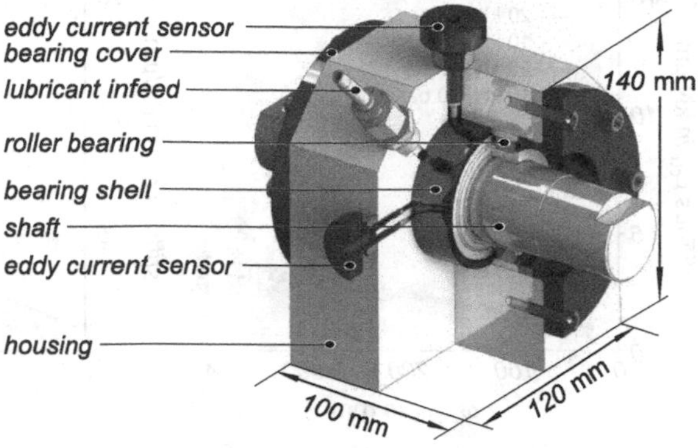

Fig. 5.42 Sensor-equipped combined roller and plain bearing, according to Sinz [247]

For this purpose, tests are carried out in full rotation and in pivoting operation mode to investigate the influence of the speed and radial displacement of the shaft on the overall stiffness of the bearing combination developed. The roller bearings are designed both as cylindrical roller bearings and angular contact ball bearings. During the test runs with various operating forces, it has been shown that in full rotation an increasing rotational speed leads to a continuously increasing load ratio. Above a certain speed, the plain bearing takes over the major part of the load and presses the shaft into the centre of the plain bearings, resulting in a higher resistance to load shocks at high speeds [247]. For the movement sequence in pivoting mode, a characteristic forward and backward stroke movement is used. The assumption of a constant roller bearing stiffness leads to a constant load ratio, so that the maxima of the forces of both bearing types are superimposed [247].

Semi-active manipulation of the bearing behaviour

On the basis of the knowledge gained on the behaviour and design of combined roller and plain bearings, the following section considers possibilities for semi-active process manipulation (cf. Secti. 3.4) during operation to react specifically to disturbing influences during operation. Two approaches are possible for this purpose: Firstly, a control of the feed pressure p_{in} of the lubricant can be used to influence the bearing behaviour; Secondly, additional actuators can be integrated and used for an adaptive adjustment of the roller bearing preload.

The lubricating oil supply pressure p_{in} can be used to preload the combined bearing via the plain bearing. The resulting preload can be interpreted as an increase in the effective operating force F_{op}, which leads to an increased radial displacement of the shaft. However, the effect on the overall stiffness of the combined bearing is small. It should be mentioned that in the calculation model described above, the outer rings of the roller bearings and the plain bearing shell are assumed to be concentric. Due to occurring inaccuracy during manufacturing and assemblage of the combined bearing, the centres of roller bearings and plain bearing do not match exactly in the

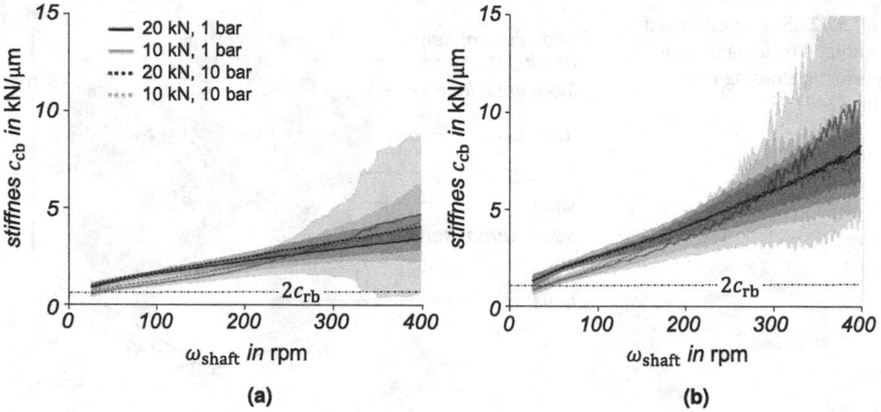

Fig. 5.43 Stiffness of combined bearings determined by carrying out experiments in full rotation mode, according to Sinz [246]: **a** roller bearings design as angular contact ball bearings, **b** roller bearings design as cylindrical roller bearings

conducted experimental investigations leading to an initial displacement of the shaft in the plain bearing.

In Fig. 5.43 the stiffness c_{cb} of the combined bearing is shown as a function of the rotational speed in full rotation operation for both angular contact ball and cylindrical roller bearings. The non-linear stiffness behaviour of the roller bearings leads to a non-linearity in the stiffness curves, especially at low speeds. If the speed is increased, the pressure build-up in the lubricant in the plain bearing gap centres the shaft in the plain bearing shell, which leads to an increase in the bearing stiffness. As the shaft displacement e in the bearing converges towards a very small value, the stiffness, determined by $c_{\text{cb}} = F_{\text{op}}/e$, reacts extremely sensitive to changes of e, resulting in an oscillation in the stiffness curves at high speeds. However, due to the better damping properties of the plain bearing, when increasing p_{in} from 1 bar to 10 bar, the preload via the lubricant pressure causes both a smoothing of the stiffness curves and a lower standard deviation, which increases operational reliability [246].

The preload of the roller bearings has a decisive influence on the behaviour of the combined bearing in different load and operating conditions. As already described, the influence of the plain bearing increases continuously with increasing speed. In the case of a radial external load perpendicular to the shaft axis, this curve flattens out when a preload is applied to one or both roller bearings, so that the load ratio with increasing preload is below that of the non-preloaded condition. Conversely, this means that, depending on the external load, the load ratio $F_{\text{pb}}/F_{\text{rb}}$ can be increased in a targeted manner by reducing the roller bearing preload, i.e. the main load can be shifted to the plain bearing.

The overall stiffness of the bearing, i.e. the resistance against a displacement of the shaft centre due to an applied operating force, grows with increasing speed, especially in full rotation mode, due to the higher load carrying capacity of the plain bearings. The stiffness can be additionally increased by preloading the roller bearings.

To make use of this knowledge, it is conceivable to equip the bearing with active components that can change the preload of the roller bearings during operation. The described effects could be validated in special tests. To change the preload, paraffin wax actuators were used. They were mounted between the bearing pedestal and the bearing covers. Through the supply of heat, a force is exerted on the outer rings of the angular contact ball bearings and thus reduces the preload [132].

Conclusion

The combination of roller and plain bearings offers high potentials for the application in servo presses. Designed properly, the roller bearings bear the largest share of the load at low circumferential speeds and operating forces, while the plain bearings carry a larger ratio of the force with increasing speeds and loads. This load transfer has already been proven experimentally [131, 245].

In-depth investigations of the operational behaviour of the combined bearing and the evaluation of its industrial applicability have been carried out on a special test rig. Due to the resulting reduction of the roller bearing load in the combined bearing compared to pure roller bearings, the nominal lifetime of the roller bearings increases significantly despite a reduction of the required inner diameter. The use of combined roller and plain bearings is currently limited to prototypes, since the maturity for widespread industrial use is not yet given [131]. However, the potential resulting from the combination have already been proven [247].

Monitoring the operation state of bearings offers new potentials for the application in modern production machines. Examining a newly developed combined roller and plain bearing, we showed the possibility to use the bearings as both a sensor and actor in the drive train of the 3D Servo Press. Sensor-based condition monitoring of bearings allows conclusions on the current operation conditions of the press system, while the implementation of active components in the bearing makes it possible to influence the system's behaviour depending on the particular load scenario.

Condition monitoring makes it possible to detect detuned load conditions in the process, such as asymmetrical process forces, which can lead to ram tilting, as was shown in [132]. With the knowledge about their influence on the bearing stiffness, damping as well as the bearing clearance and load ratio of combined bearings, the controllability of the machine as well as the operating limits of the bearing can be extended by active bearing components, which leads to a reduction of tool wear, whereby a reduction of process uncertainty as well as an increase in component quality can be expected as well.

5.4.3 Development of a Sensor-Integrated Compensation Chuck for Semi-active Control of the Tapping Process

Tuğrul Öztürk, Matthias Weigold, and Eberhard Abele

The tapping process is a widely used machining process for internal thread making and usually takes place at the end of the value chain, so that non-compliance of required thread quality, or even the destruction of the thread due to uncertainty, is therefore associated with high economic costs [273]. Possible uncertainty during thread cutting are axis offset, concentricity error, synchronisation error, sloped pre-drill bore, faulty pre-drill bore diameter and tool wear [2, 185]. Here, we are facing two types of uncertainty both classified as incertitude. As introduced in Sect. 2.1 there is data uncertainty hidden in state variables, e.g. feed rate or motor spindle speed. Structural uncertainty as described in Sect. 2.3 results from the upstream process of machining the pre-drilling, since the tapping process is a combined process. In addition to the methods and technologies to master uncertainty during the design and production phase presented in Chap. 5 an alternative strategy is the semi-active process manipulation described in Sect. 5.4. for mastering uncertainty during the production phase, Sect. 1.2. One of the core components of such a semi-active system for process manipulation are appropriate sensors to detect uncertainty. For the detecting of uncertainty within the tapping process, various approaches are being developed. In [185] and [266] the motor spindle currents were used as a data source to detect uncertainty while tapping by a two-stage pair-wise feature selection and classification algorithm based on wavelet transformations [185] and statistical process control [266]. The detection of progressive wear of an M12 tap by examining the frequency spectrum of vibration measurements of the workpiece and the motor spindle during tapping is shown in [200]. The authors in [60] and [188] used dynamometers as data source for diagnosis of the tapping process by using an information and multiple probability voting scheme [60] and artificial neural networks [188].

The company Bilz Werkzeugfabrik has developed a sensor-integrated tapping chuck which detects the occurrence of length compensation [39], thus resulting in a binary information only, which is not appropriate for semi-active control. In this subsection, we introduce the concept for a semi-active control of the tapping process on the basis of the length compensation chuck (LCC) Softsynchro 3 from the tool manufacturer Emuge Franken [88] shown in Fig. 5.44a. This is a purely mechanical system for handling synchronisation errors only. The Softsynchro 3 consists of a hollow shank taper with an attached hollow cylindrical part in which a piston clamping the tapping tool can move maximum 1.5 mm in both z-axis directions. Two pairs of axial ball bearings offset by 180° are guiding the piston and transmitting the process torque from the hollow shaft taper to the tapping tool. The axial system stiffness in z-axis direction is determined by two polymer spring elements offset by 180°. Further, we present three sensor concepts to measure the close-to-tool vibrations, the process torque and the axial length compensation for uncertainty detection during the tapping process, as well as the integration of these sensors into the LCC. Compared

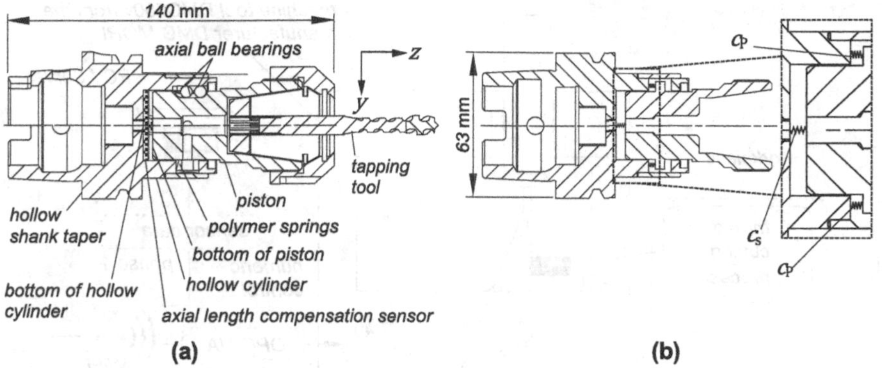

Fig. 5.44 **a** Sectional view of Softsynchro 3 [88] and **b** modelling of stiffness

to [185, 266], the process torque is measured directly on the LCC and not indirectly via the motor spindle currents. In comparison to [200], the vibration sensor introduced in this work measures the vibrations close to the tapping tool. The concept we present here to measure the axial length compensation is unique and allows to quantify the synchronisation error. The use of a dynamometer as in the approaches [60, 188] is deliberately avoided, since dynamometers have a number of disadvantages, such as high costs and a smaller workspace. The three sensor concepts measure the corresponding measurement quantities close to the process and therefore provide an appropriate data source for semi-active control to master the uncertainty of the tapping process.

Concept of semi-active control of the tapping process

Within a semi-active control approach, the production process of tapping is not directly influenced by actuators but by the machine tool. In Fig. 5.45 we present a schematic diagram of the setup and the functional principle for semi-active control during the tapping process. To detect uncertainty while tapping, the LCC Softsynchro 3 is equipped with the three sensor concepts mentioned above. For signal pre-processing, data acquisition (A/D converter) and data processing we developed a rotating telemetry unit. The data acquisition of the sensor data while tapping as well as the communication with the machine tool for semi-active control is made via the wireless radio standard WiFi. For mastering uncertainty while tapping by semi-active control, it is necessary to define appropriate control variables within the machining process containing process parameters, e.g. motor spindle speed and feed rate. Beside definition of such control variables by process parameters, it is necessary to influence them within the numerical control (NC) of the machine tool. As mentioned in Fig. 5.45, the NC of the selected machine tool contains a OPC-UA (Open Platform Communications—Unified Architecture) server, which makes a communication of the machine tool with other production machines possible. By linking

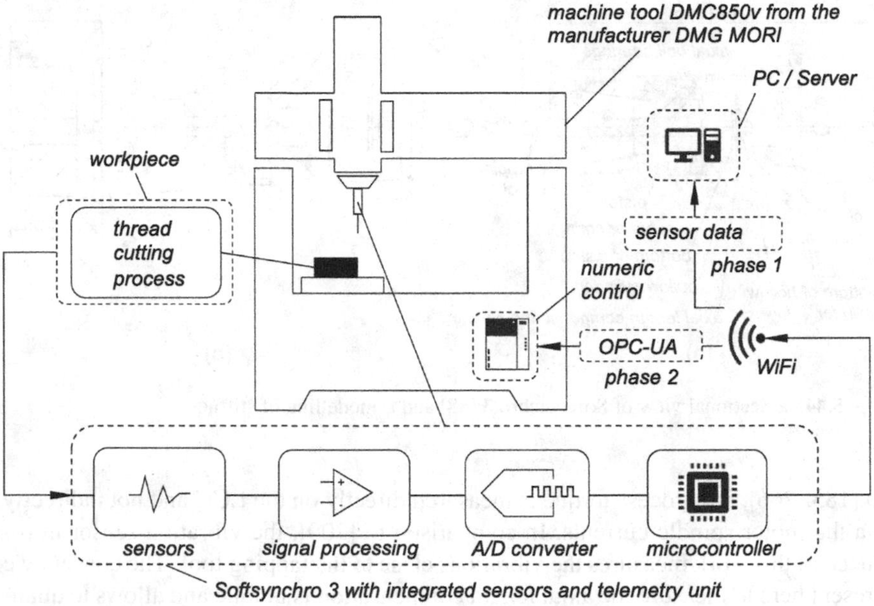

Fig. 5.45 Schematic diagram of the semi-active control concept

this ability with synchronised actions [244] within the NC program makes process manipulation during machining processes possible, since synchronised actions can influence process parameters synchronously to the NC program execution.

Sensor integration

A sensor concept has been developed to measure lateral vibrations close to the tapping tool shown in Fig. 5.46a. The vibration sensor is screwed onto a sleeve which is glued onto the tapping tool, allowing a non-destructive use of the vibration sensor in case of tool breakage. Two accelerometers based on Microelectromechanical system (MEMS) measure the acceleration in x- and y-axis direction. The process torque is determined with a strain gauge full bridge, which for design reasons can only be applied to the outer surface of the cylindrical part (Fig. 5.44a) of the LCC. By means of a Finite Element Method calculation we localised the position of the maximum mechanical shear stress illustrated in Fig. 5.46b, and thus the appropriate application positions of the strain gauges. The measured strain ε_{45} is converted into the process torque M_T by assuming a linear-elastic behaviour by Eq. (5.22)

$$M_T = \varepsilon_{45} \frac{E}{1+\nu} \frac{\pi}{16} \frac{d_a^4 - d_i^4}{d_a}, \tag{5.22}$$

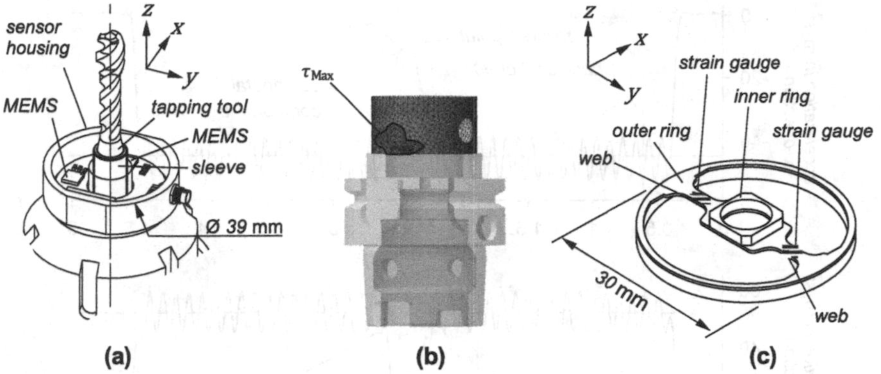

Fig. 5.46 Integrated sensors in the LCC to measure **a** close-to-tool vibrations, **b** process torque and **c** axial length compensation

where d_a is the outer and d_i the inner diameter of the cylindrical part (Fig. 5.44a), the modulus of elasticity is represented by E and the Poisson's ratio by ν.

To measure the axial length compensation a sensor based on strain gauges has been developed shown in Fig. 5.46c. This sensor consists of an outer ring and an inner ring, which are connected by two webs. Due to this design, the sensor behaves like a spring element, whereby the spring stiffness is largely determined by the lower side of the two webs forming a full bridge configuration allowing a sensitive measuring of the strain due to displacement of the inner ring in the z-axis direction. By integrating the axial length compensation sensor into the LCC (Fig. 5.44a) and connecting the outer ring with the bottom side of the cylindrical part (Fig. 5.44a) and the inner ring with the bottom side of the piston (Fig. 5.44a), the axial length compensation can be quantified indirectly. The original axial system stiffness mainly determined by the two polymer spring rings c_P in Fig. 5.44b is increased by the sensor stiffness c_S in Fig. 5.44b resulting in a new axial system stiffness which is given by

$$c_{\text{System}} = 2c_P + c_S \tag{5.23}$$

Therefore, the geometry of the sensor was designed in a way that the sensor stiffness c_S is less than ten percent of original axial system stiffness $c_S < 0.2c_P$.

Evaluation of the close-to-tool vibration sensor

For evaluation of the vibration sensor, we carried out thread cutting tests with thread dimension M8×1.25 mm on a GROB G350 machine tool. The thread has the dimension M8×1.25 mm and the material type 42CrMo4. The spindle speed was 9.7 Hz and the cutting speed 148.8 m/min. The vibration signals of both MEMS accelerometers while cutting a single thread are shown in Fig. 5.47. It can be seen that each signal contains a constant acceleration component, on which a low-frequency harmonic

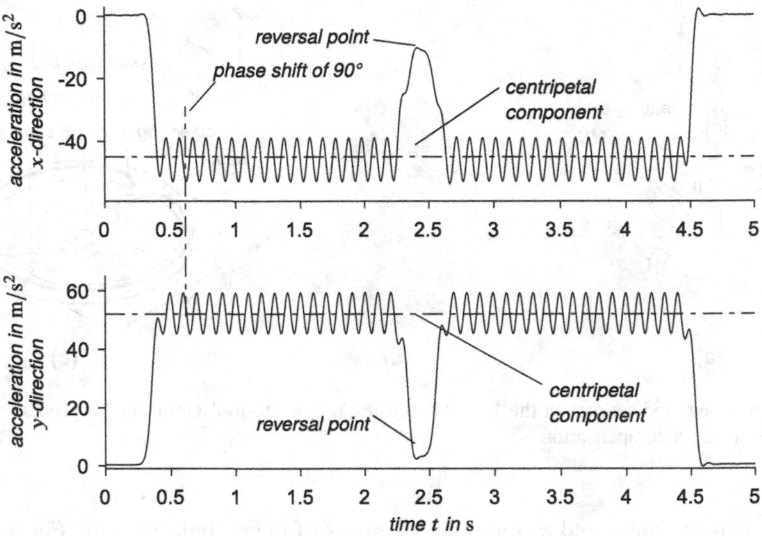

Fig. 5.47 Close-to-tool vibration signals during the cutting of a single M8 × 1.25 mm thread, sampled at 10.5 kHz and low pass filtered at 50 Hz

oscillation of 9.7 Hz is superimposed. The constant acceleration corresponds to the centripetal acceleration, which is included in the vibration signals due to the measuring direction of the MEMS accelerometers in x- and y-axis, thus radial direction. In this case, the low-frequency harmonic oscillation corresponds to the spindle speed signal, because the machine tool used to perform the experiments is a horizontally mounted motor spindle. In this arrangement, the acceleration due to gravity is no longer perpendicular to the plane of rotation (x-y plane), so that the gravity is also measured by the MEMS accelerometers depending on the angle of rotation of the motor spindle. Since the two measuring directions of the MEMS accelerometers are orthogonal to each other, the harmonic components of both MEMS accelerometers caused by gravity are phase-shifted by 90° to each other. Furthermore, the reversal point of the tapping cycle can be seen in both vibration signals and is also marked in Fig. 5.47.

Conclusion and outlook

The evaluation of the developed vibration sensor provides plausible results, so that the measurement data of the vibration sensor can be used to detect uncertainty during tapping process. Next, the sensors for measuring the process torque as well as the axial length compensation must also be validated for plausibility and calibrated within thread cutting tests. Within the scope of further thread cutting tests, uncertainty during the tapping process, e.g. exceeding threshold values of corresponding process quantities measured by the integrated sensors have to be detected. Subsequently, appropriate control variables are to be derived for the semi-active control of the tapping process and the control algorithm is to be implemented in the micro-controller of the telemetry unit. The presented technology enables the mastering of uncertainty of the tapping process within the production phase of mechanical components used in structural dynamic systems such as the MAFDS presented in Sect. 3.6.1.

5.4.4 Shock Absorber with Integrated Hydraulic Vibration Absorber to Improve Driving Dynamics

Nicolas Brötz and Peter F. Pelz

A car driving on an uneven road is subject to vibrations that are handled by suspension struts similar to the MAFDS, see Sect. 3.6.1. Their main task is to reduce vibrations. This can be done by (i) passive elements like spring, damper and dynamic vibration absorber, (ii) semi-active systems like adjustable damper, and (iii) active systems, such as the Active Air Spring, see Sect. 3.6.2. In the following, we focus on the usage of a passive dynamic vibration absorber, see Sect. 1.3, and discuss a semi-active option.

The essential requirements for the suspension strut of modern vehicles are to ensure high driving safety and high driving comfort at the same time. The driving safety is affected by data uncertainty of the wheel mass and tire stiffness, which can be classified as incertitude, Sect. 2.1. A higher wheel mass, for example a tire covered with mud, leads to a loss of driving safety. The driver's control of the vehicle can only be ensured, if there is wheel-ground contact. The suspension strut links the car's body to the wheel. The vertical suspension strut force, therefore, has to be sufficient to ensure the best transmission behaviour of horizontal manoeuvres. According to Mitschke [198], the driving dynamics are measured by the dynamic wheel load fluctuation. If the fluctuation is higher than the static load, the tire loses contact to the ground. Driving comfort is measured by accelerations on the occupants. The suspension strut as a link between body and wheel affects both, driving dynamics and driving safety.

However, when tuning the system, these two objectives are in conflict. As an example, we consider a quarter-car simulation of a reference car with adjustable damper and a constant body spring stiffness using an excitation according to a federal highway at 100 km/h. The quarter car has the parameters of a middle class car also used in Sect. 3.6.2. As a reference for following investigations, the grey line in the conflict diagram in Fig. 5.48 depicts the corresponding Pareto-front for varying damping coefficient, where the standard deviations are extracted from a time signal of wheel load and body acceleration. By increasing the damping coefficient, the body accelerations are also increased, whereas the wheel-load fluctuation decreases.

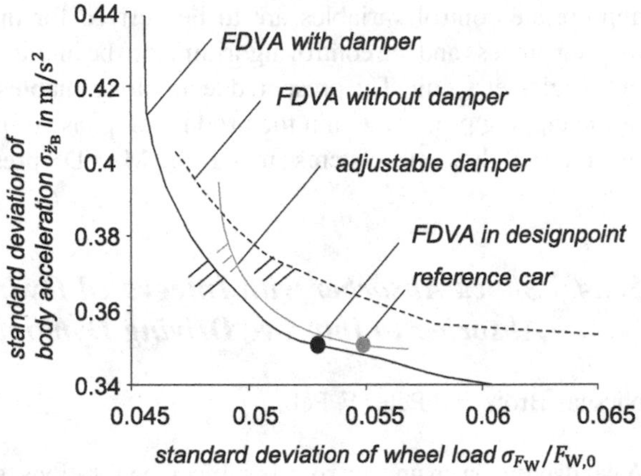

Fig. 5.48 Conflict diagram for a reference car with constant body spring stiffness, FDVA with damper and FDVA without damper

Fluid dynamic vibration absorber

To improve upon this boundary, a structural expansion, a dynamic vibration absorber with a hydraulic translated fluid mass, the Fluid Dynamic Vibration Absorber (FDVA) has been developed [52]. Its advantages can be seen in Fig. 5.48. The black solid line shows the Pareto-front of the FDVA with damper. The dashed black line refers to the FDVA without damper. It can be seen that this more simple configuration is only useful when the comfort is neglectable, e.g. for sport cars. In combination with a damper the FDVA is able to reduce wheel load fluctuations with the same comfort compared to the reference. In the following, we present the design and operation principle of the FDVA, as well as the validation of the used models and the discussion of simulations leading to Fig. 5.48.

A classic dynamic vibration absorber consists of a capacity and a heavy mass, which contradicts the goal of lightweight construction. In comparison to a classic dynamic vibration absorber, the Fluid Dynamic Vibration Absorber (FDVA) [52], reduces the dynamic mass by means of hydraulic transmission. By using the continuity equation, Bernoulli's equation and principle of linear momentum, the possible reduction of mass can be derived from the axiomatic model of the FDVA

$$\left[m_P + (2\alpha\beta + \alpha^2)m_F\right]\ddot{z} = k(z_0 - z) + p_L A \tag{5.24}$$

with the piston mass m_P, the ratio $\alpha = A/a$ between the surface of the piston A and the ducts a, the relation of duct length to cylinder length β, the duct fluid mass m_F the absorber motion z, the excitation motion z_0, the vibration absorber spring stiffness k, and the pressure loss p_L.

Figure 5.49 shows the principle of lightweight design at the hydraulic transmission. The duct fluid mass m_F is transmitted by α^2, which we see at the deviation between inertia and mass, if the piston mass m_P moves towards zero. The ratio

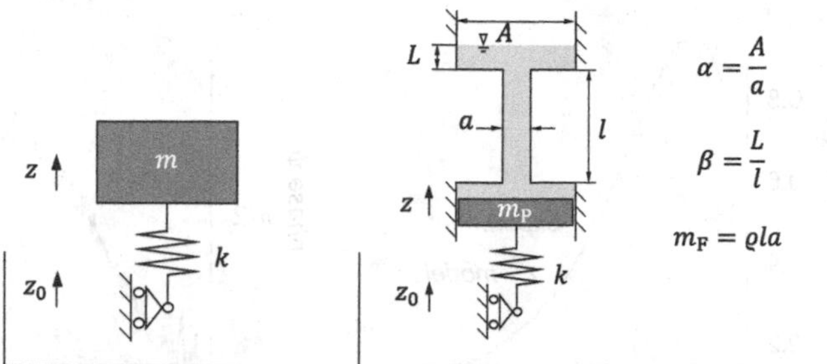

mass	m	$m_P + m_F(2\alpha\beta + 1)$
inertia θ	m	$m_P + m_F(2\alpha\beta + \alpha^2)$
inertia / mass	1	$\frac{m_P + m_F(2\alpha\beta + \alpha^2)}{m_P + m_F(2\alpha\beta + 1)}$

Fig. 5.49 Lightweight design: A comparison between standard dynamic vibration absorber and FDVA

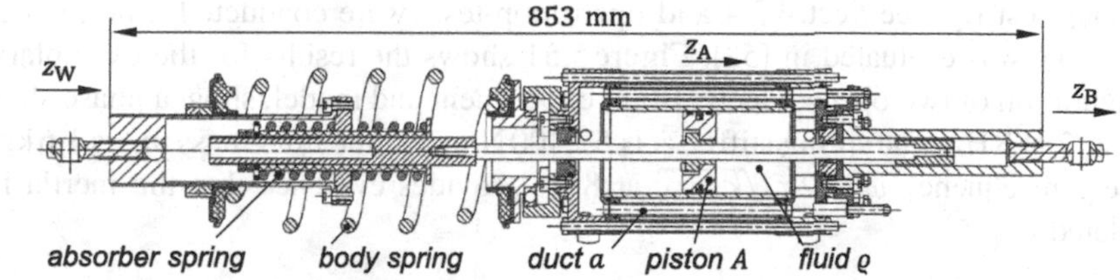

Fig. 5.50 Sectional view of the FDVA functional demonstrator [52]

$\alpha = A/a$ is determined by the geometry. The duct fluid mass m_F is negligible in terms of weight, but translated it has the largest contribution to inertia.

We built a prototype of the FDVA to validate the axiomatic model, Eq. (5.24), see Fig. 5.50. It consists of a double-acting hydraulic cylinder where the chambers are connected via twelve ducts on the outside. The ducts can be closed by mechanical valves in order to change the ratio α. A spring connects the piston rod of the hydraulic cylinder with the wheel axle. The body spring is connected parallel to the FDVA and links the wheel axle to the housing of the FDVA, which in turn is attached to the chassis.

FDVA model validation

The characteristics of a dynamic vibration absorber are used to validate the axiomatic model. Therefore, we evaluate the transmissibility V and the phase shift ψ of foot excitation z_S to piston motion z_{FDVA} at a frequency band with frequencies from 0.1

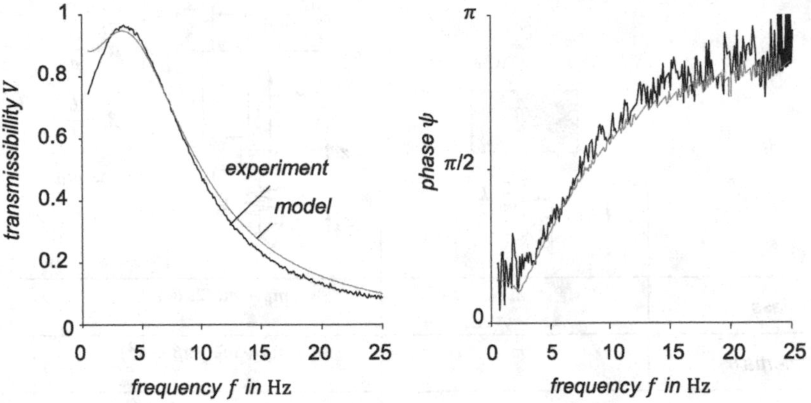

Fig. 5.51 Model validation with experimental results for the FDVA

to 25 Hz, since the quarter car, the basis of our investigations, is usually applied in this range [198]. We mounted the functional demonstrator into the Hardware-in-the-Loop test rig, see Sect. 4.3.4 and open-loop-tests were conducted. The model, Eq. (5.24), was evaluated in [53]. Figure 5.51 shows the results for the exemplary configuration of two opened ducts. Both, experiment and model, show a phase shift $\psi = \pi/2$ at 8 Hz. The spring stiffness is 57.400 N/m and the fluid mass m_F is 1.6 kg. The eigenfrequency $\omega = 2\pi\sqrt{k/m_\mathrm{A}}$ at 8 Hz provides evidence that the inertia is translated.

Improvement by use of the FDVA

We use the validated model to estimate the improved driving dynamics. The quarter car, a simple dual mass oscillator, is equipped with the model of the FDVA. The three equations of motion for body (index B), wheel (index W) and FDVA according to Fig. 5.50 are then

$$\begin{bmatrix} m_\mathrm{B} + (1+\alpha)^2 m_\mathrm{F} & 0 & -\left(\alpha+\alpha^2\right) m_\mathrm{F} \\ 0 & m_\mathrm{W} & 0 \\ -\left(\alpha+\alpha^2\right) m_\mathrm{F} & 0 & \left[m_\mathrm{P} + \left(2\alpha\beta+\alpha^2\right) m_\mathrm{F}\right] \end{bmatrix} \begin{bmatrix} \ddot{z}_\mathrm{B} \\ \ddot{z}_\mathrm{W} \\ \ddot{z}_\mathrm{FDVA} \end{bmatrix} + \begin{bmatrix} b & -b & 0 \\ -b & b & 0 \\ 0 & 0 & 0 \end{bmatrix} \begin{bmatrix} \dot{z}_\mathrm{B} \\ \dot{z}_\mathrm{W} \\ \dot{z}_\mathrm{FDVA} \end{bmatrix}$$
$$+ \begin{bmatrix} k_\mathrm{B} & -k_\mathrm{B} & 0 \\ -k_\mathrm{B} & k_\mathrm{B} + k_\mathrm{T} + k_\mathrm{FDVA} & -k_\mathrm{FDVA} \\ 0 & -k_\mathrm{FDVA} & k_\mathrm{FDVA} \end{bmatrix} \begin{bmatrix} z_\mathrm{B} \\ z_\mathrm{W} \\ z_\mathrm{FDVA} \end{bmatrix} = \begin{bmatrix} -p_\mathrm{L} a \alpha \\ k_\mathrm{T} z_0 \\ p_\mathrm{L} a \alpha \end{bmatrix} \tag{5.25}$$

with mass m, motion z, damping coefficient b, stiffness k, tire stiffness k_T and road excitation z_0. The body stiffness k_B contains the secondary stiffness of the suspension strut for a realistic simulation. These equations can be used to calculate the driving dynamics and comfort. For the FDVA with duct fluid mass of $m_F = 72$ g translated by a translation factor of $\alpha = 59.45$ the results are given in Fig. 5.48.

For the FDVA with damper in the design point an increase to the reference car of 5% in driving dynamic at a constant driving comfort is possible. The peak of wheel load fluctuation in the frequency response reduces by 15%.

Now, we discuss the possibility to master uncertainty using the passive FDVA. The passive FDVA is designed to absorb the eigenfrequency of the unsprung mass, which is formed by the mass of break, wheel carrier, upper and lower control arm and wheel. The wheel mass depends on the brand and the wear condition and therefore is uncertain.

The tire is another uncertain factor. While driving, the tire is able to heat up and the stiffness increases. Thus, the eigenfrequency of the wheel increases, too. Investigations show that in realistic changes of wheel mass, tire stiffness and road conditions the improvement through the FDVA in comparison to a standard shock absorber is independent of a change of eigenfrequency.

Semi-active FDVA

Compared to a passive FDVA, a semi-active FDVA is able to adopt its eigenfrequency to a change of the system's eigenfrequency. A system's eigenfrequency can variate by a change of inductance or compliance. In the case of a high system eigenfrequency change, e.g. for a sprung foundation, where the change of mass depends on the assembly of the test bench and thus can be in the range of several hertz, the semi-active FDVA is useful. For the adaptation to the system eigenfrequency the transmission factor α or the length of the duct and the damping of the FDVA has to be changed.

The wheel of a car has only a small change of unsprung mass and tire stiffness, thus showing a neglectable system eigenfrequency change. The extra effort to design a semi-active FDVA for a car on the one hand is not necessary, and on the other hand is too expensive.

Conclusion

A standard shock absorber equipped with the passive FDVA is able to maintain the driving dynamics. This is equivalent to the driving safety in a larger area of application, when the road, tire or loading conditions of the vehicle are unknown. If we know the boundary of the parameter changes and can classify them with tolerances, passive systems can be sufficient. Semi-active and active systems require a higher effort, but they are able to adopt a wider range of parameter changes.

5.4.5 Active Air Spring for Vibration Reduction in Vehicle Chassis

Manuel Rexer, Philipp Hedrich, and Peter F. Pelz

In autonomously driven cars, passengers are able to spend time on activities other than driving [151, 248]. The highest possible level of driving comfort is increasingly important, as it is the case with today's luxury class vehicles. The suspension system of the vehicle has the greatest influence on driving comfort [147]. When designing the spring and damper system, driving comfort and driving safety are two conflicting objectives [198]. It is not possible to optimise both at the same time, however, a compromise between the two is feasible.

The simplest model for the vertical dynamics of a car is the quarter car introduced in Sect. 3.6.2 [198]. With this two-mass oscillator, driving comfort can be evaluated by the standard deviation of body acceleration $\sigma_{\ddot{z}_b}$. The standard deviation of the force between wheel and road σ_{F_w} describes the driving safety. As shown in Sect. 3.6.2 there are boundary lines, so-called Pareto lines, for passive suspensions consisting of springs and dampers which cannot be improved upon by any passive system. To further increase the driving comfort, active systems must be applied. The active system increases the function "driving comfort" of the vehicle. This gain in function can be used to master uncertainty that occurs during the usage phase of the entire system. The active suspension is able to compensate varying parameters (data uncertainty Sect. 2.1), e.g. a changed wheel mass. Since the distribution of the variation is not known, this is an incertitude. The active module is also able to master uncertainty in excitation as we show later in this section. Since the excitation is unknown in the design phase, this is ignorance. We have developed an Active Air Spring (cf. Sect. 3.6.2) to demonstrate the capabilities of an active suspension system [20, 147]. This subsection gives a short review on experimental results of the Active Air Spring which actively increases driving comfort compared to the passive system.

All investigations are carried out on a hardware-in-the-loop (HiL) test rig (cf. Sect. 4.3.4). Figure 5.52 shows the layout of the test rig, which consists of a uniaxial test rig that deflects the air spring and measures its reaction force. The dynamics of the quarter car and the controller of the active system are simulated in a real-time simulation (for parameters see Sect. 3.6.2). Hardware and simulation are connected via the deflection $\Delta z = z_b - z_w$ and the force measurement signal F.

In principle, the quarter car can be excited by any base excitation z_r. In the following, the rides over (i) a single obstacle and (ii) on a road are examined. The ride over a single obstacle (i) demonstrates the active manipulation of the quarter car. The quarter car rides at a speed of $v = 10\,\mathrm{km/h}$ over an obstacle created as a cosine shaped bump with a height of $h = 5\,\mathrm{cm}$ and a total width of $l = 100\,\mathrm{cm}$. Maximum driving comfort is reached, if the body is at rest ($\ddot{z}_b = 0$). It is to be taken into account that the controller has been designed for the stochastic excitations of a road and is not optimised for the crossing of a single obstacle. The excitation therefore is uncertain. This data uncertainty occurs in the use phase of the Active Air Spring.

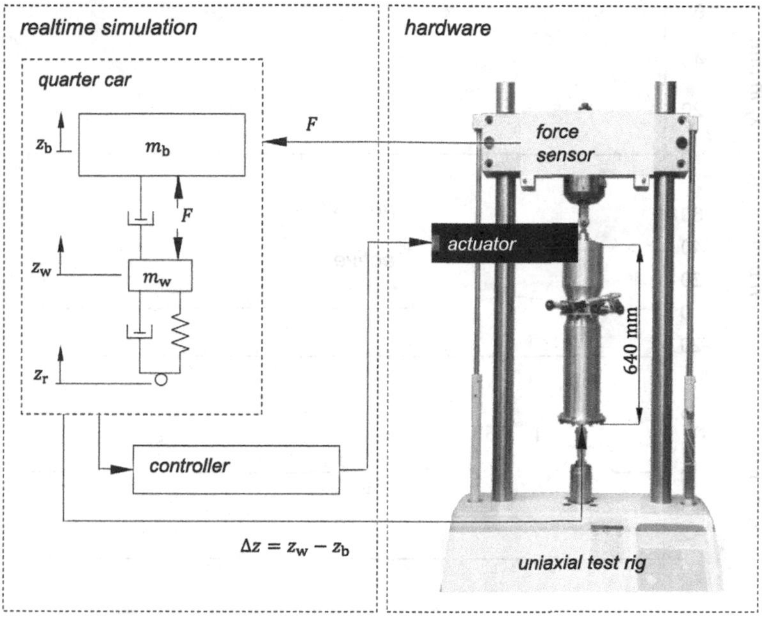

Fig. 5.52 Basic layout of the hardware-in-the-loop test rig with signal flows

Figure 5.53 shows the result of the HiL test for a passive and an active system. The time sequences of the excitation z_r, the body movement z_b and acceleration $\ddot{z}_b$, the wheel force fluctuation ΔF_w, the force between wheel and road and the deflection Δz are shown. The results for the passive system were also determined with the HiL test rig, by deactivating the actuator.

The active suspension system comes much closer to the target of keeping the body at rest than the passive system, since the body movements z_b are smaller. The lower body acceleration $\ddot{z}_b$ results in increasing driving comfort and the smaller wheel force fluctuation ΔF_w increases driving safety. It is possible to reduce the body accelerations by 53%. The active system requires more suspension stroke than the passive system, since the deflection Δz increases. The deflection of the Active Air Spring is almost equal to the excitation, which is obvious, since the body is almost at rest. Hence, increasing driving comfort is only possible with larger suspension stroke. Furthermore, these results demonstrate that the designed controller is robust against ignorance in the excitation of the system.

The ride on a road (ii) is examined by excitation with a stochastically generated signal. For this purpose, white noise n is filtered so that the contained spectrum corresponds to riding on a highway with 100 km/h [147]. Figure 5.54 shows the frequency response of the performance indicators driving safety F_w and driving comfort $\ddot{z}_b$ during such excitation. Frequencies above 25 Hz are not considered, since they are negligible for chassis applications [198]. As before, the response of the passive chassis is used as a reference.

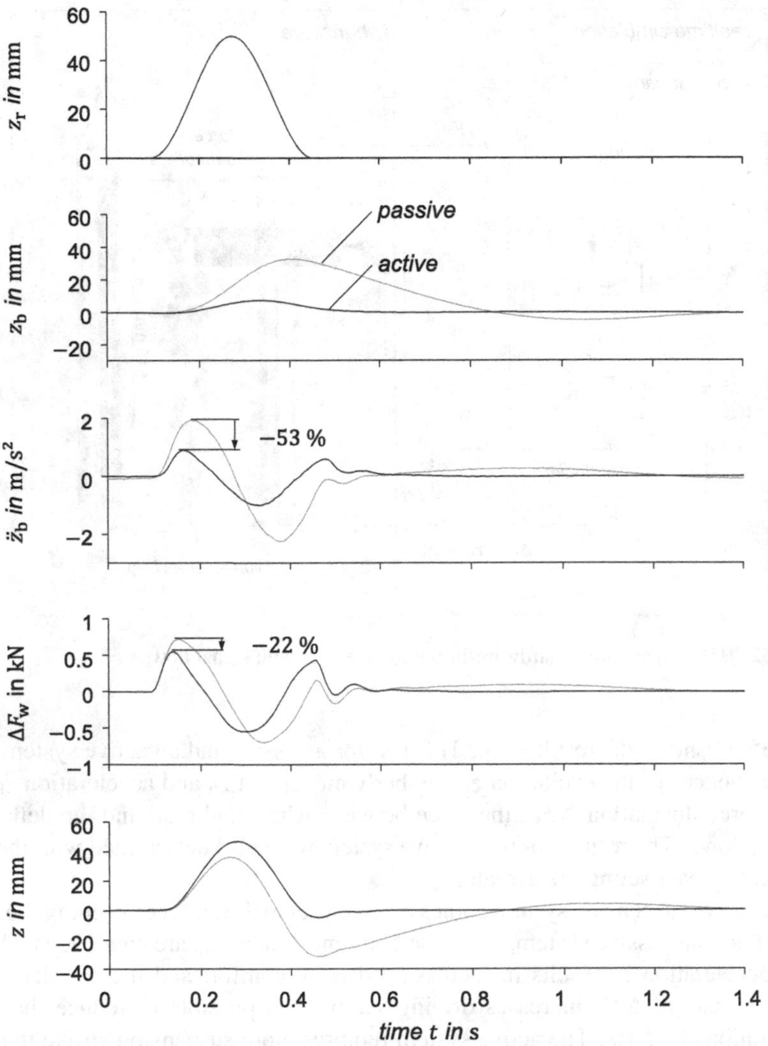

Fig. 5.53 Time sequences of the HiL test when crossing a single obstacle ($l = 1\,\text{m}$, $h = 5\,\text{cm}$, $v = 10\,\text{km/h}$) with the passive and the Active Air Spring controlled by controller with preview function

It can be seen that the active system causes a significant vibration reduction in terms of driving safety and driving comfort in the range up to 5 Hz. In the range above 5 Hz no significant improvements are achieved. However, no deterioration can be seen either. The reason is the speed of the actuator which has a transfer function that corresponds approximately to a 1st order low pass filter with a cut-off frequency of 5 Hz [183]. At frequencies above 5 Hz, the influence of the actuator is therefore

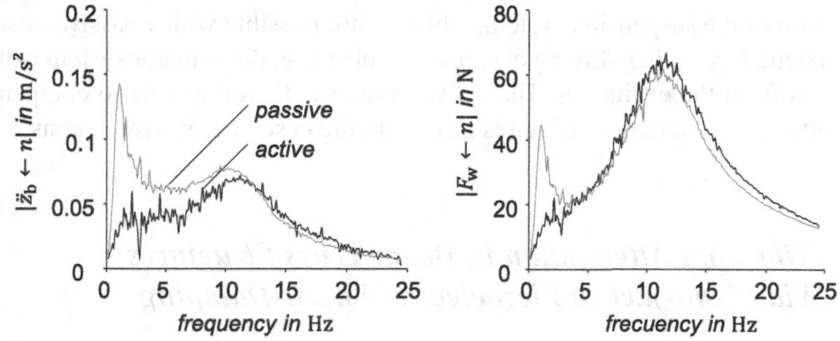

Fig. 5.54 Frequency response of passive and active system from HiL test riding on a highway with 100 km/h

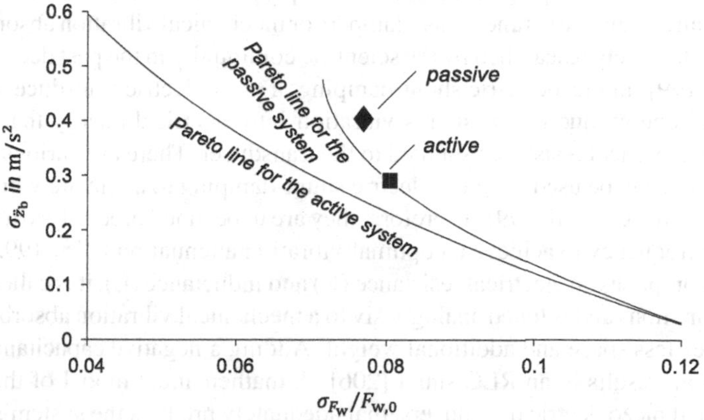

Fig. 5.55 Conflict diagram for driving safety and driving comfort of passive and Active Air Spring from HiL test riding an a highway with 100 km/h

limited. However, it should be noted that the possible comfort margin above 5 Hz is also small.

The conflict diagram, cf. Fig. 5.55, shows the performance indicators, standard deviation of body acceleration $\sigma_{\ddot{z}_b}$ and wheel load $\sigma_{F_w}/F_{w,0}$ in relation to the static wheel load can be shown in a comparative manner. Figure 5.55 shows the Pareto line for the passive and active system and the results for the passive and active chassis. The Pareto lines are the results of the H2 optimisation as shown in Sect. 3.6.2.

The active system will increase comfort by 28% with loosing 4.7% of driving safety. The active system comes much closer to the active Pareto front. We have to consider that the results for both, the passive and the active system, are influenced by the HiL test rig (see Sect. 4.3.4). This degrades the test results in comparison to the limit lines which can be determined by simulating the influence of the HiL test rig.

In summary, the active system is suitable for increasing driving comfort by up to 28% while nearly maintaining the level of driving safety. The active system increases the function of the suspension system, which is not possible with a passive or semi-active system. Even when driving over a single obstacle, the actuator system isolates the chassis from the excitation. The active system is therefore able to compensate the uncertainty in excitation, which is limited in the case of a passive chassis.

5.4.6 Vibration Attenuation in Beam Truss Structures Via (Semi-)active Piezoelectric Shunt-Damping

Jonathan Lenz, Benedict Götz, Tobias Melz, and Roland Platz

Piezoelectric shunt-damping as a (semi-)active process manipulation, see Sects. 3.2 and 3.4, is an alternative to tuned mass dampers or mechanical vibration absorbers and has been extensively researched by the scientific community in the past decades [104, 118, 138, 199]. In piezoelectric shunt-damping, a piezoelectric transducer converts mechanical energy due to a system's vibration into electrical energy that is dissipated in an electrical resistance shunted to the transducer. There are various possible shunts, which can be used for piezoelectric shunt-damping to attenuate vibrations in a narrow frequency band [199]. Therefore, they are to be tuned precisely to a system's resonance frequency to achieve an optimal vibration attenuation [118, 199, 206]. If the shunt comprises an electrical resistance (R) and inductance (L), it is called an RL-shunt. It functions and is tuned analogously to a mechanical vibration absorber [206], but requires less space and additional weight. Adding a negative capacitance (C) to the RL-shunt results in an RLC-shunt [206]. A mathematical model of the system with shunted piezoelectric transducers, that adequately predicts the system's outputs is required to tune the shunt parameters within a model-based approach. However, if data uncertainty occurs, see Sect. 2.1, e.g. due to a variation of model parameters or boundary conditions, the model output varies, which leads to detuned shunt parameters as well as to a decrease in the achievable vibration attenuation [118, 184]. In this section, we therefore discuss the effects of aleatoric and epistemic data uncertainty from different sources on the vibration attenuation of a single beam and a beam truss structure in the usage phase via piezoelectric shunt-damping, see Sect. 1.2.

Uncertainty in vibration attenuation in a single beam test setup

In [118], Götz conducted an extensive numerical and experimental study on the attenuation of lateral vibrations of a single beam at the first resonance frequency via piezoelectric shunt-damping. Figure 5.56a shows the corresponding single beam test setup with the slender aluminum beam *1* that is of the same design as the beams comprising the MAFDS described in Sect. 3.6.1. It is connected to a slide bearing *2* and a fixed bearing *3* via the novel piezo-elastic supports *A* and *B*. A static axial load F_{ax} can be applied to the beam using the spindle type lifting gear *4*, which is measured with the force sensor *5*. Support *A* is used to excite the beam in the *y*-

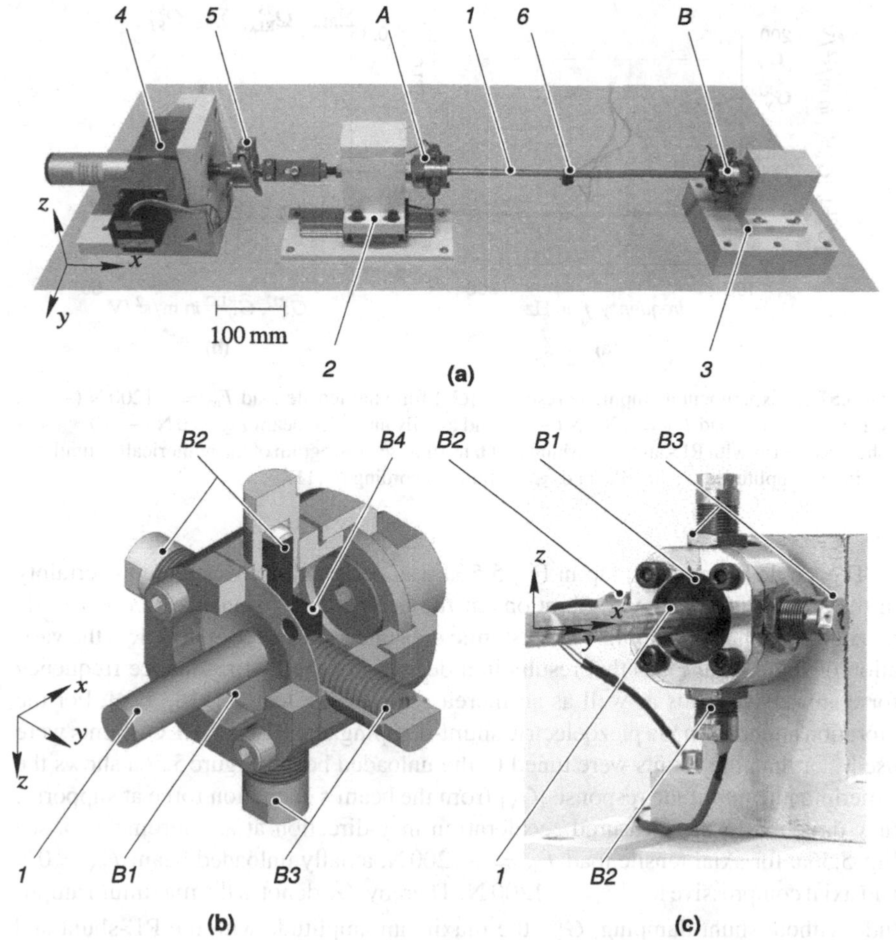

Fig. 5.56 **a** Single beam test setup, **b** sectional view and **c** photo of the piezo-elastic support [118]

and z-direction, while support B is shunted to an RL- or RLC-shunt to attenuate the lateral vibrations, which are measured by two accelerometers *6*. Figure 5.56b shows the sectional view of the piezo-elastic support in Fig. 5.56c, with the membrane spring *B1* that restricts lateral and axial displacement but allows rotation of the beam's end. Two piezoelectric stack transducers *B2* are positioned opposite two helical disc springs *B3* mechanically prestressing the stack transducers. Both are connected to the beam via the axial extension *B4*, which connects both springs to the beam, and transforms the beam's lateral vibration into axial deflections of the stack transducers. The concept of the piezo-elastic support was patented in [91] and, apart from piezoelectric shunt-damping, is also used for active buckling control, see Sect. 5.4.7 [235].

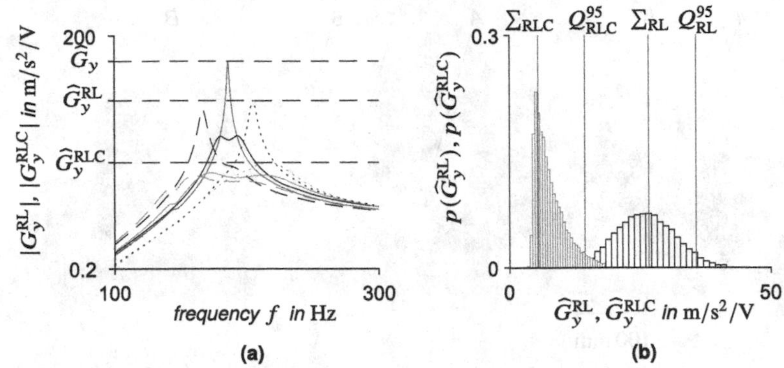

Fig. 5.57 **a** Experimental amplitude response $|G_y|$ for axial tensile load $F_{ax} = -1200$ N (- - - -), axial compressive load $F_{ax} = 1200$ N (— —) and axially unloaded beam $F_{ax} = 0$ N (——) without a shunt as well as with RL- and RLC-shunt and **b** normalised histogram of the numerically simulated maximum amplitudes $\hat{G}_y$ for RL- and RLC-shunt, according to [118]

The single beam test setup in Fig. 5.56a was designed to investigate uncertainty in the attenuation of lateral vibrations at the beam's first resonance frequency via piezoelectric shunt-damping. An epistemic data uncertainty, see Sect. 2.1, is the variation of static axial loads that results in a decrease of the first resonance frequency for compressive loads as well as an increase for tensile loads, [118, 120]. For the vibration attenuation via piezoelectric shunt-damping, the RL- and RLC-shunts were used. For that, the shunts were tuned to the unloaded beam. Figure 5.56a shows the experimental amplitude response $|G_y|$ from the beam's excitation force at support A in y-direction to the measured acceleration in y-direction at accelerometer 6, see Fig. 5.56a, for axial tensile load $F_{ax} = -1200$ N, axially unloaded beam $F_{ax} = 0$ N and axial compressive load $F_{ax} = 1200$ N. Thereby, $\hat{G}_y$ denotes the maximum amplitude without shunt-damping, $\hat{G}_y^{RL}$ the maximum amplitude with the RL-shunt and $\hat{G}_y^{RLC}$ the maximum amplitude with the RLC-shunt in Fig. 5.57a. Vibration attenuation is achieved with RL- and RLC-shunts, but is reduced for the uncertain axial loads compared to the unloaded beam. However, the attenuation with the RLC-shunt is always higher than with the RL-shunt with the maximum amplitude of the RL-shunt $\hat{G}_y^{RL} = 29.26\,\text{m/s}^2/\text{V}$ being higher than the maximum amplitude of the RLC-shunt $\hat{G}_y^{RLC} = 4.68\,\text{m/s}^2/\text{V}$. Furthermore, the attenuation deviates less for the RLC-shunt, suggesting that the RLC-shunt is less sensitive to uncertainty in the axial load compared to the RL-shunt [118, 120].

Figure 5.57b shows the histogram of the numerically simulated maximum amplitudes $\hat{G}_y$ with epistemic uncertainty from the axial load, aleatoric uncertainty from the manufacturing and assembly of the piezo-elastic support as well as from the variation of shunt components and transducer parameters for RL- and RLC-shunts, according to [118]. The 95% quantile of the maximum amplitude Q^{95} and the most likely maximum amplitude $\sum$ are used to compare the uncertainty in vibration attenuation when using the RL- and RLC-shunts. The most likely maximum ampli-

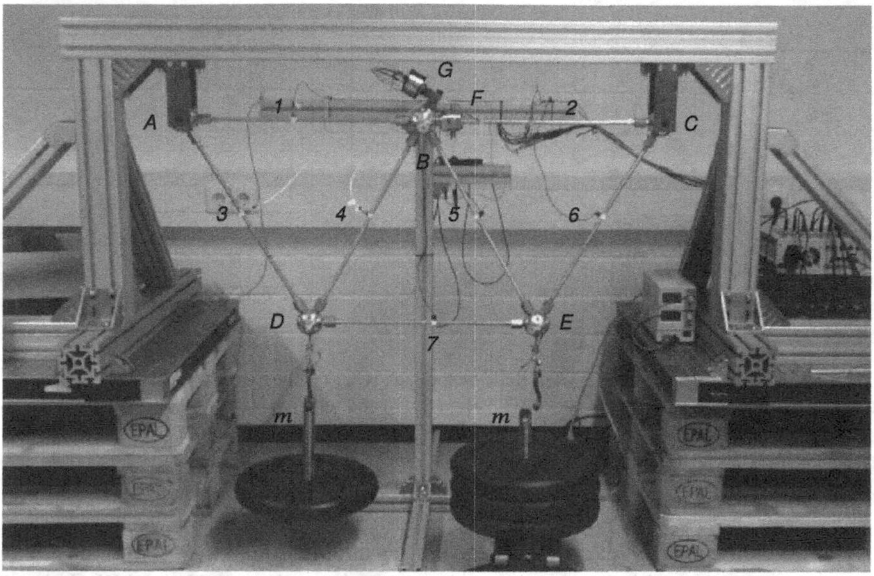

Fig. 5.58 Test setup of the two dimensional beam truss structure

tude $\sum_{RL} = 26.77\,\mathrm{m/s^2/V}$ as well as the 95% quantile of the maximum amplitude $Q_{RL}^{95} = 35.81\,\mathrm{m/s^2/V}$ of the RL-shunt are higher than the RLC-shunt's most likely maximum amplitude $\sum_{RLC} = 5.42\,\mathrm{m/s^2/V}$ and 95% quantile of the maximum amplitude $Q_{RL}^{95} = 35.81\,\mathrm{m/s^2/V}$.

Numerical investigation of piezoelectric shunt-damping in a beam truss structure

After the extensive investigation of piezoelectric shunt-damping in a single beam test setup subject to data uncertainty, piezoelectric shunt-damping with the piezo-elastic support is investigated in the two-dimensional beam truss structure shown in Fig. 5.58. It is derived from the upper truss structure of the MAFDS in Sect. 3.6.1. The beam truss structure comprises seven beams *1* to *7* connected to each other via five spheres *A* to *E* with connector elements. Additional masses *m* can be attached at spheres *D* and *E* to introduce epistemic uncertainty in the static load of the beam truss structure. This results in a variation of the axial loads of the individual beams shifting the structural eigenfrequencies of the two-dimensional beam truss structure, see Fig. 5.56a. The piezo-elastic support *F* connects beam *2* with sphere *B*, where an electrodynamic shaker *G* is attached that dynamically excites the beam truss structure. Seven accelerometers are each positioned slightly off-centred at beams *1* to *7* to determine the local lateral vibrational behaviour.

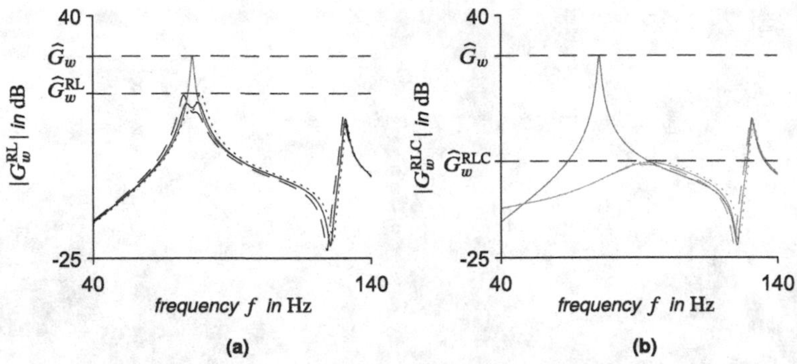

Fig. 5.59 Numerically simulated amplitude responses $|G_\mathrm{w}|$ of the two-dimensional beam truss structure a shunt and additional mass $m_0 = 50\,\mathrm{kg}$ (——) and with **(a)** RL-shunt and **(b)** RLC-shunt for various additional masses $m_- = 0\,\mathrm{kg}$ (----), $m_0 = 50\,\mathrm{kg}$ (——) and $m_+ = 100\,\mathrm{kg}$ (— —) at spheres *D* and *E*, Fig. 5.58

Here, we numerically investigate the vibration attenuation of the two-dimensional beam truss structure with the piezo-elastic support shunted to an RL- and RLC-shunt without additional masses $m_- = 0\,\mathrm{kg}$ as well as with an additional mass $m_0 = 50\,\mathrm{kg}$ and $m_+ = 100\,\mathrm{kg}$ at each sphere *D* and *E* using a finite-element model derived in [184]. Figure 5.59 shows the numerically simulated amplitude response $|G_\mathrm{w}|$ with maximum amplitude $\hat{G}_\mathrm{w} = 29.06\,\mathrm{dB}$ from the excitation with the electrodynamic shaker to the accelerometer at beam 2, Fig. 5.58. The RL- and RLC-shunts are tuned to the two-dimensional beam truss structure's first eigenfrequency at additional mass m_0 and used for all three uncertain load cases. Analogous to the single beam test setup, the maximum amplitude with RL-shunt $\hat{G}_\mathrm{w}^\mathrm{RL} = 19.08\,\mathrm{dB}$ is higher than the RLC-shunt's maximum amplitude $\hat{G}_\mathrm{w}^\mathrm{RL} = 0.81\,\mathrm{dB}$. Furthermore, the vibration attenuation with the RLC-shunt is less sensitive to the uncertain loads than with the RL-shunt.

Conclusion

In this section, we presented piezoelectric shunt-dampings with the piezo-elastic support as a possible (semi-)active process manipulation for vibration attenuation in the single beam shown in Fig. 5.56a, if non-stochastic and stochastic data uncertainty are present [118]. For a two-dimensional beam truss structure, numerical investigations show a successful vibration attenuation with piezoelectric shunt-damping, if data uncertainty is present. In general, we are able to demonstrate that a higher and less sensitive vibration attenuation of a single beam and a beam truss structure are achieved with the RLC-shunt in comparison to the RL-shunt. The numerical results for the beam truss structure will be experimentally validated in the future at the test setup of the two-dimensional beam truss structure, see Fig. 5.58. Furthermore, piezoelectric shunt-damping will be further investigated in the complex structural dynamic system of the MAFDS, Sect. 3.6.1.

5.4.7 Active Buckling Control of Compressively Loaded Beam-Columns and Trusses

Maximilian Schaeffner, Roland Platz, and Tobias Melz

Lightweight mechanical truss structures that comprise slender beam-columns and stiff nodes are commonly used in mechanical engineering applications to distribute both (quasi-)static and dynamic loads within the usage phase, see Sect. 3.1, e.g. in aircraft landing gears or vehicle suspension struts [280]. However, the slender beam-columns are sensitive to failure by buckling when loaded by compressive axial loads F_x and their maximum bearable axial loads $F_{x,\max}$ are considerably reduced by data uncertainty in the material, geometry, loading or support properties [262], see Sect. 2.1. In order to increase the maximum bearable axial load, active buckling control of rather academic beam-column systems with rectangular cross-sections has been investigated numerically and experimentally [25, 93, 94, 196, 261, 275, 286]. In contrast, we apply active buckling control to a practical beam-column system with a circular cross-section in order to demonstrate the mastering of uncertainty in beam-column buckling by active process manipulation, which was motivated in Sect. 3.2 and Sect. 3.4. The beam-column system is later integrated into the Modular Active Spring-Damper System (MAFDS) presented in Sect. 3.6.1. In the following, first the concept of active buckling control and the application to a single beam-column system is presented. Then, the application to a three-dimensional tetrahedron truss structure is discussed.

Active buckling control in a single beam-column system

The concept of active buckling control was experimentally investigated first for the single beam-column system shown in Fig. 5.60a [235, 236]. The beam-column used for active buckling control is a slender beam-column with length $l_\mathrm{b} = 400$ mm and a circular solid cross-section with constant radius $r_\mathrm{b} = 4$ mm. It is made from high-strength aluminium alloy EN AW-7075 with Young's modulus $E_\mathrm{b} = 71$ GPa and density $\varrho_\mathrm{b} = 2850\,\mathrm{kg/m^3}$ to avoid plastic deformation due to beam-column buckling [262]. Four strain gauge sensors in the beam-column centre are used to measure the surface strains due to bending to calculate the deflection of the beam-column in the local y- and z-directions.

The lower and upper beam-column ends are connected to piezo-elastic supports with integrated piezoelectric stack actuators depicted in Fig. 5.60b; these are fixed to a baseplate and a parallel guidance to allow the introduction of compressive axial loads F_x, respectively. The novel concept of the piezo-elastic support was patented in [91] and, other than for active buckling control, is also used for vibration attenuation with shunted piezoelectric transducers, as discussed in Sect. 5.4.6 and [118, 120]. The piezo-elastic supports provide elastic boundary conditions for the beam-column and include the piezoelectric stack actuators, which are mechanically prestressed by

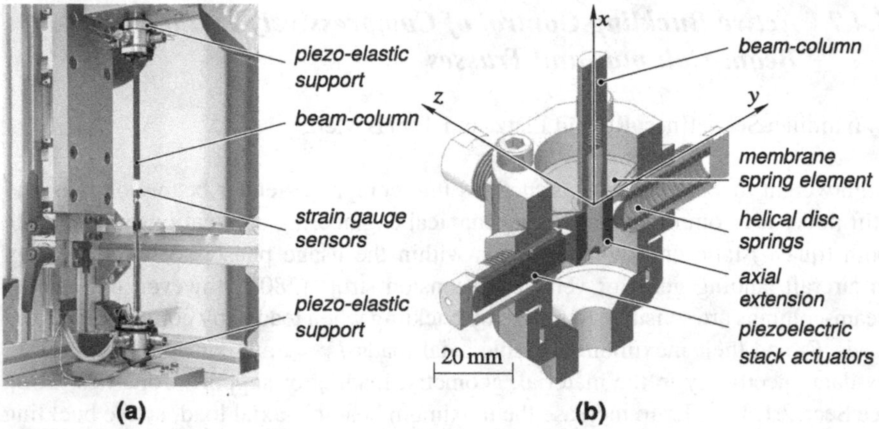

Fig. 5.60 **a** Single beam-column system for active buckling control, **b** sectional view of piezo-elastic support with local x-, y- and z-directions [235]

allocated helical disk springs. The piezoelectric stack actuators are integrated in the lateral load path via axial extensions at a distance l_{ext} from the beam-column ends and exert active lateral forces in the local y- and z-directions, so that they may influence the lateral beam-column deflections. The central element of the piezo-elastic supports are two differently shaped membrane spring elements that are manufactured by a single point incremental forming process with the 3D Servo Press presented in Sects. 3.6.3 and 5.4.1 and [155]. The membrane spring elements bear the axial and lateral loads and allow rotations in any plane perpendicular to the beam-column's local x-axis.

The piezo-elastic supports with integrated piezoelectric stack actuators stabilise the beam-column in arbitrary lateral direction by active bending moments at each end of the beam-column. A linear parameter-varying (LPV) controller, in particular a gain-scheduled $\mathcal{H}_\infty$ controller, which guarantees stability and performance for arbitrary trajectories of dynamic axial loads $F_x(t)$, is used for the active buckling control. For controller design, the beam-column including the piezo-elastic supports is modelled by an axial load-dependent finite element (FE) model for the FE degrees of freedom $r(t)$ with equation of motion and output equation

$$M\,\ddot{r}(t) + D\,\dot{r}(t) + K\big(F_x(t)\big)\,r(t) = B\,u(t) \tag{5.26}$$

$$y(t) = C\,r(t) \tag{5.27}$$

where M is the mass matrix, D is the damping matrix, $K\big(F_x(t)\big)$ is the axial load-dependent stiffness matrix, B is the voltage input matrix and C is the output matrix. The actuator voltages of the piezoelectric stack actuators V_{pz} and the beam-column bending strains ε_{s} are combined in the beam-column input and output vectors

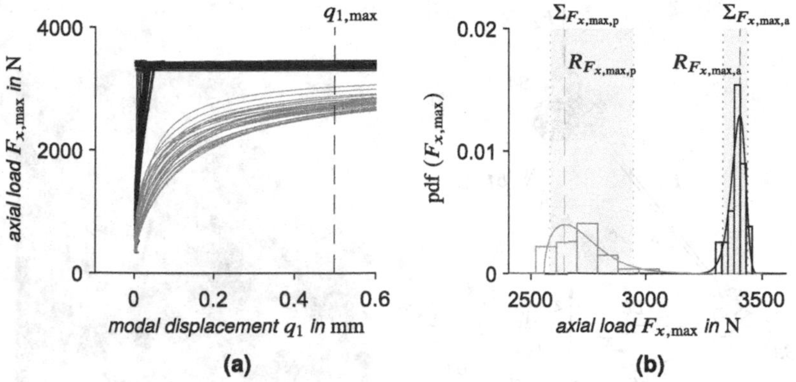

Fig. 5.61 Passive (——) and active (——) beam-column systems with quasi-static axial load $F_x(t)$, **a** absolute modal displacement q_1 versus axial load F_x with maximum admissible displacement $q_{1,\max}$ (— —), **b** normalised histograms and fitted three-parameter WEIBULL distributions $p_W(F_{x,\max})$ of maximum bearable axial loads $F_{x,\max}$ with most likely value $\Sigma_{F_{x,\max}}$ and interpercentile range $R_{F_{x,\max}}$ [235]

$$u(t) = \begin{bmatrix} V_{pz,y}(t) \\ V_{pz,z}(t) \end{bmatrix} \quad \text{and} \quad y(t) = \begin{bmatrix} \varepsilon_{s,y}(t) \\ \varepsilon_{s,z}(t) \end{bmatrix}. \tag{5.28}$$

After Laplace transformation, the 2×2 matrix of transfer functions

$$G(F_x, s) = \frac{y(s)}{u(s)} \tag{5.29}$$

describes the transfer behaviour from the actuator voltages u to the beam-column strains y (5.28) [249].

The most sensitive parameters of the axial load-dependent FE model (5.26) and (5.27) of the beam-column system, which is augmented by the dynamic transfer behaviour of electrical components used for signal conditioning, are calibrated with experimental data and then used to design the LPV controller. Subsequently, the passive (without controller) and active (with controller) beam-column systems are loaded by quasi-static and dynamic axial loads $F_x(t)$ and the absolute lateral modal displacement $q_1(t)$ is measured. Figure 5.61a shows the experimental load-displacement curves of passive buckling and active buckling control for the single beam-column system in Fig. 5.60a subjected to a slowly increasing quasi-static axial load $F_x(t)$. We conducted the experiments with a representative sample of 30 nominally identical passive and active beam-column systems in order to quantify stochastic data uncertainty in the maximum bearable axial loads according to the classification of uncertainty presented in Chap. 2.

Normalised histograms and three-parameter WEIBULL distribution fits $p_W(F_{x,\max})$ for the maximum bearable loads $F_{x,\max}$ of the passive and active beam-column systems are shown in Fig. 5.61b. For the active beam-column system, the most likely

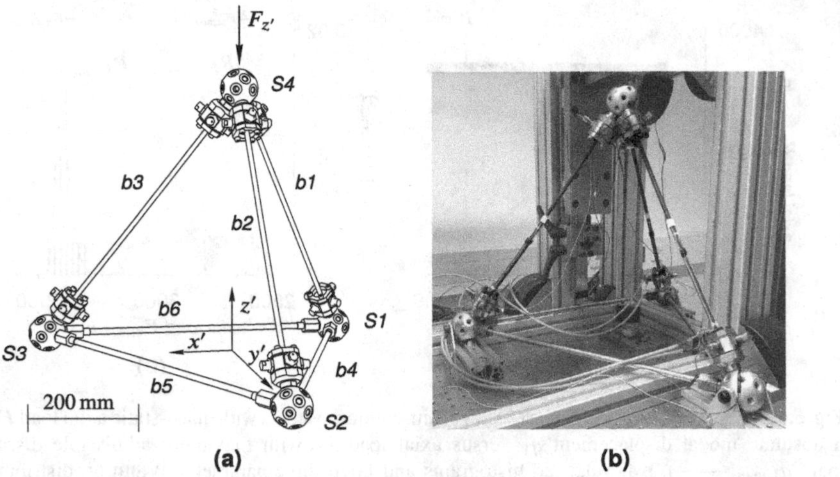

Fig. 5.62 Experimental tetrahedron truss structure, **a** CAD sketch of test setup with global x'-, y'- and z'-directions, **b** photo of test setup

maximum bearable axial load $\Sigma_{F_{x,\max}}$ increases by 29% and the variability expressed by the interpercentile range $R_{F_{x,\max}}$ reduces by 70% in comparison to the passive beam-column system. Thus, stochastic uncertainty in the maximum bearable axial loads is mastered by active process manipulation using active buckling control for the single beam-column system.

Active buckling control in a tetrahedron truss structure

As an intermediate step to the integration of active buckling control in the MAFDS, which represents a realistic load-bearing structure, Sect. 3.6.1, we investigate active buckling control in an experimental three-dimensional tetrahedron truss structure. Figure 5.62 shows the experimental tetrahedron truss structure that comprises three beam-columns *b1–b3* with piezo-elastic supports, in the following called active beam-columns, and three passive beams *b4–b6*. They are connected to each other via the spheres *S1–S4*, where the spheres *S1–S3* are clamped and sphere *S4* is free and may be used to introduce vertical compressive loads $F_{z'}$ into the tetrahedron truss structure in global z'-direction.

The lateral dynamic behaviour of the three active beam-columns, which is essential for the model-based controller synthesis for active buckling control, was investigated in [237]. Here, we used the experimental beam-column transfer functions (5.29) of the unloaded active beam-columns *b1–b3* in the tetrahedron truss structure in Fig. 5.62 to calibrate the FE model of the single beam-column (5.26) and (5.27). The lateral dynamic behaviour of all active beam-columns *b1–b3* are very similar and their boundary conditions in the tetrahedron truss structure are adequately described by the numerical beam-column transfer function (5.29) originating from the single beam-column test setup in Fig. 5.60a. Thus, no additional model of the tetrahedron truss structure is necessary and the calibrated FE beam-column models will be used for the model-based controller synthesis as well as active buckling control of the tetrahedron truss structure in future investigations.

In conclusion, active buckling control may be used to master uncertainty in the maximum bearable load of mechanical load-bearing structures prone to buckling. The effectiveness of this form of active process manipulation was shown for a single beam-column system subject to quasi-static and dynamic axial loads. Furthermore, it is currently tested for a tetrahedron truss structure and prepared for the integration in the MAFDS.

5.4.8 Load Redistribution Via Semi-active Guidance Elements in a Kinematic Structure

Christopher M. Gehb, Roland Platz, and Tobias Melz

In many mechanical engineering applications, withstanding external loads is one of the key tasks within the usage phase, see Sect. 3.1. In most cases, the load is transmitted from one or more points of load application via a predetermined load path to the structural supports. Additionally, defined kinematics are often an important part of the functional performance in a load-bearing structure to enable a specified relative displacement of structural components. An example is the compression stroke of a landing gear or a suspension strut in airplanes or vehicles. A spring-damper often determines the main kinetic properties [183]. The compression stroke, in turn, is enabled by guidance elements, such as torque-links or other suspension links [65, 152]. In general, the guidance elements kinematically connect two or more parts of a load-bearing structure in order to achieve defined relative displacements. In most cases, the load path going through the load-bearing structure is predetermined in the design phase and, mostly, is not subject to any changes during the structure's lifetime [218]. However, if parts of the load-bearing structure become weak or suffer damage, e.g. due to deterioration or overload, the load capacity may become lower than designed leading to uncertainty [112]. In this case, semi-active process manipulation, as introduced in Sects. 3.2 and 3.4, in form of load redistribution can be an option to master this uncertainty by adjusting the load path and, thus, reducing the effects of damage or prevent further damage, compare Sect. 3.6.1. Also, a desired support reaction force ratio achieved by load redistribution during operation might be useful if the predetermined load path is not suitable anymore [115]. In the following, first the concept of load redistribution and the investigated test setup are presented. Then, semi-active guidance elements for load redistribution are introduced. Finally, numerical and experimental results are discussed.

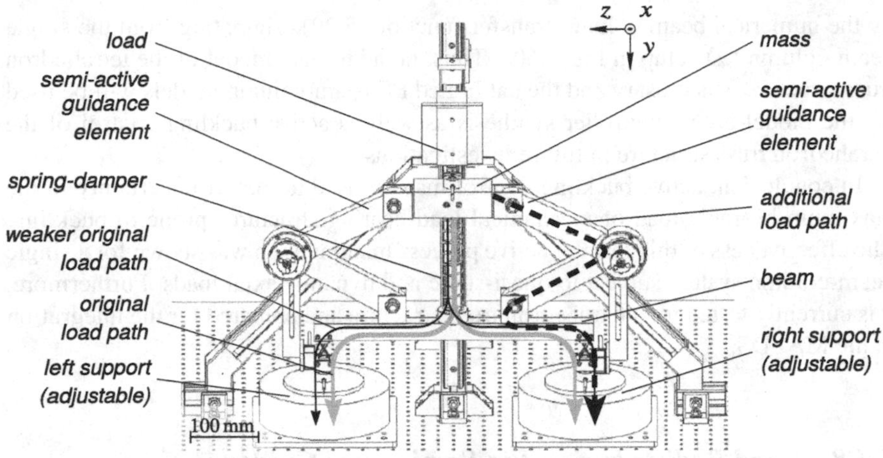

Fig. 5.63 Semi-active load redistribution test setup with original and additional load path via the right semi-active guidance element [113]

Concept of load redistribution

Figure 5.63 depicts a schematic representation of an exemplary load-bearing structure to demonstrate load redistribution, compare Sect. 4.1.2. It represents a simplified surrogate version of the MAFDS, Sect. 3.6.1. A spring-damper and two semi-active guidance elements connect a mass and a beam forming a load-bearing structure. The beam distributes the load to two supports at its ends. Varying support stiffness simulates varying load capacity in an academic and reproducible way and is a manifestation of data uncertainty, see Sect. 2.1. Load redistribution according to [114, 115] entails the redistribution of loads between the left and right support in case of present or anticipated damage. The loads previously passed through the spring-damper solely (grey line in Fig. 5.63) are partly redistributed via the controlled semi-active guidance elements. If, for example, the left support is assumed to be damaged or weak, it can be relieved via the right semi-active guidance element providing a load path in addition to the spring-damper. Hence, parts of the loading are bypassed through the right semi-active guidance element towards the undamaged right support. The additional load path is depicted in Fig. 5.63 by a dashed black line passing through the right semi-active guidance element compared to a thin black line indicating the corresponding weaker load path through the spring-damper. Load redistribution during operation can be attributed as semi-active process manipulation and is part of Structural Health Control SHC [194]. SHC combines structural health monitoring, assessing the structural health condition, e.g. the load-bearing capacity, and an adequate semi-active or active process manipulation in order to load or unload load-bearing components, compare Sect. 3.4.

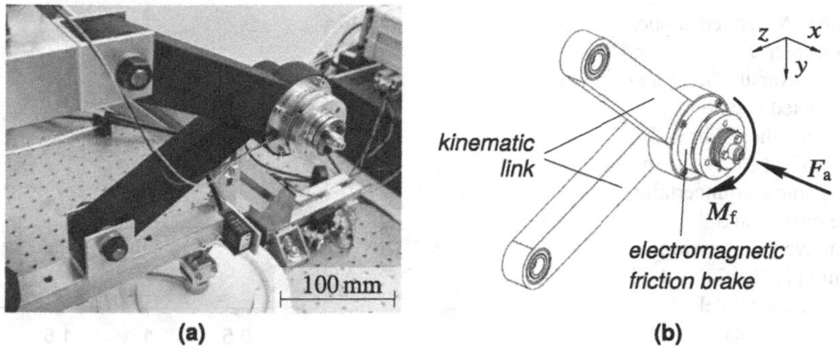

Fig. 5.64 Semi-active guidance element, **a** realisation and **b** CAD illustration according to [113]

Semi-active guidance elements

Figure 5.64a depicts a close-up view of the right semi-active guidance element mounted to the test setup, compare Sect. 4.1.2. The already existing components of the load-bearing structure, i.e. the guidance elements, are augmented with electromagnetic friction brakes for semi-active process manipulation to adapt the structure's load path via an induced friction moment. Figure 5.64b depicts a CAD illustration of the semi-active guidance element. It consists of two kinematic links connected by a middle joint, which is equipped with an electromagnetic friction brake. Within the friction brake, the electromagnetically induced normal force F_a acts on the friction lining in negative x-direction and is controlled via a controlled input voltage applied to the brake's electromagnet. The normal force F_a, in turn, causes a friction moment M_f and changes the load transmitting properties of the semi-active guidance element by increasing or decreasing the possible amount of the loading passing through the joint. The mathematical relation between the electromagnetic normal force F_a and the related friction moment M_f is presented in [112, 113] applying the LuGre friction model.

Numerical and experimental results of load redistribution

The uncertainty within the load path, assumed to be a support with reduced stiffness, is supposed to be mastered by means of semi-active load redistribution. The supports' stiffness is adjustable for the experiments in order to introduce uncertainty in a repeatable and measurable way. Thereby, we can numerically and experimentally investigate and evaluate the load redistribution capability within the exemplary load-bearing structure from Sect. 4.1.2. The stiffness reduction, which is achieved via the adjustable supports in Fig. 5.63 according to [144], causes uncertain dynamic behaviour and uncertain load capacity. In our example, this is misalignment of the beam and undesired support reaction force ratio among the two supports, for details see [112]. Here, we exemplarily present the resulting load path when trying to reduce

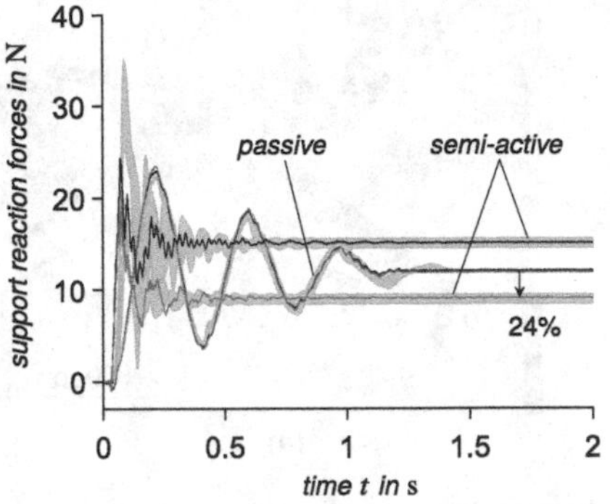

Fig. 5.65 Measured support reaction force F_L (——) and F_R (——) versus time t and the predicted uncertainty ranges with the calibrated math. model ( ) due to model parameter uncertainty for the passive and semi-active load-bearing structure [112], cf. Fig. 4.7 in Sect. 4.1.2 for model calibration results

the undesired misalignment caused by a damaged support with reduced stiffness. The load path is evaluated by means of the left and right support reaction forces F_L and F_R. A comprehensive case study comprising different control strategies and different levels of damage is presented in [112].

Figure 5.65 depicts the simulated and measured load path of the load-bearing structure due to an external force excitation by the dropped load mass resulting in a step load applied to the mass, compare Fig. 5.63. A time series of two seconds is simulated and measured to analyse the load path and load redistribution capability with and without semi-active control. For the numerical results, we consider the remaining model prediction uncertainty by conducting Monte Carlo (MC) simulations with the calibrated parameter ranges, compare Sect. 4.1.2. Instead of ignoring the uncertainty by stating only one deterministic curve, the uncertainty in the model prediction caused by the remaining parameter uncertainty is indicated by shaded areas in Fig. 5.65. Solid lines represent the measured load paths, averaged for 10 measurements [112].

For the experimental results in Fig. 5.65, the excitation load is equally distributed to both supports in case of no semi-active load redistribution. This leads to almost identical measured support reaction forces. In case of a support with reduced stiffness and, hence, load-bearing capacity, the load redistribution results from the semi-active guidance elements. The excitation load is no longer equally distributed to the supports, but depends on the support stiffness. The steady state load reduction of the assumed to be damaged right support is about 24%. The shaded area of the MC simulation results show similar dynamic behaviour for the entire time scale. Thus, the numerical results widely encompass the experimental results for both support reaction forces with and without semi-active load redistribution. The measured steady state left and right support reaction forces F_L and F_R are within the range of the model predictions. Taking into account the overall time scale, the load paths and the load redistribution capability is predicted sufficiently accurate using the calibrated parameter ranges from Sect. 4.1.2 and considering the remaining parameter uncertainty.

The load is partly redistributed towards the undamaged support and the damaged support is relieved. Since we cannot redistribute loads unlimitedly, there remains a steady state control deviation for the numerical and experimental results. This is most probably bot not exclusively due to the limitations of the semi-active approach, which cannot introduce energy into the structure via the electromagnetic friction brakes to completely eliminate the control deviation, compare Sect. 3.4. Overall, a successful semi-active process manipulation with significant load redistribution is numerically and experimentally proved [112, 114, 115].

For a more practical application of load redistribution, the concept of semi-active guidance elements as shown in Fig. 5.64 is transferred into up-scaled semi-active guidance elements for the MAFDS in future work, see Sect. 3.6.1. Thus, the planar load redistribution problem transforms into a spatial load redistribution problem. The load path of the MAFDS will become adaptable in order to master uncertainty in the spatial structure application.

References

1. Abele E, Bölling C (2014) Modellierung der geometrischen Eingriffsbedingungen bei der Ventilsitzbearbeitung am Zylinderkopf. wt Werkstattstechnik online 104(H. 11/12):735–740
2. Abele E, Geßner F (2018) Spanungsquerschnittmodell zum Gewindebohren. wt Werkstattstechnik. Online 108(1–2):2–6
3. Abele E, Hauer T, Haydn M (2011) Modellierung der Prozesskräfte beim Reiben mit Mehrschneidenreibahlen—Implementierung eines Kraftmodells für die Simulation der Reibbearbeitung. wt Werkstattstechnik online 101(6):407–412
4. Ackoff RL (1989) From data to wisdom. J Appl Syst Anal 16(1):3–9
5. Affenzeller J, Gläser H (1996) Lagerung und Schmierung von Verbrennungsmotoren, Die Verbrennungskraftmaschine Neue Folge, vol 8. Springer, Vienna. https://doi.org/10.1007/978-3-7091-6568-3
6. Al-Baradoni N, Krech M, Groche P (2020) In-process calibration of smart structures produced by incremental forming. In: Production engineering, pp 1–9. https://doi.org/10.1007/s11740-020-01002-6
7. Ali S, Hinduja S, Atkinson J, Bolt P, Werkhoven R (2008) The effect of ultra-low frequency pulsations on tearing during deep drawing of cylindrical cups. Int J Mach Tools Manuf 48(5):558–564. https://doi.org/10.1016/j.ijmachtools.2007.06.013
8. Allwood JM, Duncan SR, Cao J, Groche P, Hirt G, Kinsey BL, Kuboki T, Liewald M, Sterzing A, Tekkaya AE (2016) Closed-loop control of product properties in metal forming. CIRP Ann 65(2):573–596. https://doi.org/10.1016/j.cirp.2016.06.002
9. Altherr LC, Joggerst L, Leise P, Pfetsch ME, Schmitt A, Wendt J (2019) On obligations in the development process of resilient systems with algorithmic design methods. In: Pelz PF, Groche P (eds) Uncertainty in mechanical engineering III, applied mechanics and materials, vol 885. Trans Tech Publications, pp 240–252. https://doi.org/10.4028/www.scientific.net/AMM.885.240
10. Archibald RD (1994) Uncertainty and change require a new management model. PM Netw. 8(5):6–10
11. Arentoft M, Eriksen M, Wanheim T (2000) Determination of six stiffnesses for a press. J Mater Process Technol 105(3):246–252. https://doi.org/10.1016/s0924-0136(00)00559-8
12. Arentoft M, Wanheim T (2005) A new approach to determine press stiffness. CIRP Ann 54(1):265–268. https://doi.org/10.1016/s0007-8506(07)60099-7
13. Bates H, Holweg M, Lewis M, Oliver N (2007) Motor vehicle recalls: Trends, patterns and emerging issues. Omega 35(2):202–210

14. Bauer B, Devroye L, Kohler M, Krzyżak A, Walk H (2017) Nonparametric estimation of a function from noiseless observations at random points. J Multivar Anal 160:93–104. https://doi.org/10.1016/j.jmva.2017.05.010
15. Bauer B, Heimrich F, Kohler M, Krzyżak A (2019) On estimation of surrogate models for multivariate computer experiments. Ann Inst Stat Math 71(1):107–136. https://doi.org/10.1007/s10463-017-0627-8
16. Bauer W (2016) Planung und Entwicklung änderungsrobuster Plattformarchitekturen. Dissertation, TU München
17. Baumberger GC (2007) Methoden zur kundenspezifischen Produktdefinition bei individualisierten Produkten. Produktentwicklung. Verl. Dr. Hut, München
18. Becerril L, Sauer M, Lindemann U (2016) Estimating the effects of engineering changes in early stage product development. In: DSM 2016: sustainability in modern project management—proceedings of the 18th international DSM conference, pp 125–135. https://doi.org/10.19255/JMPM-DSM2016
19. Beck U (1986) Risikogesellschaft: Auf dem Weg in eine andere Moderne. Suhrkamp, Frankfurt a. M
20. Bedarff T, Hedrich P, Pelz PF (2014) Design of an active air spring damper. In: Murrenhoff H (ed) 9th international fluid power conference (9th IFK). HP - Fördervereinigung Fluidtechnik, Aachen, pp 356–365
21. Beder S (1998) The new engineer: management and professional responsibility in a changing world. Macmillan Education
22. Beetz JP (2018) Modelle und Methoden zur systematischen Entwicklung hygienegerechter Produkte. Dissertation, TU Darmstadt
23. Behrens BA, Yilkiran T, Vahed N, Frischkorn C (2014) Einfluss des Materialflusses auf die Verdichtung beim Sinterschmieden: Pulvermetallurgie und Umformtechnik. Whitepaper, Umformtechnik.net. https://umformtechnik.net/content/location/72775
24. Bender M (2016) Forschungsumgebungen in den Digital Humanities: Nutzerbedarf, Wissenstransfer, Textualität, Sprache und Wissen, vol 22. De Gruyter, Berlin
25. Berlin AA, Chase JG, Yim MH, Maclean JB, Olivier M, Jacobsen SC (1998) MEMS-based control of structural dynamic instability. J Intell Mater Syst Struct 9(7):574–586. https://doi.org/10.1177/1045389X9800900709
26. BGH: Urteil vom 02. März 2010—VI ZR 223/09
27. BGH: Urteil vom 02. Oktober 2012—VI ZR 311/11
28. BGH: Urteil vom 07. Oktober 1975—VI ZR 43/74
29. BGH: Urteil vom 08. November 2005—VI ZR 332/04
30. BGH: Urteil vom 09. Dezember 1986—VI ZR 65/86
31. BGH: Urteil vom 09. September 2008—VI ZR 279/06
32. BGH: Urteil vom 16. Juni 2009—VI ZR 107/08
33. BGH: Urteil vom 17. Juni 1997—VI ZR 156/96
34. BGH: Urteil vom 17. März 1981—VI ZR 286/78
35. BGH: Urteil vom 17. März 2009—VI ZR 176/08
36. BGH: Urteil vom 26. November 1968—VI ZR 212/66
37. BGH: Urteil vom 28. April 1952—III ZR 118/51
38. Bhattacharyya O, Kapoor SG, DeVor RE (2006) Mechanistic model for the reaming process with emphasis on process faults. Int J Mach Tools Manuf 46(7–8):836–846. https://doi.org/10.1016/j.ijmachtools.2005.07.022
39. Bilz Werkzeugfabrik GmbH and Co. KG: HFP—wireless monitoring for multispindle tapping (2020). https://www.bilz.de/produkte/messtechnik-smarttools/prozessueberwachung-fuer-gewindebearbeitung. (Visited 12/2020)
40. Birkhofer H (2011) The future of design methodology. Springer, Berlin
41. Birkhofer H, Waeldele M (2008) Properties and characteristics and attributes and...—an approach on structuring the description of technical systems. In: DS 57: Proceedings of AEDS 2008 workshop

42. Böhle F, Heidling E, Neumer J, Kuhlmey A, Winnig M, Trobisch N, Kraft D, Denisow K (2016) Umgang mit Ungewissheit in Projekten. Technical Report, GPM Deutsche Gesellschaft für Projektmanagement e. V., Nuremberg
43. Boller C (2001) Composites for sensors and actuators. In: Buschow KJ, Cahn RW, Flemings MC, Ilschner B, Kramer EJ, Mahajan S, Veyssière P (eds) Encyclopedia of materials: science and technology. Elsevier, Oxford, pp 1376–1382. https://doi.org/10.1016/B0-08-043152-6/00256-4
44. Bölling C (2019) Simulationsbasierte Auslegung mehrstufiger Werkzeugsysteme zur Bohrungsfeinbearbeitung am Beispiel der Ventilführungs- und Ventilsitzbearbeitung. Dissertation, TU Darmstadt. http://tubiblio.ulb.tu-darmstadt.de/111353/
45. Bölling C, Güth S, Abele E (2015) Control of uncertainty in high precision cutting processes: reaming of valve guides in a cylinder head of a combustion engine. In: Pelz PF, Groche P (eds) Uncertainty in mechanical engineering II, applied mechanics and materials, vol 807. Trans Tech Publications, pp 153–161
46. Bölling C, Hoppe F, Geßner F, Knoll M, Abele E, Groche P (2018) Fortpflanzung von Unsicherheit in Prozessketten. wt Werkstattstechnik Online 108(1–2):82–88
47. Bonnardel N, Marmèche E (2005) Towards supporting evocation processes in creative design: A cognitive approach. Int J Hum Comput Stud 63(4–5):422–435
48. Brenneis M, Groche P (2012) Smart components through rotary swaging. In: Material forming ESAFORM 2012, key engineering materials, vol 504. Trans Tech Publications, pp 723–728. https://doi.org/10.4028/www.scientific.net/KEM.504-506.723
49. Brenneis M, Ibis M, Duschka A, Groche P (2014) Towards mass production of smart products by forming technologies. In: WGP congress 2012, advanced materials research, vol 907. Trans Tech Publications, pp 113–125. https://doi.org/10.4028/www.scientific.net/AMR.907.113
50. Brenner DP (2014) Einsatzmöglichkeiten der zusätzlichen Stößelfreiheitsgrade der 3D-Servo-Presse in Prozessen der Pulvermetallurgie: Potential applications using the additional degrees of freedom of the 3d-servo-press in powder metallurgy processes. Diplomarbeit, TU Darmstadt
51. Breunig A, Hoppe F, Groche P (2019) Localized blank-holder pressure control in cup drawing through tilting of the ram. In: Proceedings of NUMIFORM, pp 583–586
52. Brötz N, Hedrich P, Pelz PF (2018) Integrated fluid dynamic vibration absorber for mobile applications. In: 11th international fluid power conference, vol 1, pp 14–24
53. Brötz N, Pelz PF (2020) Bayesian uncertainty quantification in the development of a new vibration absorber technology. In: IMAC XXXVIII, vol 1
54. Budiansky B, Hutchinson JW, Slutsky S (1982) Void growth and collapse in viscous solids. In: Hopkins HG (ed) Mechanics of solid. Pergamon Press, Oxford, pp 13–45. https://doi.org/10.1016/B978-0-08-025443-2.50009-4
55. Bührig-Polaczek A (ed) (2013) Handbuch Urformen: Handbuch der Fertigungstechnik, second edn. Hanser, München. https://doi.org/10.3139/9783446434066
56. Calmano S (2017) Methodik zur Regelung von Bauteileigenschaften in Umformprozessen. Dissertation, TU Darmstadt
57. Calmano S, Hesse D, Hoppe F, Groche P (2015) Evaluation of control strategies in forming processes. In: Qin Y, Dean TA, Lin J, Yuan SJ, Vollertsen F (eds) Proceedings of the 4th international conference on new forming technology (ICNFT), Article 04002. https://doi.org/10.1051/matecconf/20152104002
58. Calmano S, Hesse D, Hoppe F, Traidl P, Sinz J, Groche P (2015) Orbital forming of flange parts under uncertainty. In: Pelz PF, Groche P (eds) Uncertainty in mechanical engineering II, applied mechanics and materials, vol 807. Trans Tech Publications, pp 121–129. https://doi.org/10.4028/www.scientific.net/AMM.807.121
59. Calmano S, Schmitt S, Groche P (2013) Prevention of over-dimensioning in light-weight structures by control of uncertainties during production. In: International conference on "New Developments in Forging Technology". MAT INFO Werkstoff-Informationsgesellschaft mbH, Fellbach, pp 313–317

60. Chen YB, Sha JL, Wu SM (1990) Diagnosis of the tapping process by information measure and probability voting approach. J Eng Ind 112(4):319–325. https://doi.org/10.1115/1.2899594
61. Chodnikiewicz K, Balendra R (2000) The calibration of metal-forming presses. J Mater Process Technol 106(1–3):28–33. https://doi.org/10.1016/s0924-0136(00)00633-6
62. Chucholowski N, Langer S, Ferreira M, Forcellini F, Maier A (2012) Engineering change management report 2012: Survey results on causes and effects, current practice, problems, and strategies in Brazil. Universidade Federal de Santa Catarina
63. Clark KB, Fujimoto T (1991) Product development performance: Strategy, organization, and management in the world auto industry. Harvard Business School Press
64. Conrat Niemerg JI (1997) Änderungskosten in der Produktentwicklung. Dissertation, TU München
65. Currey NS (1988) Aircraft landing gear design: Principles and practices, 4th edn. AIAA education series. American Institute of Aeronautics and Astronautics, Washington, DC
66. Deutscher Bundestag: Drucksache 17/6276 (2001)
67. Deutsches Institut für Normung (1979) DIN 199–4. Begriffe im Zeichnungs- und Stücklistenwesen - Zeichnungen. Beuth, Berlin
68. Deutsches Institut für Normung (1988) DIN 55189-1:1988-12. Werkzeugmaschinen; Ermittlung von Kennwerten für Pressen der Blechverarbeitung bei statischer Belastung; Mechanische Pressen [Machine tools; determination of the ratings of presses for sheet metal working under static load; mechanical presses]. Beuth, Berlin
69. Deutsches Institut für Normung (2009) DIN 69901-5:2009-01. Projektmanagement – Projektmanagementsysteme – Teil 5: Begriffe. Beuth, Berlin. https://doi.org/10.31030/1498911
70. Deutsches Institut für Normung (2011) DIN EN ISO 376:2011–09 Metallische Werkstoffe—Kalibrierung der Kraftmessgeräte für die Prüfung von Prüfmaschinen mit einachsiger Beanspruchung [Metallic materials—Calibration of force-proving instruments used for the verification of uniaxial testing machines]. Beuth, Berlin
71. Deutsches Institut für Normung (2012) DIN 1988-200:2012-05. Technische Regeln für Trinkwasser-Installationen – Teil 200: Installation Typ A (Geschlossenes System) – Planung, Bauteile, Apparate, Werkstoffe; Technische Regel des DVGW. Beuth, Berlin. https://doi.org/10.31030/1887421
72. Deutsches Institut für Normung (2014) DIN IEC 60050–351:2014–09 Internationales Elektrotechnisches Wörterbuch: Teil 351: Leittechnik [International electrotechnical vocabulary - Part 351: Control technology]. Beuth, Berlin
73. Deutsches Institut für Normung: DIN EN 62198, VDE 0050-6. Managing risk in projects – Application guidelines. Beuth, Berlin (2014-08)
74. Deutsches Institut für Normung (2015) DIN 60812. Fehlzustandsart- und -auswirkungsanalyse (FMEA): (IEC 56/1579/CD:2014), vol. 03.120.01, 21.020, 29.020. Beuth, Berlin. https://doi.org/10.31030/2328411
75. Deutsches Institut für Normung (2017) DIN 31652-1. Gleitlager – Hydrodynamische Radial-Gleitlager im stationären Betrieb: Teil 2: Funktionen für die Berechnung von Kreiszylinderlagern. https://doi.org/10.31030/2588236
76. Deutsches Institut für Normung (2020) DIN 820-2. Standardization – Part 2: Presentation of documents (ISO/IEC Directives – Part 2:2018, modified). 820-2. Beuth, Berlin
77. Devroye L (1986) Non-uniform random variate generation. Springer, New York
78. Doege E, Behrens BA (2010) Handbuch Umformtechnik: Grundlagen, Technologien, Maschinen, 2nd edn. Springer, Berlin, Heidelberg. https://doi.org/10.1007/978-3-642-04249-2
79. Doege E, Lange K (1980) Static and dynamic stiffness of presses and some effects on the accuracy of workpieces. CIRP Ann 29(1):167–171. https://doi.org/10.1016/s0007-8506(07)61316-x
80. Doege E, Silberbach G (1990) Influence of various machine tool components on workpiece quality. CIRP Ann 39(1):209–213. https://doi.org/10.1016/s0007-8506(07)61037-3
81. Dowlatshahi S (2000) Designer-buyer-supplier interface: Theory versus practice. Int J Prod Econ 63(2):111–130

82. Drechsler S (2016) Kompetenzbedarfe von Maschinenbauingenieuren in Bezug auf Richtlinien, Normen und Standards zur Ausübung ihrer beruflichen Tätigkeit. Dissertation, Karlsruher Institut für Technologie (KIT). https://doi.org/10.5445/IR/1000059347
83. Dsuban A, Lohn J, Brüggemann JP, Kullmer G: Ein Qualitätssicherungskonzept für die additive Fertigung. In: H.A. Richard, B. Schramm, T. Zipsner (eds.) Additive Fertigung von Bauteilen und Strukturen, pp. 23–34. Springer, Wiesbaden and Cham (2019). DOI https://doi.org/10.1007/978-3-658-27412-2_2
84. Eckert C, Clarkson PJ, Zanker W (2004) Change and customisation in complex engineering domains. Res Eng Des 15(1):1–21
85. Ehrlenspiel K, Meerkamm H (2013) Integrierte Produktentwicklung: Denkabläufe, Methodeneinsatz. Hanser, Zusammenarbeit
86. Eifler T (2014) Modellgestützte Methodik zur systematischen Analyse von Unsicherheit im Lebenslauf technischer Systeme. Dissertation, TU Darmstadt
87. Eifler T, Enss GC, Haydn M, Mosch L, Platz R, Hanselka H (2012) Approach for a consistent description of uncertainty in process chains of load carrying mechanical systems. In: Hanselka H, Groche P, Platz R (eds) Uncertainty in mechanical engineering, applied mechanics and materials, vol 104. Trans Tech Publications, pp 133–144. https://doi.org/10.4028/www.scientific.net/AMM.104.133
88. EMUGE-Werk Richard Glimpel GmbH & Co. KG (2019) Softsynchro 3. https://www.emuge.com/sites/default/files/literature/Emuge-Catalog-520.pdf
89. Engelhardt R (2012) Uncertainty mode and effects analysis—Heuristische Methodik zur Analyse und Beurteilung von Unsicherheiten in technischen Systemen des Maschinenbaus. Dissertation, TU Darmstadt (2012)
90. Engelhardt R, Birkhofer H, Kloberdanz H, Mathias J (2009) Uncertainty-mode-and effects-analysis—an approach to analyze and estimate uncertainty in the product life cycle. In: Norell Bergendahl M (ed) Design theory and research methodology. Design Society, Glasgow, pp 191–202. https://www.designsociety.org/publication/28585/Uncertainty-Mode-+and+Effects-Analysis+ Visited 12/2020
91. Enss GC, Gehb CM, Götz B, Melz T, Ondoua S, Platz R, Schaeffner M (2015) Device for bearing design elements in lightweight structures (Festkörperlager). Patent DE 10 2015 101 084 A1
92. Enss GC, Kohler M, Krzyżak A, Platz R (2016) Nonparametric quantile estimation based on surrogate models. IEEE Trans Inform Theory 62(10):5727–5739. https://doi.org/10.1109/TIT.2016.2586080
93. Enss GC, Platz R (2016) Evaluation of uncertainty in experimental active buckling control of a slender beam-column with disturbance forces using Weibull analysis. Mech Syst Signal Process 79:123–131. https://doi.org/10.1016/j.ymssp.2016.02.066
94. Enss GC, Platz R, Hanselka H (2012) Uncertainty in loading and control of an active column critical to buckling. Shock Vib 19(5):929–937. https://doi.org/10.3233/SAV-2012-0700
95. European Commission: Commission Implementing Decision (EU) 2019/417 of 8 November 2018 laying down guidelines for the management of the European Union Rapid Information System 'RAPEX' established under Article 12 of Directive 2001/95/EC on general product safety and its notification system – C(2018) 7334
96. European Commission: Proposal for a Regulation of the European Parliament and of the Council on consumer product safety and repealing Council Directive 87/357/EEC and Directive 2001/95/EC: COM(2013) – 78 final (Feb. 13, 2013)
97. European Parliament, Council of the European Union: Directive 2006/42/EC of the European Parliament and of the council of 17 may 2006 on machinery, and amending directive 95/16/EC. Official Journal of the European Union (2006). URL https://eur-lex.europa.eu/eli/dir/2006/42/oj
98. Faleskog J, Gao X, Shih CF (1998) Cell model for nonlinear fracture analysis - I. Micromechanics calibration. Int J Fracture 89(4):355–373. https://doi.org/10.1023/A:1007421420901
99. Fang KT, Li R, Sudjianto A (2006) Design and modeling for computer experiments. Chapman & Hall/CRC, Boca Raton, FL

100. Fankhauser P, Kupietz M (2017) DeReKoVecs. http://corpora.ids-mannheim.de/openlab/derekovecs/. (Visited 12/2020)
101. Felber T, Kohler M, Krzyżak A (2015) Adaptive density estimation from data with small measurement errors. IEEE Trans Inform Theory 61(6):3446–3456. https://doi.org/10.1109/TIT.2015.2421297
102. Feldhusen J, Grote KH (eds) (2013) Pahl/Beitz Konstruktionslehre: Methoden und Anwendung erfolgreicher Produktentwicklung, eighth edn. Springer Vieweg, Berlin, Heidelberg. https://doi.org/10.1007/978-3-642-29569-0
103. Foerste U (2012) §24. In: Foerste U, Westphalen F (eds) Produkthaftungshandbuch. Beck, München
104. Forward RL (1979) Electronic damping of vibrations in optical structures. Appl Opt 18(5):690–697
105. Frame JD (1994) The new project management: tools for an age of rapid change, complexity, and other business realities. Jossey-Bass, Chichester
106. Freeman RE, McVea J (2001) A stakeholder approach to strategic management. The Blackwell Handbook of Strategic Management pp. 189–207 (2001)
107. Fricke E (1998) Der Änderungsprozeß als Grundlage einer nutzerzentrierten Systementwicklung. Dissertation, TU München, München
108. Fricke E, Gebhard B, Negele H, Igenbergs E (2000) Coping with changes: causes, findings, and strategies. Syst Eng 3(4):169–179
109. Friedman M (2007) The social responsibility of business is to increase its profits. In: W.C. Zimmerli, M. Holzinger, K. Richter (eds.) Corporate Ethics and Corporate Governance. Springer, Berlin, Heidelberg, pp 173–178. https://doi.org/10.1007/978-3-540-70818-6_14
110. Funtowicz SO, Ravetz JR (1993) Science for the post-normal age. Futures 25(7):739–755
111. Gantar G, Kuzman K, Filipič B (2005) Increasing the stability of the deep drawing process by simulation-based optimization. J Mater Process Technol 164–165:1343–1350. https://doi.org/10.1016/j.jmatprotec.2005.02.099
112. Gehb CM (2019) Uncertainty evaluation of semi-active load redistribution in a mechanical load-bearing structure. Dissertation, TU Darmstadt
113. Gehb CM, Atamturktur S, Platz R, Melz T (2020) Bayesian inference based parameter calibration of the LuGre-friction model. Exp Tech 44:369–382. https://doi.org/10.1007/s40799-019-00355-7
114. Gehb CM, Platz R, Melz T (2017) Global load path adaption in a simple kinematic load-bearing structure to compensate uncertainty of misalignment due to changing stiffness conditions of the structure's supports. In: Barthorpe RJ, Platz R, Lopez I, Moaveni B, Papadimitriou C (eds) Model validation and uncertainty quantification, Volume 3, conference proceedings of the society for experimental mechanics series. Springer, Cham, pp 133–144. https://doi.org/10.1007/978-3-319-54858-6_14
115. Gehb CM, Platz R, Melz T (2019) Two control strategies for semi-active load path redistribution in a load-bearing structure. Mech Syst Signal Process 118:195–208. https://doi.org/10.1016/j.ymssp.2018.08.044
116. Gemmerich M (1995) Technische Produktänderungen. Betriebswirtschaftliche und empirische Modellanalyse. https://doi.org/10.1007/978-3-663-09030-4
117. Giesberts L, Gayger M (2019) Sichere Produkte ohne technische Normen: Anforderungen an neuartige Produkte wie autonome Systeme. Neue Zeitschrift für Verwaltungsrecht 38(20):1491–1495
118. Götz B (2019) Evaluation of uncertainty in the vibration attenuation with shunted piezoelectric transducers integrated in a beam-column support. Dissertation, TU Darmstadt. https://tuprints.ulb.tu-darmstadt.de/8608
119. Götz B, Kersting S, Kohler M (2018) Estimation of an improved surrogate model in uncertainty quantification by neural networks. Submitted for publication
120. Götz B, Platz R, Melz T (2018) Effect of static axial loads on the lateral vibration attenuation of a beam with piezo-elastic supports. Smart Mater Struct 27(3). https://doi.org/10.1088/1361-665X/aaa937

121. Groche P, Brenneis M (2014) Manufacturing and use of novel sensoric fasteners for monitoring forming processes. Measurement 53:136–144. https://doi.org/10.1016/j.measurement.2014.03.042
122. Groche P, Calmano S, Felber T, Schmitt S (2015) Statistical analysis of a model based product property control for sheet bending. Prod Eng 9(1):25–34
123. Groche P, Fritsche D, Tekkaya E, Allwood J, Hirt G, Neugebauer R (2007) Incremental bulk metal forming. CIRP Ann 56(2):635–656. https://doi.org/10.1016/j.cirp.2007.10.006
124. Groche P, Hoppe F, Hesse D, Calmano S (2016) Blanking-bending process chain with disturbance feed-forward and closed-loop control. J Manuf Process 24:62–70. https://doi.org/10.1016/j.jmapro.2016.07.005
125. Groche P, Hoppe F, Kessler T, Kleemann A (2018) Industrial working environment 2025. In: Liewald M (ed) New Developments in Sheet Metal Forming. institute for metal forming technology. Stuttgart, pp 125–136
126. Groche P, Hoppe F, Sinz J (2017) Stiffness of multipoint servo presses: Mechanics versus control. CIRP Ann 66(1):373–376. https://doi.org/10.1016/j.cirp.2017.04.053
127. Groche P, Krech M (2017) Efficient production of sensory machine elements by a two-stage rotary swaging process - relevant phenomena and numerical modelling. J Mater Process Technol 242:205–217. https://doi.org/10.1016/j.jmatprotec.2016.11.034
128. Groche P, Scheitza M, Kraft M, Schmitt S (2010) Increased total flexibility by 3D servo presses. CIRP Ann 59(1):267–270. https://doi.org/10.1016/j.cirp.2010.03.013
129. Groche P, Schneider R (2004) Method for the optimization of forming presses for the manufacturing of micro parts. CIRP Ann 53(1):281–284. https://doi.org/10.1016/s0007-8506(07)60698-2
130. Groche P, Sinz J (2016) Innovation durch Kombination: Wälz-Gleit-Lagerung für Servopressen. MM Maschinenmarkt 41
131. Groche P, Sinz J, Felber P (2018) Kombinierte Wälz-Gleitlager: Anforderungsgerechter Funktionsübergang. Konstruktionspraxis. Sonderheft SH1/2018
132. Groche P, Sinz J, Germann T (2020) Efficient validation of novel machine elements for capital goods. CIRP Ann 69(1):125–128. https://doi.org/10.1016/j.cirp.2020.03.004
133. Groche P, Türk M (2010) Integration of adaptive components by incremental forming processes. In: SPIE Smart Structures/NDE—For the Latest Research on Smart Sensors, NDE, Aerospace Systems, Energy Harvesting, San Diego, USA (2010)
134. Groche P, Türk M (2011) Smart structures assembly through incremental forming. CIRP Ann 60(1):21–24. https://doi.org/10.1016/j.cirp.2011.03.003
135. Groche P, Wohletz S, Brenneis M, Pabst C, Resch F (2014) Joining by forming—a review on joint mechanisms, applications and future trends. J Mater Process Technol 214(10):1972–1994. https://doi.org/10.1016/j.jmatprotec.2013.12.022
136. Gurson AL (1977) Continuum theory of ductile rupture by void nucleation and growth: Part I—yield criteria and flow rules for porous ductile media. J Eng Mater Technol 99(1):2–15. https://doi.org/10.1115/1.3443401
137. Haberhauer H, Bodenstein F (2011) Maschinenelemente: Gestaltung, Berechnung, Anwendung, sixteenth edn. Springer, Berlin. https://doi.org/10.1007/978-3-642-14290-1
138. Hagood NW, Crawley EF (1991) Experimental investigation into passive damping enhancement for space structures. J Guidance, Control, Dyn 14(6):1100–1109
139. Hales C, Gooch S (2004) Managing engineering design. Springer, Berlin
140. Hamp B, Feldweg H (1997) GermaNet—a lexical-semantic net for German. In: Proceedings of ACL workshop automatic information extraction and building of lexical semantic resources for NLP applications, pp 9–15
141. Hamraz B, Caldwell NH, Clarkson PJ (2013) A holistic categorization framework for literature on engineering change management. Syst Eng 16(4):473–505
142. Hanselka H, Engelhardt R, Koenen JF, Enss GC, Sichau A, Platz R, Kloberdanz H, Birkhofer H (2010) A model to categorise uncertainty in load-carrying systems. In: Heisig P (ed) Proceedings of the 1st international conference on modelling and management engineering processes. University of Cambridge, pp 53–64

143. Hanselka H, Platz R (2010) Ansätze und Maßnahmen zur Beherrschung von Unsicherheit in lasttragenden Systemen des Maschinenbaus. Konstruktion 2010(11–12):55–62
144. Hansmann J, Kaal W, Seipel B, Melz T (2012) Einstellbares Federelement für Adaptierbaren Schwingungstilger. ATZ - Automobiltechnische Zeitschrift 114(3):242–247. https://doi.org/10.1365/s35148-012-0295-1
145. Hauer T (2012) Modellierung der Werkzeugabdrängung beim Reiben – Ableitung von Empfehlungen für die Gestaltung von Mehrschneidenreibahlen. Schriftenreihe des PTW. Shaker, Aachen. TU Darmstadt, Dissertation
146. Hauschka C, Klindt T (2007) Eine Rechtspflicht zur Compliance im Reklamationsmanagement. Neue Juristische Wochenschrift 60(38):2726–2729
147. Hedrich P (2018) Konzeptvalidierung einer aktiven Luftfederung im Kontext autonomer Fahrzeuge, Forschungsberichte zur Fluidsystemtechnik, vol 20. Shaker, Aachen
148. Heid U, Schierholz S, Schweickard W, Wiegand HE, Gouws RH, Wolski W (eds) (2010) Das Digitale Wörterbuch der Deutschen Sprache (DWDS). Lexicographica. de Gruyter, Berlin, New York
149. Heidemann B (2001) Trennende Verknüpfung: Ein Prozessmodell als Quelle für Produktideen, Fortschritt-Berichte VDI Reihe 1, Konstruktionstechnik, Maschinenelemente, vol 351. VDI Verlag, Düsseldorf (2001). TU Darmstadt, Dissertation
150. Heidling E (2016) Erscheinungsformen und Typen von Ungewissheit in Projekten. Technical Report, GPM Deutsche Gesellschaft für Projektmanagement e. V., Nuremberg. Chapter in Böhle. Umgang mit Ungewissheit, Fritz et al. in Projekten
151. Heinrichs D (2015) Autonomes Fahren und Stadtstruktur. In: Maurer M, Gerdes, JC, Lenz B, Winner H (eds) Autonomes Fahren: Technische, rechtliche und gesellschaftliche Aspekte. Springer, Berlin, Heidelberg, pp 219–239. https://doi.org/10.1007/978-3-662-45854-9_11
152. Heißing B, Ersoy M (eds) (2007) Fahrwerkhandbuch: Grundlagen, Fahrdynamik, Komponenten, Systeme, Mechatronik, Perspektiven. Vieweg & Sohn, GWV Fachverlage, Wiesbaden. https://doi.org/10.1007/978-3-8348-9151-8
153. Henrich V (2015) Word sense disambiguation with GermaNet. Dissertation, Universität Tübingen. https://doi.org/10.15496/publikation-4706
154. Henrich V, Hinrichs E (2010) GernEdiT—the Germanet editing tool. In: Proceedings of the seventh international conference on language resources and evaluation (LREC' 10). European Language Resources Association (ELRA), Valletta, Malta
155. Hesse D, Hoppe F, Groche P (2017) Controlling product stiffness by an incremental sheet metal forming process. Procedia Manuf 10:276–285. https://doi.org/10.1016/j.promfg.2017.07.058
156. Hiller F (1997) Ein Konzept zur Gestaltung von Änderungsprozessen in der Produktentwicklung. Dissertation, TU Kaiserslautern
157. Hinrichs EW, Hinrichs M, Zastrow T (2010) WebLicht: Web-based LRT services for German. In: Proceedings of the ACL 2010 system demonstrations, pp 25–29
158. Holland J (1996) Beitrag zur Erfassung der Schmierungsverhältnisse in Verbrennungskraftmaschinen, VDI-Forschungsheft, vol 475. VDI Verlag, Düsseldorf
159. Hoppe F, Hohmann J, Knoll M, Kubik C, Groche P (2019) Feature-based supervision of shear cutting processes on the basis of force measurements: Evaluation of feature engineering and feature extraction. Procedia Manuf 34:847–856. https://doi.org/10.1016/j.promfg.2019.06.164
160. Hoppe F, Knoll M, Götz B, Schaeffner M, Groche P (2018) Reducing uncertainty in shunt damping by model-predictive product stiffness control in a single point incremental forming process. In: Pelz PF, Groche P (eds) Uncertainty in mechanical engineering III, applied mechanics and materials, vol 885. Trans Tech Publications, pp 35–47. https://doi.org/10.4028/www.scientific.net/AMM.885.35
161. Hoppe F, Pihan C, Groche P (2019) Closed-loop control of eccentric presses based on inverse kinematic models. Procedia Manuf 29:240–247. https://doi.org/10.1016/j.promfg.2019.02.132

162. Jarratt T, Clarkson J, Eckert C (2005) Engineering change. In: Design process improvement. Springer, Berlin, pp 262–285
163. Joggerst L, Knoll M, Hoppe F, Wendt J, Groche P (2018) Autonomous manufacturing processes under legal uncertainty. In: Pelz PF, Groche P (eds) Uncertainty in mechanical engineering III, applied mechanics and materials, vol 885. Trans Tech Publications, pp 227–239. https://doi.org/10.4028/www.scientific.net/AMM.885.227
164. Jürgens U (2000) New product development and production networks: global industrial experience. Springer, Berlin
165. Kapoor A, Klindt T (2012) Das neue deutsche Produktsicherheitsgesetz (ProdSG). Neue Zeitschrift für Verwaltungsrecht 31(12):719–724
166. Kennedy MC, O'Hagan A (2001) Bayesian calibration of computer models (with discussion). J R Stat Soc: Ser B (Statistical Methodology) 63(3):425–464. https://doi.org/10.1111/1467-9868.00294
167. Kier M, Wegner S, Bey R (2011) Neuartige Lagerungstechnik des Kurbeltriebs von Verbrennungsmotoren zur Verbesserung der Effizienz und Minderung der Emissionen: Abschlussbericht. Technical Report. 25518-24/0, Deutsche Bundesstiftung Umwelt AZ, Herzogenrath
168. Kitayama S, Natsume S, Yamazaki K, Han J, Uchida H (2016) Numerical investigation and optimization of pulsating and variable blank holder force for identification of formability window for deep drawing of cylindrical cup. Int J Adv Manuf Technol 82:583–593. https://doi.org/10.1007/s00170-015-7385-7
169. Klammt A (2019) Erfolgreicher Start von neuem Verbundprojekt CLARIAH-DE. https://dhd-blog.org/?p=11593. (Visited 12/2020)
170. Kleedörfer RW (1998) Prozess-und Änderungsmanagement der integrierten Produktentwicklung. Dissertation, TU München
171. Kleinaltenkamp M (1996) Customer integration - Kundenintegration als Leitbild für das Business-to-Business Marketing. In: Kleinaltenkamp M, Fließ S, Jacob F (eds) Customer integration. Gabler Verlag, Wiesbaden, pp 13–24
172. Klie JC, Bugert M, Boullosa B, de Castilho RE, Gurevych I (2018) The INCEpTION Platform: Machine-assisted and knowledge-oriented interactive annotation. In: Proceedings of the 27th international conference on computational linguistics: system demonstrations. Association for Computational Linguistics, pp 5–9
173. Klindt T (2014) § 8 Produktsicherheitsrecht – Marktüberwachungsrecht. In: Lenz T (ed) Produkthaftung. C. H. Beck, München, pp I–III
174. Klindt T (2015) Produktsicherheitsrecht, quo vadis? Zeitschrift für Innovations und Technikrecht, pp 37–38
175. Kloberdanz H, Engelhardt R, Mathias J, Birkhofer H (2009) Process based uncertainty analysis—an approach to analyze uncertainties using a process model. In: Norell Bergendahl M (ed) Design theory and research methodology, ds/design society. Design Society, Glasgow, pp 465–474. https://www.designsociety.org/publication/28608/Process+Based+Uncertainty+Analysis+
176. Kohler M, Krzyżak A (2017) Nonparametric regression based on hierarchical interaction models. IEEE Trans Inform Theory 63(3):1620–1630. https://doi.org/10.1109/TIT.2016.2634401
177. Kohler M, Krzyżak A (2018) Adaptive estimation of quantiles in a simulation model. IEEE Trans Inform Theory 64(1):501–512. https://doi.org/10.1109/TIT.2017.2743177
178. Kohler M, Tent R (2019) Nonparametric quantile estimation using surrogate models and importance sampling. Metrika. https://doi.org/10.1007/s00184-019-00736-3
179. Koppka F (2008) A contribution to the maximization of productivity and workpiece quality of the reaming process by analyzing its static and dynamic behavior: An analysis with focus on automotive powertrain production. Schriftenreihe des PTW. Shaker, Aachen (2009). TU Darmstadt, Dissertation
180. Krech M (2020) Adaptiver Umformprozess zur Herstellung und simultanen Kalibrierung von metallischen Strukturen mit bauteilintegrierten Kraft- und Drehmomentsensoren [Adaptive forming process for the production and simultaneous calibration of metallic structures with component-integrated force and torque sensors]. Dissertation, TU Darmstadt

181. Krech M, Trunk A, Groche P (2017) Controlling the sensor properties of smart structures produced by metal forming. Procedia Eng 207:1415–1420. https://doi.org/10.1016/j.proeng.2017.10.906. International Conference on the Technology of Plasticity, ICTP 2017
182. Lee BJ, Mear ME (1994) Studies of the growth and collapse of voids in viscous solids. J Eng Mater Technol 116(3):348–358. https://doi.org/10.1115/1.2904298
183. Lenz E, Hedrich P, Pelz PF (2018) Aktive Luftfederung – Modellierung, Regelung und Hardware-in-the-Loop-Experimente. Forschung im Ingenieurwesen 82(3):171–185. https://doi.org/10.1007/s10010-018-0272-2
184. Lenz J, Holzmann H, Platz R, Melz T (2019) Vibration attenuation of a truss structure with piezoelectric shunt-damping for varying static axial loads in the truss members. In: Proceedings of international conference on structural engineering dynamics (ICEDyn), pp 1–11
185. Li W, Li D, Ni J (2003) Diagnosis of tapping process using spindle motor current. Int J Mach Tools Manuf 43(1):73–79. https://doi.org/10.1016/S0890-6955(02)00142-6
186. Lichtenberg S (2000) Proactive management of uncertainty using the successive principle: a practical way to manage opportunities and risks. Polyteknisk Press
187. Liu G, Yuan SJ, Wang ZR, Zhou DC (2004) Explanation of the mushroom effect in the rotary forging of a cylinder. J Mater Process Technol 151(1–3):178–182. https://doi.org/10.1016/j.jmatprotec.2004.04.035
188. Liu T, Ko E, Sha S (1991) Diagnosis of tapping operations using an AI approach. In: Proceedings. 1991 IEEE international conference on robotics and automation. IEEE Comput. Soc. Press, pp 1556–1561. https://doi.org/10.1109/ROBOT.1991.131838
189. Luhmann N (1997) Die Gesellschaft der Gesellschaft, vol 2. Suhrkamp, Frankfurt a. M
190. Mahlamäki K, Ström M, Eisto T, Hölttä V (2009) Lean product development point of view to current challenges of engineering change management in traditional manufacturing industries. In: 2009 IEEE international technology management conference (ICE). IEEE, pp 1–8
191. Marciniak Z, Duncan J, Hu J (2002) Mechanics of sheet metal forming. Elsevier
192. Mathias J (2016) Auf dem Weg zu robusten Lösungen. Dissertation, TU Darmstadt
193. Meister JC (forthcoming) From TACT to CATMA or a mindful approach to text annotation and analysis. In: Rockwell G, Sinclair S (eds) Festschrift for John Bradley
194. Melz T, Hanselka H, Matthias M (2006) Adaptronische Systeme für automotive Anwendungen am Beispiel eines modularen, aktiven Strukturinterfaces [Smart Structures for Automotive Applications with the Example of a Modular Active Interface]. at – Automatisierungstechnik 54(6). https://doi.org/10.1524/auto.2006.54.6.284
195. Menz S (2015) §4. In: Klindt T (ed) Produktsicherheitsgesetz ProdSG. C. H. Beck, München
196. Meressi T, Paden B (1993) Buckling control of a flexible beam using piezoelectric actuators. J Guidance, Control, Dyn 16(5):977–980. https://doi.org/10.2514/3.21113
197. Merkel J (2008) Untersuchungen zum Einfluss von Kugelstrahl- und Festwalzbehandlungen auf den Randschichtzustand und die Schwingfestigkeit von Sintereisenwerkstoffen, Schriftenreihe Werkstoffwissenschaft und Werkstofftechnik, vol 45. Shaker, Aachen. Karlsruhe University, Dissertation
198. Mitschke M, Wallentowitz H (2014) Dynamik der Kraftfahrzeuge. Springer Vieweg, Wiesbaden. https://doi.org/10.1007/978-3-658-05068-9
199. Moheimani SOR, Fleming AJ (2006) Piezoelectric transducers for vibration control and damping. Advances in industrial control. Springer, London. https://doi.org/10.1007/1-84628-332-9
200. Monka P, Monkova K, Uban M, Hruzik L, Vasina M (2018) Vibrodiagnostics as the tool of a tap wear monitoring. Procedia Struct Integrity 13:959–964. https://doi.org/10.1016/j.prostr.2018.12.179
201. Moré JJ (1978) The Levenberg–Marquardt algorithm: implementation and theory. In: Numerical analysis. Springer, Berlin, pp 105–116
202. Mori K (2011) Application of servo presses to sheet metal forming. Key Eng Mater 473:27–36. https://doi.org/10.4028/www.scientific.net/KEM.473.27
203. Morris PWG (1997) The management of projects. T. Telford, London

204. Naber D (2005) OpenThesaurus: Ein offenes deutsches Wortnetz. In: Fisseni B, Schmitz HC, Schroeder B, Wagner P (eds) Sprachtechnologie, mobile Kommunikation und linguistische Ressourcen: Beiträge zur GLDV-Tagung 2005 in Bonn, no. 8 in Sprache, Sprechen und Computer; Computer studies in language and speech. Peter Lang, Frankfurt am Main, New York
205. Needleman A (1972) A numerical study of necking in circular cylindrical bar. J Mech Phys Solids 20(2):111–127. https://doi.org/10.1016/0022-5096(72)90035-X
206. Neubauer M, Oleskiewicz R, Popp K, Krzyzynski T (2006) Optimization of damping and absorbing performance of shunted piezo elements utilizing negative capacitance. J Sound Vib 298(1–2):84–107. https://doi.org/10.1016/j.jsv.2006.04.043
207. Niwa K, Matsumura H (1993) Gleit-wälz-lager. Patent DE4327543 C2
208. Noy NF, McGuinness DL (2000) Ontology development 101: A guide to reating your first ontology
209. Nusskern P, Hoffmeister J, Schulze V (2014) Powder metallurgical components: Improvement of surface integrity by deep rolling and case hardening. Procedia CIRP 13:192–197. https://doi.org/10.1016/j.procir.2014.04.033
210. Obara RM (1998) Kombiniertes Lager. Patent DE69927078 T2
211. Ogden CK, Richards IA (1923) The meaning of meaning. K. Paul, Trench, Trubner and Company, New York
212. Okereke C (2015) Business. In: Bäckstrand K, Lövbrand E (eds) Research handbook on climate governance. Edward Elgar Publishing, pp 262–272. https://doi.org/10.4337/9781783470600.00034
213. OLG Frankfurt: Urteil vom 05. Juli 2018—6 U 28/18
214. Osakada K, Mori K, Altan T, Groche P (2011) Mechanical servo press technology for metal forming. CIRP Ann 60(2):651–672. https://doi.org/10.1016/j.cirp.2011.05.007
215. Pahl G, Beitz W (1977) Konstruktionslehre: Handbuch für Studium und Praxis. Springer, Berlin
216. Pahl G, Beitz W (1997) Konstruktionslehre: Methoden und Anwendung, 4th edn. Springer, Berlin, Heidelberg, New York
217. Pahl G, Beitz W (2013) Konstruktionslehre: Methoden und Anwendung. Springer, Berlin
218. Pahl G, Beitz W, Feldhusen J, Grote KH (2005) Konstruktionslehre: Grundlagen erfolgreicher Produktentwicklung Methoden und Anwendung, 6th edn. Springer, Berlin, Heidelberg. https://doi.org/10.1007/b137606
219. Pahl G, Beitz W, Feldhusen J, Grote KH (2007) Engineering design: a systematic approach, 3rd edn. Springer, London, Cham. https://doi.org/10.1007/978-1-84628-319-2
220. Parteder E (2000) Ein Modell zur Simulation von Umformprozessen pulvermetallurgisch hergestellter hochschmelzender Metalle, Umformtechnische Schriften, vol 94. Shaker, Aachen (2000). RWTH Aachen, Dissertation
221. Perminova O, Gustafsson M, Wikström K (2008) Defining uncertainty in projects - a new perspective. Int J Project Manag 26(1):73–79
222. Peterson RA (1964) Combination bearing. Patent US3305280 A
223. Platz R, Ondoua S, Habermehl K, Bedarff T, Hauer T, Schmitt SO, Hanselka H (2010) Approach to validate the influences of uncertainties in manufacturing on using load-carrying structures. In: ISMA2010 international conference on noise and vibration engineering, pp 5319–5334
224. Pohl K (2010) Requirements engineering: fundamentals. Principles and Techniques. Springer, Heidelberg
225. Pohl K, Rupp C (2015) Basiswissen Requirements Engineering: Aus- und Weiterbildung zum Certified Professsional for Requirements Engineering, fourth edn. dpunkt.verlag, Heidelberg
226. Ravetz JR (1986) Usable knowledge, usable ignorance: Incomplete science with policy implications. In: Clark C, Munn RE (eds) Sustainable development of the biosphere. Cambridge University Press, Cambridge, pp 415–432
227. Renn O (1981) Wahrnehmung und Akzeptanz technischer Risiken. Dissertation, University of Köln. https://doi.org/10.18419/opus-7503

228. Reynolds O (1886) IV. On the theory of lubrication and its application to Mr. Beauchamp tower's experiments, including an experimental determination of the viscosity of olive oil. Philoso Trans R Soc Lond 177:157–234. https://doi.org/10.1098/rstl.1886.0005
229. Riviere A, Féru F, Tollenaere M (2003) Controlling product related engineering changes in the aircraft industry. In: DS 31: Proceedings of ICED 03, the 14th international conference on engineering design. Stockholm
230. Rowley J (2007) The wisdom hierarchy: Representations of the DIKW hierarchy. J Inform Sci 2(33):163–180. https://doi.org/10.1177/0165551506070706
231. Ruder S (2016) An overview of gradient descent optimization algorithms. arXiv:1609.04747
232. Santner TJ, Williams BJ, Notz WI (2018) The design and analysis of computer experiments, 2nd edn. Springer series in statistics. Springer, New York
233. SASIG (2009) ECM Recommendation Part 0. Verband der Automobilindustrie
234. Sauer M (2013) Vorgehensmodell zur Implementierung eines meilensteingestützten Kundenänderungsmanagements. No. 21 in Stuttgarter Beiträge zur Produktionsforschung. Fraunhofer-Verlag, Stuttgart
235. Schaeffner M (2019) Quantification and evaluation of uncertainty in active buckling control of a beam-column subject to dynamic axial loads. Dissertation, TU Darmstadt. https://tuprints.ulb.tu-darmstadt.de/8652
236. Schaeffner M, Platz R (2018) Gain-scheduled H_∞ buckling control of a circular beam-column subject to time-varying axial loads. Smart Mater Struct 27(6). https://doi.org/10.1088/1361-665X/aab63a
237. Schaeffner M, Platz R, Melz T (2020) Adequate mathematical beam-column model for active buckling control in a tetrahedron truss structure. In: Proceedings of the 38th IMAC, conference proceedings of the society for experimental mechanics. Springer, Berlin, pp 323–332. https://doi.org/10.1007/978-3-030-47638-0_35
238. Schatt W, Wieters KP, Kieback B (eds) (2007) Pulvermetallurgie: Technologien und Werkstoffe, second edn. VDI-Buch. Springer, Berlin. https://doi.org/10.1007/978-3-540-68112-0
239. Scheitza M (2010) Konzeption eines flexiblen 3D-Servo-Pressensystems und repräsentative Basisanwendungen: Matthias Scheitza. Berichte aus Produktion und Umformtechnik. Shaker, Aachen
240. Scheitza M, Schmitt SO, Emde S (2013) Potential and challenges of combined roller and plain bearings for servo presses. Adv Mater Res 769:285–292. https://doi.org/10.4028/www.scientific.net/AMR.769.285
241. Schmitt SO (2015) Vorgehensmodell der realen iterativen Produktentwicklung. Shaker, Aachen
242. Schmitt SO, Scheitza M, Groche P (2015) A model for improving the applicability of design methodologies to mechanical engineering design routines. J Eng Des 26(10–12):302–320
243. Schneider W, Heinrich B (2008) Praktische Regelungstechnik, vol 3. Vieweg+Teubner, Wiesbaden
244. Siemens AG: SINUMERIK 840D sl/828D Synchronaktionen. Funktionshandbuch (2015). https://cache.industry.siemens.com/dl/files/521/109481521/att_862872/v1/FBSYsl_1015_de_de-DE.pdf
245. Sinz J, Groche P (2019) Effekte der Größenskalierung auf die Funktionsfähigkeit kombinierter Wälz-Gleitlager. In: 13. VDI-Fachtagung Gleit- und Wälzlagerungen 2019: Gestaltung – Berechnung – Einsatz, VDI-Berichte Band 2348. VDI Verlag
246. Sinz J, Knoll M, Groche P (2019) Operational effects on the stiffness of combined roller and plain bearings. Procedia Manuf 41:650–657. https://doi.org/10.1016/j.promfg.2019.09.054
247. Sinz J, Niessen B, Groche P (2018) Combined roller and plain bearings for forming machines: Design methodology and validation. In: Schmitt R, Schuh G (eds) Advances in production research: proceedings of the 8th congress of the german academic association for production technology (WGP). Springer, Berlin, pp 126–135
248. Sivak M, Schoettle B (2015) Motion sickness in self-driving vehicles. Technical Report. UMTRI-2015-12, Transportation Research Institute, University of Michigan

249. Skogestad S, Postlethwaite I (2001) Multivariable feedback control. Wiley, New York
250. Smith PG, Reinertsen DG (1992) Shortening the product development cycle. Res Technol Manag 35(3):44–49
251. Sommer S (2020) Wertberichtigung. Brand eins 10:94
252. Spath D, Koch S (2009) Grundlagen der Organisationsgestaltung. In: Handbuch Unternehmensorganisation. Springer, Berlin, pp 3–24
253. Spindler G (2004) IT-Sicherheit und Produkthaftung – Sicherheitslücken, Pflichten der Hersteller und der Softwarenutzer. Neue Juristische Wochenschrift 57(44):3145–3150
254. Stegmeier J, Hartig J, Bartsch S, Leštáková M, Logan K, Rapp A, Pelz PF (2021) Semantic uncertainty in norms. In: Groche P, Pelz PF (eds) Uncertainty in mechanical engineering: selected, peer reviewed papers from the 4th international conference on uncertainty in mechanical engineering (ICUME 2021). Springer, Berlin
255. Stekolschik A (2016) Engineering change management method framework in mechanical engineering. In: IOP conference series: materials science and engineering, vol 157. https://doi.org/10.1088/1757-899x/157/1/012008
256. Strauß D (2015) Examination regarding the redensification of porous sintering parts respecting the additional movement capabilities of the 3D servo press. Master-thesis, TU Darmstadt
257. Taguchi G, Chowdhury S, Taguchi S (2000) Robust engineering: learn how to boost quality while reducing costs and time to market. McGraw-Hill, New York. http://www.loc.gov/catdir/bios/mh041/99041786.html
258. Taguchi G, Yano H, Chowdhury S, Taguchi S (2005) Taguchi's quality engineering handbook. Wiley, Hoboken, N.J and Livonia, Mich. https://doi.org/10.1002/9780470258354
259. Takemasu T, Koide T, Shinbutsu T, Sasaki H, Takeda Y, Nishida S (2014) Effect of surface rolling on load bearing capacity of pre-alloyed sintered steel gears with different densities. Procedia Eng 81:334–339. https://doi.org/10.1016/j.proeng.2014.10.002
260. Thamhain H (2013) Managing risks in complex projects. Project Manag J 44(2):20–35
261. Thompson SP, Loughlan J (1995) The active buckling control of some composite column strips using piezoceramic actuators. Compos Struct 32:59–67. https://doi.org/10.1016/0263-8223(95)00048-8
262. Timoshenko SP, Gere JM (1961) Theory of elastic stability. McGraw-Hill
263. Türk M, Groche P (2010) Integration of adaptive components by incremental forming processes. In: Ghasemi-Nejhad MN (ed) Active and passive smart structures and integrated systems 2010, vol 7643. International Society for Optics and Photonics, SPIEM, pp 447–455. https://doi.org/10.1117/12.847280
264. Tvergaard V (1981) Influence of voids on shear band instabilities under plane strain conditions. Int J Fracture 17(4):389–407. https://doi.org/10.1007/BF00036191
265. Unsworth J (2000) Scholarly primitives: What methods do humanities researchers have in common, and how might our tools reflect this? In: Symposium on humanities computing: formal methods, experimental practice, vol 13. London. https://johnunsworth.name/Kings.5-00/primitives.html (Visited 12/2020)
266. Val AGD, Fernández J, Arizmendi M, Veiga F, Urízar J, Berriozábal A, Axpe A, Diéguez P (2013) On line diagnosis strategy of thread quality in tapping. Procedia Engineering 63:208–217. https://doi.org/10.1016/j.proeng.2013.08.196. The Manufacturing Engineering Society International Conference, MESIC 2013
267. Value Balancing Alliance e. V (2020). https://www.value-balancing.com
268. Verband der Automobilindustrie (VDA) (2015) Automatisierung von Fahrerassistenzsystemen zum automatisierten Fahren. VDA Magazin – Automatisierung
269. Verein Deutscher Ingenieure (1986) VDI 2221:1986–11 Methodik zum Entwickeln und Konstruieren technischer Systeme und Produkte [Systematic approach to the development and design of technical systems and products]. Beuth, Berlin
270. Verein Deutscher Ingenieure (1992) VDI 2204 Blatt 2:1992–09 Auslegung von Gleitlagerungen; Berechnung [Design of plain bearings; calculation]. Beuth, Berlin
271. Verein Deutscher Ingenieure (1993) VDI 2221:1993–05 Methodik zum Entwickeln und Konstruieren technischer Systeme und Produkte [Systematic approach to the development and design of technical systems and products]. Beuth, Berlin

272. Verein Deutscher Ingenieure (2005) VDI/VDE 3682:2005–09 Formalisierte Prozessbeschreibungen [Formalised process descriptions]. Beuth, Berlin
273. Verein Deutscher Ingenieure (2015) VDI 3334:2015–03 Maschinelle Innengewindefertigung [Machining of internal threads]. Beuth, Berlin
274. Wagner G (2020) § 823. In: F.J. Säcker, R. Rixecker, H. Oetker, B. Limperg (eds.) Münchener Kommentar zum Bürgerlichen Gesetzbuch, vol 7, eighth edn. C. H. Beck, München
275. Wang QS (2010) Active buckling control of beams using piezoelectric actuators and strain gauge sensors. Smart Mater Struct 19(6):1–8. https://doi.org/10.1088/0964-1726/19/6/065022
276. Wehling P (2007) Wissen und Nichtwissen. In: Schützeichel R (ed) Handbuch Wissenssoziologie und Wissensforschung. UVK, Konstanz, pp 485–494
277. Wehling P (2012) Gibt es Grenzen der Erkenntnis? Von der Fiktion grenzenlosen Wissens zur Politisierung des Nichtwissens. In: Wengenroth U (ed) Grenzen des Wissens - Wissen um Grenzen. Velbrück, Weilerswist, pp 90–117
278. Wendt J, Oberländer M (2016) Product compliance: Neue Anforderungen an sichere Produkte. Zeitschrift für Energie- und Technikrecht, pp 62–70
279. Wideman R (1986) Risk management. Project Manag J17(8):20–26
280. Wiedemann J (2007) Leichtbau. Springer, Elemente und Konstruktion
281. Wong RKW, Storlie CB, Lee TCM (2017) A frequentist approach to computer model calibration. J R Stat Soc Ser B Stat Methodol 79(2):635–648. https://doi.org/10.1111/rssb.12182
282. Wu WH, Fang LC, Wang WY, Yu MC, Kao HY (2014) An advanced CMII-based engineering change management framework: the integration of PLM and ERP perspectives. Int J Prod Res 52(20):6092–6109
283. Würtenberger J (2018) Ein Beitrag zur Identifikation und Beherrschung von Unsicherheit bei der Modellierung technischer Systeme. Dissertation, TU Darmstadt
284. Yu X, Yang Z, Wang G, Lu J (2013) Study on design change review for small and medium-sized enterprises. In: 2013 IEEE international conference on industrial engineering and engineering management. IEEE, pp 556–560
285. Zech H (2019) Künstliche Intelligenz und Haftungsfragen. Zeitschrift für die Privatrechtswissenschaft 5(2):198–219
286. Zenz G, Humer A (2015) Stability enhancement of beam-type structures by piezoelectric transducers. Acta Mechanica 226(12):3961–3976. https://doi.org/10.1007/s00707-015-1445-9
287. Zirn C, Nastase V, Strube M (2008) Distinguishing between instances and classes in the Wikipedia taxonomy. In: Bechhofer S, Hauswirth M, Hoffmann J, Koubarakis M (eds) The semantic web: research and applications, vol 5021. Springer, Berlin, Heidelberg, pp 376–387. https://doi.org/10.1007/978-3-540-68234-9_29

Chapter 6
Strategies for Mastering Uncertainty

Marc E. Pfetsch, Eberhard Abele, Lena C. Altherr, Christian Bölling, Nicolas Brötz, Ingo Dietrich, Tristan Gally, Felix Geßner, Peter Groche, Florian Hoppe, Eckhard Kirchner, Hermann Kloberdanz, Maximilian Knoll, Philip Kolvenbach, Anja Kuttich-Meinlschmidt, Philipp Leise, Ulf Lorenz, Alexander Matei, Dirk A. Molitor, Pia Niessen, Peter F. Pelz, Manuel Rexer, Andreas Schmitt, Johann M. Schmitt, Fiona Schulte, Stefan Ulbrich, and Matthias Weigold

Abstract This chapter describes three general strategies to master uncertainty in technical systems: robustness, flexibility and resilience. It builds on the previous chapters about methods to analyse and identify uncertainty and may rely on the availability of technologies for particular systems, such as active components. Robustness aims for the design of technical systems that are insensitive to anticipated uncertainties. Flexibility increases the ability of a system to work under different situations. Resilience extends this characteristic by requiring a given minimal functional performance, even after disturbances or failure of system components, and it may incorporate recovery. The three strategies are described and discussed in turn. Moreover, they are demonstrated on specific technical systems.

In this chapter, we eventually come to the final key topic of this book, namely strategies to master uncertainty in technical systems. The underlying concepts and ideas of this chapter have already been introduced in Sect. 3.5.

M. E. Pfetsch (✉) · T. Gally · P. Kolvenbach · A. Kuttich-Meinlschmidt · A. Matei · A. Schmitt · J. M. Schmitt · S. Ulbrich
Department of Mathematics, TU Darmstadt, Darmstadt, Germany
e-mail: pfetsch@mathematik.tu-darmstadt.de

E. Abele · C. Bölling · N. Brötz · I. Dietrich · F. Geßner · P. Groche · F. Hoppe · E. Kirchner · H. Kloberdanz · M. Knoll · P. Leise · Dirk A. Molitor · P. Niessen · P. F. Pelz · M. Rexer · F. Schulte · M. Weigold
Department of Mechanical Engineering, TU Darmstadt, Darmstadt, Germany

L. C. Altherr
Faculty of Energy, Building Services and Environmental Engineering, Münster University of Applied Sciences, Münster, Germany

U. Lorenz
Chair of Technology Management, Universität Siegen, Siegen, Germany

P. F. Pelz et al. (eds.), *Mastering Uncertainty in Mechanical Engineering*, Springer Tracts in Mechanical Engineering,
https://doi.org/10.1007/978-3-030-78354-9_6

It is useful to recall that several prior steps are necessary to master uncertainty. This is illustrated in Fig. 1.12, where the methods of this chapter are addressed on the top layer, with the layers below corresponding to the preceding chapters. In the first step, one needs to be aware of the existence of uncertainty and the different types of uncertainty as described in Chap. 2. The next step is to analyse, quantify and evaluate uncertainty as presented in Chap. 4. After the identification of uncertainty in a particular system, the legal requirements are determined (Sect. 5.1), before the technological options have to be reviewed, created and evaluated. In Chap. 5 technologies and methods with focus on product design and process chains are introduced (Sect. 5.3). Moreover, it might be possible to use (semi-)active components to master uncertainty in the system (Sect. 5.4).

The first strategy to master uncertainty described in the following is *robustness*, see Sect. 6.1. The goal is to design a robust system that not only fulfils its function at the design point, but also in the surrounding neighbourhood, see Sect. 3.5. This is achieved by anticipating uncertainty in the design phase, following robust design principles or by applying robust optimisation. These general methods are described and illustrated on several technical systems and processes, such as presses, as well as tapping and reaming, see Sects. 6.1.7 and 6.1.8, respectively. The description of such applications highlights the fact that the general approach needs to be adapted to the particular circumstances.

The second strategy is *flexibility*, see Sect. 6.2. The objective is to design flexible systems that can react to uncertain conditions during the usage phase. Hence, even unpredicted disturbances might be mastered.

The third strategy is *resilience*, see Sect. 6.3. Here a technical system is designed in such a way that it fulfils a given predetermined minimal functional performance, even when disturbances and failures of system components occur and may include recovery, see the definition introduced in Sect. 3.5. As motivated in the latter section, both flexibility and resilience try to handle ignorance (see Chap. 2). This topic is depicted and detailed in Sect. 6.3, including several measures for resilience and demonstrating the practical application in systems, such as truss topologies (Sect. 6.3.4) and fluid systems (Sect. 6.3.8).

Many of the sections in this chapter combine knowledge from mechanical engineering and mathematics, e.g. by combining technological and domain knowledge with mathematical optimisation. Moreover, the presented strategies to master uncertainty of this chapter connect the different product life phases from system design to usage, see Fig. 3.1. Overall, this chapter provides a broad discussion of general strategies to master uncertainty, including a discussion of specific technical systems.

6.1 Robustness

Hermann Kloberdanz, Alexander Matei, Marc E. Pfetsch, Andreas Schmitt,
Johann M. Schmitt, and Stefan Ulbrich

In all life cycle processes, robust systems prove to be insensitive or only insignificantly sensitive to deviations in system properties or varying usage. In this section, we consider robustness as a strategy to master uncertainty from the different perspectives of mathematical optimisation, product or system design and production.

As an example, to further illustrate our understanding of robustness as introduced in Sect. 3.5, we first examine how robustness is incorporated in mathematical optimisation, before we give a short overview of the Sects. 6.1.1–6.1.8.

Robust optimisation is a mathematical approach that seeks solutions with guaranteed worst-case behaviour, provided that the uncertain data comes from a known *uncertainty set* $\mathcal{U}$. Let an optimisation program be described in the form

$$\min_x f^0(x, p) \quad \text{s.t.} \quad f^i(x, p) \leq 0, \quad \text{for } i \in I, \tag{6.1}$$

where x is the optimisation (or design) variable, p is a vector of uncertain parameters, f^0 is the scalar objective function, and f^i, $i \in I$, are finitely many scalar constraint functions. Provided that $p \in \mathcal{U}$ is known and fixed, the Problem (6.1) reduces to a classic optimisation program. In practice, however, this assumption does not hold. The parameters p are not exactly known, but we assume that they are contained in the given uncertainty set $\mathcal{U}$.

The robust approach eliminates the unknown p from Problem (6.1) by using a pessimistic assumption on the objective function and by requiring the constraints to hold regardless of the value of p, i.e. for its worst-case realisation, which leads to

$$\min_x \max_{p \in \mathcal{U}} f^0(x, p) \quad \text{s.t.} \quad f^i(x, p) \leq 0, \quad \text{for } i \in I, \quad \text{for } p \in \mathcal{U}.$$

Due to its bilevel (min-max) structure, this problem is difficult to solve in this general setting. In the following subsections, we therefore present different solution strategies which exploit the specific problem structure, such as the f^i being linear or nonlinear, time-variant or time-invariant, and also the analytic structure of $\mathcal{U}$. The latter could be in ellipsoidal form or consist of finitely many elements, for example.

In more detail, we exemplify robust optimisation techniques, models and applications in the first four sections. In Sect. 6.1.1 we introduce a robust truss topology optimisation framework in which we particularly consider dynamic models, beam elements and discrete decision variables. Furthermore, in Sect. 6.1.2 the employment of active elements is discussed for static and dynamic bar models, and demonstrated at the examples of active buckling control and shunt damping. In Sect. 6.1.3, we present robust optimisation techniques for problems involving partial differential equations and apply these techniques to the optimal design of a truss structure under uncertain dynamic load as well as the optimal design of a sensor element. Sect. 6.1.4 is concerned with quantified programs, which extend the robust optimisation approach to more than two stages.

In the subsequent sections, we move away from the mathematical point of view to investigate design principles and present control strategies to achieve robustness in a technical system. Sects. 6.1.5 and 6.1.6 describe possibilities of robust design of mechatronic systems. First, the mastering of disturbing influences in the early phases of the design process by process model-based analysis and synthesis strategies is presented. The process-oriented robust design then focuses mainly on the design of the mechanical components in the force flow. The design for clarity is recognised as particularly effective in robust design.

If measures regarding the mechanical system are limited, the control of the system offers additional possibilities to master uncertainty during the production phase. Potentials and effectiveness of nonlinear robust closed-loop control systems are shown in Sect. 6.1.7.

In Sect. 6.1.8, the robust design of process chains is explained using the linked production processes of drilling and reaming as well as drilling and tapping. The robustness of process chains is achieved by tool design, by optimising process parameters, and by additional adaptation process steps.

6.1.1 Robust Topology Optimisation of Truss Structures

Tristan Gally, Philip Kolvenbach, Anja Kuttich-Meinlschmidt, Alexander Matei, Marc E. Pfetsch, Johann M. Schmitt, and Stefan Ulbrich

The goal of truss topology design is to determine truss structures that are both, stable and lightweight. Stability here means that data uncertainty in the form of incertitude in the inputs, cf. Sect. 2.1, is taken into account by a robust approach in the system design phase, see Sect. 3.5. In the following, we concentrate on a particular approach via a semidefinite program (SDP), which was originally introduced by Ben-Tal and Nemirovski [17]. Exemplary alternatives to our treatment of uncertainty in truss topology design are described in [81, 101, 176]. For an overview of non-robust topology optimisation we refer to [18]. The approach, as presented here and in [111], is unique in the sense that dynamic uncertainty is mastered in robust truss topology design with SDP. Furthermore, another extension is the usage of binary variables for trusses introduced by Mars [122].

We first introduce the basic model and the corresponding optimisation problem for truss topology design. Then, we discuss the concept of robust optimisation as adopted in Sect. 6.1 with regard to this model. In the following paragraphs, we extend the basic optimisation problem to beam elements and dynamic truss models for vibration attenuation following [75, 111]. Finally, all these different models are compared using the example of the upper truss of the Modular Active Spring-Damper System, see Sect. 3.6.1.

Basic model

In this paragraph, we present the basic model of a truss which uses a so-called ground structure, i.e. a simple directed graph $\mathcal{D} = (\mathcal{V}, \mathcal{E})$ with n nodes $\mathcal{V} = \{v_1, \dots, v_n\} \subseteq \mathbb{R}^d$. The edges $\mathcal{E}$ represent possible bars. A subset $\mathcal{V}_f \subset \mathcal{V}$ of size n_f of the nodes is freely movable, while the remaining ones are fixed. At each of the n_f free nodes d-dimensional forces are applied, which are contained in the vector $f \in \mathbb{R}^{d_f}$ with $d_f = d \cdot n_f$. These forces cause displacements $u \in \mathbb{R}^{d_f}$, which are determined by the equilibrium constraint $A(x)\, u = f$, where $x \in \mathbb{R}_+^{\mathcal{E}}$ represents the cross-sectional areas of the possible bars in $\mathcal{E}$. Here, $A(x) = \sum_{e \in \mathcal{E}} A_e\, x_e$ is the *stiffness matrix* with $A_e = b_e b_e^\top$, where $b_e = (b_e(v))_{v \in \mathcal{V}_f} \in \mathbb{R}^{d_f}$ and

$$b_e(v) = \begin{cases} \sqrt{E}\, \frac{v_i - v_j}{\|v_i - v_j\|_2^{3/2}}, & \text{if } v = v_i, \\ \sqrt{E}\, \frac{v_j - v_i}{\|v_i - v_j\|_2^{3/2}}, & \text{if } v = v_j, \\ 0 & \text{otherwise,} \end{cases} \qquad \text{for } e = (v_i, v_j),\ v \in \mathcal{V}_f,$$

where E is Young's modulus of the used material. One possible aim is to find the stiffest truss under the restriction of a total volume bound $V_{\max} \in \mathbb{R}_+$. We measure the stiffness of the structure by the *compliance* $c = \frac{1}{2} f^\top u$, which represents the potential energy stored in the deformed truss and has to be minimised to maximise stiffness. This yields the optimisation problem

$$\min_{x \in \mathbb{R}^{\mathcal{E}}} \tfrac{1}{2} f^\top u \quad \text{s.t.} \quad A(x)\, u = f, \ \sum_{e \in \mathcal{E}} l_e x_e \leq V_{\max}, \quad x \geq 0, \tag{6.2}$$

where l_e denotes the length of the edge $e \in \mathcal{E}$. This optimisation problem can be reformulated as a semidefinite program (SDP)

$$\min_{\tau \in \mathbb{R}_+, x \in \mathbb{R}^{\mathcal{E}}} \tau \quad \text{s.t.} \quad \begin{pmatrix} 2\tau & f^\top \\ f & A(x) \end{pmatrix} \succeq 0, \ \sum_{e \in \mathcal{E}} l_e x_e \leq V_{\max}, \quad x \geq 0, \tag{6.3}$$

cf. [17]. Here, a symmetric, positive semidefinite matrix M is denoted by $M \succeq 0$. Analogously, we can also minimise the volume of the truss for a given upper bound on the compliance $c_{\max}$ which leads to a similar optimisation problem (6.3).

Robustness

In mechanical structures, uncertainty often appears in the form of parameters that are not exactly known, e.g. loads acting on a truss. One main topic of this section is the modelling and the mathematical treatment of data uncertainty, see Sect. 2.1. In the context of truss topology optimisation, even small changes of the considered load scenario may lead to severe instabilities. In order to cope with this problem, we use robust optimisation, see Sect. 6.1, where we consider uncertainty sets instead of fixed parameters. We consider the given force f to be uncertain. Then the robust optimisation problem corresponding to Problem (6.2) consists of finding a vector $x \in \mathbb{R}_+^{\mathcal{E}}$ to such an extent that the compliance is minimal under the worst-case load scenario, i.e.

$$\min_{x\in\mathbb{R}^{\mathcal{E}}} \max_{f\in\mathcal{U}} \ \tfrac{1}{2} f^\top u \quad \text{s.t.} \quad A(x)\,u = f, \ \sum_{e\in\mathcal{E}} l_e x_e \leq V_{\max}, \ \ x \geq 0, \tag{6.4}$$

where $\mathcal{U}$ denotes an uncertain set of forces f. Note that (6.4) is of the same form as the robust formulation in Sect. 6.1 and can be reformulated as an SDP via the techniques of Ben-Tal and Nemirovski [17]. This SDP can be solved efficiently. Therefore, it is desirable to reformulate these robust problems as SDPs. Nevertheless, the results of Ben-Tal and Nemirovski [17] strongly rely on the special structure of the inner maximisation problem. For dynamic problems, which are also addressed in this section, these techniques cannot be applied, since we have to deal with ordinary differential equations as constraints. However, using the Bounded Real Lemma we can still reformulate the robust problem as an SDP.

There are different ways to choose the uncertainty set of forces $\mathcal{U}$. One could for example consider *polyhedral uncertainty sets* $\mathcal{U}$ which are given by the convex hull of $s \in \mathbb{N}$ many forces $f_1, \dots, f_s$. These can be integrated into the problem by adding an additional SDP constraint for each force. A second possibility is to work with *ellipsoidal uncertainty sets*, where $\mathcal{U} = \{f = Qa : a^\top a \leq 1\}$ for some scaling matrix $Q \in \mathbb{R}^{d_f \times d_f}$. A common choice for Q is given by $Q = [f_1, \dots, f_s, \theta e_1, \dots, \theta e_{n_U}]$ with a scaling factor $\theta > 0$ and $n_U = d_f - s$. The scenario set $\mathcal{F} = \{f_1, \dots, f_s\}$ describes the "most important loads" and $\{\theta e_1, \dots, \theta e_{n_U}\}$ the "occasional loads". Here, $\{e_1, \dots, e_{n_u}\}$ is chosen as an orthonormal basis of the orthogonal complement to $\mathcal{L}(\mathcal{F})$ in $\mathbb{R}^{d_f}$, where $\mathcal{L}(\mathcal{F}) \subset \mathbb{R}^s$ denotes the linear span of $\mathcal{F}$, see [17]. In case of ellipsoidal uncertainty sets, a major result of [17] is that (6.4) is equivalent to

$$\min_{\tau\in\mathbb{R}_+, x\in\mathbb{R}^{\mathcal{E}}} \ \tau \quad \text{s.t.} \quad \begin{pmatrix} 2\tau I & Q^\top \\ Q & A(x) \end{pmatrix} \succeq 0, \ \sum_{e\in\mathcal{E}} l_e x_e \leq V_{\max}, \ \ x \geq 0. \tag{6.5}$$

In practice, often only a finite set $\mathcal{A}$ of cross-sectional areas is available. Then, we introduce binary variables x_e^a, which have value 1 if and only if the cross-sectional area of bar $e \in \mathcal{E}$ is equal to $a \in \mathcal{A}$. Integrating these binary decisions in our model, we obtain a *mixed-integer SDP* (MISDP) formulation

$$\min_{\substack{\tau\in\mathbb{R}_+,\\ x\in\{0,1\}^{\mathcal{E}\times\mathcal{A}}}} \tau \quad \text{s.t.} \quad \begin{pmatrix} 2\tau I & Q^\top \\ Q & A(x) \end{pmatrix} \succeq 0, \quad \sum_{e\in\mathcal{E}}\sum_{a\in\mathcal{A}} a\, l_e x_e^a \le V_{\max}, \quad \sum_{a\in\mathcal{A}} x_e^a \le 1, \quad \text{for } e\in\mathcal{E}$$

as independently shown in [106, 122].

Beam elements

The Truss Topology Design Problem (6.5) uses an idealised model of pin-connected bars. This can be extended to beam elements which can also represent bending. The new stiffness matrix, which depends nonlinearly on x, can be computed by using a finite element approach and inserted into (6.5), see [63]. The obtained non-convex SDP can be solved by a sequential SDP method based on [37], in which the nonlinear SDP constraint is linearised and iteratively solved by applying a suitable step length rule. In addition to rigid connections, one can also model pin-connected beams by introducing binary variables, which are coupled via linear constraints and indicate the connection type. In this approach, the stiffness matrix has to be further modified and the resulting nonlinear mixed-integer SDP can be solved as in [75] by using a sequential SDP method which is embedded in a Branch-and-Bound algorithm solving a non-convex SDP in each node.

Solving mixed-integer semidefinite programs

As described before, solving the problems arising from robust optimisation models, possibly incorporating integer decisions like a discrete choice of truss thicknesses or placing actuators (see Sect. 6.1.2), results in mixed-integer SDPs. For this class of problems, only very few software packages are available. Therefore, SCIP-SDP, a software system based on the framework SCIP [67] was created, which is publicly available [147]. SCIP-SDP contains interfaces to several SDP-solvers, such as Mosek, DSDP and SDPA. Moreover, it contains a variety of presolving techniques, branching rules and primal heuristics. The paper [65] describes some of the used techniques and provides an analysis of the preservation of strong duality when working in a branch-and-cut framework. More details are given in [62, 122, 123]. A parallel version of SCIP-SDP is also available, see [149].

Dynamic model

As an extension to the static approach above, we consider a dynamic truss model which can be used for example to describe and reduce structural vibrations in mechanical systems resulting from time-dependent uncertain loads $f : \mathbb{R}_+ \to \mathbb{R}^{d_f}$. Within the dynamic model, the displacements $u(t)$ are given by the solution of the ordinary differential equation system

$$\begin{aligned} &M(x)\,\ddot{u}(t) + D(x)\,\dot{u}(t) + A(x)\,u(t) = f(t), \quad t > 0, \\ &u(0) = 0, \;\; \dot{u}(0) = 0, \end{aligned} \tag{6.6}$$

where $M(x) \in \mathbb{R}^{d_f \times d_f}$, $D(x) \in \mathbb{R}^{d_f \times d_f}$ and $K(x) \in \mathbb{R}^{d_f \times d_f}$ denote the mass matrix, the damping matrix and the stiffness matrix, respectively. We use the mean squared displacement

$$J(u) = \int_0^\infty \|u(t)\|_2^2 \,\mathrm{d}t = \|u\|^2_{L^2(\mathbb{R}_+;\mathbb{R}^{d_f})}$$

as a measure of stability and stiffness. The time-dependent load $f \in \mathcal{U}$ is again uncertain. Here, the uncertainty set is of ellipsoidal form

$$\mathcal{U} = \{f = Qa : \|a\|_{L^2(\mathbb{R}_+;\mathbb{R}^{d_f})} \leq 1\},$$

where Q is chosen as explained before. Then the robust dynamic truss topology design problem reads

$$\min_{x \in \mathbb{R}^{\mathcal{E}}} \max_{f \in \mathcal{U}} \; J(u) \quad \text{s.t.} \quad u \text{ solves (6.6)}, \; \sum_{e \in \mathcal{E}} l_e x_e \leq V_{\max}, \quad x \geq 0. \tag{6.7}$$

Rewriting (6.6) as a system of first order differential equations

$$\begin{aligned} &\dot{y}(t) = P(x)\,y(t) + B(x)\,Q\,a(t), \; u(t) = L\,y(t), \quad t > 0, \\ &u(0) = 0, \;\; \dot{u}(0) = 0, \end{aligned} \tag{6.8}$$

and using the Bounded Real Lemma, see [11], we can reformulate the optimisation problem (6.7) as

$$\begin{aligned} \min_{x \in \mathbb{R}^{\mathcal{E}}, \gamma \in \mathbb{R}_+, Y \in \mathbb{R}^{d_f \times d_f}} \quad & \gamma + \varepsilon_Y \,\|Y\|_{\mathrm{F}}^2 \\ \text{s.t.} \quad & \begin{pmatrix} P(x)^\top Y + Y P(x) & Y B(x) Q & L^\top \\ Q^\top B(x)^\top Y & -\gamma I & 0 \\ L & 0 & -\gamma I \end{pmatrix} \preceq 0, \\ & -Y \preceq 0, \;\; \sum_{e \in \mathcal{E}} l_e x_e \leq V_{\max}, \; x \geq 0, \end{aligned} \tag{6.9}$$

where a penalisation term $\varepsilon_Y \,\|Y\|_{\mathrm{F}}^2$ is added to the cost function, see [111]. Here, $\|\cdot\|_{\mathrm{F}}$ denotes the Frobenius norm and ε_Y is some positive constant. In [111], the optimisation problem (6.9) is solved by using a sequential SDP algorithm, see also [113].

Comparison of the models

In the following, we compare the models from this section using the example of the upper truss of the Modular Active Spring-Damper System, see Sect. 3.6.1 and Fig. 6.1a, where one possibly uncertain force is acting on the centre node. In Table 6.1, the cost function values for the solutions of the optimisation problems introduced in the subsections above are evaluated for the nominal case, i.e. one fixed force f is considered, and for the worst case scenario, where the acting force is uncertain. Moreover, we compare the corresponding solving times in the last column of Table 6.1. As we can see, in the non-robust static case (b) the compliance for the nominal force is quite small, whereas in the worst-case scenario it is much larger. This shows that the truss becomes unstable, if the force acting on the truss changes. In contrast to this behaviour, the values of the cost function in the nominal case and in the worst-case almost coincide for the robust problems, where the compliances in the robust static case with (d) and without discrete cross-sectional areas (c) are only slightly larger than in the non-robust static case. Nevertheless, we note that restricting the available cross-sectional areas to a discrete set or considering the dynamic model (f) leads to longer solving times.

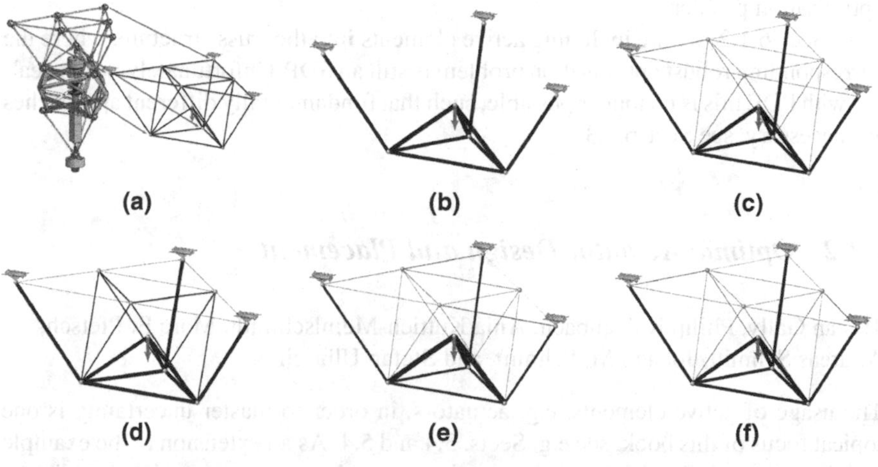

Fig. 6.1 Results of different optimisation models for the upper truss of the Modular Active Spring-Damper System: **a** basic truss structure, **b** non-robust static model, **c** robust static model, **d** discrete cross-sectional areas, **e** robust static model with beam elements and **f** robust dynamical truss. Varying truss thicknesses are recognisable upon close inspection; see also Table 6.1 for the differences in performance

Table 6.1 Comparison of solution characteristics for the different models

Problem	Nominal	Worst-case	Time in s
Non-robust static (b)	0.1124	$4.328 \cdot 10^5$	0.64
Robust static (c)	0.1162	0.1181	0.98
Robust static with discrete cross-sectional surfaces (d)	0.1260	0.1275	74.77
Robust static with beam elements (e)	5.6624	5.8609	8.18
Robust dynamic (f)	65.0858	66.0157	2199.01

Conclusion

In this section, we discussed robust truss topology design via an SDP-approach. We introduced a basic truss model and reformulated the minimum compliance problem for the robust case as a (mixed-integer) SDP. Furthermore, we compared static and dynamic models regarding the worst-case behaviour of the solution to the respective optimisation problem.

In Sect. 6.1.2, we are including active elements into the truss structure, where the corresponding robust optimisation problem is still an SDP. Unfortunately, when dealing with PDE this is no longer possible, such that fundamentally different approaches are necessary, see Sect. 6.1.3.

6.1.2 Optimal Actuator Design and Placement

Tristan Gally, Philip Kolvenbach, Anja Kuttich-Meinlschmidt, Marc E. Pfetsch, Andreas Schmitt, Johann M. Schmitt, and Stefan Ulbrich

The usage of active elements, e.g. actuators, in order to master uncertainty is one topical focus of this book, see e.g. Sects. 3.4 and 5.4. As an extension to the example of robust truss design in Sect. 6.1.1, we integrate abstract actuators inspired by the technologies introduced in Sects. 5.4.6 and 5.4.7. The resulting active truss structures can employ additional forces f^α in each of the different load scenarios. These forces form an additional design parameter, besides the cross-sectional areas of the bars. Thus, the robustness of the mechanical structure with respect to uncertain input data, e.g. the loads acting on the truss, can be further improved by the optimal design and placement of actuators. In Sect. 6.1.1 the robust optimisation approach as introduced in Sect. 6.1 is used to obtain robust truss topology designs. The incorporation of actuators into the model leads to more complex formulations compared to Sect. 6.1.1, e.g. certain actuators increase the number of stages in the robust optimisation problem. Nevertheless, we show that these can be reformulated as semidefinite programs (SDP). The objectives of this section are to integrate actuators into the truss topology models introduced in Sect. 6.1.1 and to reformulate the corresponding robust optimisation problems as SDPs to make these accessible to optimisation methods.

For this purpose, we present four models for the optimal design of active trusses, i.e. trusses incorporating active elements, such as actuators under uncertain loads. In the first model, the actuators generate a counterforce $f^\alpha \in \mathcal{F}^{\text{act}}$ acting on each free node of the truss for each load scenario. Here, the set $\mathcal{F}^{\text{act}}$ depends on the choice of the bars equipped with actuators. We call bars with integrated actuators *active* and those without actuators *passive*. This approach is considered in the paragraph below, see also [75]. Another possibility is the operation of the actuators via a parameterised algorithm, for example feedback controllers, where the parameters are considered as optimisation variables. This second model is examined in the next paragraph, cf. [111]. In the third model, we apply actuators with the aim to achieve an improvement of the buckling resistance of trusses, which is based on [64]. In the last paragraph, we introduce the fourth model, namely the optimal design of shunt damping for vibration attenuation as presented in [112].

Another approach for truss topology design with active elements was made by [116] using genetic algorithms but neglecting uncertainty. Further methods for the reduction of vibrations in mechanical systems by actuator placement can be found in [86, 140]. Buckling control has also been addressed by [16, 34, 139].

Robust optimisation of active trusses via mixed-integer semidefinite programming

In the following, we extend the robust static truss topology design problem from Sect. 6.1.1 with actuators of the first type. We introduce the binary variables $z \in \{0, 1\}^{\mathcal{E}}$ with $z_e = 1$, if bar $e \in \mathcal{E}$ is active and $z_e = 0$ otherwise. For all bars e, let $f_e^{\max} \in \mathbb{R}^{d_f}$ be the maximal force that can be applied to the truss by the actuator and let $\alpha_e \in [0, 1]$ describe the exposure of the actuator, if the bar is active. By

$$\mathcal{F}^{\text{act}}(z) = \left\{ f^\alpha = \sum_{e \in \mathcal{E}} z_e \, \alpha_e \, f_e^{\max} : \alpha_e \in [0, 1] \text{ for } e \in \mathcal{E} \right\}$$

we denote the set of all possible forces which the actuators given by z can implement. The goal is to choose the cross-sectional areas $x \in \mathbb{R}_+^{\mathcal{E}}$, $z \in \{0, 1\}^{\mathcal{E}}$ and the counterforces $f^\alpha \in \mathcal{F}^{\text{act}}(z)$, in such a way that the compliance $\frac{1}{2} f^\top u$ is minimal under the worst-case scenario with respect to the uncertainty of the force f. This leads to

$$\begin{aligned} \min_{\substack{z \in \{0,1\}^{\mathcal{E}},\\ x \in \mathbb{R}_+^{\mathcal{E}}}} \; \max_{f \in \mathcal{U}} \; \min_{f^\alpha \in \mathcal{F}^{\text{act}}(z)} \quad & \tfrac{1}{2}(f + f^\alpha)^\top u \\ \text{s.t.} \quad & A(x)u = f + f^\alpha, \; \sum_{e \in \mathcal{E}} l_e x_e \le V_{\max}, \; \sum_{e \in \mathcal{E}} z_e \le N, \end{aligned} \tag{6.10}$$

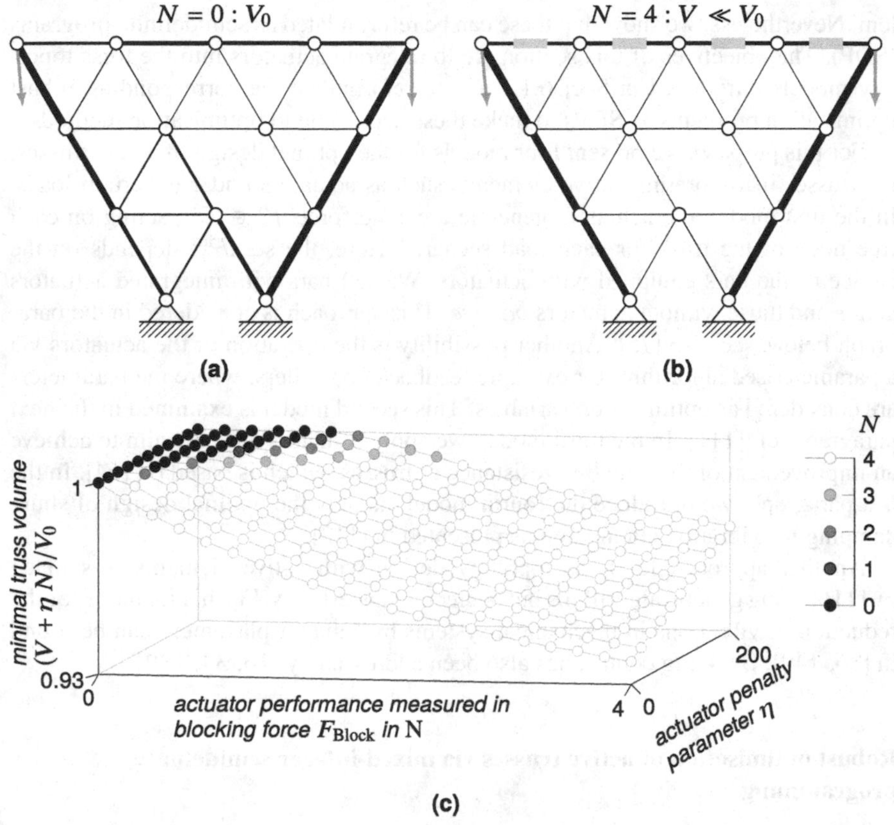

Fig. 6.2 Exemplary reduction of material for an optimised active/passive truss: **a** optimal passive, robust truss, **b** optimal active, robust truss with $N = 4$ grey active beams, **c** ratio of the total volume of optimised active trusses including penalty term and passive ones, depending on blocking force F_{Block} and penalty parameter η

where N is an upper bound for the number of active bars. In order to solve this problem, a key step in [75] is the splitting of (6.10) into an *inner* and an *outer problem*. The outer problem, which involves the decision for the binary variables $z \in \{0, 1\}^{\mathcal{E}}$, is solved by a Branch-and-Bound-type method. The remaining inner problem can be reformulated as a nonlinear SDP similarly to Sect. 6.1.1. This problem is solved by a sequential SDP algorithm, see [75]. We can analogously minimise the volume of a truss, where the compliance has to stay below an upper bound $c_{\max}$. This leads to a similar optimisation problem as (6.10).

Finally, we present the example of Fig. 6.2 also given in Sect. 3.4 to show that the usage of actuators can lead to substantial material savings. This also leads to significantly different trusses with minimal volume as shown by an optimal truss without actuators ($N = 0$), Fig. 6.2a, and with ($N = 4$) actuators, Fig. 6.2b. However, this does not consider the trade-off between potential actuator costs and material savings. Therefore, we show results in which we use the objective

$$\min_{z\in\{0,1\}^{\mathcal{E}},x\in\mathbb{R}_+^{\mathcal{E}}} \sum_{e\in\mathcal{E}} l_e x_e + \eta \sum_{e\in\mathcal{E}} z_e$$

for some actuator penalty parameter η. In the computations we model piezoelectric actuators for which $f_e^{\max}$ is given by

$$f_e^{\max} = \frac{2Ex_e}{2Ex_e + l_e\kappa_{\text{act}}} F_{\text{Block}},$$

where E is the Young's modulus of the used material, x_e denotes the cross-sectional area and l_e the length of bar e, κ_{act} denotes the stiffness of the actuator and F_{Block} the blocking force. This problem can be solved with the methods described above. In Fig. 6.2c, we illustrate the results for varying blocking force F_{Block} and penalty parameter η using the ratio of the total volume for the active case including penalty term and the total volume for the passive case. We see again that material savings are possible. However, high actuator costs can balance these savings.

Optimal feedback controller design

If we consider dynamic load scenarios, the reduction of structural vibrations with uncertain inputs, such as uncertain loads acting on a truss, becomes important. The usage of adaptive elements is a possibility to achieve this; it makes mechanical structures safer and more resistant against effects of external disturbances. These components consist of a sensor, a control unit and an actuator, thus compensating external forces acting on the structure. In [111], feedback controllers for the adaptive components are considered, which produce a response $f^\alpha(t)$ to external inputs based on a measurement of the current state $y(t)$, i.e. $f^\alpha(t) = Ky(t)$ for a matrix $K \in \mathbb{R}^{d_f\times d_f}$. Then the system of first order differential equations from Sect. 6.1.1 reads

$$\dot{y}(t) = P(x)\,y(t) + B(x)(Q\,a(t) + K\,y(t)), \quad u(t) = L\,y(t), \quad \text{for } t > 0,$$
$$u(0) = 0, \quad \dot{u}(0) = 0.$$

Here, $Q\,a(t) \in \mathbb{R}^{d_f}$ describes the uncertain loads lying in an ellipsoidal uncertainty set $\mathcal{U} = \{f = Qa : \|a\|_{L^2(\mathbb{R}_+;\mathbb{R}^{d_f})} \le 1\}$. For more details concerning the matrices $P(x)$, $B(x)$, $L \in \mathbb{R}^{d_f\times d_f}$ see Sect. 6.1.1. The matrix K is added to the Robust Dynamic Truss Topology Design Problem (6.9) as an additional optimisation variable. The corresponding optimal control problem is then given by

$$\min_{\substack{x\in\mathbb{R}^{\mathcal{E}}_+,\gamma\in\mathbb{R}_+,\\ K,Y,W\in\mathbb{R}^{d_f\times d_f}}} \quad \gamma+\varepsilon_Y\|Y\|_F^2+\varepsilon_K\|K\|_F^2+\varepsilon_W\|W\|_F^2$$

$$\text{s.t.} \quad \begin{pmatrix} (P(x)+B(x)K)^\top Y+Y(P(x)+B(x)K) & YB(x)Q & L^\top \\ Q^\top B(x)^\top Y & -\gamma I & 0 \\ L & 0 & -\gamma I \end{pmatrix} \preceq 0,$$

$$-Y,-W\preceq 0,\ \sum_{e\in\mathcal{E}} l_e x_e \le V_{\max},$$

$$(P(x)+B(x)K)^\top W+W(P(x)+B(x)K)\prec 0,$$

where ε_Y, ε_K and ε_W are positive constants. This optimisation problem can be solved by using a sequential SDP algorithm, cf. [111].

Active buckling control

An additional critical failure mode to be considered is buckling due to axial loads. In the following, we show its inclusion presented in [64], for the case of discrete cross-sectional areas for each bar indicated by binary variables x_e^a, where x_e^a is 1 if and only if bar $e\in\mathcal{E}$ has the cross-sectional area $a\in\mathcal{A}$, which we assume to be of circular shape, see also Sect. 6.1.1.

In order to model buckling, we need to know the bar force q_e for each bar e. Thus, we assume in the following a so-called statically determined truss, which allows to compute q using the invertible geometry matrix B and the equilibrium condition $Bq=f$ for a given force f. A compressed bar ($q_e<0$) buckles if the bar force exceeds the critical buckling load. For the pinned-pinned Euler buckling case this load is given by $\pi Ea^2/4\ell^2$, see e.g. [168], where E and ℓ are the Young's modulus and the bar's length, respectively. For a bar with area a under tension ($q_e>0$) an upper bound is given by $\sigma_0\, a$, where σ_0 is the proportional limit after which the stress-strain curve deviates from the linearity, see [44].

Buckling can be avoided by increasing the diameters of the bars. Alternatively, actuators, as presented in Sect. 5.4.7, can be used to avoid buckling by increasing the critical buckling load of a bar. To optimally place these active elements into the truss structure, we assume their buckling load increase ρ to be additive and independent of the bar area. Then the binary variables z_e, indicating whether the bar e is active, can be combined with the discussed buckling constraints in the inequalities

$$-\sum_{a\in\mathcal{A}}\frac{\pi Ea^2}{4\ell_e^2}x_e^a-\rho\, z_e\le q_e\le\sigma_0\sum_{a\in\mathcal{A}} a\, x_e^a.$$

If one minimises the volume of the truss for a given upper bound on the compliance $c_{\max}$, a bound N on the number of active bars and a polyhedral uncertainty set with forces $\mathcal{S}=\{f_1,\dots,f_s\}$, we obtain the following mixed-integer SDP

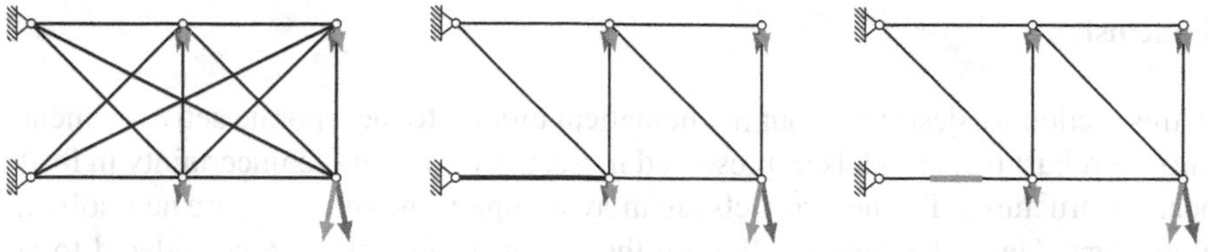

Fig. 6.3 Example truss for an active buckling control for two load scenarios: **a** ground-structure, **b** $r = 0$ active bars and **c** $r = 2$ active bars [64]

$$
\begin{aligned}
\min_{\substack{x\in\{0,1\}^{\mathcal{E}\times\mathcal{A}},\\ z\in\{0,1\}^{\mathcal{E}},\\ q^1,\dots,q^s\in\mathbb{R}^{\mathcal{E}}}} \quad & \sum_{e\in\mathcal{E}}\sum_{a\in\mathcal{A}} a\, l_e x_e^a \\
\text{s.t.} \quad & \begin{pmatrix} 2\, c_{\max} I & f_s^\top \\ f_s & A(x) \end{pmatrix} \succeq 0, \quad \text{for } f_s \in \mathcal{S}, \\
& B\, q^s = f_s, \quad \text{for } f_s \in \mathcal{S}, \\
& -\sum_{a\in\mathcal{A}} \frac{\pi E a^2}{4\ell_e^2} x_e^a - \rho\, z_e \le q_e^s \le \sigma_0 \sum_{a\in\mathcal{A}} a\, x_e^a, \quad \text{for } e \in \mathcal{E},\ f_s \in \mathcal{S}, \\
& z_e \le \sum_{a\in\mathcal{A}} x_e^a \le 1, \quad \text{for } e \in \mathcal{E}, \\
& \sum_{e\in\mathcal{E}} z_e \le N.
\end{aligned}
$$

We solve this problem using SCIP-SDP [147], see also Sect. 6.1.1. Exemplary, Fig. 6.3 shows optimal solutions, when minimising the total bar volume for two load-scenarios using no or at most two actuators on the ground-structure given by Fig. 6.3a. Without actuators one bar has to be bigger than the others, see Fig. 6.3b. Replacing this bar by an active one, the solution can be improved to use the smallest diameter everywhere, see Fig. 6.3c.

Optimal design of shunt damping

In addition to the methods described above, one can use shunted piezoelectric transducers to attenuate structural vibrations in mechanical systems, cf. Sect. 5.4.6. The attenuation of vibration strongly depends on the choice of the shunt parameters, where uncertainty in design and application may lead to a loss of attenuation performance. In [112], the Bounded Real Lemma is applied to formulate the corresponding optimisation problem as a nonlinear SDP, which is solved using a sequential SDP method for a demonstrating example.

Conclusion

In this section we described four mathematical models to incorporate active elements into the robust truss models as presented in Sect. 6.1.1 to handle uncertainty in load-bearing structures. The new models are more complex and often require new solving algorithms. Here, the forces acting on the nodes of the truss are considered to be uncertain. Section 6.3.4 extends this idea by also considering arbitrary bar-failures.

6.1.3 Mathematical Optimisation in Robust Product Design

Philip Kolvenbach, Alexander Matei, and Stefan Ulbrich

Optimisation tasks in engineering applications are usually described by (nonlinear) mathematical models and are often based on partial differential equations (PDE). However, these models depend on uncertain data in virtually every real-world application, for example, in the form of parameters that are not known exactly, cf. Sect. 2.1. In the context of mathematical optimisation, it is well known that optimal solutions are often sensitive to the problem data to such an extent that even small perturbations in the uncertain data can severely reduce the quality of a solution and might even cause a solution to violate important design and safety constraints. As a consequence, it is very important to take uncertainty into account during the optimisation process in early stage product design, cf. Sect. 3.1. One approach to achieve this is through robust optimisation, see also Sect. 6.1. In this subsection, a mathematical method to deal with robustness in the PDE-setting is depicted and applied to a truss structure that is derived from the Modular Active Spring-Damper System, see Sect. 3.6.1, and an integrated sensor element.

This exposition is based on [107, 150], which substantially extends the methods presented in [43, 178] by using a second order approximation instead of a linearisation of the worst-case function. Further methods to deal with nonlinear robust optimisation are investigated by [20, 36, 93, 121].

Robust optimisation as a two-level problem

Recall the nominal optimisation problem from Sect. 6.1

$$\min_x f^0(x, p) \quad \text{s.t.} \quad f^i(x, p) \leq 0, \quad \text{for } i \in I, \tag{6.11}$$

and its robust counterpart

$$\min_x \max_{p \in \mathcal{U}} f^0(x, p) \quad \text{s.t.} \quad f^i(x, p) \leq 0, \quad \text{for } i \in I, \quad \text{for } p \in \mathcal{U}, \tag{6.12}$$

where x is the optimisation (or design) variable, $p \in \mathcal{U}$ is a vector of uncertain parameters from an uncertainty set $\mathcal{U}$, f^0 is the scalar objective function, and f^i, $i \in I$, are finitely many scalar constraint functions.

In the case at hand, the objective and constraint functions depend on the physical state of the system, which in the following is the solution of a PDE. Therefore, the methods presented in Sects. 6.1.1 and 6.1.2 are no longer applicable and we develop alternative techniques. Problem (6.12) has infinitely many constraints but is easily seen to be equivalent to a two-level problem with finitely many constraints, i.e.

$$\min_{x} \max_{p \in \mathcal{U}} f^0(x, p) \quad \text{s.t.} \quad \max_{p \in \mathcal{U}} f^i(x, p) \leq 0, \quad \text{for } i \in I.$$

The optimisation problem can be further simplified by using the worst-case functions $\Phi^i(x) = \max\{f^i(x, p) : p \in \mathcal{U}\}$, $i \in I_0 = I \cup \{0\}$, which yields

$$\min_{x} \Phi^0(x) \quad \text{s.t.} \quad \Phi^i(x) \leq 0, \quad \text{for } i \in I. \tag{6.13}$$

While quite similar to (6.11) in form, the so-called *robust counterpart* (6.13) is decisively more difficult to solve because of its two-level structure and, in particular, because the lower-level maximisation problems are generally non-convex, but need to be solved globally in order to evaluate the worst-case functions. In addition, the worst-case functions are generally non-smooth functions, which hinders the application of efficient gradient-based optimisation methods. Even so, there has been considerable progress in the field of non-smooth constrained optimisation in recent years, such that suitable optimisation methods for (6.13) exist and are openly available; for example, see [38, 39].

An alternative way to deal with the non-smoothness is to lift optimality conditions, if available, of the lower-level problems to the upper-level problem, thereby obtaining an often smooth, single-level mathematical program with complementarity conditions (MPCC) eligible to tailored sequential quadratic programming (SQP) methods, see [114, 150].

The other difficulty—having to globally solve non-convex programs—is much more severe, especially in applications that involve PDE that make function evaluations of f^i, $i \in I_0$, extremely expensive. One approach is to approximate the functions $p \mapsto f^i(x, p)$, $i \in I_0$, by Taylor models of first or second order; see [6, 43, 107–109, 150, 151]. Once the models are built, their global maxima (i.e. their worst cases) can be computed very efficiently without further function evaluations of $p \mapsto f^i(x, p)$ and, hence, without further solving expensive PDE. Since Taylor models can only be expected to be locally accurate, strategies have been investigated to iteratively move the model expansion point in the course of the optimisation in order to increase the quality of approximation by closing the gap between the model and the modelled functions [6, 114]. In the following, we apply this approach to two examples of shape optimisation problems from structural mechanics.

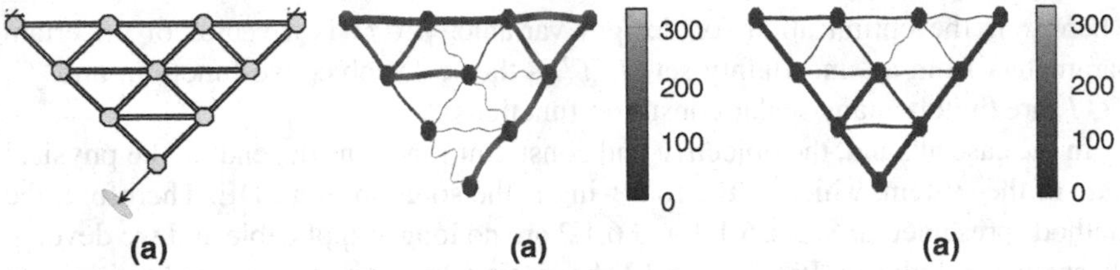

Fig. 6.4 **a** Initial truss with the position of the uncertain load, **b–c** snapshot of the displacements y in the optimal non-robust truss (**b**) and the optimal robust truss (**c**) under their respective worst-case dynamic load [107]

Example 1: a truss structure subject to uncertain dynamic load

As a first example, we consider a truss structure with 18 bars and ten connector nodes, see Fig. 6.4a, also considered in [107, 108]. The truss supported at two of its outer nodes is subject to an uncertain time-dependent diagonal load at the bottom, indicated by an arrow and an ellipse. The goal is to redistribute the volume between the 18 bars so as to minimise the L^2-norm of the displacement over space and time, with identical upper and lower bounds for all bar volumes. The physical behaviour is modelled by an equation of motion with linear elasticity. This PDE is discretised in space with a standard finite element method and a Newmark method in time, which leads to more than four million degrees of freedom in total. A scaled 10 % ellipsoid in the space-time domain, which is centred around a constant diagonal force, is chosen as the uncertainty set for the load.

Figures 6.4b, c show the optimal non-robust and optimal robust truss structures under their respective worst-case loads at the time point of maximum displacement; the greyscale indicates the von-Mises stress in MPa. The dynamic behaviour of the non-robust and the robust truss structures is displayed in the plots of Fig. 6.5. From both figures we infer that the non-robust structure not only has a considerably larger scale on the displacement $y(t_k)$, but it is also clearly susceptible to resonance on the fixed time interval, unlike the robust structure. These observations illustrate that the worst-case behaviour of the robust structure is much better—in terms of the objective function by a factor of 13 in this example, compare Table 6.2.

Table 6.2 also demonstrates the cost of robustness in terms of the number of additional PDE that have to be solved (PDE s.). Even so, the increase can be mitigated by a factor of 10 by utilising specialised, i.e. so-called matrix-free, second-order optimisation methods for the lower-level problems; see row "Robust-2".

Example 2: sensor element with manufacturing tolerance

When uncertainty is taken into account during the design phase of a mechanical structure, the designer usually has to have specific knowledge on the source of uncertainty. Since it is often too demanding to assume all potential sources of uncertainty

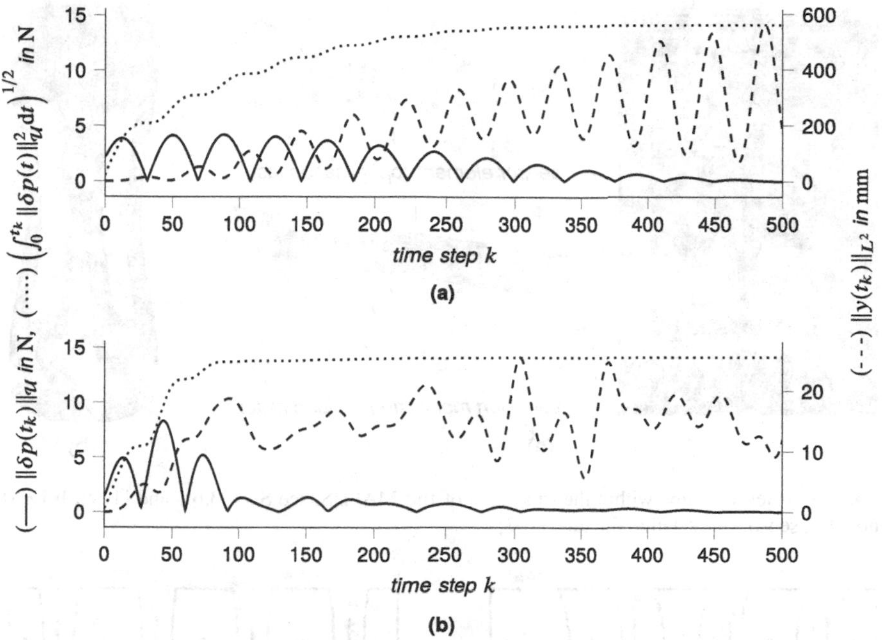

Fig. 6.5 Worst-case dynamic behaviour of **a** the non-robust and **b** the robust truss structures. In both plots, on the left axis, (—) is the magnitude of the worst-case load over time in N, (·····) is the accumulated L^2-norm of the worst-case load over time and on the right axis, (- - -) is the accumulated L^2-norm of the displacement in space in mm [107]

Table 6.2 Relative worst-case objective, number of optimisation iterations (it), number of steps that were fully accepted by the line search (fsteps) and number of solved PDE (PDE s.) for different truss optimisation methods [107]

Method	Rel. worst-case objectiveit		fsteps	PDE s.
Non-robust	13.25	146	142	450
Robust	1.00	143	127	339928
Robust-2	1.02	67	61	3672

are known, it is crucial to monitor safety-related structures during their usage phase. As an example, we consider a novel manufacturing process that integrates sensors into the inside of metallic structures such as bars enabling such monitoring; see Fig. 6.6. Specifically for bars, the sensor element needs to be sensitive to axial loads, but insensitive to transverse forces. At the same time, it should not be too sensitive, because otherwise the signal-to-noise ratio becomes unfavourable. In a collaboration between mathematicians and mechanical engineers [107, 109], we modelled this task as a shape optimisation problem. A robust optimisation approach has been applied in order to preserve the desirable sensor properties despite manufacturing tolerances. The physics are modelled by the three-dimensional equations of linear

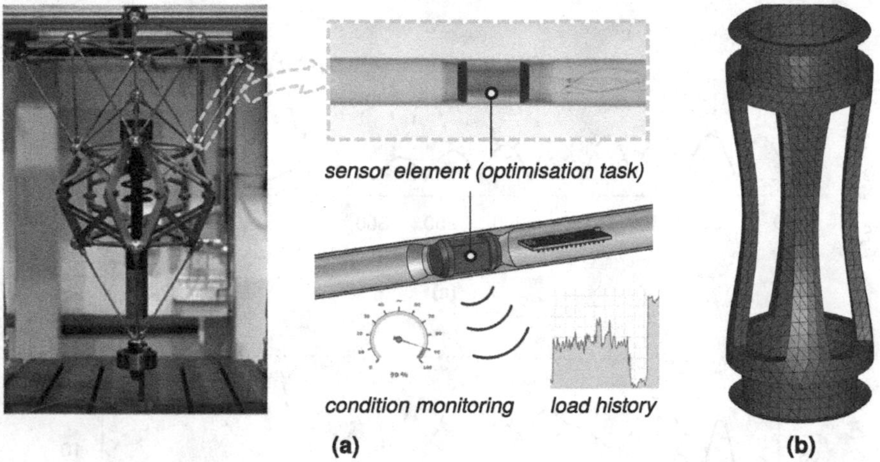

Fig. 6.6 **a** Sensory tube within the truss bars of the MAFDS, see Sect. 3.6.1 and [109], **b** FEM model of sensor body within the tube [107]

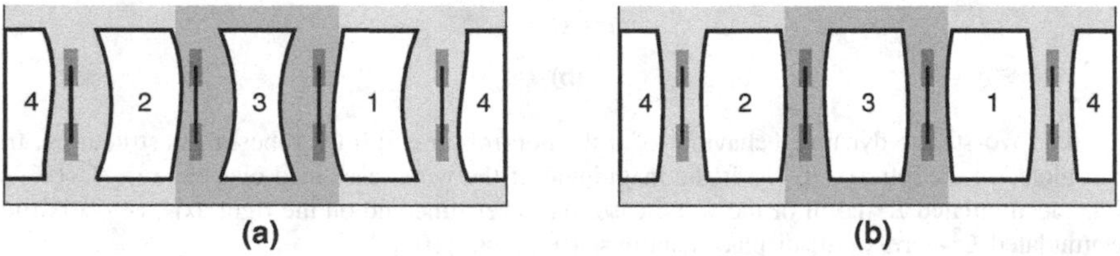

Fig. 6.7 Mantle view of the sensor element for **a** the optimal non-robust design and **b** the optimal robust design [107]

elasticity, which are solved with a finite element method with about 750 000 degrees of freedom in total.

Figure 6.7 shows the mantle of the cylindrically shaped sensor elements for the non-robust and the robust solution, respectively. The white areas are holes cut into the steel, compare to Fig. 6.6b; their shape is parameterised by cubic Bézier curves and chosen during optimisation. The black vertical lines show the desired but uncertain position of the four strain gauges used to measure the axial loads. The optimisation results are given in Table 6.3, where q stands for the ratio between transverse and tensile sensitivity (smaller is better) and $\|c_+\|$ for the relative constraint violation of the sensitivity bounds. In both cases, the prefix "max" indicates the respective value in the worst case under manufacturing tolerance. It can be seen that the robust solution is feasible in every scenario and even has a better worst-case sensitivity ratio than the non-robust solution, at the cost of a slightly worse sensitivity ratio in the undisturbed case, and greater computational effort (see columns it and eval). The optimal robust design, Fig. 6.7b, shows a perfect symmetry even though there was no symmetry constraint given. This result can be explained by the observation that tensile strains

Table 6.3 Optimisation results for the different problems and methods. The table displays the objective function value (q), the constraint violation at the respective solution ($\|c_+\|$) and their robust counterpart function value ($\max q$, $\max \|c_+\|$), both for the non-robust/undisturbed and for the robust problem. Also the number of iterations (it) and the number of objective function and constraint evaluations (eval) are given

Problem	q	$\max q$	$\|c_+\|$	$\max \|c_+\|$	it	eval
Non-robust	1	1	0	0.7369	41	153
Robust	1.0274	0.9566	0	0	192	467

at one bar cause compression strains at the opposing bar. Thus, the solver tends to solutions in which opposing bars have nearly the same geometry. Furthermore, the cross-section of the bars has been reduced by the optimisation algorithm. This leads to an increase in the observed strains which seems very plausible since the solver has the task to increase the sensitivity of the sensor.

Conclusion

In conclusion, we have seen that second order approximations of the robust counterpart with moving model expansion point perform well in PDE-constrained optimisation of mechanical structures. This method is indeed able to increase the robustness and efficiency of structural components in various applications.

6.1.4 Quantified Programs

Ulf Lorenz, Marc E. Pfetsch, and Andreas Schmitt

Uncertainty is ubiquitous in the production phase, as well as in the usage phase of products, see Sect. 1.2. On the one hand, random fluctuations in the properties of the semi-finished parts or the raw material occur; on the other hand, uncertainty results from unpredictable process behaviour, or due to the fact that the behaviour of the end customers is difficult to predict, see also Sect. 3.2. In this section, we showcase a mathematical modelling and optimisation method, which was developed to master uncertainty in process chains.

The goal is to provide a general framework to optimally solve multi-stage decision-making problems under uncertainty. To model this structure, the class of optimisation programs termed quantified problems (QP) by Subramani [156] is used. In these problems, each variable is associated with the existential or universal quantifier. This corresponds to the mentioned multi-stage structure, in which variables depend on the variables of previous stages. Therefore, we are able to model problems with a dynamic structure, e.g. decisions have to be made before an uncertain outcome is revealed.

Quantified programming has close ties to robust optimisation as introduced in Sect. 6.1. In fact, the decision version of every (integer) linear robust optimisation problem with an interval uncertainty set $\mathcal{U}$ can be modelled as a quantified linear program consisting of existential quantifiers followed by universal quantifiers. However, due to the possibility of a mixed appearance of the two kinds of quantifiers leading to more than two stages, but also due to the integrality constraints, more general methods than the ones presented in Sects. 6.1.1, 6.1.2 and 6.1.3 are needed. In this section we present theoretical results and extensions to the QP-framework.

Quantified programs

A quantified linear program (QLP) is the problem to decide whether the following logic formula in real variables $x_1, \ldots, x_n$ holds:

$$\exists x_1 \in [\ell_1, u_1]\ \forall x_2 \in [\ell_2, u_2]\ \ldots\ \exists x_{n-1} \in [\ell_{n-1}, u_{n-1}]\ \forall x_n \in [\ell_n, u_n]\ :\ Ax \leq b.$$

Here, n is even, $\ell, u \in \mathbb{Z}^n$ are lower and upper bounds, and the coefficient matrix $A \in \mathbb{Q}^{m\times n}$ as well as the vector $b \in \mathbb{Q}^m$ define the linear constraints $Ax \leq b$. An arbitrary sequence of universally and existentially quantified variables is possible by including dummy variables. A quantified integer program (QIP) additionally restricts the variables to attain integral values.

One interpretation of quantified programs is given via two-person zero-sum games: The existential player plays against the universal player. During the game, the values of the existential ($\exists$) and universal ($\forall$) quantified variables are chosen in order 1 to n by the corresponding player in the given variable bounds $[\ell, u]$. In iteration i, the previous values x_1 to x_{i-1} are known. The existential player wins if the condition $Ax \leq b$ holds. The question is whether there is a winning-strategy for the existential player, i.e. can this player win the game independently of the universal player's actions?

An illustration is the design and time-discretised operation of a technical system under an uncertain and time-varying load. If lower and upper bounds to the load are known, the first block of existential variables could model the selection of components, whereas the t-th following pair of universally and existentially quantified variable blocks models the worst-case load and the corresponding system's operation at time t. In this way, the solution design will be able to handle every possible combination of loads.

Properties and extensions

It is known that QLP is coNP-hard and QIP is PSPACE-complete [157], i.e. their solution is theoretically hard, but QIPs make it possible to model a wide range of applications. The expressiveness of QIPs has been shown by using it to model the classical job-shop and car-sequencing scheduling problems in [49] and the PSPACE-complete game Go-Moku in [50]. Polyhedral properties have been researched in [118].

If there even exist several winning strategies for the existential player, a best choice with respect to a given measure can be considered. One example, see [119], is a min-max-objective for a vector of objective coefficients $c \in \mathbb{Q}^n$:

$$\min_{x_1 \in [\ell_1, u_1]} \left(c_1 x_1 + \left(\max_{x_2 \in [\ell_2, u_2]} c_2 x_2 + \left(\min_{x_3 \in [\ell_3, u_3]} c_3 x_3 + \left(\dots \max_{x_n \in [\ell_n, u_n]} c_n x_n + F(x) \right) \dots \right) \right) \right),$$

where $F(x) = 0$ if x satisfies $Ax \leq b$ and $F(x) = \infty$ otherwise. Thus, an optimal solution will be a winning strategy of the existential player with minimal cost and minimal worst-case cost of the universal player. This extension has greater expressive power. Problems with objective functions composed of costs or efficiency can be treated.

In recent works [77, 80], the above framework is extended to bound also the universally quantified variables using a polytope. Here, a player loses, if she is the first not being able to satisfy her system of inequalities. This makes it easier to formulate problems involving more constraints on universal variables, e.g., the maintenance of a machine (∃-variable) prevents its failure (∀-variable). Instances of this extension can be reduced to greater but polynomial sized instances of the interval case. Thus, obstacles when formulating application-problems as QIP are simplified without raising theoretical complexities.

To solve quantified programs, different techniques can be used. As it has been noted early in the literature, see [48], one possibility is to use the so-called deterministic equivalent problem (DEP), which contains the existential variables and includes the universal variables placed at their bounds for QLP or their integer feasible values for QIP. The multi-stage character of problems makes it possible to use a specialised nested Benders decomposition to solve the DEP of a QLP [48]. The interpretation as a game motivates the usage of the Alphabeta algorithm as shown in [51, 120]. By reordering the quantifiers in a given quantified program, the so called quantifier shifting, an efficient relaxation can be formed [174]. Further, a pruning technique is shown to be computationally efficient in [79].

These and more techniques are applied in the QIP-solver Yasol [47], which is used, e.g. to solve a resilient booster design problem [78].

Conclusion

Quantified programming is a framework to model multi-stage structure in optimisation problems. Many QIPs were solved by using the DEP, as implicitly done in [8] and [141]. Future research will hopefully allow to replace algorithms based on DEPs by improved methods, equally leading to a wider applicability of QIPs.

6.1.5 Mastering of Disturbing Influences in Early Phases of Product Development

Fiona Schulte and Hermann Kloberdanz

In principle, robust design follows the same objective as sustainable design. This means that products should be developed with regard to functionality, costs and availability to ensure that acceptance is by and large guaranteed, as described in the Sects. 1.6 and 3.5. This equally includes very different product usages and future changes in the entire product life cycle. The challenge for developers is particularly high when products are newly developed and when they have little experience with the planned product usage. This section presents how robust design can be applied to meet increased expectations, especially for new types of usage with intensive disturbance effects.

In view of rapidly changing technologies, markets and customer needs, innovative products are of great importance for the sustainable success of companies. New production processes and process chains as well as new types of usage processes and environments offer opportunities for the successful marketing of innovative products. However, their development is also associated with a high risk due to uncertainty.

Frequently, a high degree of innovation can only be achieved by developing a product from scratch. A lack of experience and missing reference products as well as working at a high level of abstraction at the beginning of the development represent great challenges. In particular, developers have to make far-reaching decisions in the early stages of the development process, even though the product is still widely unknown. Overall, the situation is characterised by a lack of reliable information, which correlates with a high degree of uncertainty as introduced in Sect. 1.3 Therefore, mastering of uncertainty in the early phases of the development of innovative systems is of great importance. In addition, the new development of products and systems is very complex, since almost all properties have to be defined depending on different requirements, cf. [12].

To master the complexity, developments of new systems are performed systematically and supported by methods. Development processes according to the guideline VDI 2221 [172] are widely used. This guideline recommends a discursive development process, which is structured according to phases and work steps in which defined results are achieved. The basic phases are (i) task clarification and project definition, (ii) concept development, (iii) embodiment design and (iv) detail design.

Robust design in the early phases of product development

The first two phases 'task clarification and project definition' as well as 'concept development' are called the early phases of product development [22]. In these phases the basic characteristics of the products and systems to be developed are defined.

Especially in innovation projects, the early phases are intensified, as the developers are able to greatly influence the subsequent production processes and usage properties here, thus making a significant contribution to the success of the product. On an abstract level, models of functional structures, physical effects and working principles are developed. Deviations from the ideal function, disturbance parameters and their influences are only rudimentarily known and initially not taken into account in solution synthesis. The current robust design methods therefore focus mainly on the 'embodiment design' and 'detail design', where more concrete models of the developed product are already available [52, 124].

However, the full potential for mastering uncertainty can only be exploited if robustness is considered as a central criterion from the beginning of the development process [12]. Therefore, we developed further robust design methods for the early phases of product development [124, 125]. These methods provide a decisive way to master uncertainty in innovation projects.

In the early phases of product development, products and systems are only modelled in the form of process models and functional structures, see Sects. 5.2.3 and 5.1.2. Since these models contain little specific information about the system to be developed, they are poorly suited for mathematical modelling and simulation. Therefore, methods that support developers in the synthesis of robust concepts are more important than analytical methods. Methods to support system syntheses ideally complement the uncertainty analysis according to the UMEA methodology from Sect. 5.2.1 and support robust design by providing additional models and tools.

In the following we focus on two essential elements of the robust design methodology for early phases:

- assessment of sources of uncertainty based on physical effects: Sources of uncertainty can be identified comprehensively with checklists mainly based on physical effects. The checklists are compatible with the robustness evaluation of principal solutions during concept development.
- strategies of mastering uncertainty caused by disturbance parameters: The strategies describe the principle mastering of uncertainty caused by disturbances and serve as an orientation for the development of solution approaches and their prioritisation.

Assessment of uncertainty source in early phases of product development

In case of new product developments, the systems are considered as a whole in the 'task clarification and project definition' phase. Mainly, the process model is used to analyse the overall system in detail with regard to the expected benefits, the fulfilment of functions and the corresponding relationships with the system environment as shown in Sect. 5.2.3. On one hand, the planned use of resources and operation of the system, and on the other hand, disturbance parameters from the environment, as well as disturbing side effects on the system environment are considered.

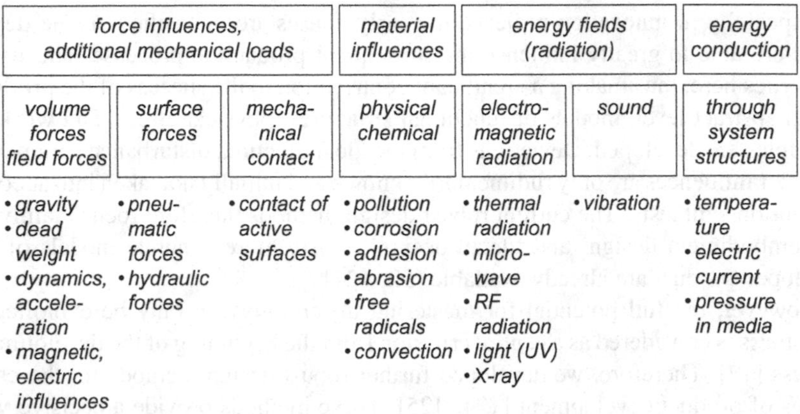
force influences, additional mechanical loads
material influences
energy fields (radiation)
energy conduction
volume forces field forces
surface forces
mecha-nical contact
physical chemical
electro-magnetic radiation
sound
through system structures
• gravity
• dead weight
• dynamics, accele-ration
• magnetic, electric influences
• pneu-matic forces
• hydraulic forces
• contact of active surfaces
• pollution
• corrosion
• adhesion
• abrasion
• free radicals
• convection
• thermal radiation
• micro-wave
• RF radiation
• light (UV)
• X-ray
• vibration
• tempera-ture
• electric current
• pressure in media

Fig. 6.8 Structure of checklists for disturbance identification in early phases of product development based on physical effects

With regard to robust design, especially potential disturbances and side effects have to be recognised, since they are the main identifiable sources of uncertainty in this phase. Side effects may not be acceptable, if they do not comply with restrictions, while disturbances can reduce the system's performance as described in Sect. 5.2.3. In the 'task clarification and project definition' phase, the developers can be supported especially by checklists for the determination of potential disturbance parameters. The usefulness of such checklists is mainly determined by their applicability. The applicability of the checklist is based on the checklist's structure, completeness and handling due to their scope and versatility listing the sources of uncertainty. The checklist's focus on physical effects is purposeful, since a substantial part of the causes of disturbances is considered and can be structured in a well-founded way [124].

In addition, detected disturbance influences can be assigned to principle solutions during concept development, which allows a simple robustness evaluation of the solutions as shown below. Furthermore, the sole consideration of physical effects is not sufficient for the complete detection of potential disturbance parameters as Mathias states [124]. For example, uncertainty due to contamination or other external influences must be added based on experience. For load-bearing systems, in addition, it is purposeful to emphasise aspects of force flow. We therefore propose the structure shown in Fig. 6.8, based on the proposal of Mathias.

These checklists primarily support the identification of relevant disturbance parameters as safety-relevant requirements and their documentation in requirement lists. These serve as a basis for decision-making in the entire development process as discussed in Sect. 5.1.

Strategies of mastering uncertainty in early phases of product development

Furthermore, developers can be supported effectively in their search for solutions using reference objects as orientation. In the phase of 'concept development', very basic strategies and principles prove to be suitable as work is done at a high level of abstraction. The analysis of the system's vulnerability follows the analysis of the system environment with regard to disturbance parameters as indicated in Fig. 6.9. Vulnerability describes the possibility of serious functional disorders of the system due to external disturbances. Both, the exposure and sensitivity of the system to the disturbance parameters, must be taken into account. Therefore, we derived three basic strategies for the development of robust concepts from the chain of effects of disturbance parameters as shown in Fig. 6.9: (i) eliminate disturbance parameters, (ii) reduce (or eliminate) the influence of disturbance parameters, (iii) avoid the impact of disturbance parameters, cf. [124, 125].

The elimination of disturbance parameters means to use the system only in an environment where no relevant disturbance parameters are present. Solutions restricted to these conditions are known as, for example, air-conditioned measuring rooms and particle-free clean rooms. For load-bearing systems, such solutions usually are impracticable or not very effective, since their area of application would be restricted.

The reduction of the influence of disturbance parameters is a frequently used strategy. In most cases shields, insulations, seals, housings or surface coatings reduce the influence of radiation, dirt or mechanical impact. These approaches can be understood as robust design in a broader sense. However, the application of this strategy is mostly associated with additional measures or components. Protective measures are often not an optimal solution in terms of additional effort, limited effectiveness and additional uncertainty.

By contrast, solutions that are based on the strategy to avoid the impact of disturbance parameters are understood as robust design in a narrower sense [12]. Typical robust solution approaches select functional principles based on physical effects that are basically not or only slightly influenced by the expected disturbance parameters [125]. For example, extreme temperatures have less effect on mechanical solution principles than on electronic solutions. Conversely, electronic components e.g.

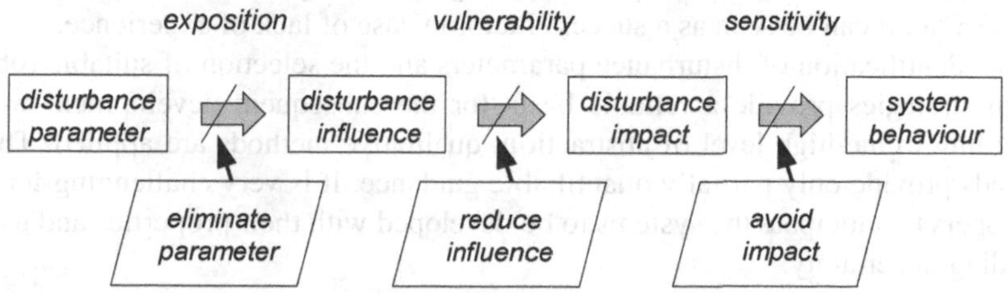

Fig. 6.9 Robust design strategies of mastering uncertainty caused by disturbance parameters in early phases of the product development, cf. [126]

engine control units are less sensitive to strong accelerations. The strategy to avoid the impact of disturbance parameters must be prioritised because in most cases the overall solution is more cost effective and incorporates less uncertainty. In most cases, additional effort can be avoided from the beginning by considering the corresponding uncertainty.

The strategy to avoid the impact of disturbance parameters can also be applied in thc 'cmbodiment design' phase. For example, a symmetrical design of components or materials with a high thermal conductivity can avoid component distortion due to heat impacts and the associated functional impairment as shown in [124].

In the 'concept development' phase basic solutions are determined. The principle solutions are described by the underlying physical effects, working principles and working structures, and are presented as simple sketches [22]. Due to the high degree of abstraction, uncertainty regarding the flow of forces cannot be estimated at this stage of the development process.

However, potential uncertainty influences can be estimated based on a principal evaluation of the robustness [124]. A rough estimation of uncertainty can be done by evaluating the principal correlations between the identified disturbances and the physical effects on which the intended overall solution is based. Thus, to assess the robustness of solutions on the principle level, only the influences of disturbance parameters on physical effects need to be known. This method abstracts the approach of Taguchi [162], that is based on the signal-to-noise ratio for the evaluation of the robustness. However, an exact calculation of the signal-to-noise ratio representing the robustness properties of the system, as Taguchi strives for as discussed in Sect. 3.5, is neither possible nor necessary at this stage of the development. An estimation of the basic sensitivity to disturbance parameters is adequate for assessing uncertainty in this early phase. The sensitivity of physical effects to disturbances was summarised by Mathias in tables [124]. In addition, equations for the calculation of principal robustness values have been developed, which, however, are only suitable for a relative comparison of alternative solutions of the same system.

Conclusion

In the early phases of the development of new innovative products, the robust design approach contributes significantly to mastering uncertainty from the very beginning. In particular, it can be seen as a success factor in case of lack of experience.

The identification of disturbance parameters and the selection of suitable robust design strategies provide a reliable basis for the subsequent development work. According to the high level of abstraction, qualitative methods are applied. These methods provide only partially quantifiable guidance. It is very challenging for the developers to anticipate the systems to be developed with their properties and usage including uncertainty.

6.1.6 Uncertainty-Based Product Design in Robust Design

Hermann Kloberdanz, Fiona Schulte, and Eckhard Kirchner

A technical system is said to be robust, when it does not only fulfil its predefined function at the design point, but also in the surrounding neighbourhood, the so-called uncertainty set. The accepted functional quality is guaranteed even under uncertain resources or disturbances by uncertain external influences as defined in Sect. 3.5. Both, the constructional design and the development process of such systems, are referred to as robust design. The basic idea was developed by Genichi Taguchi [161, 162] and refined several times over decades, e.g. by Ulrich and Eppinger [170]. The characteristic of the system perceived by the user during usage under the influence of disturbance parameters is decisive and is determined by uncertainty [76]. Therefore, uncertainty has to be considered in the context with all phases of the product life cycle as described in Sect. 3.2. In this section, we show how and by which measures in the design of parts and components uncertainty can be mastered over all phases of the product life cycle in the sense of robust design.

Uncertainty in load-bearing systems mainly affects the performance of the force transmission function. Unacceptable deformations or damage to the system or its components are typical functional deviations. The load carrying capacity of such systems is mainly determined by the life cycle phases prior to product use Sect. 1.2. The product components are manufactured and assembled in production processes. These processes are performed by work equipment e.g. machine tools. Similar to usage processes, uncertainty also influences the production processes. Uncertainty of the network of manufacturing and assembly processes accumulates in deviations from product properties as explained in Sect. 5.2.3. A comprehensive mastering of uncertainty in robust design must therefore take all processes in the product life cycle into account when designing products and their components.

As a consequence, robust design demands that three basic requirements must be considered when developing load-bearing products:

- the products must prove to be insensitive to disturbance parameters in usage processes,
- uncertainty of the production processes may only accumulate to a small extent in the process chains or must even be able to be reduced,
- it must be possible to produce the components of the products with low uncertainty in critical processes.

The basic approach of robust design is to master uncertainty in the entire life cycle process network by designing the products. Therefore, we refer to this approach as process oriented robust design. This means that all life cycle processes and their interactions are not directly but indirectly defined by the embodiment design properties of the product components. In other words, process oriented robust design strives to reduce uncertainty of the properties of parts and components through their design by producing them in the tightest possible tolerances without additional effort or even with reduced effort. Process orientated robust design can therefore make an essential contribution to mastering uncertainty of load-bearing systems as part of life cycle engineering.

Mastering uncertainty caused by usage process influences

Uncertainty of system functions in usage processes comprises three types of sources:

- direct or indirect effect of external influences on the system or components of the system in usage processes (external usage uncertainty),
- mutual influence of the components of the system through side effects and interactions in usage processes (internal usage uncertainty of the system)
- deviation of the properties of the product acting as work equipment in the usage process which result from the previous life cycle processes (internal uncertainty of the system).

Well-known approaches of robust design e.g. Ulrich and Eppinger [170] assume that external and internal uncertainty in load-bearing systems during usage can be mastered mainly by the design of the mechanical components. The mechanical components form the essential part of load-bearing products in mechanical engineering. They guide and transfer the forces in usage processes and are referred to collectively as mechanical system. Therefore, the comprehensive mastering of uncertainty by the design of the components of load-bearing products in the embodiment design phase is in the focus of the consideration.

While robust design in early phases considers the product as a whole, see Sect. 6.1.5, the consideration of uncertainty in the embodiment design requires more detailed models as the solutions become more specific. Therefore, we have further detailed the model of technical processes presented in Sect. 5.2.3 for depicting the mechanical system in order to categorise the sources of uncertainty like in Fig. 6.10. In addition to the external disturbance parameters in the usage processes outlined in Sect. 6.1.5, sources of uncertainty can be depicted in this model due to varying functions of the components with regard to the force flow and the interacting influence of the components. In this way it is possible to locate typical sources of uncertainty regarding power transmission within the mechanical system.

In the example of the 3D Servo Press introduced in Sect. 3.6.3 servo motors generate the drive forces. However, the heat loss that is generated can heat up the transformer components and thus influence the power transmission as well as the ram movement and position. These dependencies can be illustrated and analysed analogous to the robust design approach in the early phase of the specialisation of the detailed generic process model according to Fig. 6.10.

Uncertainty due to external influences and mutual influence of the components can be mastered in embodiment design analogous to the strategies in the early phases. In particular, ensuring an unambiguous force transmission can reduce the effect of uncertainty influences [57, 60]. This approach is also known as design for clarity [59].

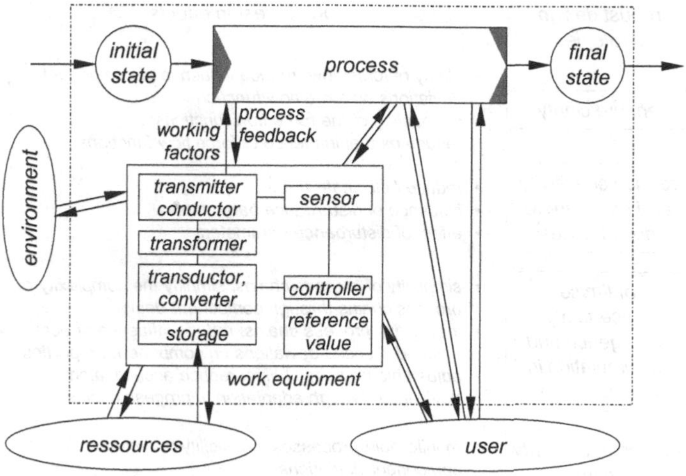

Fig. 6.10 Extended process model of technical systems, cf. [58]

Ensuring clarity is of significant importance in robust design. The basic character of clarity has already been recognised by Pahl [136] and formulated as the basic rule of design. Pahl even demands mandatory compliance with the basic rule of clarity. Kirchner has reworked these basic rules recently [103]. Here, clarity means the wide degree of independence of the force transmission and the resulting component stress from external influences as well as from component and part tolerances. For example, in the Modular Active Spring-Damper System, see Sect. 3.6.1, the load is clearly initiated via three points. In this way, a deviating force can be clearly recorded and the stress on the system components can be clearly calculated.

Mastering of uncertainty caused by production process influences

As stated above, robust design strives to minimise the accumulation of uncertainty in production processes to reduce deviations in usage processes. Uncertainty especially of critical production processes is mainly mastered by the design of the components. To cope with the high level of complexity of design and process-related interrelationships, we developed systematically structured procedures and assistances to support the robust design of load-bearing systems [57, 60]. For that, we structured the design advices and strategic procedures that serve as an orientation for the development of robust load-bearing systems. They are described based on a set of eleven robust design effects, see Fig. 6.11.

The robust design effects were determined based on a systematic analysis of the relationships between the properties of the product and in particular its components on the one hand and the life cycle processes on the other. These interrelationships summarise effective ways of influencing all life cycle processes by product and

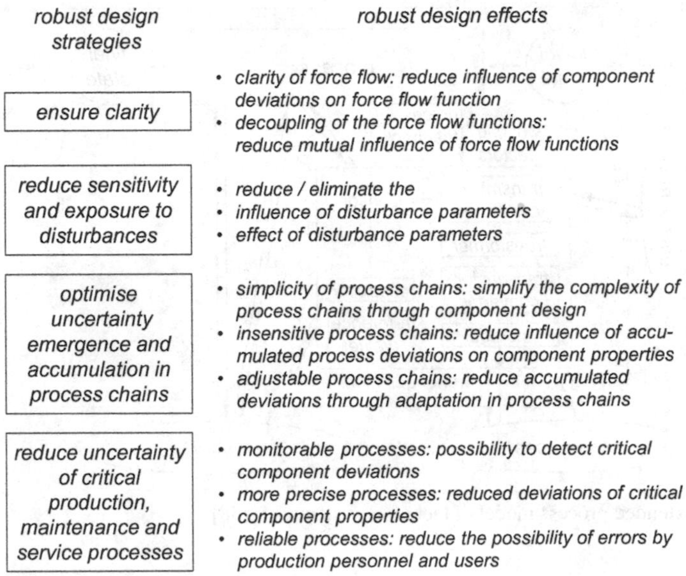

Fig. 6.11 Robust design strategies and effects for mastering uncertainty in process oriented robust design, following [57]

component design in terms of uncertainty. Thus robust design contributes directly and indirectly to mastering uncertainty of product usage. Furthermore, strategies were formulated to categorise the robust design effects. The order of the robust design strategies and effects recommends a prioritisation of their usage.

As stated above, ensuring clarity is of significant importance in robust design. This applies to mastering uncertainty in both usage and production processes. For example, there is ambiguity in force flow, if the mechanical system is kinematically overdetermined by multiple power transmission paths with high rigidity. Deviations in critical component dimensions usually lead to considerable constraining forces. Typical consequences are reduced functionality and unexpectedly high component stresses in the system that cannot be reliably calculated and lead to increased component wear. Dimensional deviations are often caused by uncertainty in component production, by elastic component deformation or by component deformation as a result of temperature influences. Ambiguous designs are generally only acceptable by considerable additional effort to guarantee narrow component tolerances. The clarity of robust structures can be supported by decoupling of power transmitting functions in order to avoid mutual influence of the power transmitting elements.

The strategy to reduce the sensitivity and exposure of components to disturbances corresponds to basic strategies for reducing uncertainty caused by disturbance parameters described in Sect. 6.1.5. While disturbances caused by the usage environment can only be reduced or eliminated to a limited extent in the total system, robust design offers the possibility of reducing influences (disturbances) from one system element to another. Typical examples are the prevention of waste, heat loss and wear particles, which reduce the component strength or can influence the functionality of other components through pollution. The influence of most disturbance parameters can be reduced or eliminated by robust design, for example by shielding or insulation. However, this requires additional components and system elements. Therefore, generally design solutions according to the robust design effect 'reduce or eliminate effects of disturbance variables' should be aimed at. As a rule, this can be achieved by an appropriate choice of material or component geometry that is insensitive to disturbance parameters. In order to master uncertainty, unambiguity and low sensitivity as prerequisites for robust design should be given the highest priority.

A detailed description of how robust design effects and strategies can be integrated in the development of products and components was provided by Freund [57]. As an aid for the application in robust design, he provides a comprehensive catalogue with examples ('RopEx catalog: Design notes for mastering uncertainty in usage processes' (German)) [57]. Typical examples for part and component design clarify the abstract robust design strategies and effects.

The formulation of the robust design strategies and effects presented here is mainly based on the power transmission functions of passive mechanical systems. The presented procedures, principles and instructions can be transferred analogously to semi-active and active systems.

Conclusion

The design of the components and parts, which is mainly determined in the embodiment design phase of product development, significantly defines the production processes and the interaction of the parts in usage. Thus, the process oriented robust design allows to master uncertainty, which arises during product usage due to the mutual influence of components and ambiguous force flow. Appropriately structured and prioritised robust design strategies and effects effectively support product developers by providing orientation. The applicability is considerably improved by demonstrative examples.

6.1.7 Non-linear Robust Closed-Loop Control of Presses with Geometric Singularities

Florian Hoppe, Dirk A. Molitor, and Peter Groche

Servo presses like the 3D Servo Press presented in Sect. 3.6.3 fulfil the purpose of producing metal-formed parts with high accuracy. Besides increasing the passive stiffness, active compensation measures in terms of closed-loop control have been established. Control laws are based on knowledge about the machine model, e.g. robot control mainly focuses on inverse kinematic models [152]. Especially kinematic singularities are very sensitive to uncertainty and require a high model accuracy. A kinematic model is typically based on the assumption of rigid bodies that are connected by rigid joints. Uncertainty affects the whole lifecycle of a product, cf. Sect. 1.2. In the case of a press, uncertainty occurs in form of inaccuracies during production and assembly, as well as thermal or elastic expansion during its use. Therefore, the geometric dimensions of the machine components are not exactly known. This data uncertainty as well as model uncertainty arising from ignored relevant physical phenomena, cf. Sects. 2.1 and 2, result in instable regions at the kinematic singularities, i.e. the top dead centre (TDC) and bottom dead centre (BDC). While methods based on stiffness models presented in Sect. 5.4.1 are able to reduce this uncertainty, they are not able to eliminate it [71]. An alternative approach is to accept the uncertainty and to increase the robustness of a control. Robust methods seek to reduce the impact of uncertainty on the worst-case scenario, which is the control near a singularity.

Servo presses are electromechanical systems consisting of a controlled servo drive and a mechanical gear. The most common gear kinematics are eccentric and knuckle-joint kinematics, both containing singularities. Despite the fact that servo drives allow for real-time adaptions and therefore to control the ram of the tool centre point (TCP), the closed-loop control of the TCP still faces unresolved challenges. Therefore, in industrial applications, only the drives are controlled in a closed-loop way, and the kinematics are open-loop.

Dulger et al. have drawn a parallel to robotics and have adopted the approach of a model-based control [46]. It was possible to demonstrate the feasibility of the control qualitatively, but the question of stability at the singularities remains open. In the investigation of a master-slave approach, Kirchner et al. have shown an instability in the dead centres [104]. This stability problem also occurs in robot kinematics and is typically avoided by limiting the joint space [152]. However, the operating point of presses is close to the BDC, since the highest possible transmission ratio of the force is achieved here. Current robust control methods focus on linear systems, and are not able to cope with non-linearities that emerge especially at kinematic singularities [94]. To investigate the influence of uncertainty, the press system is subdivided into the drive and the kinematics. The drive control has the task of adapting the actual drive speed $\dot{\varphi}$ to the setpoint speed $\dot{\varphi}_{\text{ctrl}}$. Since synchronous motors allow an almost step-like change of the drive torque, the controlled drive can be approximated as a time-invariant first-order transfer function

$$\ddot{\varphi}\tau + \dot{\varphi} = \dot{\varphi}_{\text{ctrl}} \tag{6.14}$$

with the settling time τ. Since all quantities of the differential equation can be measured or derived on the drive side, τ can be identified by means of the step response as described in [92]. The transmission behaviour, however, depends largely on the load torque, which is governed by the coupled inertia of the machine, friction and the forming force. Since the eccentric angle φ changes the transmission ratio of the machine and thus the feedback of the mass inertias, τ is dependent on φ. While the smallest mass moment of inertia is applied to the drive in the dead centres, as

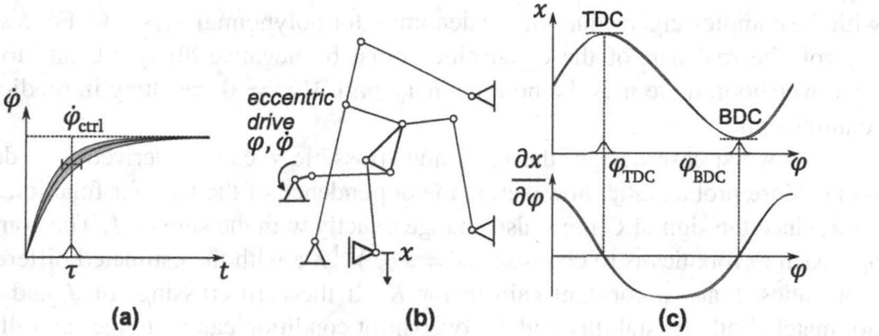

Fig. 6.12 Uncertainty in a closed-loop control of a servo eccentric press: **a** effect of uncertain time constant τ on the drive speed dynamics, **b** kinematic diagram of the 3D Servo Press, **c** kinematic function $f(\varphi)$ with uncertain model parameters causing uncertain dead centres φ_{TDC}, φ_{BDC}

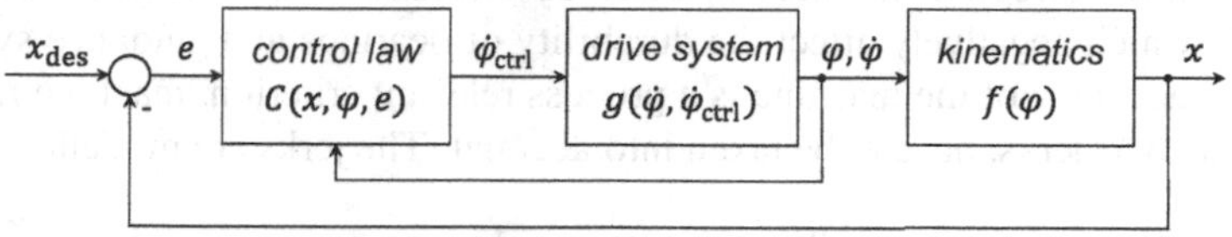

Fig. 6.13 Control loop for a servo press ram control

the power transmission is theoretically infinite here, the mass moment of inertia is greatest between the dead centres. The interval of τ can be obtained by identifying it in both points.

The position of the input φ and the kinematics of the machine $f(\varphi)$ result in the position at the output x, which is the control variable. To derive a control law, the differential kinematics, i.e. the Jacobian J

$$\dot{x} = \frac{\partial f(\varphi)}{\partial \varphi} \dot{\varphi} = J(\varphi) \cdot \dot{\varphi} \tag{6.15}$$

is required. These can be determined from the nominal geometric values of the rigid bodies, but in practice are subject to uncertainty as shown in Fig. 6.12.

For the press system consisting of drive and kinematics, we seek to design a control law C shown in Fig. 6.13 that stabilises the control loop while maximising its performance. Performance criteria are the settling time of control deviations e and the overshoot, which are represented by the integral of $e(t)$ in a time interval T. To investigate the stability, we assume J and C to be approximately constant and thus a linear ordinary differential equation can be used. This results in the transfer function of the closed control loop in the Laplace s-domain

$$G(s) = \frac{x(s)}{x_{des}(s)} = \frac{CJ}{\tau s^2 + s + CJ} \tag{6.16}$$

with the complex eigenvalues of the denominator polynomial $\lambda_{1/2} \in \mathbb{C}$. For a stable control, the real part of the eigenvalues must be negative $\Re\{\lambda\} < 0$ and to prevent overshoot, there must be no imaginary part $\Im\{\lambda\} = 0$, resulting in predictable requirements.

For a worst-case design, the maximum possible τ can be derived as a design point. More problematic, however, is the dependence of the transfer function $G(s)$ on J, since the sign of C must also change exactly with the sign of J. The common approach in robotics is to choose $\dot{\varphi}_{\text{ctrl}} = \hat{J}(\varphi)^{-1} K e$ with the estimated differential kinematics $\hat{J}$ and a constant gain factor K. If the zero crossings of J and $\hat{J}$ do not match, both the stability and the overshoot condition can no longer be fulfilled, resulting in very high $\dot{\varphi}_{\text{ctrl}}$. The resulting motion again leads to a change of J, back into the stable area. Hence, the instability is locally limited in a region around the TDC and BDC of φ.

These instable regions are accompanied by undesirable acceleration peaks and thus jerk, which negatively affect the durability of bearings and guidance systems. While the accuracy of the machine is a process relevant criterion, machine relevant criteria, such as jerks, have to be taken into account. The jerks are modelled as

$$\dddot{\varphi} = -\tau^{-1}\ddot{\varphi} + \tau^{-1}\ddot{\varphi}_{\text{ctrl}} \tag{6.17}$$

$$\dddot{x} = \ddot{J}\dot{\varphi} + 2\dot{J}\ddot{\varphi} + J\dddot{\varphi}. \tag{6.18}$$

A common approach in robot applications is to lock the dead centres by setting $\hat{J}^{-1}$ to zero in a region around φ_{TDC} and φ_{BDC}. However, since the dead centres of a press are a relevant operating point, this approach is not practical for presses. Another approach is the damped least-squares, which is based on a regularised inverse Jacobian [32]. This causes the inverse Jacobian to become increasingly damped near the singularity. While this reduces the effect of the instability and allows to control positions near the dead centres, the speed starts to creep, increasing the settling time dramatically. The result is getting stuck at the dead centres. Therefore, we investigate both, process and machine relevant criteria, in our proposed control.

As the desired motion of the machine typically is known before, it is possible to combine closed-loop control with a feedforward control. The task of motion control can then be split into the feedforward path, which generates the control speed for the open-loop kinematics $\dot{\varphi}_0(t)$, and the closed-loop control, which only compensates for disturbances. Using the generalised regularisation of Tikhonov to find an optimal control law

$$\dot{\varphi}_{\text{ctrl}} = \operatorname{argmin}_{\dot{\varphi}} \|J\dot{\varphi} - K(x_{\text{des}} - x)\|_2^2 + \|\gamma(\dot{\varphi} - \dot{\varphi}_0)\|_2^2 \tag{6.19}$$

allows to include a null motion in the regularisation term, which takes effect when approaching the singularities [166]. This null motion can be used to deduce a combined closed-loop and open-loop control law C_4, which smoothly switches to open-loop via $\dot{\varphi}_0 = \hat{J}^{-1}\dot{x}_{\text{des}}$ close to the singularities. It corresponds to an additional zero

Table 6.4 Simulation results for different control laws using $K = 50, \gamma = 0.02$

Type	Control law for $\dot{\varphi}_{\text{ctrl}}$	$\mu(\|e\|/h)$	$\|e\|_{\max}/h$	$\mu(\|\dddot{\varphi}\|)$ in 2ß/s^3	$\mu(\|\dddot{x}\|/h)$ in 1/s^3
C_1	$\hat{J}^{-1}\dot{x}_{\text{des}}$	$21.9 \cdot 10^{-3}$	$36.0 \cdot 10^{-3}$	0.00	0.65
C_2	$K\hat{J}^{-1}e$	$6.3 \cdot 10^{-3}$	$10.5 \cdot 10^{-3}$	5491.84	7.98
C_3	$(\hat{J}^2+\gamma^2)^{-1}\hat{J}e$	$9.6 \cdot 10^{-3}$	$28.2 \cdot 10^{-3}$	3.66	6.66
C_4	$(\hat{J}^2+\gamma^2)^{-1}(\hat{J}e+\gamma^2\dot{\varphi}_0)$	$6.2 \cdot 10^{-3}$	$10.8 \cdot 10^{-3}$	0.04	0.32

in the transfer function and speeds up the settling time for changes in the setpoint x_{des}.

Based on the 1600 kN version of the 3D Servo Press presented in Sect. 3.6.3, we examine different control approaches in a simulation with uncertain parameters. The uncertainty is chosen according to the machine's actual manufacturing tolerances, stiffness, actual sensor errors and noise.

The performance criteria for the control are the absolute error $|e| = |x_{\text{des}} - x|$, but also the magnitude of jerks $\dddot{\varphi}$, $\dddot{x}$. Table 6.4 shows the results for an open-loop control C_1, a closed-loop control with inverse kinematics C_2, a closed-loop control with regularised inverse kinematics C_3 and the combined control C_4.

Because of the singularities, a large dependence on the operating point φ is to be expected. Therefore a trajectory for all control laws is chosen to cover the complete operating area of $\varphi \in [0, 360°]$. To extract one feature for each performance criterion, the mean $\mu(\cdot)$ of the time series is calculated. The variance of the error e is normalised to the stroke height h, which is the difference between the dead centres $h = x_{\text{TDC}} - x_{\text{BDC}} = 100\,\text{mm}$.

The simulations show that an open-loop control C_1 allows to smoothly control the machine with little jerks but neglects disturbances and uncertainty leading to a significant normalised position error. While an inverse kinematics control C_2 increases the accuracy in x, it also leads to higher jerks, mainly at the dead centres. According to (6.18), high jerks in the drive $\dddot{\varphi}$ have little effect on jerk in the ram $\dddot{x}$ at the dead centres where $J \approx 0$. Therefore, control design C_2 leads to high jerks in the drive and moderate jerks in the ram. C_3 finds a trade-off between accuracy and jerk in $\dddot{\varphi}$. The only minor reduction in $\dddot{x}$ shows, that the contributing jerks do not occur near the dead centres. All closed-loop controls share the drawback that they only react to position and errors are not able to anticipate. C_4 combines the ability to compensate errors while anticipating trajectory changes. The result is a position error comparable to C_2 and a drastically smoothed motion whose jerks are comparable to C_1.

A typical approach to master uncertainty in the control of machines is closed-loop control. However, especially in servo presses, stability issues occur at their dead centres of motion, i.e. at their kinematic singularities. These lead to undesired jerks that affect the durability of bearings. Therefore, we presented and evaluated open-loop and closed-loop control methods taking into account process accuracy as well as machine criteria, i.e. jerk. The investigations have shown that a combination of open-loop and robust closed-loop resolves the conflicting objectives of high process accuracy and low machine jerk to the greatest possible extent.

6.1.8 Mastering Uncertainty in Tapping and Reaming by Robust Tools and Processes

Christian Bölling, Felix Geßner, Eberhard Abele, and Matthias Weigold

As described in Sect. 4.1.3, machining processes that are used in the production of technical systems are generally affected by data uncertainty in form of incertitude, see Sect. 2.1. For the manufacturing of the final component geometry, several individual processes are linked to form a process chain. Since the output of one process is the input for the following process, uncertainty propagates through the process chain, as shown in Sect. 3.2. Therefore, regarding the final process of reaming or tapping, the uncertainty can have its origin in the preceding process (pre-drill geometry), the current process (runout error or synchronisation error), or the combination of the two individual process steps (positioning errors).

For reaming operations, a comparison of different uncertainty factors shows that an axis offset between the tool and the bore hole centre has major influence on the final geometry [82]. In the simulation of tool displacement in reaming an axis offset of 30 μm shows a resulting tool displacement of 2.8 μm. Due to the lack of radial guidance of the tool, the beginning of the penetration phase is a decisive point for the control of uncertainty. Tool deflection during the first cut leads to an inclination of the tool, which, due to the continuously increasing radial tool guidance, remains constant over the entire reaming depth [82]. This applies equally to tapping. For both processes uncertainty can be mastered by using a robust tool design, or by adapting the manufacturing strategies. Within this subsection different approaches based on the robust tool design and robust manufacturing strategies are presented for tapping and reaming.

Mastering uncertainty based on the robust tool design

In earlier investigations, uncertainty factors during reaming were eliminated by adjusting the macro geometry of a reaming tool. Schmalz [143] suggests an unequal distribution of the cutting edges in order to reduce roundness errors. This adaptation became state of the art in the industry. In further investigations on reaming, axis offsets are compensated by using tools with a chamfer angle of $\kappa_R = 90°$. By means of a simulation model for passive forces during reaming [82], we can show that, for a cutting speed of 60 m/min, a chamfer angle of 90° and different variations of cutting depths and tooth feeds, the simulated forces are smaller compared to the original reaming tool with a smaller chamfer angle. This results in a lower radial force, and in turn in lower tool deflection, during the penetration phase.

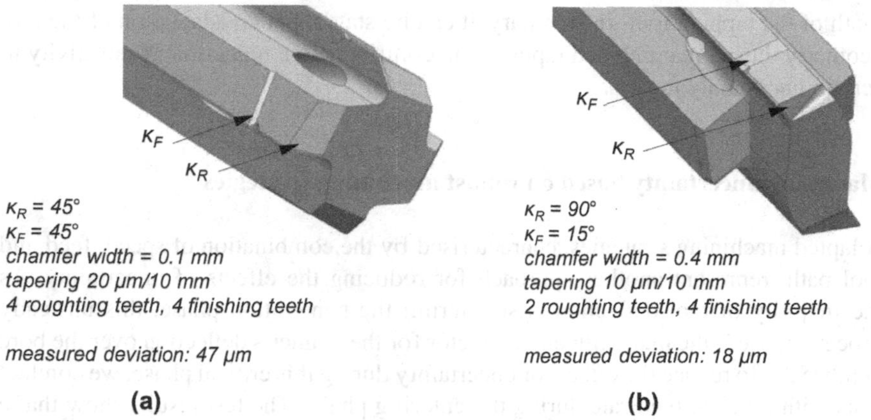

Fig. 6.14 **a** Original reaming tool geometry and **b** adapted reaming tool geometry with measured radial deviation for cutting speed of 60 m/min, feed of 0.4 mm/U and axis offset of 0.06 mm [82]

Based on further investigations on the influences of geometry elements, we modify the original tool geometry (Fig. 6.14a) to an adapted tool geometry (Fig. 6.14b). The key modification is a larger κ_R of the roughing cutting edge. In addition, two of the roughing teeth in the chamfer area are removed, since a lower number of teeth showed less deviation of the reaming tool in the simulation, especially if the tool is not yet guided through the secondary cutting edges. By applying these modifications, which are based on a mechanistic model (see Sect. 4.1.3), we can reduce the measured radial deviation by over 60 % [82].

The tool shown in Fig. 6.14b is adapted to reduce the effects of axis-offset errors of the pre-bore and the reaming tool. Under the assumption of an even drilling surface, an inclined pre-drilling has no effect on the radial course. It was therefore not considered in the tool design phase. However, if the drilling surface is not even, i.e. if the whole component is inclined, a chamfer angle of $\kappa_R = 90°$ has a negative effect due to the resulting lateral forces. Considering this effect, we reduce the chamfer angle to $\kappa_R = 65°$ and add a pilot reaming process. This leads to better results in terms of the lateral tool displacement and reduces the sensitivity of the process concerning uneven drilling surfaces [23].

When considering positioning inaccuracies in tapping, geometric modelling shows that these deviations can lead to an interrupted cut. This results in simultaneous removal of long chips as well as short thin chips, which increases the risk of chips jamming between tool and workpiece. We can detect the critical teeth using a geometric model [1]. One approach to increase the tool's robustness and to ensure the removal of long chips, even under positioning inaccuracies, is to remove the first tooth that potentially leads to an interrupted cut. The adapted tool geometry cuts the thread with fewer teeth, reducing the number of chips per flute, while increasing the load on the subsequent teeth. Using the simulation tool from [1], we show that for axial-offsets of 0.1 mm, a disrupted cut can be prevented by removing the first cutting tooth of the tapping tool. In summary, it can be stated that an adaptation of the tool geometry during reaming and tapping can contribute to a reduction in sensitivity to certain uncertainty factors.

Mastering uncertainty based on robust machining strategies

Adapted machining strategies, characterised by the combination of speed, feed and tool path, represent another approach for reducing the effects of uncertainty. As the majority of tool deflection arises during the penetration phase, this unsteady process phase is the main influencing factor for the reamer's deflection over the bore depth [82]. To reduce the effects of uncertainty during this critical phase, we conduct tests with reduced feed rate during the entering phase. The test results show that a higher feed rate results in a higher deflection of the reamer. This matches theoretical considerations. A reduced feed rate leads to a decrease of forces, which deflect the tool in the unsteady phase. However, the application of special entry strategies, such as a cubic increase in the feed rate, shows no significant reduction of the medium deflection [74].

An alternative approach based on the active process control to reduce the effects of uncertainty in reaming is to apply suitable machining strategies [28]. The axis-offset is determined by the forces during the penetration phase with a sensor-integrated reaming tool. After the penetration phase, the reaming process is interrupted and the reaming tool is removed from the bore. This procedure can be compared to a countersinking process. Based on the integrated sensor data the axial misalignment is detected and quantified using a neural network. Thus, we can compensate the axial misalignment, before the actual reaming process is conducted. Despite a disrupted cut in the entry phase of the reamer, due to the compensation of the axial misalignment the countersinking process does not lead to a significant deterioration of the bore [28]. The proposed method is therefore suitable for compensating axial displacements and thus reaming the bores at the desired location.

Since the feed rate per revolution is much higher, this approach cannot be easily transferred to tapping. However, we can adapt the idea of an upstream countersinking process. To obtain threads that are true to gauge, the feed must fit the pitch of the thread. If the feed per revolution is smaller than the thread pitch, the tap would drill out the existing bore. Using simulations based on [1], we can show that by using an M8 tap with a feed of 0.09 mm/rev we can produce a chamfer to a pre-drilled bore that has the exact same angle as the chamfer angle of the tapping tool. This pilot process step could be used to provide a pilot bore with a defined chamfer that allows several teeth to be engaged with the workpiece material at the same time. As a result, the lateral forces partly compensate each other, which leads to a lateral support of the tap. To show the general feasibility of this approach, we carry out experimental investigations in 42CrMo4 steel on a 5-axis machining centre "Grob G350". The pilot process is realised with an M8 machine tap, a feed of 0.09 mm/rev and a cutting speed of 60 m/min. The drilling depth of the countersinking process is varied between 1.25 and 3.75 mm, which equals 1 to 3 times the pitch of the tap. We carry out the

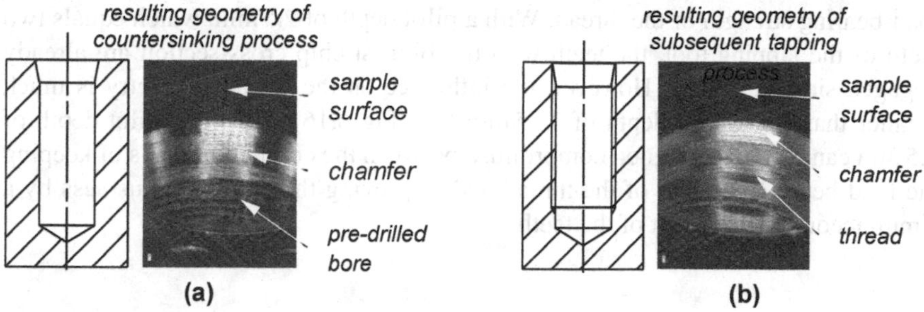

Fig. 6.15 Resulting geometry of **a** the countersinking process and **b** the subsequent tapping process with a pilot depth of 2.5 mm

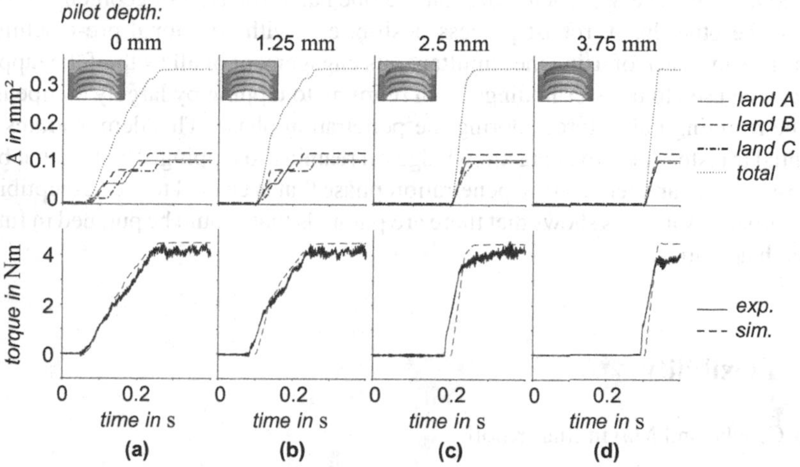

Fig. 6.16 Simulated and experimental results of the countersinking process in tapping for pilot depths of **a** 0 mm, **b** 1.25 mm, **c** 2.5 mm and **d** 3.75 mm

subsequent tapping process using the same tool with a feed of 1.25 mm/rev and a cutting speed of 15 m/min. The resulting geometry is shown in Fig. 6.15. This is due to the removed material causing an air cut at the entrance to the bore. Furthermore, the effect of multiple teeth engaging with the material at the same time is visualised by the steps of the calculated chip cross-section A of each land of the tapping tool. With an increasing number of teeth being engaged simultaneously, the number of visible steps declines until the stationary process phase, when all teeth are engaged, is reached. For the pilot depth of 3.75 mm, which equals the chamfer length of the used tool, all of the chamfered teeth are engaged with the material at the same time.

By increasing the pilot depth, the starting time of the first material engagement is shifted backwards (see Fig. 6.16).

Since the depth of the pilot process affects the inner diameter of the thread in the penetration phase, a large pilot depth could lead to negative influences on the load-bearing strength of the thread. With a pilot depth of 2.5 mm, which equals two teeth of the tapping tool, the teeth with the biggest chip cross-section are already engaged simultaneously. However, the influence on the thread geometry is much smaller than for a pilot depth of 3.75 mm (see Fig. 6.16). Thus, the pilot depth of 2.5 mm can be seen as a good compromise between the conflicting goals of keeping the load-bearing strength of the thread and improving the process robustness by a simultaneous engagement of the teeth.

Conclusion

In this section, we demonstrate a number of approaches to master uncertainty in reaming and tapping. The approaches are, on the one hand, robust tool geometry design, and on the other hand, robust process design, e.g. with additional pre-machining steps. The approach of using the simultaneous engagement of all teeth of the tapping tool can be used to transfer findings from reaming to tapping by largely compensating the resulting radial forces during the penetration phase. This demonstrates the potential transfer of the existing knowledge on reaming to tapping. The fact that both processes are characterised by a penetration phase that is critical to the susceptibility to disturbance variables shows that there are parallels that should be pursued in future research activities.

6.2 Flexibility

Peter Groche and Maximilian Knoll

In addition to the strategies already presented for mastering uncertainty, a strategy for mastering uncertainty by means of increasing machine flexibility is presented below. In Sect. 3.5 we defined: A flexible system is characterised by the fact that it fulfils $i = 1, \dots, N$ predefined functions g_i with accepted functional quality δg_i. Flexible manufacturing systems are advantageous with respect to mastering uncertainty in all phases of a product life cycle (cf. Sect. 1.2). An example of flexibility in product design is given in Sect. 3.5. In the following, we focus on the production phase.

The planning and selection of manufacturing systems are associated with uncertainty. Uncertainty may occur if a future event for the manufacturing system considered is not known, or if future events are probabilistic. Manufacturers have to expect four variants of uncertainty which are the market acceptance of product types, length of product life cycles, specific product properties and aggregated product demand [66].

One approach to counter uncertainty in the area of production is to increase the flexibility of the used manufacturing systems and processes. According to [153] flexibility is divided into time and range. A possible solution, which copes with the uncertainty is called "bank flexibility", which is a financial buffer built up for future needs. Investments financed from this can, for example, absorb unforeseen changes in market conditions. Furthermore, investments to cope with uncertainty can be achieved by new flexible manufacturing systems. One manufacturing system is considered more flexible than another, if it can handle a wider range of products, processes and tools. [66].

6.2.1 *Total Flexibility in Forming Technology*

Peter Groche and Maximilian Knoll

Total flexibility can be further divided into four types, namely equipment flexibility, product flexibility, process flexibility and demand flexibility [154]. Equipment flexibility is defined as the ability of a system to integrate new products and variants of existing products. Product flexibility is the ability of a production system to adapt to the changing product spectrum. Process flexibility characterises the adaptability of the system to changes in parts processing, e.g. caused by changes in technology. Demand flexibility describes the ability of a production system to respond to changes in the demand of the market. Flexible manufacturing systems can cope with occurring fluctuations. Based on the four types of flexibility, different concepts for flexible assembly and cutting machines have already been intensively investigated in the past. However, uncertainty also influences the future market value of a specific forming machine, so the focus will be on forming machines [72].

Today's forming machines are used for large series production with fixed selected forming processes, or for small series production with predetermined special tool movements. Consequently, the available flexibility in terms of adaptation to changing market conditions is limited [144]. The implementation of servo technology (see Sect. 3.6.3) is a first step towards increased flexibility by an extended process control of forming processes [73].

Forming machines and the associated processes differ in the number and type of degrees of freedom (DoF) in their driven movements. A DoF is defined as an independent way of moving a body. A three-dimensional space thus has three translational and three rotational degrees of freedom. This means that an input with one DoF can provide a maximum of one independent degree of process freedom at the output. In the following discussion, a forming system is considered, which is described only by the drives made available in the machine and thus the DoF.

The influence of the additional degrees of freedom on the flexibility types is analysed using continuous and discontinuous forming processes, see Fig. 6.17. It becomes obvious that with increasing degrees of freedom the flexibility of tool, product and process flexibility increases. However, due to longer tool paths, the demand flexibility decreases with the increasing number of degrees of freedom. Especially large series production is based on forming technologies, which are characterised by a one degree of freedom movement. This type of movement leads to low tool-to-workpiece contact times and thus high productivity.

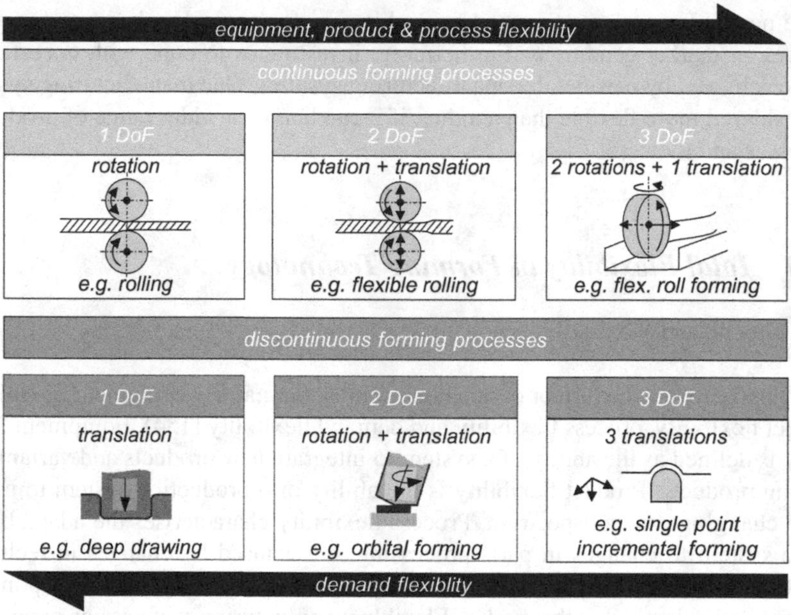

Fig. 6.17 Classification of continuous and discontinuous forming processes by DoF, in accordance with [72]

Considering forming machines or processes with one rotational or translational DoF, such as rolling or deep drawing, the result is a low equipment flexibility. This results from the fact that the shape of the component cross section to be achieved is given by the shape of tools (Fig. 6.17). Due to the rigid tools, machines and processes with only one DoF in their movements possess a smaller product flexibility compared to processes with higher degrees of freedom. Additionally, they require more capacity for tool production and setup. The process flexibility is also low compared to processes with higher degrees of freedom, because of the limited usability of existing tools dedicated to previously executed production routes. In contrast, the demand flexibility of processes with few degrees of freedom is high since even a large number of parts can be produced on short notice. A control of process fluctuations is only possible to a limited extent, which is why the product quality is strongly dependent on the fluctuations in the semi-finished product. Conventional mechanical presses as well as servo presses allow translatory movements. Servo presses allow freely programmable speed-time sequences, which contribute to improved material flow, shorter setup times or increased product quality [73].

Forming processes with two degrees of freedom are, for example, flexible rolling or orbital forming. These processes not only allow the production of new products or variants by the development of new tools, they also allow the independent controllable speed-time curves for the respective DoF to react to the uncertainty of semi-finished product variations. By increasing the degrees of freedom, the tool contact time is increased in comparison to a movement with only one DoF. At the same time, the complexity of the control system is increased, which reduces the flexibility with respect to demand [73].

The increase to three DoF in the tool movement provides the so far conventionally established highest flexibility in machine, product and process design. By using three degrees of freedom, complex component operations can be produced in continuous and discontinuous processes. These include the processes of flexible roll forming and the Single Point Incremental Forming process. The Single Point Incremental Forming process was used for the production of spring elements in Sect. 5.4.8. These processes offer the possibility of producing a variety of three-dimensional geometries with a fixed tool geometry. Specific workpiece geometries are achieved by specific tool movements. Due to the flexibility of the tool, the set-up time is reduced to a minimum. The increased control effort and the associated increased manufacturing time reduce the flexibility of demand. Thus, the ability to react to increasing sales figures is significantly reduced [73].

Processes with one DoF offer the highest degree of demand flexibility, especially when considering the costs related to the batch size and the setup and maintenance effort. In the case of a smaller batch size, these production processes are not economically feasible due to their high fixed costs. Forming operations with two or three DoFs are in most cases designed for more complex products so that a smaller batch size of workpieces can be achieved. An increase in demand can usually only be achieved by using additional machines. The machines used for this purpose are mostly designed for special operations and are only capable for series production in a very limited way. Based on these findings, a forming machine has so far shown either high product, process and equipment flexibility or high productivity [73].

Frequently, used conventional presses provide a pure translatory relative movement of tool and workpiece. The investigation of forming processes shows that forming machines with flexible tool movements can have economic disadvantages in production technology (Sect. 3.6.3). In order to achieve the high productivity necessary for mass products in addition to complex products, the drive system of a forming machine should be usable for simple linear movements without reducing the production speed. This makes it possible to combine a flexible forming machine with high batch sizes. To realise a press with translatory stroke motion with additional degrees of freedom, three drive points of a plane are necessary. These three drive points can be driven by three axis-parallel translatory drives, see 3D Servo Press Sect. 3.6.3. By the simultaneous linear displacement of the three points, a translatory movement is realised similar to conventional presses. In contrast, the asymmetrical control of the three drive points results in a tilting movement of the point plane. The Tool Centre Point (TCP), which is located in the centre of the three drive points, shifts exclusively in the vertical plane, assuming small tilting angles. Based on these considerations, they can be transferred to a new press type [73].

In the following, a short description of the design is given. For the movement of the ram, a servo motor and a crank mechanism are used at each of the three drive points of the plane. These drives have a good controllability, and high stroke rates can be realised with them. Furthermore, the drives can be controlled independently

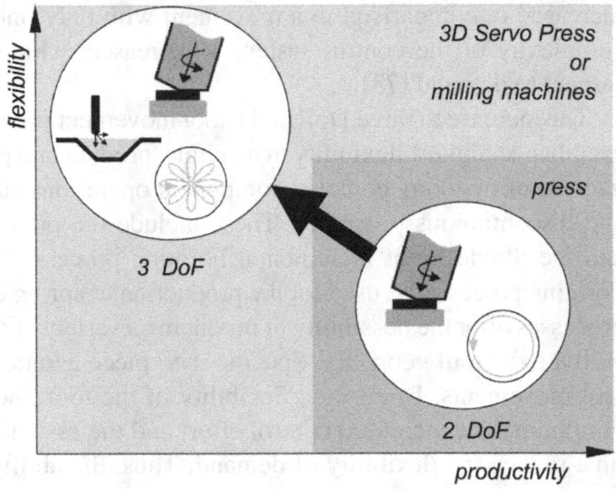

Fig. 6.18 Increase of the process control strategy through additional degrees of freedom in orbital forming and Single Point Incremental Forming

of each other, so that the translation and tilting movements of the TCP can be controlled freely. The kinematics of the 3D Servo Press enables processes with one DoF of the ram such as punching, embossing and deep drawing. Furthermore, combined flexibly controlled processes, which are presented in [14, 70], are possible. Due to the additional degrees of freedom, the multi-technology machine can be used to investigate new and existing processes to extend existing process limits. The additional degrees of freedom in orbital forming processes offer the possibility of using the process strategy for a targeted control of product properties independently of the machine or tool kinematics (Fig. 6.18) [31]. Furthermore, Single Point Incremental Forming can be investigated for higher sheet thicknesses and high-strength steels. Previous investigations on these processes are limited to applications on special or milling machines. In comparison, the use of the 3D Servo Press offers the potential to explore new process limits and process control strategies.

Production technologies are confronted with uncertainty based on unknown material, product and demand influences. One way to master the uncertainty is to increase the flexibility of manufacturing systems. This can be done as described by increasing the degrees of freedom in machines and processes. Flexible manufacturing machines have advantages in comparison to conventional manufacturing systems, especially in uncertain demand scenarios. It becomes clear that the 3D Servo Press with its implemented control system is able to achieve a consistent product quality through different adopted process parameters. The total flexibility in forming technology shows a promising approach to cope with different types of uncertainty.

6.3 Resilience of Technical Systems

Marc E. Pfetsch

Resilience is a topic that is currently in the focus of many different research areas: psychology, sociology, safety-critical infrastructure, and many others. In this section, we concentrate on the resilience of technical systems. Thus, we define the following understanding of resilient technical systems in mechanical engineering, cf. [7, p. 189]:

> A resilient technical system guarantees a predetermined minimum of functional performance even in the event of disturbances and failures of system components, and a subsequent possibility of recovering.

Disturbances and/or failures can lead to a loss of functionality in a technical system; in particular, the disturbances or failures can be severe and might not be anticipated. The resulting ignorance can then be mastered by a technical system that is resilient. For an additional motivation and discussion of resilience see Sect. 3.5.

This section investigates resilience as a strategy to master uncertainty in its many facets: from a structural discussion of resilience characteristics, over methods to guarantee resilience, to an experimental evaluation for specific technical systems. The discussion also includes and combines contributions from mathematical optimisation and different areas of mechanical engineering.

Section 6.3.1 starts with a discussion of the definition and the differences to robustness as presented in Sect. 6.1 and introduced in Sect. 3.5. Moreover, it presents several metrics to quantify resilience. Section 6.3.2 covers so-called adaptive resilience and the difference to flexibility. It characterises different methods to obtain adaptivity. Sect. 6.3.3 describes the role of human interaction on and in resilient systems. In Sect. 6.3.4, mathematical optimisation methods to design resilient trusses are presented. The effect of different buffering capacities and its influence on the performance range are discussed. Sect. 6.3.5 continues with optimisation methods for designing water supply networks. An adaptive method for computing networks with different buffering capacities is presented and evaluated. Section 6.3.6 considers drop tests using a Fluid Dynamic Vibration Absorber. It experimentally demonstrates the increased resilience of a system incorporating this technology. The topic of Sect. 6.3.7 is the interplay of the production and usage phase of a hydraulic actuator. Experiments evaluate the effect of production disturbances in the usage phase. Finally, Sect. 6.3.8 considers a real resilient fluid system test rig. The incorporation of algorithmic models for learning and the applicability of the resilience triangle are evaluated experimentally.

Designing a completely resilient technical system requires high effort and costs. Nevertheless, the virtual examples in Sects. 6.3.4 and 6.3.5 demonstrate how such systems can be approached and allow for a quantification of resilience costs. Moreover, the basic research example in Sect. 6.3.8 shows how the control of a technical system can be adapted to achieve a higher resilience. This is complemented by the other sections, which focus on the resilient design process, resilience metrics and human factors.

6.3.1 Resilience as a Concept to Master Uncertainty

Lena C. Altherr and Philipp Leise

The concept of *resilience* has found its way into different disciplines where it is commonly used to describe the ability of an individual or a system to withstand and adapt to changes in its environment, cf. [55, 97, 148, 163]. In order to address the resilience of technical systems, a tailored definition and understanding, as well as suitable metrics to quantify resilience are required. In this subsection, we show first results on adapting the concept of resilience to technical systems, and address the following questions: (i) How to differentiate between robustness and resilience? (ii) How to quantify the resilience of a technical system?

What is resilience of technical systems? how is it possible to differentiate between robustness and resilience?

Prior to engineering, resilience was introduced in the domain of human factors. Within this domain, special attention is paid to resilience for the design of socio-technical systems and safety management. According to Hollnagel, "a system is resilient if it can adjust its functioning prior to, during, or following events (changes, disturbances, and opportunities), and thereby sustain the required operations under both expected and unexpected conditions" [91]. Given this understanding, resilience of a technical system can be seen as a concept to master the unexpected and thus uncertainty.

While the human factor research community focuses on analysing the socio-technical interaction between humans and technology (cf. Sect. 6.3.3), in the following subsections of Sect. 6.3, we focus on transferring the idea of resilience to technical systems. We explicitly consider each product life phase as introduced in Sect. 3.1 and show resilience principles and approaches that can be transferred to a broad range of systems, as for instance the example reference systems introduced in Sect. 3.6. In the context of product development, resilience—as compared to robustness—can be regarded as a paradigm shift. According to Taguchi, [159–161], robust design refers to functional characteristics of a system perceived by the user as being unaffected by disturbances and failures. Robust Design may be achieved by mathematical optimisation methods dealing with uncertainty, cf. [69]. Robust optimisation allows to optimise technical systems regarding user-specific objectives, while guaranteeing robustness of the design against uncertain input parameters, e.g. feasibility even in case of uncertain load parameters. The generated robust solutions fulfil their purpose not only at the design point, but also in a surrounding neighbourhood, the so-called uncertainty set, cf. [15]. For further details and examples of robust design and robust optimisation see Sects. 6.1.1–6.1.4.

New approaches to design products considering not only robustness, but also resilience, allow us to master *ignorance* or *nescience* (see Chap. 2 and Sect. 3.5): Compared to a robust system, a resilient system is not only able to withstand expected disturbances and failures, but is also able to handle unexpected disturbances and failures which were not explicitly taken into account during the design phase. Robust systems guarantee performance for a known range of uncertain parameters (the uncertainty set), however, once outside this range, they might break down completely. Yet, resilient systems are characterised not only by their ability to withstand specific disturbances and/or failures, but are also "safe-to-fail" [2], yielding a minimum performance even in case of a failure and a subsequent ability to recover.

Thus, we use the definition already mentioned in Sect. 6.3: "A resilient technical system guarantees a predetermined minimum of functional performance even in the event of disturbances and failures of system components, and a subsequent possibility of recovering.", cf. [7, p. 189].

In socio-technical or technical systems, the ability to recover is connected to maintenance measures, e.g. replacement of damaged components and/or capacity adaptations. In some rare cases, the technical system itself is able to recover to some extent, as shown for example by Bongard and Lipson in [27]. In this example, they describe an intelligent starfish-shaped robot that recovers autonomously from removing parts of its legs.

Given the ability of ideally resilient systems to be "safe-to-fail" [2] and to recover, the next logical step is to characterise resilient technical systems and to investigate how it is possible to specifically design the resilience.

One approach to master uncertainty by building resilient systems is given by Hollnagel [87–89]. Hollnagel distinguishes four abilities/functions (*monitoring*, *responding*, *learning*, *anticipating*) that resilient systems should contain. They are shown in a systematic way in Fig. 6.19. We will address the transfer of this approach to the mechanical engineering domain in more detail in Sect. 6.3.8.

How is the resilience of technical systems quantified?

A first prerequisite to design the resilience properties of a system is to be able to assess them. Different qualitative and quantitative concepts for measuring resilience are proposed in the literature. For an overview see, e.g. [54]. While some metrics in the literature are either described in very general terms, or are not applied to technical systems, others are very specific to the system under consideration. For instance, for water distribution systems alone, more than 20 tailored resilience metrics were proposed by different authors [148]. Since many of the critical infrastructures that are examined with regard to their resilience, such as roads or supply systems for energy or water, are network-like, graphs can be used to describe them. Thus, also graph theoretical metrics (e.g. average path length, link density, central point dominance or k-shortest path length) have been proposed to assess resilience, see e.g. [85, 127].

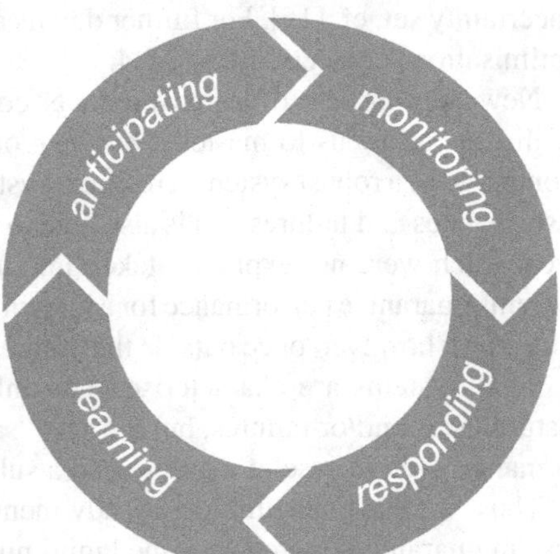

Fig. 6.19 Four abilities/functions to derive a resilient system, based on [83, 87]

Regarding general technical systems that are not necessarily network-like, a catalogue of design principles is proposed in [97], and the first author states that "the measurability of resilience should be a top priority topic for further research" [98, p. 35]. Since we characterise a resilient system by the fact that it can sustain the required operations even in the event of disturbances and failures, we use its performance under varying external and internal influencing factors i as the basis for assessing its resilience. Based on this understanding, we discuss a collection of different metrics, which we exemplify qualitatively. While there may be multiple external and internal influencing factors, which may influence the performance, we choose an univariate depiction for reasons of clarity.

In [7] we defined the *performance range p* of a technical system. It describes the subset of influencing factors for which the system is able to maintain a predefined minimum performance $f_{\min}$. Figure 6.20a depicts the performance range for an univariate function. Mathematically, the performance range can be expressed by the so-called 'superlevel set',

$$\mathcal{L}_f^{\geq}(f_{\min}) = \left\{x \in X \,|\, f(x) \geq f_{\min}\right\}.$$

Here, $f : X \to \mathbb{R}$ is the system's performance for varying arbitrary influencing factors $x \in X$, cf. [7, p. 189]. Based on the performance range, several metrics can be introduced, e.g. the area under the curve above $f_{\min}$, or the performance range times a problem-specific weighting factor that gets smaller with a growing distance from the design point. As a general metric, we propose in [7] the *radius of performance* r_{p}, cf. Fig. 6.20a. It measures the minimum distance between the design point and a realisation of an influencing factor for which the minimum performance can no longer be maintained.

In addition to the radius of performance, also the *margin m* can be used to assess the resilience of a technical system. It describes "how closely or how precarious

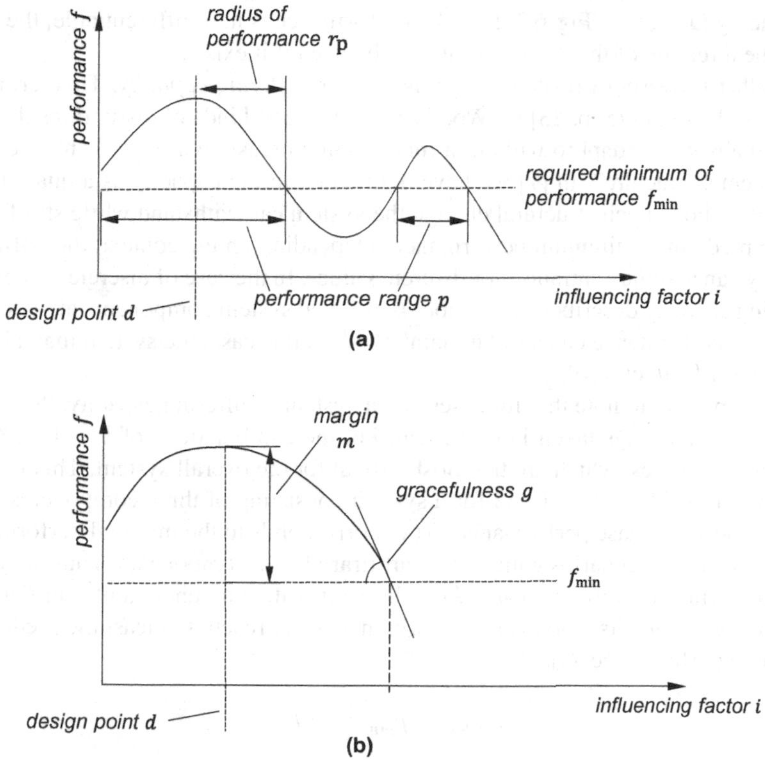

Fig. 6.20 Resilience metrics: **a** performance range p and radius of performance r_p, **b** margin m and gracefulness g [7]

the system is currently operating relative to one or another kind of performance boundary", cf. [175, p. 23]. In [7], we proposed to quantify the *margin* by measuring the distance between the performance at the design point and the required minimum performance, see Fig. 6.20b for an illustration. Thus, it can be calculated by

$$m = f(d) - f_{min},$$

where m is the margin, $f(d)$ is the performance of the system at the design point d, and f_{min} is the predefined minimum required performance of the system.

Resilient systems have been characterised to be safe-to-fail, cf. [2]. For this, it is important "how a system behaves near a boundary—whether the system gracefully degrades as stress/pressure increase or collapses quickly when pressure exceeds adaptive capacity.", cf. [175, p. 23]. As such, resilience includes *graceful degradation* [68], once the system reaches its performance limit. In [7], we defined the *gracefulness* g of a system mathematically, being the directional derivative of the performance f in the direction of a given influencing factor or a vector of multiple influencing factors, cf. Fig. 6.20b. If the performance is non-differentiable, the limit from the direction of the design point may be used if it exists.

Another property of resilient systems is their *buffering capacity*. This term was first described in [175, p. 23] by Woods as "the size or kinds of disruptions the system can absorb or adapt to without a fundamental breakdown in performance or in the system's structure". In [7] we have defined buffering capacity as a quantitative measure of how much structural change the system can withstand while still fulfilling the predefined minimum performance. Depending on the context, the buffering capacity can assume continuous or discrete values. In the case of discrete values, the buffering capacity describes the number k of failed system components at which the minimum performance can still be maintained. In this case, the system may also be called *k-resilient*, cf. [10].

It is important to note that for assessing a system's buffering capacity, the worst-case failure is always taken into account, i.e. the combinations of $k = 1, 2, 3, \ldots$ component failures, which are the most critical for the overall system. This is illustrated in Fig. 6.21 for $k = 1$ and for a system consisting of three components A, B and C. The worst-case performance $f_k(x)$ corresponds to the minimal performance over the set of all scenarios with up to k arbitrary failed components. Thus, a system has a buffering capacity of k (or is k-resilient) if, within an uncertainty set U of the influencing factors, its worst-case performance $f_k(x)$ reaches at least the predefined minimum performance f_{min}, i.e.

$$f_k(x) \geq f_{\text{min}} \; \forall x \in U,$$

where U is the uncertainty set. The smallest possible uncertainty set corresponds to the design point d.

While the above-mentioned metrics can be used to measure static characteristics of resilient systems, also its dynamic behaviour should be taken into account. The metric *rapidity* r_t was proposed for measuring the system's capacity to recover its functionality in a timely way, [165]. For this purpose, the time period between the

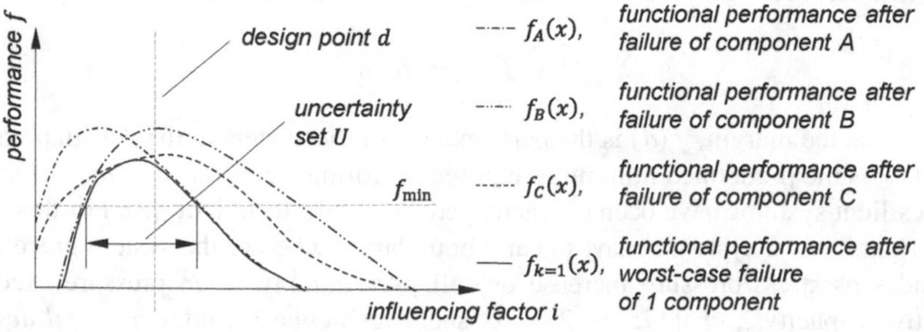

Fig. 6.21 Worst-case functional performance f of a system with failure of one of its components A, B or C, [7]

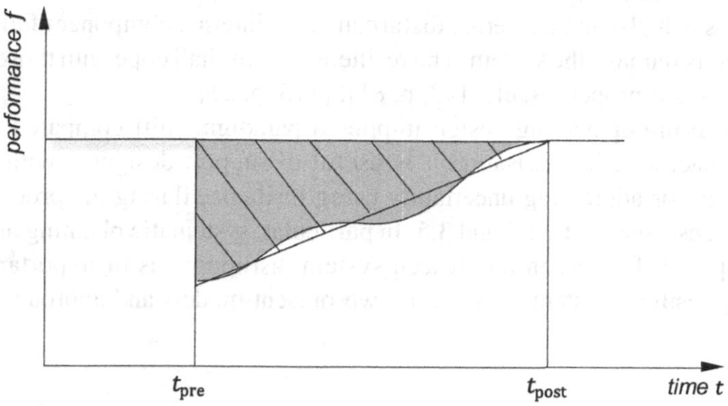

Fig. 6.22 Metrics for measuring the system's ability to recover: *rapidity* and *resilience triangle R*, cf. [7, 165]

occurrence of the disturbance and/or failure and the restoration of functionality is measured, i.e. $t_{post} - t_{pre}$, cf. Fig. 6.22.

Another measure for the dynamic aspects of resilience, also proposed in [165], is the so-called *resilience triangle*. Here, the total losses (e.g. in utility, revenue or performance) are measured until the system is recovered. These losses can either be approximated by a triangle, or calculated by

$$R = \int_{t_{pre}}^{t_{post}} \max\{0, f_{pre} - f(t)\}\, dt,$$

see, e.g. [163]. The triangle approximation is shown in Fig. 6.22 with hatched lines, while R is shown in grey.

While the presented metrics are intended to be suitable for the resilience assessment of general technical systems, their application in practical usage has to be proven. The next subsections show the application of the proposed metrics to practical engineering examples.

6.3.2 *Mastering Uncertainty in Engineering Design by Adaptive Resilience*

Fiona Schulte, Hermann Kloberdanz, and Eckhard Kirchner

We consider resilient system properties as an extension of robustness to handle uncertainty caused by nescience, see Sects. 2.3 and 6.3.1. A central aspect of resilient system behaviour is the adaptivity of the system. A purposive adaption is required to make the system continuously usable under changed internal or external system conditions, as introduced in Sect. 6.3.1. The condition changes also include disruptive changes, which can be external disturbances or internal component failures, that could severely damage the system. The resilient system shall cope with those changes using its resilient properties, cf. [142, p. 81]; [175, p. 21].

Resilience in engineering design implies a paradigm shift compared to robust design. To facilitate its realisation it is useful to support designers with a design methodology for addressing uncertainty using resilience during the product development process, see Sects. 1.2 and 3.5. In particular, systematic planning and design of the adaptivity for certain unforeseen system disruptions is of importance when developing resilient systems. Therefore, we present models and approaches in this Section.

Adaptivity

Adaptivity in the resilience context is defined as the system's ability to adjust to changing purposes or conditions in a suitable way. The desired adjustment aims at approaching a predetermined behaviour under the new conditions and indicates that the adaption of the system allows the maintenance of elementary system functionalities instead of a system failure and possible consequential damages, cf. [98, p. 107]; [96, p. 7].

Adaptivity in load-bearing systems can either be realised autonomously or externally induced as distinguished in Fig. 6.23. Autonomous or internal adaptivity includes all adjustments that happen within the system simply triggered by a change of conditions. Externally induced adaptivity usually requires a human operator or an additional external system that influences the system towards the desired adaption, cf. [142, p. 81]; [90, p. 224 et seq.].

The kinds of adaptions vary in the way of realisation and timing. Autonomous adaptions normally apply quickly because they are triggered by the disruption itself or correlating signals within the system, though in many cases the measures are only effective short-term. They necessitate the system's ability to improvise. This can, e.g., be realised by physically or functionally redundant structures [142, p. 81]. A prompt

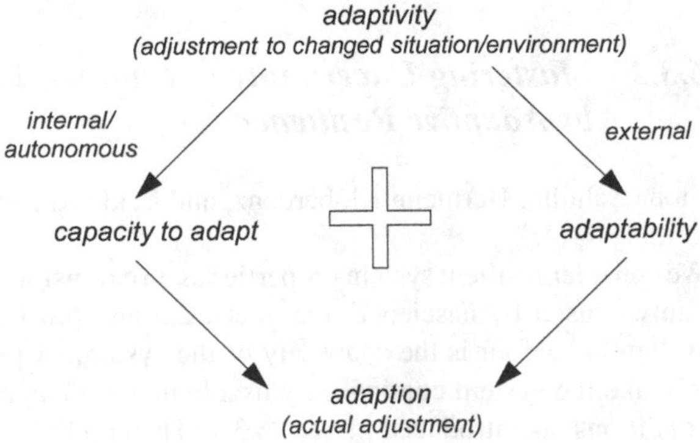

Fig. 6.23 Adaptivity in resilient systems is divided into autonomous and externally induced reactions

reaction realised by autonomous adaptivity is often required due to sudden rapid disruptions. However, the disruptions can last for a long term, which makes additional measures necessary, if the autonomous measure is only applicable short-term. In many cases, the additional measures are externally induced. Externally induced adaptions can usually not be applied promptly, as a human operator's response time is interjacent. In return, externally induced measures are effective for a longer period of time. Thus a combination of a prompt autonomous and a long-term externally induced adaptivity is recommendable. Externally induced adaptivity relies on the ability to convert, i.e. a *replacement* of the damaged subsystem by an identical one. Alternatively, the subsystem can be exchanged or even extended. We define the term *exchange* as the incorporation of a partly improved subsystem. *Extension* describes a significant improvement of the subsystem and requires an innovative capability, which usually implies the contribution of a human operator's development work, cf. [142, p. 81].

An adaption is realised by implementing the resilience functions monitoring, responding, learning and anticipating, as introduced in Sect. 6.3.1 and evaluated in Sect. 6.3.8, cf. [90, p. 227]. Depending on the system's complexity and the necessity of resilient functionalities, only one or several of the functions can be combined. The central resilience function is 'responding' because it describes the execution of the purposeful system adaption instead of an arbitrary reaction of the system. This means responding is always required in case a disruption, which is either an external disturbance or an internal failure, occurs. More sophisticated systems additionally use the resilience functions monitoring and anticipating. Monitoring means that parameters that have an influence on the system or quantities correlating to the influencing factors are measured. The anticipating function then describes the interpretation of the monitored data, which allows the system to foresee upcoming or potential disruptions and thus to react and apply measures before the disruption actually occurs, while responding only describes a reaction towards a disruption, cf. [90, p. 224 et seq.]; [175, p. 121 et seq.].

The resilience function learning exceeds anticipating by interpreting the data not only according to upcoming disruptions but also regarding the success or failure of the applied measure. Depending on the measure's result, a learning system is able to adjust the reaction for the case of a repeating or similar disruption. The implementation of learning relies on an artificial intelligence (AI) or a human operator included within the system as contemplated in Sect. 6.3.3. The AI or human operator is able to take on the processing of the monitored data regarding the measure and its particular success or failure, respectively. As an AI or a human operator cannot be presumed in most technical systems, especially technical subsystems, the realisation of learning in technical systems is rather an outlook for further development. Nonetheless it is beneficial to design resilient subsystems, because such subsystems support the realisation of resilient properties in the superordinate system, cf. [90, p. 224 et seq.]. In the following, we focus on the resilience functions monitoring, responding and anticipating.

The resilience application model

We developed the resilience application model to describe the interdependencies between the resilience characteristics and behaviour, as well as the disruption and the signal, respectively. The resilience application model is based on the definition of resilience for load-bearing systems, in particular, and on the definition of metrics regarding the system resilience characteristics and behaviour (Sect. 6.3.1), cf. [7, p. 189 et seq.]. The application model considers the resilience characteristics and behaviour as supplied before in Sect. 6.3.1 and adds the consideration of the disruption and a correlating signal progression. According to Jackson [96, p. 6] considering the disruption is of high importance for the realisation of resilience and the four resilience functions. In addition, the correlating signal is especially important for the functions monitoring, anticipating and learning, cf. [146, p. 1406 et seq.]; [145, p. 3 et seq.].

The resilience application model offers support for analysis and synthesis during the development process of resilient load-bearing systems. The analysis approach starts with the identification of the disruption and the determination of its temporal progression shown in Fig. 6.24c. If monitoring and anticipating are used within the system, the identification of a correlating signal, such as in Fig. 6.24d, and its progression are crucial, as monitoring is responsible for gathering data of a correlating signal, and anticipating relies on the gathered information. With knowledge of the resilience characteristics as shown in Fig. 6.24a, which provide the interrelation of the functional performance and different influencing factors, the impact of a disruption can be determined, as also described in Sect. 6.3.1. Based on the disruption, respectively signal progression and the resilience characteristics, the expected dynamic resilience behaviour can be determined as in Fig. 6.24b (see Sect. 6.3.1). During the synthesis, the aspired system properties can be described using the resilience application model. First, the aspired resilient behaviour is defined. Afterwards the required resilience characteristics for realising the behaviour can be deduced, e.g., the value of the required minimum performance or the system's gracefulness. Furthermore, the necessity of monitoring the disrupted influencing factor or correlating signals for realising the required characteristics can be examined, cf. [146, p. 1406 et seq.]; [145].

For the depiction in Fig. 6.24 we chose the example of a system that is disrupted and applies a booster, which increases the possible performance of the system for a certain time. It shows how the four graphs of the resilience application model could look like for a particular case. A boosted system could, e.g., be a car using snow chains in case of the disruption of sudden black ice represented by a jump in the disruption graph. Figure 6.24 shows the exemplary progressions of the system in the resilience application model. Without snow chains the car is not able to deliver the required performance. Hence a measure is required to restore the functional performance to the minimum level $f_{\min}$. The booster is represented by the upper grey graph in comparison to a system not applying a measure depicted in black. Applying snow chains increases the traction performance of the car on black ice. As soon as the disruption of black ice disappears, the snow chains can be removed and

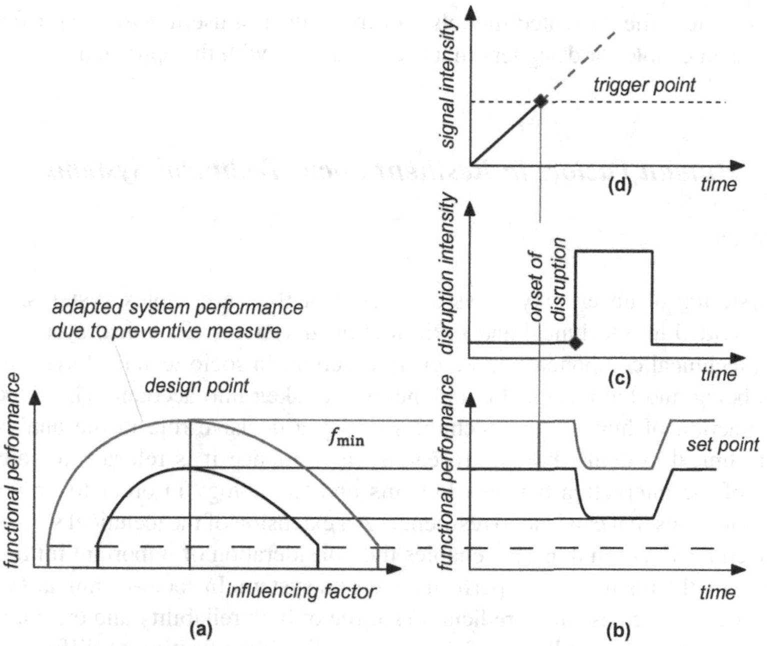

Fig. 6.24 Resilience application model with exemplary progressions for a system applying a booster **a** resilience characteristics, **b** resilience behaviour, **c** disruption progression, **d** signal progression following [146, p. 1407]

the car resumes its functional performance without additional measure, as shown by the upper grey curve which reaches its original value again, cf. [146, p. 1410 et seq.]. In case of the snow chains, it becomes apparent that this booster functionality is only useful during the disturbance phase because the snow chains increase traction and therefore allow to maintain the system's essential ability to drive. However, they reduce the achievable velocity significantly, which is not desired under undisturbed conditions. Beyond conflicts of several desired system properties, it often also saves resources to apply the booster only for a certain period of time, like in case of emergency generators with limited energy capacity.

The central aspect for realising resilient behaviour is that a system is able to purposively adapt to new conditions. The purposive adaption is realised implementing the four resilience functions, while responding is the most important resilience function, as it describes the application of the adaption. The interrelations of the resilience properties, which comprise the resilience characteristics and the resilience behaviour, as well as the disruption progression and possible correlating signals are describable using the resilience application model. The model is also able to describe a desired behaviour and the complementing resilience characteristics for the system synthesis. The desired behaviour can be formulated as an ideal system and then be refined to an actual technical solution. The methodological approach is not fully composed yet, but we consider the presented models and methods as a useful support for making resilience amenable for designers that are unfamiliar with the approach.

6.3.3 Human Factors in Resilient Socio-Technical Systems

Pia Niessen

The mastering of uncertainty plays an important role in complex systems, which can be divided into technical and socio-technical systems. Technical systems only include technical components and their interaction. In socio-technical systems, the human being and the technical components are taken into account. The effects of the interaction of humans and technology play a decisive role in the analysis of socio-technical systems. For the research on resilience it is relevant to integrate aspects of the interaction between humans and technology in order to implement certain functions that can lead to resilience. An extension of the technical system to a socio-technical system thus also enables the consideration of important influencing variables on the resilience and performance of a system. In socio-technical systems humans can be seen as an unpredictable source of both reliability and errors, which has an impact on the resilience of the system. The identification of different components of resilience is complex, especially the question, to what extent humans are involved as an actor in a resilient system. Most of the research in this field is taking place in the area of Resilience Engineering. The resilience of the systems here, is linked to safety management, faced with known or unknown situations. The context of the studies are sectors where humans and machines work together in critical situations, for example in aviation, healthcare, chemical and petrochemical industry, nuclear power plants, and railways [138]. In the following section of this overview, the possibility to integrate the human into a resilient system is outlined.

Modelling human behaviour in socio-technical systems

To classify the human in a socio-technical system, biological, cognitive, emotional, motivational or dispositional aspects can be considered. The consequences of these aspects are observable actions or decisions. In the case of modelling, it is therefore relevant to determine which part of the human being is to be investigated. The analysis of behaviour leads to behaviour models while the analysis of decisions leads to decision-making models. Behavioural models explain human behaviour as an outcome and model the upstream processes, for example cognitions. They can explain how a particular disposition leads to a particular behaviour of humans. In a socio-technical system, these variables have an influence on the entire system output. By manipulating the upstream processes, a certain human behaviour can be simulated and promoted. Examples for this can be found in numerous disciplines analysing and predicting behaviour in social systems [3]. A further class of human models are

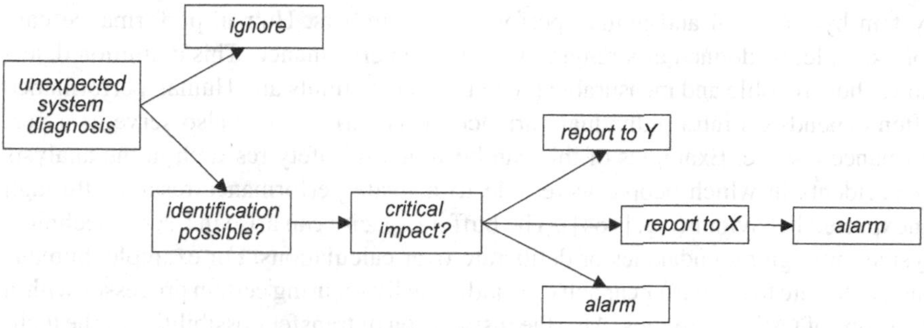

Fig. 6.25 Example for a decision-making model in a safety-critical context with an unexpected system diagnosis and the following decision possibilities

decision-making models. These can be used as process models to simulate resilience in human decisions [117]. Thereby criteria are defined, for example redundancy, and afterwards such a criterion can be applied to the human decision, understood as a process chain. This modelling is mainly used in safety-critical contexts. Figure 6.25 depicts an example of a decision-making model which shows the possibilities for the development of resilience triggered by human beings. Both types of models can be tested in various ways, for example with the fuzzy logic [40].

Application of resilience metrics in a socio-technical system

In order to measure resilience, metrics are established in Sect. 6.3.1. These metrics can be equally applied to the socio-technical system, depending on the type of the human modelling. In his Functional Resonance Analysis Method (FRAM), Hollnagel [89] outlines the application of four resilience functions in an organisational context. All four functions (learning, responding, anticipating, monitoring) are seen as cross-sectional claims. This means that people are also required to learn, anticipate, monitor and respond in order to create a resilient socio-technical system. They can, of course benefit from the resilience of the system as well, which is important, especially in safety contexts. These interactions are different for each system. A few authors have already assessed this framework (see [4, 35, 164]). Especially for the factors anticipating and learning, it makes sense to consider the human being as a source of resilience. While the responding and monitoring within socio-technical systems is often performed by the technical components, the human being is able to take over tasks which serve the learning and anticipation ability of the entire system. Human operators are able to gain experience and to learn by repeating their control tasks, thus improving their behaviour [134]. They can adjust themselves according to the dynamic changes of the socio-technical systems.This requires adaptive and proactive behaviour (i.e. resilient behaviour) to control the system performances, especially when faced with unexpected situations. The metrics introduced in Sect. 6.3.1 can also be applied to human capabilities, for example the performance range or the buffering capacity. The performance limits and reserves are defined in a socio-technical system by technical and human performance numbers. Human performance can, for example, be defined as strength or cognitive performance. This definition determines how reliable and measurable the performance limits are. Human performance often depends on intra-individual variance. This variance can also serve as a performance reserve. Examples of this can be found in safety research in the analysis of accidents in which people were able to activate performance reserves through unexpected behaviour (e.g. [164]). The buffer capacity can arise in a socio-technical system through redundancies or deliberate over-calculations. For example, humans can contribute to the buffer capacity by additionally securing certain processes within the scope of testing activities. Also the installation of transfer possibilities of the technical system by humans increases the buffer capacity of the entire socio-technical system.

Conclusion

In order to design systems resiliently, it makes sense to strive for a socio-technical modelling. In order to promote certain properties of the system human behaviour and decisions can be presented as a source of increasing resilience. Human modelling is also necessary to rule out opposing effects, such as a reduction of resilience because of human decisions or behaviour that can increase errors or uncertainty.

6.3.4 Truss Topology Optimisation Under Aspects of Resilience

Tristan Gally, Philip Kolvenbach, Anja Kuttich-Meinlschmidt, Marc E. Pfetsch, Andreas Schmitt, Johann M. Schmitt, and Stefan Ulbrich

Truss structures are load-bearing systems that are found in many applications of mechanical engineering. This includes the Modular Active Spring-Damper System presented in Sect. 3.6.1. As introduced in Sect. 6.1.1, a typical truss design problem is to find a truss topology that is as light-weight as possible while being stable enough to withstand certain load scenarios. In these design problems, there are various sources of uncertainty that need to be accounted for. For instance, typical optimisation methods lead to truss designs that are stable only for a small, predetermined set of external forces; even small deviations from these forces can lead to extremely poor performance or failure of the structure. This issue is well studied in the field of robust optimisation, where the worst-case behaviour of a structure over a given uncertainty set of forces is decisive [17].

In Sects. 6.1.1, 6.1.2 and 6.1.3, we used robust optimisation to control uncertain loads for different models of truss topology problems. Robust optimisation can also be applied to other kinds of uncertain parametric dependency, e.g. material properties or manufacturing tolerances. However, sometimes sources of uncertainty are unknown or cannot be quantified. In these cases, it can be worthwhile to move the focus from the source of the uncertainty to its impact on the truss structure. One possible approach is to design *resilient* truss structures in the sense that the truss remains stable even if a predetermined number of bars fail, for whatever reason, i.e., we use resilience to master ignorance, see Sect. 3.5. Following Sect. 6.3.1 we therefore also evaluate truss structures with respect to their *buffering capacity*. This strategy can be combined with robust parametric optimisation leading to light-weight robust and resilient trusses. To master the inherent structural uncertainty, cf. Sect. 2.3, in the truss topology design, we use mathematical optimisation, which takes the complete solution space into account.

The consideration of complete bar failures in truss topology optimisation has started only recently except for [158], which considers only a small failure set. Continuous topology optimisation problems are considered by [99, 179]. Redundancy from coding theory applied to truss design is considered in [131]. Kanno [100] also designs resilient trusses according to our definition, however displacement constraints are not included in the model. Non-robust truss topology design under bar-failure is considered in [155]. In the following we will extend the robust truss topology design problems from Sect. 6.1.1 and 6.1.2 to include resilience as presented in detail in [7, 64].

Resilient truss topology design via semidefinite programming

In this section we extend the basic robust truss topology design problem to also consider bar failures as shown in [7]. This will allow the computation of a resilient and robust truss with minimal volume. The base is formed by the robust truss topology optimisation problem

$$\min_{x \in \mathbb{R}^{\mathcal{E}}} \sum_{e \in \mathcal{E}} l_e \, x_e \quad \text{s.t.} \quad \begin{pmatrix} 2C_{\max} & Q \\ Q & A(x) \end{pmatrix} \succeq 0, \; x \geq 0,$$

which finds an optimal cross-sectional area x_e for each bar e under a semidefinite stiffness constraint. Here, l_e denotes the length of bar e, $A(x) = \sum_{e \in \mathcal{E}} A_e \, x_e$ is the stiffness matrix, Q describes uncertain forces on the nodes of the truss and $C_{\max}$ is a bound on the compliance of the truss, which must not be exceeded. For more details see Sect. 6.1.1. A truss has a buffering capacity of k if the above semidefinite constraint still holds, even after up to k arbitrary bars have failed. This condition can be incorporated into the above formulation with the help of the set of failure scenarios $\mathcal{Z} = \{z \in \{0, 1\}^{\mathcal{E}} \colon \sum_{e \in \mathcal{E}} z_e \leq k\}$. For a failure scenario $z \in \mathcal{Z}$, bar e fails if and only if z_e is 1. The failure of a bar can be represented by removing its influence on the corresponding stiffness matrix. Therefore, the stiffness matrix for a given failure scenario z is given by $A(x, z) = \sum_{e \in \mathcal{E}} A_e x_e (1 - z_e)$. The resilient design problem then reads

Table 6.5 Statistics for the computation of resilient trusses with buffering capacity k

k	$\vert\mathcal{Z}\vert$	Relative volume	Runtime in s
0	–	1	0.19
1	137	2.38	21.65
2	9316	4.03	2240.18

$$\min_{x \in \mathbb{R}^{\mathcal{E}}} \sum_{e \in \mathcal{E}} l_e x_e \quad \text{s.t.} \quad x \geq 0, \quad \begin{pmatrix} 2C_{\max} & Q \\ Q & A(x, z) \end{pmatrix} \succeq 0, \text{ for } z \in \mathcal{Z}. \tag{6.20}$$

The cardinality of $\mathcal{Z}$, and thus the number of semidefinite constraints, increases exponentially with k, which makes this approach feasible only for small values of k, see [64]. For a different dynamic approach, see also the design of resilient water supply networks in Sect. 6.3.5. The following examples, however, use the complete set $\mathcal{Z}$.

Our approach is in the spirit of robust optimisation, see Sect. 6.1, and the presented solution methods in Sects. 6.1.1 to 6.1.3, as $\mathcal{Z}$ may be identified as an uncertainty set. However, a different solution approach is needed due to the discrete $\mathcal{Z}$ in contrast to the continuous ellipsoidal uncertainty sets considered there.

Figure 6.26 shows three optimal crane truss structures with different buffering capacities k [7]. The nominal forces are displayed as arrows, the uncertainty sets as ellipsoids around them. In Table 6.5, the size of the failure set $\mathcal{Z}$, the objective function value, and the runtime of the optimisation are displayed for three values of k. It is apparent and not unexpected that resilience does not come without a significant cost: Requiring the truss to be stable even after failure of any two bars, increases the volume by a factor of four and also the computational cost. Figure 6.27 shows the maximal increase of the nominal forces for different attack angles that the three trusses can sustain (0° means the forces face downward). It indicates a relationship between the different metrics to assess resilience defined in Sect. 6.3.1 as a greater buffering capacity leads to a greater performance range and margin, as well.

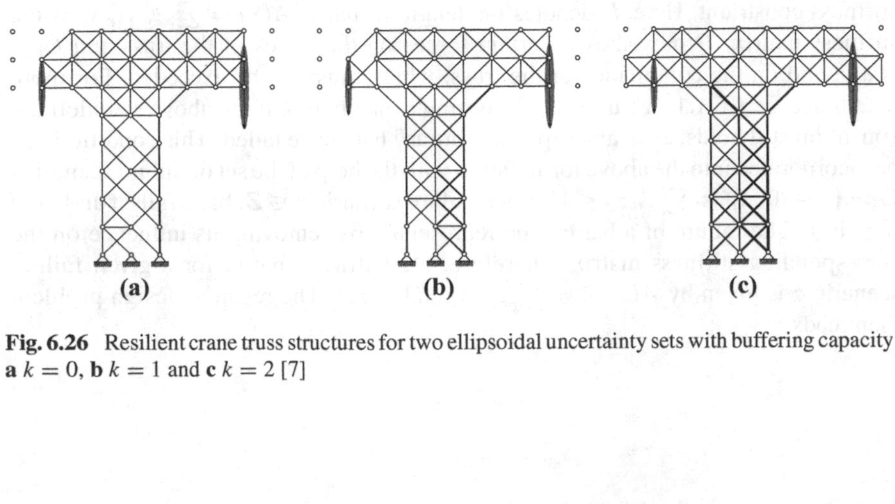

Fig. 6.26 Resilient crane truss structures for two ellipsoidal uncertainty sets with buffering capacity **a** $k = 0$, **b** $k = 1$ and **c** $k = 2$ [7]

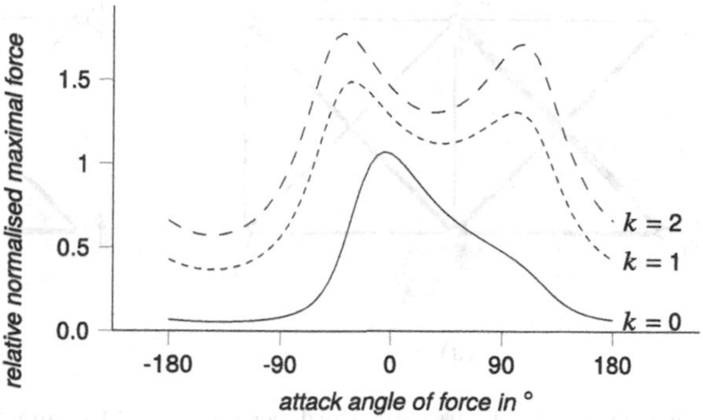

Fig. 6.27 Performance curves for trusses given in Fig. 6.26 with buffering capacity k [7]

Buckling control in truss structures under bar failures

An important cause of failure for truss structures is the buckling of individual bars, which is caused by excessive axial compressive loads and which cannot be detected through the compliance condition alone, see Sect. 6.1.2. As shown in [64], the optimisation problem (6.20) can be augmented with buckling constraints to prevent bars to buckle in the optimal design. These buckling constraints also have to be copied for each failure scenario in $\mathcal{Z}$ in order to obtain a resilient structure with buffering capacity. Furthermore, variables for the bar forces for each failure scenario must be added. These are determined by so-called indicator constraints, in contrast to the equilibrium conditions used in Sect. 6.1.2, as the geometry matrix after the failure of bars is possibly singular. For more details see [64]. A passive option to avoid bar buckling is to increase the width of vulnerable bars. An alternative that helps to reduce the mass of the truss is presented in Sect. 5.4.7. It introduces active buckling control by integrating piezoelectric stack actuators in compact piezo-elastic supports at the bar ends and has been integrated in the robust problem in Sect. 6.1.2.

Figures 6.28 and 6.29 show resilient truss structures with different buffering capacity and different number of active bars for the example presented in Sect. 6.1.2. The objective function values and computational costs are given in Table 6.6. It can be seen that active bars indeed help to decrease the volume of the truss, but at the price of increased cost due to the actors. Note that the added weight of the actuators is not considered in the model.

Conclusion

In this section we have shown how to include bar failures into truss topology optimisation to design light-weight but robust and resilient trusses. Future research could investigate, whether the increased size by additional variables and constraints for each failure scenario can be avoided.

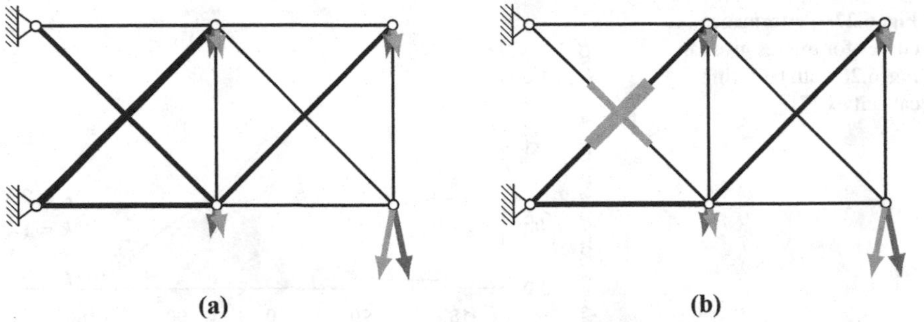

Fig. 6.28 Resilient truss for two load scenarios, $k = 1$ bar failures and different numbers r of active bars: **a** $r = 0$ active bars and **b** $r = 2$ active bars [64]

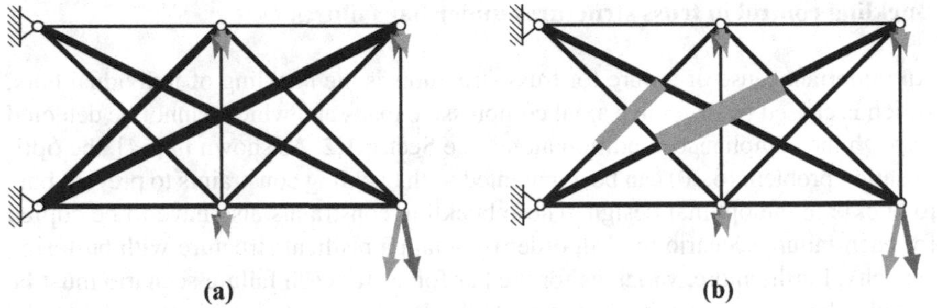

Fig. 6.29 Resilient truss for two load scenarios, $k = 2$ bar failures and different numbers r of active bars: **a** $r = 0$ active bars and **b** $r = 2$ active bars [64]

Table 6.6 Optimal objective values and solving times for different combinations of failure scenarios k and maximal number of active bars r [64]: **(a)** volume in 10^5 mm^3 and **(b)** solving times in s

(a)			
Volume	$k = 0$	$k = 1$	$k = 2$
$r = 0$	1.9457	3.6088	7.4511
$r = 2$	1.7657	3.0205	6.4644
(b)			
Time	$k = 0$	$k = 1$	$k = 2$
$r = 0$	7.47	52.26	890.23
$r = 2$	6.64	86.49	581.77

6.3.5 *Optimal Design of Resilient Systems on the Example of Water Supply Systems*

Lena C. Altherr, Philipp Leise, Marc E. Pfetsch, and Andreas Schmitt

This section presents optimisation methods to consider resilience, as introduced in Sects. 3.5 and 6.3.1, in the design phase of a technical system. In order to master uncertainty, our goal is to find an optimal combination of different components constituting a resilient system structure, i.e. a structure which is able to tolerate and react to failing components. To assess and optimise resilience, we use the concept of *buffering capacity* described in Sect. 6.3.1: If a system has a buffering capacity of k, any k components can fail and a previously defined minimum system functionality is still fulfillable. We also say the system is *k-resilient.*

In [9, 10] we have analysed a Mixed-Integer Nonlinear Programming (MINLP) model to design a cost-optimal but k-resilient water supply system for a high-rise building and presented a solution approach. A similar model has been validated in [133] with the help of a test rig. In the following, we briefly summarise the model and the solution algorithm presented. Furthermore, we review the found characteristics of the resilient designs.

In comparison to Sect. 6.3.4, which presents the computation of k-resilient trusses, the basic models differ. Therefore, the computation of worst-case failures also differs and needs to be treated differently. In both cases, the design of a topology is considered, and structural uncertainty is present, see Sect. 2.3. The consideration of buffering capacity further increases this uncertainty.

For an overview on using MINLP to optimise water distribution networks (WDN) we refer to [41]. The literature on resilient WDN focuses on measures to quantify resilience and testing these on existing networks, see e.g. the well know resilience index by Todini [169] and the overview article [148]. For an example for an optimisation of resilience using a surrogate measure in the context of WDN see [171]. The inclusion of component failures in layout optimisation can also be regarded as a defender-attacker-defender game model, see e.g. [5], in which the defender designs a layout. The attacker interdicts components in this layout, whereupon the defender reacts to these contingencies. These games can be understood as multi-stage optimisation/adjustable robust optimisation with integer variables, see [19, 102, 177]. In some cases the tri-level structure can be reformulated as a structure with only two levels, see e.g. [33] and the general bilevel solution approaches [53, 105, 129].

Optimisation model

In high-rise buildings, pumps are used for pressure-boosting in order to supply all floors with water. In each pressure zone, a given pressure and volume flow demand has to be fulfilled. The aim of the optimisation model is to design a pump and pipe system which fulfils these demands and minimises operating and investment

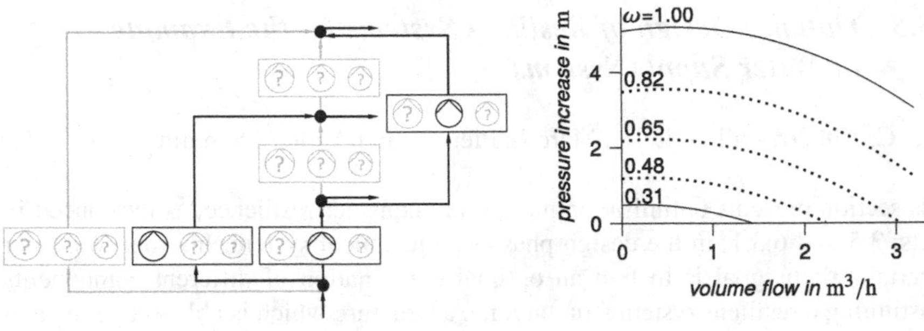

Fig. 6.30 Features of decentralised water supply systems: **a** possible discrete decisions (grey) and exemplary solution (black), **b** exemplary characteristic diagram

costs. In order to do so, each floor has to be connected to the ground floor either by connecting it directly or by connecting it to a lower floor. Thus, the topology as well as the diameters of the resulting pipe network have to be determined. These possible pipe layouts are restricted to be tree-shaped, i.e. each floor is connected to exactly one lower floor. Further discrete decisions concern the placement of pumps. The model equally determines the cheapest pump operation to provide water. Altogether, the model contains discrete decision variables, such as pipe and pump placement, as indicated in Fig. 6.30b, and nonlinear non-convex constraints to approximate the pump characteristic diagrams, as shown in Fig. 6.30b. This leads to a complex mixed-integer nonlinear problem which is already strongly NP-hard to solve for $k = 0$.

Computation of resilient solutions

To obtain solutions which are robust against uncertain pump failures, we developed a method to find a k-resilient system which minimises the investment and operating costs in [9, 10]. In order to integrate this robustness we define the set of failure scenarios

$$\mathcal{Z} = \left\{ z \in \{0,1\}^n : \sum_{i=1}^{n} z_i \leq k \right\},$$

where we have enumerated all possible pumps in the building from 1 to n. Thus, for scenario $z \in \mathcal{Z}$, the entry z_i is 1 if and only if pump i fails in the scenario. Using this set, we can guarantee resilient system structures by modelling the successful operation of the system for each failure scenario $z \in \mathcal{Z}$, even though the pumps given by z fail.

Integrating all scenarios in $\mathcal{Z}$ in the model would lead to exceedingly large solution times, due to the exponential growing cardinality of $\mathcal{Z}$ with respect to k. Therefore, an iterative strategy is used. We solve models which only consider a subset $\mathcal{Z}'$ of $\mathcal{Z}$. For a solution, we compute a worst-case failure scenario in $\mathcal{Z}$. If the solution can

Table 6.7 Shifted geometric mean of solving time in seconds and number of instances solved withing the timelimit, clustered by number of pressure zones and buffering capacity k with 36 instances

# zones		k				
		0	1	2	3	4
7	Time	412.62	847.73	1087.60	1252.01	1414.59
	Solved	36	36	36	36	36
8	Time	3315.81	6388.67	6733.21	6570.97	6451.66
	Solved	36	22	15	10	11

sustain this scenario, the optimal resilient solution is found. Otherwise, the failure scenario is added to $\mathcal{Z}'$ and the model is solved again.

This scheme is further adapted to the use case of the high-rise building. Due to the tree-shaped network topology and the usage of only parallel pumps of the same type, the volume flows in each floor and in each pump are pre-determined for a fixed pipe topology. Thus, the number of nonlinear constraints decreases. Furthermore, for the optimal placement and operation of pumps on this topology, resilience can be modelled by a set of linear inequalities for each failure scenario. These inequalities can be separated by a simple dynamic program with running time polynomially bounded in the input parameters. Thus, a branch and bound scheme which branches on the pipe connections from the bottom to the top was developed. Computational tests in [10] show the computational benefits of this approach.

To further improve running times an alternative representation of the characteristic diagrams independent of the operating speed presented in [137] can be used. This representation is convex allowing the usage of perspective cuts introduced in [56]. New computational results for the combination of the branch and bound scheme and these cuts are presented in Table 6.7 for the test instances and test environment used in [10]. The modification allows solving instances with one more pressure zone and larger buffering capacity. An increasing computational burden for increasing k and an increasing number of pressure zones is observable.

Assessment of resilience

In the following, we discuss some findings which can be useful to understand the advantages and properties of resilience. This is possible since we are able to rapidly compute resilient solutions with the above presented scheme. Thus, we can compare resilient solutions for different parameters.

Figure 6.31a shows the power consumption and investment costs of all Pareto-optimal solution topologies with respect to power consumption and investment costs of a building with six pressure zones and different levels of k-resilience. There exists no solution topology which, at the same time, is more energy-efficient and cheaper than the depicted solutions. We first note that larger investment leads to lower energy

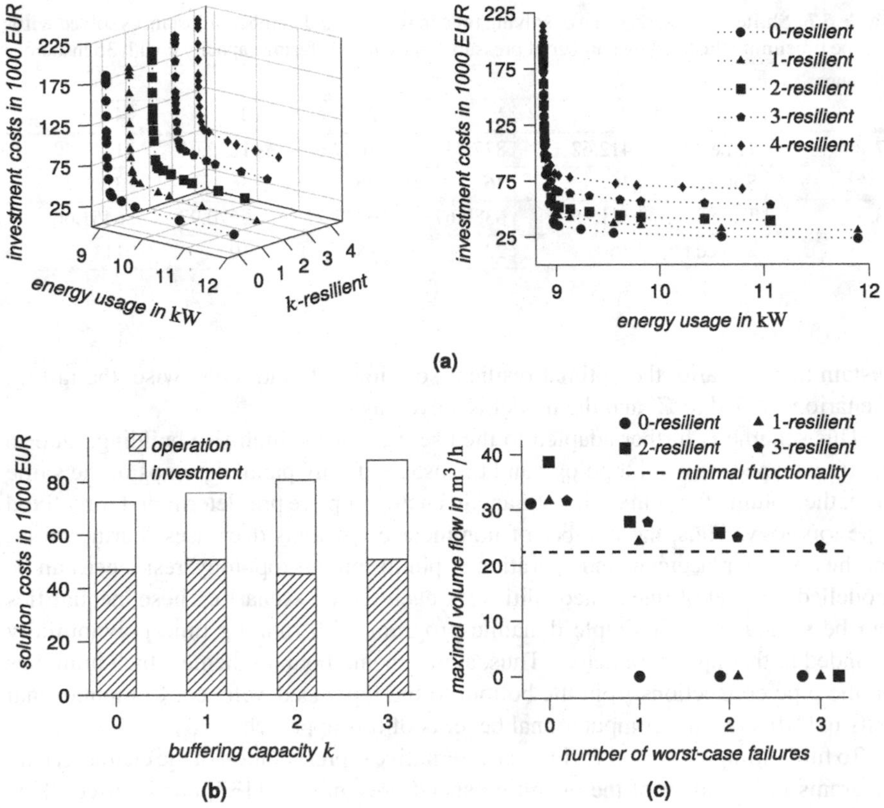

Fig. 6.31 Resilience properties: **a** three and two-dimensional depiction of power consumption and investment costs of Pareto-efficient topologies, **b** solution costs for exemplary optimal k-resilient solutions, **c** maximal volume flow these solution can maintain after worst-case failures

costs, since more pumps are built, which can then be operated more efficiently. It can be seen that the minimal investment increases, but also the worst power consumption achievable by a Pareto-optimal solution, decreases with an increasing k-resilience. This is due to the increased number of pumps needed to guarantee fault-tolerance. The overall positions of the solutions are coherent with the observation that the number of pumps and thus the investment costs increase with more emphasis on the energy costs. Interestingly, for a small power consumption ($\leq 9\,\text{kW}$) a large number of different Pareto-optimal solutions exists. Small efficiency improvements correspond to large changes in the investment costs. The best power of around 8.9 kW is for larger k only achievable with greater investment costs. However, due to the great density of solutions, resilience can be achieved for this efficiency without big disadvantages. We furthermore see that for solutions with a power consumption of at least 9 kW the investment costs tend to scale almost proportional to the resilience factor k with a proportionality constant smaller than 1. This contrasts an exemplary conventional redundant design strategy, which could build every pump of the 0-resilient solution $k + 1$ times and would scale with a proportionality constant of 1.

In [9] the characteristics of resilient designs are analysed for an exemplary application for $k \in \{0, \dots, 3\}$ with seven pressure zones. Using perspective cuts, we are able to solve this application for nine pressure zones leading to different solution topologies. Despite the difference in the number of pressure zones, we will draw very similar conclusions on resilience in the following to the ones in [9].

The price of resilience is mainly due to the increased number of needed pumps as observable by the solution costs depicted in Fig. 6.31b. Nevertheless, it is a lot more advantageous than a pure strategy of redundancy. Placing another pump in each pressure zone used of the non-resilient design increases the investment costs by more than 50 %, whereas the topology with a buffering capacity of $k = 1$ instead of $k = 0$ is only 5 % more expensive. The additional pumps which are needed for resilience are even able to decrease the operating costs for the $k = 2$ solution. It remains similar for the other levels of resilience.

Several metrics to quantify resilience have been introduced in Sect. 6.3.1, and it is not clear whether the choice of buffering capacity is preferable to other metrics in the design of a resilient high-rise water supply system. Nevertheless, it is indicated in [9] that consideration of the buffering capacity is also linked to the improvement of performance range, radius of performance, and margin. We obtain the margin for each solution, by computing the worst-case combination of one up to three failing pumps and the subsequent maximal volume flow which can be transported, compare Fig. 6.31c. In these computations, the minimal functionality after failures was defined as 80 % of the design point volume flow of $28\,m^3/h$. Thus, each k-resilient solution lies above the dotted line for up to k failures. It can be seen that resilient solutions are oversized for standard operation, since without failures they exceed the required volume flow of $28\,m^3/h$. Thus, we claim resilience is a property which has to be actively sought for. Conventional methods will seek solutions which are "just right" for the given operating point and have no reserves. We again observe that our approach for resilience is finer-grained in comparison to simple redundancy. The solution with a buffering capacity of $k = 3$ is not just obtained by including another pump to the solution which considers two pump failures. This is the case, since the latter has the largest reserves for one and no failures.

An observation specific to resilience and the design of decentralised high-rise water supply networks has been made in [10]. Here it has been shown that increasing the weight of the energy costs, i.e. shifting the importance of investment versus operating costs, leads to solutions which are branched out, whereas demanding greater resilience tends to solutions that connect the floors in series. This can be explained by the fact that in the former layout one pump can supply fewer floors than in the later scheme. Thus, it has fewer redundancies and is inferior with respect to resilience aspects.

6.3.6 Application of Resilience Metrics to the Fluid Dynamic Vibration Absorber in Drop Tests

Nicolas Brötz and Peter F. Pelz

If we want to apply resilience properties, we must be able to assess the system's resilience. For this purpose, we defined the resilience metrics in Sect. 6.3.2. We evaluate drop tests of the Modular Active Spring-Damper System (MAFDS), presented in Sect. 3.6.1, with the Fluid Dynamic Vibration Absorber (FDVA), see Sect. 5.4.4, to apply these metrics to a technical system in comparison to a conventional damper.

A vibration absorber is used to reduce vibrations from an oscillating system. A conventional dynamic vibration absorber consists of a heavy mass and a capacity. However, this additional weight counteracts the goal of a lightweight construction. In contrast, the FDVA reduces the dynamic mass by the use of hydrostatic transmission, see Sect. 5.4.4

The MAFDS, shown in Fig. 6.32 represents a dual mass oscillator and is therefore suitable for demonstrating the functionality of the FDVA. The purpose is to reduce the vibrations of the lower structure. Since this lower structure is a single mass oscillator comparable to a wheel, we will refer to it as wheel in the following. We consider a maximum wheel load for the lower structure, which is important in drop tests. The tests in the MAFDS are designed in such a way that possible influences, such as a change in the body mass, can be investigated in order to address the data uncertainty

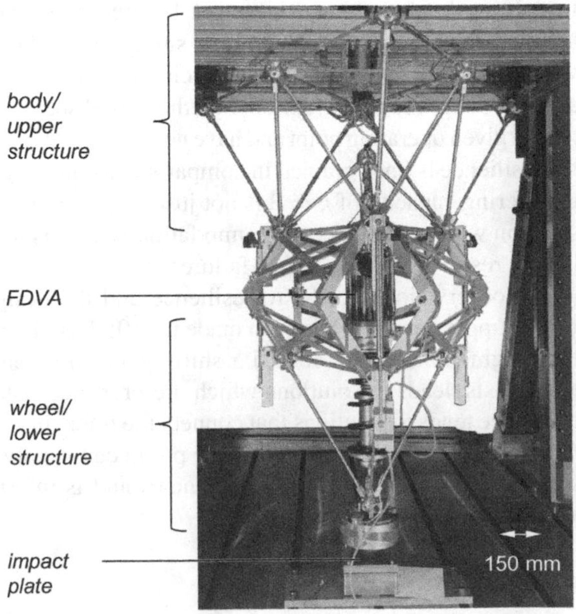

Fig. 6.32 Test rig of the MAFDS for drop tests with integrated FDVA

of the load-bearing system, see Sect. 2.1 and apply the resilience metrics, defined in Sect. 6.3.1 to the usage of the FDVA.

Drop test of a dual mass oscillator

The MAFDS, which incorporates the technologies damper or FDVA, is dropped during the measurements with a variation of additional weight on the upper structure (body) to vary the body size and represent incertitude. The additional weight addresses the unknown loading conditions and is varied in 20 kg steps from 0 to a maximum additional weight of 80 kg. The test rig is shown in Fig. 6.32.

The MAFDS is similar to a car suspension strut because both are dual mass oscillators. A drop test is an unusual but possible use case. The driver might steer the car at a high speed down a sidewalk. A scenario which is not in focus of the suspension strut adjustment.

The MAFDS is in free fall of 30 mm until impact. For each drop test, the force at the impact plate is recorded. This force is equal to the wheel load F_W. Figure 6.33 shows the wheel load F_W over time for a drop test without additional weight. The FDVA has two opened ducts to realise an eigenfrequency of 10 Hz. This is the nearest possible frequency of the FDVA to adapt the lower structure eigenfrequency. The better the eigenfrequencies match, the better the wheel load fluctuation is reduced. We calculate the lower structure eigenfrequency $\omega = \sqrt{k/m}$ by lower structure mass m and the stiffness k of the elastic foot.

The first peak for both measurements with damper and FDVA is the first lower structure contact to the impact plate. We measure the highest wheel load when the upper structure compresses the suspension. The highest wheel load is the critical load. At this point the highest force acts on the wheel and thus on the tire. The MAFDS in our case has a rubber buffer, that cannot burst, instead of a tire. Thus we are able to perform drop tests with high wheel loads at which a real tire would already burst.

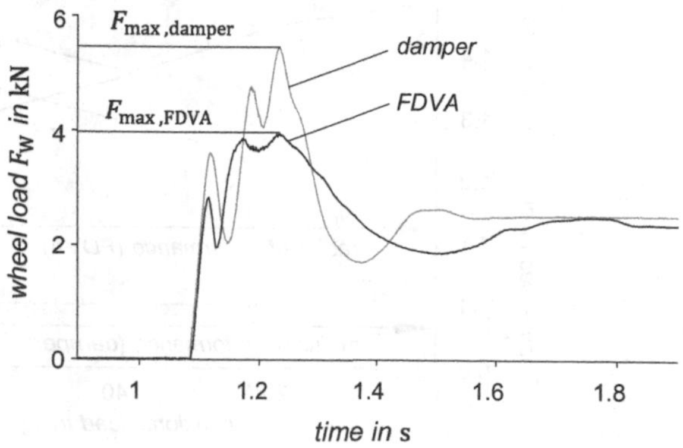

Fig. 6.33 Wheel load for FDVA and (conventional) damper for drop test with no additional weight at 30 mm height

Resilience metrics for drop test

To quantify uncertainty we consider the resilience metrics defined in Sect. 6.3.1. First, we define the functional performance. For this example the functional performance $1/F_{max}$ is the inverse of the maximum wheel load. The influencing factor is the additional weight. The design point is the MAFDS without an additional weight. To measure the resilience we need to define a minimum performance. The critical element in such a use case is the tire. The tire has a load index which defines how much weight it can carry. For this example, a tire with a load index of 91 is chosen because this load index is used for cars with body mass similar to the MAFDS. The load index 91 allows to carry 615 kg. The static load should be the maximum load the tire is exposed to. This is equivalent to the minimum functional performance f_{min}. In Fig. 6.34 the functional performance is normalised by the minimum functional performance. Every point below this minimum represents a failing system. Failing means, the tire could burst due to excessive load. With this definition, we can compare the resilience of a damper and the FDVA. We can see that the FDVA has a higher functional performance. The FDVA's margin at zero additional load is 4 % higher than the damper's margin.

The radius of performance defines the minimum distance between the design point and the point for which the functional performance undercuts the minimum functional performance f_{min}. This radius of performance is 70 kg of additional load for the damper. The radius of performance for the FDVA is higher than the performed tests. Therefore, the resilience of the MAFDS can be increased by using the FDVA: even at an additional load of 80 kg and higher the system guarantees a predetermined minimum of functional performance.

The performance range in this example is equal to the radius of performance because there is a constant decrease of functional performance with higher additional load. To evaluate the gracefulness, measurements with the FDVA at higher additional loads would have to be performed resulting in a destructive test. Thus, we do not evaluate the resilience gracefulness here.

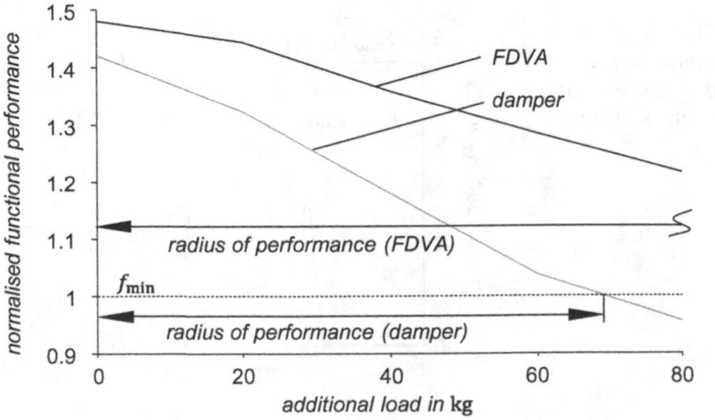

Fig. 6.34 Normalised functional performance for drop tests for FDVA compared to a damper

Conclusion

In technical systems where a critical minimum exists, which defines the minimum functional performance, we can calculate the resilience metrics margin and radius of performance. In the drop tests the radius of performance has a higher use case than the margin because it describes how uncertain the additional load can be until the system fails. But to evaluate the radius of performance, tests with failure have to be evaluated in a normal case.

The margin is useful to quantify the standard usage with no additional weight. In this drop test we have an improvement of 4 % of functional performance with FDVA in relation to a standard damper. But the value of margin on its own can not give an information about when the system fails.

6.3.7 *Concept of a Resilient Process Chain to Control Uncertainty of a Hydraulic Actuator*

Ingo Dietrich, Manuel Rexer, and Peter F. Pelz

The concept of resilience is not only applied to master uncertainty during design, but equally to connect the product life phases production and usage, (Chap. 3) by integrating the four resilience functions monitoring, responding, learning and anticipating [88], see Sect. 6.3.1. Within the product life phases production and usage, the state of the art is to establish variable process windows to master uncertainty. For example, modern cars have flexible oil changing intervals, based on numerous operating parameters of the engine. However, currently the connection between the life phases production and usage is still formed by the product design. Customer feedback or guaranteed returns are analysed individually, and the component life, described by usage and environmental parameters, is deduced. Ultimately, the product or its production are changed to cope with the findings.

Today, an increasing number of technical products offers the possibility to collect data during the usage phase. Paired with the development of technologies, such as single part tracking, the increasing modelling of production processes and process chains, resilient product life phase spanning process chains become possible. In this section, we show the general concept of this resilient process chain and apply it to the hydraulic actuator of the Active Air Spring introduced in Sect. 3.6.2. We introduce production uncertainty to individual parts of the actuator and conduct experiments to determine the effect on usage parameters.

General concept of resilient process chains

The concept of a product life phase spanning a resilient process chain was presented by Dietrich et al. in [45] and is shown in Fig. 6.35. Based on the production plan, a

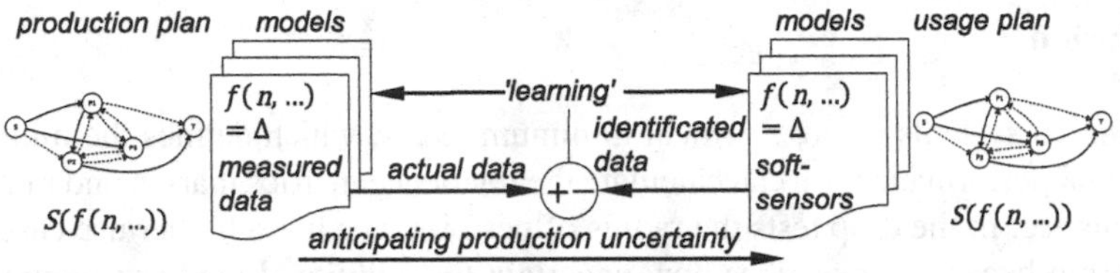

Fig. 6.35 Product life phase spanning process chain as presented by Dietrich et al. [45]

technical component is produced. The individual production steps are described by models that use measured parameters during production. Following the nomenclature introduced in Sect. 1.5, the production and usage plan are structures S, described by functional relations f, that rely on data b. (Soft-)sensors feed models that aggregate information during the usage. By a suitable selection of data that are logged during production as well as during usage, the data can be matched and compared. Mostly, the data obtained from production and usage is not the same, thus correlating models for the matching need to be developed. For example, these models may be developed by domain-specific experience or empirical correlation. By the feedback of the differences between actual data from production and identified data from usage, the models can be adapted and a learning function might be established. Using time histories and correlating single part data to the respective usage data, the component behaviour can be anticipated already in the production itself. Based on this anticipation, the usage plan of the component can be adapted. In reference to Sect. 3.5 and Fig. 3.16 'the system function is evolving', thus enabling resilience. The resilient process chain deals with structural uncertainty, according to the Sects. 1.5 and 2.3.

The concept of a resilient process chain results in four requirements for the production and the usage of the component.

1. Production parameters must have an influence on the usage.
2. Data that can be measured during production as well as during usage must be identified.
3. Data during production and usage must be collected.
4. Models that process the measured data from production and usage must exist.

Resilient process chain applied to the active air spring

In the following, we want to evaluate this concept by applying it to the Active Air Spring, which is described in detail in Sect. 3.6.2. The Active Air Spring is an active system that combines the advantages of an air suspension, such as level control or the load-independent body eigenfrequency, with those of an active system that can actively reduce vibrations and has a flexible working area. For example, it can be used to minimise kinetosis during autonomous driving of cars [83].

The active elements are two hydraulic diaphragm actuators with linear moving segments, which vary the load-bearing area of the Active Air Spring. Each segment

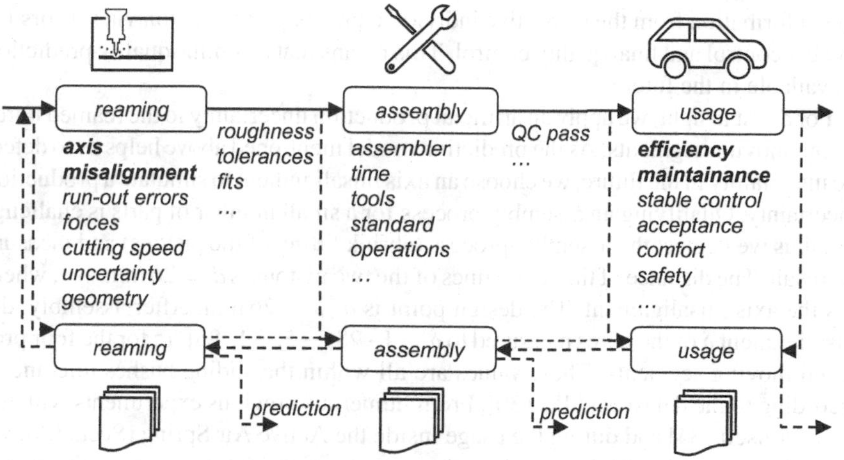

Fig. 6.36 Part of the process chain for machining, assembly and usage of the hydraulic diaphragm actuator. Solid lines represent material flow and dashed lines information flow. The parameters we focused are bold

has two pistons that run in a sliding bushing and is actuated with hydraulic oil. For a more detailed understanding of these actuators we refer to [83, 84] and Sect. 3.6.2.

To evaluate the concept of the resilient process chain, we want to focus on a rather simple mechanical property of these segments. In Fig. 6.36 the part of the process chain for the final machining, assembly and usage of the hydraulic diaphragm actuator is shown. It is the most crucial part of the production for the functional performance of the actuator. The process steps reaming, assembly and usage are labeled with examples for relevant parameters in the according step. In difference to a classical production chain we want to use information from each step in models to perform predictions on the one hand, and to improve the process steps on the other hand. As presented in Sect. 6.1.8, we have existing models for the reaming process. Typically, reaming within a process chain happens at the end of the value chain. Its purpose is to produce the shape and position of functional bores within the required tolerance range [42]. For productivity reasons, nowadays often multi-blade reamers are used where the functions "cutting" and "guiding" are combined in one geometric element. The production of precision bores using multi-bladed reamers is the subject of various scientific studies. Uncertainty in the form of disturbances regularly influences the reaming process in industrial practice. Typical disturbances are axis misalignment, run-out errors and inclined surfaces with sloped pilot holes [23, 24].

The influence of disturbances on the quality of the reamed bore has been investigated with regard to diameter, circular shape and cylindrical shape [21, 110] as well as with regard to the deflection of the tool [25, 82]. A deflection of the tool leads to an increased diameter of the casing cylinder of all bore centres of the reamed bore. This in turn leads to an increase of the casing cylinder of all bore centres [26]. The current object of research is the development of an online prediction model, which uses information from the respective individual processes production via sensors for process control and final quality control. This means that an online quality prediction is available in the future.

For a first insight, we apply an artificial production uncertainty to the reamed bores of four moving segments. As the prediction model mentioned above helps us to detect the uncertainty in the future, we choose an axis misalignment to simulate a production uncertainty. Qualifying an assembly process for a small number of parts is challenging, thus we neglect the assembly process (shrink fitting of the pistons) and measure the result. The distance of the centre lines of the two pistons is $d = 26\,\text{mm} + \delta$, where δ is the axis misalignment. The design point is $d_{DP} = 26\,\text{mm}$. After assembly, the misalignment δ of the bores measured to $\delta = [-28, -5, 32, 85]$ ¯m for the four produced moving segments. These values are all within the sliding bushes tolerances, according to the datasheet [45, 130]. From numerous previous experiments with the actuator itself [84] and during the usage inside the Active Air Spring (Sect. 3.6), we know that assembly and disassembly, changing the membranes and mounting in the test bench are robust in respect to the experimental results.

Experimental evaluation of the productions influence on the usage

In our experimental evaluation we want to investigate the influence from the production on the relevant usage parameters. To generate a viable data-set of the usage phase, we use a Hardware-in-the-Loop test rig, in which the hydraulic diaphragm actuator can be investigated without being mounted inside the Active Air Spring [45, 84]. For a more detailed insight on Hardware-in-the-Loop tests in general, we refer to Sect. 4.3.4. The test rig enables the characterisation of the actuator to calculate its efficiency, as well as the simulation of a road ride in a car equipped with the Active Air Spring. The efficiency is calculated by the input energy W_{in} and the dissipated energy W_{dis} as

$$\eta := \frac{W_{\text{in}} - W_{\text{dis}}}{W_{\text{in}}}.$$

For a detailed understanding of the characterisation we refer to [45]. To increase the wear-rate, we use an axial counter force to the moving segments of the actuator four times as high as in a typical application. To get a time series of usage data, we iterated between a characterisation cycle which allows us to calculate the desired parameters and a cycle with a road signal (national highway with a speed of 100 km/h) [128]. We perform a characterisation after each hour of a road signal. At the end of each test, we took an oil-sample and analysed it in the lab. The effort for the tests is high, due to the assembly process of the actuator and the runtime needed for results. Thus the number of experiments had to be limited.

Figure 6.37 shows the experimental results for the four specimen. Figure 6.37a shows the efficiency over the run-time. We can see a general trend for a decreasing efficiency over the run-time, which results from the wear of the sliding bushes. From $-28\,\mu\text{m}$ to $32\,\mu\text{m}$ the efficiency rises. From $32\,\mu\text{m}$ to $85\,\mu\text{m}$ it decreases. This leads

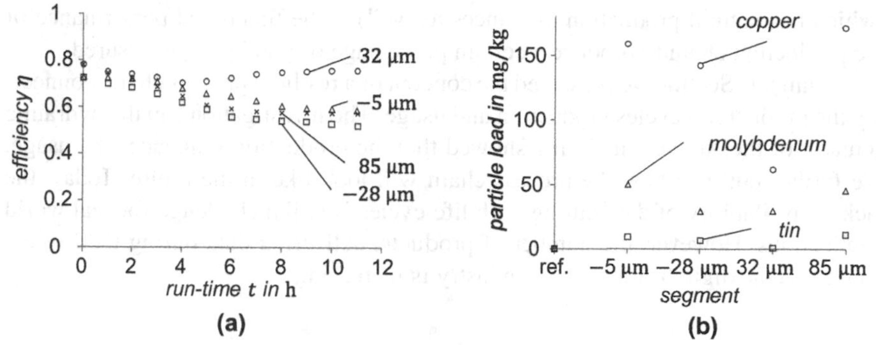

Fig. 6.37 Experimental results for the four different segments: **a** the efficiency over the run-time for the four different segments, **b** the particle load in the hydraulic oil after for each segment

to the assumption that the chain of tolerances 'bores in the actuator body—sliding bushes—pistons' has its optimum between −5 µm and 85 µm. Figure 6.37b shows the particle load in the hydraulic oil samples after the respective test was finished. In contrast to the reference sample of the oil three materials occur. Copper, molybdenum and tin are all used in the sliding bushes [130] and result from wear. The particle load correlates with the efficiency results. The 32 µm sample has the highest efficiency, thus the lowest friction and wear, which results in the lowest particle load in the oil. It should be mentioned, that these results cannot be transferred one-to-one to the real usage phase of the Active Air Spring, as the load was increased for a faster wear of the sliding bushes. The experiments show that there is an effect from the final production stage (reaming) on the usage of the Active Air Spring, or the hydraulic diaphragm actuator respectively. This effect is already measurable for production uncertainties that lie within the sliding bushes tolerances. However, the actuator could be assembled and was working as intended for all four different segments.

Conclusion

At the end of this section, we give an outline of a possible resilient process chain in the future: We use reaming models during the production to predict the quality of the bores. An empirical model correlates the bore quality to the outcome of the assembly. Another empirical model, gained by preliminary tests and real usage data, predicts the individual component's usage parameters. Resulting user stories would read: 'Based on the quality during production, the oil-changing interval for the individual actuator is determined.': Referring to Sect. 6.3 a disturbance in the production phase is mastered in the usage phase. The minimal functional performance for the actuator life time is ensured by reacting to the production disturbance with the adaptation of the oil-changing interval. 'Based on the quality during production—and the predicted efficiency, an inefficient actuator is combined with a very efficient hydraulic drive (which has natural production tolerances as well)': The functional performance of the production chain to produce a certain percentage of good parts is ensured.

Within this Section we presented the concept of a resilient process chain, connecting the product life cycles production and usage. The investigations on the hydraulic actuator of the Active Air Spring showed that the production influences the usage. We further outlined how the process chain will look like in the future. Today, the lack of availability of data during both life cycles is still a challenge for real world applications. However, the number of products, delivering data during their usage phase and the digitalisation of the industry is increasing.

6.3.8 Experimental Evaluation of Resilience Metrics in a Fluid System

Philipp Leise and Lena C. Altherr

As mentioned by Folke et al. in [55, p. 1], a resilient system has the ability to "continually change and adapt yet remain within critical thresholds". Folke et al. focused on the resilience of socio-ecological systems. Nevertheless, this concept, as already mentioned earlier in Sect. 6.3.1, can be transferred to the domain of mechanical engineering. While this concept is easily understood, the transfer to the domain of mechanical engineering is more challenging. We present a modular test rig, that is used to evaluate the applicability of the four functions (monitoring, responding, learning and anticipating) to derive resilient systems on the one hand and investigate selected resilience metrics experimentally on the other hand. We refer to Chaps. 1, 3 and Sect. 6.3.1 for a broader introduction to resilience of technical systems.

The focus of the considered resilience metrics that can be evaluated at the test rig, is set on the metrics, which correspond to an adaption of their behaviour over time. As mentioned at the beginning of this section, a resilient system has the ability to continuously adapt to external changes. These changes can be initiated by "exogenous drivers [...] and endogenous processes", as noticed by Walker et al. in [173, p. 3]. The test rig in its modular and variable design is capable to host multiple experiments for a large variety of resilience metrics that consider both "exogenous drivers" and/or "endogenous processes". It is designed to be able to derive basic implementation ideas and to present proof-of-concepts for metrics shown in Sect. 6.3.1. Therefore, we will present a brief overview of its capabilities and discuss the outcomes of the selected experiments.

The test rig with an exemplified piping is presented in Fig. 6.38. The purpose of this test rig is to supply water in each of the two acrylic water tanks which are depicted on the right side in Fig. 6.38. We are able to read out and control multiple sensors and actuators. As actuators there are up to three pumps and up to ten control valves. For instance, we can use a control valve as an *exogenous driver* as introduced in [173] to induce external disturbances on the water supply system within the test rig. We have the possibility to place multiple pressure sensors at different locations in the system.

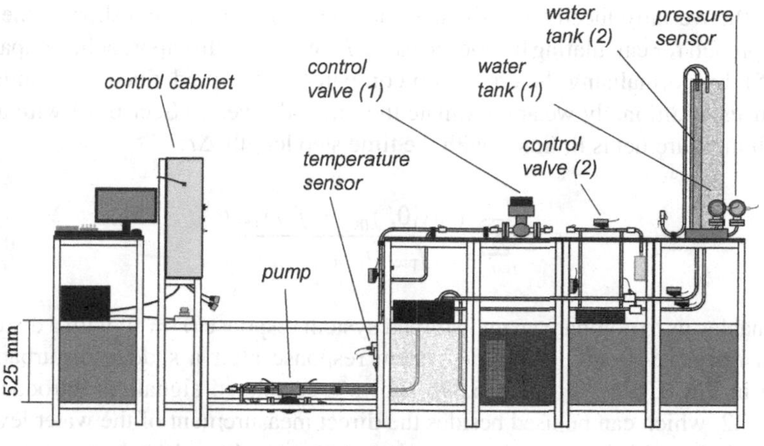

Fig. 6.38 Test rig to illustrate the application of different resilience metrics

Additionally, we can measure the power demand of each pump, the water temperature at four locations and the opening of all electrically actuated control valves. This sensor data can be evaluated in real-time on the affiliated computer system. We implemented a Python-based control system to be able to use all available sensors and actuators for each dedicated experiment.

In addition to the design and implementation of the test rig, we derived a simulation model based on Modelica, cf. [61, 167], which represents the basic structure of the test rig. This simulation model was first validated by experiments at the test rig and then used to derive a database of simulation runs, where each run represents a dedicated disruption scenario. This database can then be used to derive a system, which can automatically reconfigure itself in case of external or internal malfunctions based on the pre-calculated system behaviour. This approach eliminates the need to conduct multiple runs of actuators within the real system to derive system models in case of failures, as this is done in other domains, see e.g. [27].

We conducted multiple experiments on this test rig, to verify the usage and applicability of predefined resilience metrics, which we will briefly present in the following. Moreover, we assessed the four functions (monitoring, responding learning and anticipating) of resilience, as introduced in Chap. 3 and Sect. 6.3.1, with the help of our test rig.

Resilience triangle

We conducted experiments with specific resilience metrics, as introduced in Sect. 6.3.1 and [7]. We refer to Sect. 6.3.3 for a more detailed view on the functional performance and to Sect. 6.3.5 for more details on the buffering capacity. Within this section, we will only focus on the practical usage of the *resilience triangle*, as shown by Bruneau et al. [30]. We introduced the resilience triangle in Sect. 6.3.1 and showed the general approach for calculating the metric value R. We adapt this approach, comparable to [135], by normalising the error ratio compared to the predefined minimum performance. Additionally we approximate the integral given in Sect. 6.3.1 with a sum over all measurements in time t with the time step length Δt:

$$r = \sum_{t=t_{\text{pre}}}^{t_{\text{post}}} \frac{\max\{0, f_{\text{pre}} - f(t)\}\Delta t}{f_{\text{pre}}\left(t_{\text{post}} - t_{\text{pre}}\right)}. \tag{6.21}$$

This enables us to compare an undisturbed system response ($r = 0$) with a disturbed system response ($0 < r \leq 1$). The system response after a sudden disturbance is shown in Fig. 6.39a, while Fig. 6.39b shows a correlated signal, as introduced in Sect. 6.3.2, which can be used besides the direct measurement of the water level, to derive rule-based learning and anticipation strategies. As a disturbance, we used a square-wave (0.02 Hz, offset 70 % of maximum command signal, amplitude 30 % of maximum command signal) as a reference signal for the valve displacement. This signal was transmitted to the valve using the commercial National Instruments software as well as custom-developed Python software.

The system is able to reach the setpoint value (f_{pre}), which is in our case equivalent to the minimum performance, while the disturbance is still active. If we evaluate Eq. (6.21) on the shown example we get $r = 0.252$ if we consider $t_{\text{pre}} = t_0$ and $t_{\text{post}} = t_1$. The metric r can also be interpreted as the percentage of lost functional performance in the given time period. This loss is marked in grey in Fig. 6.39. Additionally, if we also consider the performance loss after the external failure period (hatched in Fig. 6.39a) we get $r = 0.202$ ($t_{\text{pre}} = t_0$, $t_{\text{post}} = t_2$). After stopping the disturbance signal ($t > t_1$) the considered controller tends to overshoot and produces a second performance loss before finally reaching the setpoint function. The experiment shows that an extension of the resilience triangle as given in Eq. (6.21) can be used to compare the resilience of different systems (e.g. different controllers) on common failures and common time horizons.

Four functions of resilience

Next to the quantification of resilience, we show the usage of the four functions of resilience, i.e. monitoring, responding, learning and anticipating, and evaluate the applicability of this concept on the test rig system, cf. [115]. We start to evaluate this resilience approach, by using one proportional-integral-derivative (PID) controlled pump to set the height of water in one tank. As a disturbance, we use a control valve (marked with (1) in Fig. 6.38) to disrupt the system at a given point in time, after the steady-state is reached. We try to minimise negative deviations from the desired reference water height, while positive deviations in the water height are accepted. This allows to derive a system representation that is related to classic resilience examinations, as for instance shown by [30, 165].

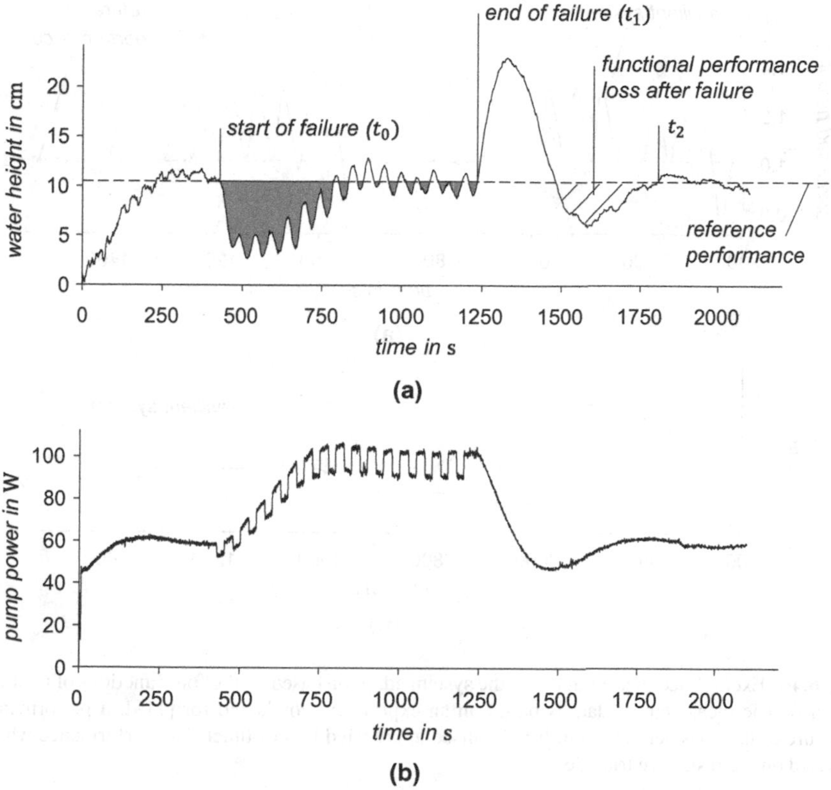

Fig. 6.39 Measurement of a sudden disturbance on the test rig system. **a** measured signal of water height **b** indirect measurement of the disturbance in the pump power signal

Resilient systems should adapt to changes continuously, as proposed in [55, 173]. This goal is promising, but difficult to reach, as the resulting system must be adaptive and not only flexible, cf. Sect. 6.3.2.

The four functions of resilience are used to build a more resilient system that can be seen as a first step towards a resilient system that can master arbitrary disturbances. If the disturbance is severe, the traditional system design with a PID controller is unable to retain the predefined functionality. As the presented system has only little possibility to adapt its behaviour, we chose to build a system that can partially anticipate future disturbances and adapt its behaviour accordingly to reduce the future loss in functionality. Therefore, we implemented among other methods (for more details we refer to [115]), an exemplified learning model, which is based on an auto-regressive (AR) model, cf. [29, 132]. It models and predicts the behaviour of the complete system based on the current system response in the time-domain. This approach is also known as a method of system identification for dynamical systems as shown in [95]. To train the AR model, we split a time-series of stored values,

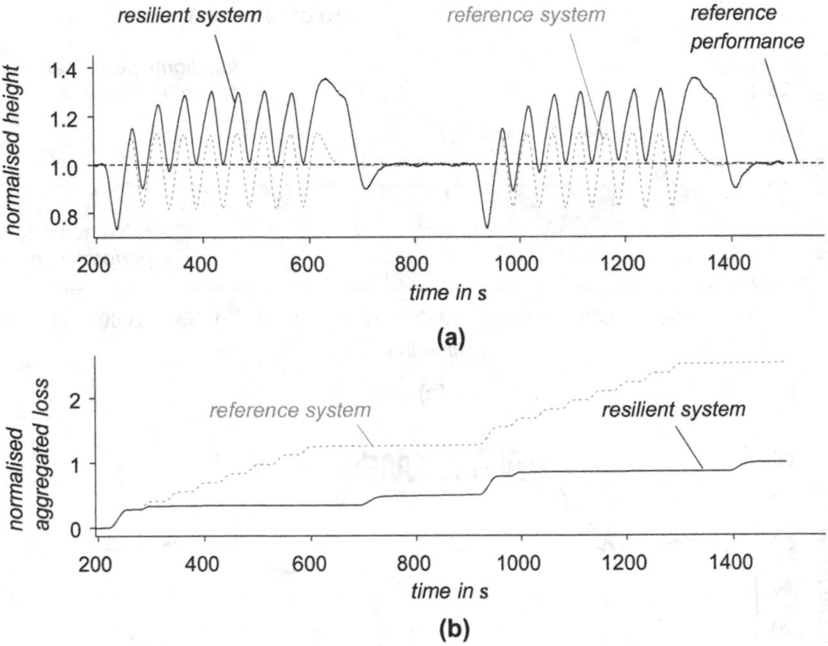

Fig. 6.40 Exemplified measurement of the system adaption based on the four functions of resilient systems. The measurement data is based on an experiment conducted for [115]. **a** performance measure of the considered system **b** normalised aggregated loss of functional performance which is based on the resilience triangle

which were measured at the test rig in a training (70 %) and a test set (30 %). The training set is then used to train the underlying AR model, while the test set serves to evaluate the performance of the trained AR model. Overall, there are five different training-test-splits within cross-validation, [13].

An exemplified system adaption is shown in Fig. 6.40. The system is adapting its behaviour based on the automatically detected deviation of the defined functionality. The system tries to recover over time by adding additional water in the reservoir when possible to avoid a severe decrease under the predefined reference performance value in the case of anticipated future losses. This approach enables the system to minimise its performance losses with a time-depending strategy that is based on the anticipated disturbances. It is important to mention that the algorithm does not use any measurement signal related to the control valve displacement that represents the "exogenous driver" as introduced in [173] within the conducted experiment. Instead it autonomously develops a model of estimated future disturbances.

The "resilient system" tries to minimise the negative deviation from the desired reference performance for all time steps within a detected disturbance in Fig. 6.40a. It learns and anticipates future losses that are caused by the changing valve opening, which simulates a severe disturbance. The reference system uses a classic design with only one PID controller, which is unable to minimise the losses over time. The "resilient system" shows a consecutive adaption capability, where comparable losses within an active disturbance only occur at the beginning. The reference system design results in a more than two times higher loss over the considered time period, as shown in Fig. 6.40b.

Conclusion

We conclude that the experiments conducted at the presented test rig show it is possible to transfer the considered resilience concepts and metrics of Sect. 6.3.1 and [7] to the mechanical engineering domain, and therefore to other technical systems demonstrating uncertainty as given in Sect. 3.6. Furthermore, the shown algorithmic approach which is based on the four functions monitoring, responding, learning, and anticipating is suitable to derive more resilient technical system designs.

References

1. Abele E, Geßner F (2018) Spanungsquerschnittmodell zum Gewindebohren: Modellierung der Auswirkung von Unsicherheit auf den Spanungsquerschnitt beim Gewindebohren. wt Werkstattstechnik online 108(1–2):2–6 (2018)
2. Ahern J (2011) From fail-safe to safe-to-fail: sustainability and resilience in the new urban world. Landscape Urban Plan 100(4):341–343
3. Ajzen I, Fishbein M (1980) Understanding attitudes and predicting social behavior. Prentice-Hall, Englewood Cliffs, N.J
4. Albrechtsen E, Besnard D (2013) Oil and gas. Technology and Humans. Assessing the Human Factors of Technological Change. Ashgate, Burlington, VT
5. Alderson DL, Brown GG, Carlyle WM, Wood RK (2011) Solving defender-attacker-defender models for infrastructure defense. In: Wood R, Dell R (eds) Operations research, computing and homeland defense, proceedings of the 12th INFORMS computing society conference, pp 28–49
6. Alla A, Hinze M, Kolvenbach P, Lass O, Ulbrich S (2019) A certified model reduction approach for robust parameter optimization with PDE constraints. Adv Comput Math 45:1221–1250. https://doi.org/10.1007/s10444-018-9653-1
7. Altherr LC, Brötz N, Dietrich I, Gally T, Geßner F, Kloberdanz H, Leise P, Pelz PF, Schlemmer P, Schmitt A (2018) Resilience in mechanical engineering—a concept for controlling uncertainty during design, production and usage phase of load-carrying structures. In: Pelz PF, Groche P (eds) Uncertainty in mechanical engineering III, applied mechanics and materials, vol 885. Trans Tech Publications, pp 187–198. https://doi.org/10.4028/www.scientific.net/AMM.885.187
8. Altherr LC, Ederer T, Pöttgen P, Lorenz U, Pelz PF (2015) Multicriterial optimization of technical systems considering multiple load and availability scenarios. In: Pelz PF, Groche P (eds) Uncertainty in mechanical engineering II, applied mechanics and materials, vol 807. Trans Tech Publications, pp. 247–256. https://doi.org/10.4028/www.scientific.net/AMM.807.247
9. Altherr LC, Leise P, Pfetsch ME, Schmitt A (2018) Algorithmic design and resilience assessment of energy efficient high-rise water supply systems. In: Pelz PF, Groche P (eds) Uncertainty in mechanical engineering III, Applied mechanics and materials, vol 885. Trans Tech Publications, pp. 211–223. https://doi.org/10.4028/www.scientific.net/AMM.885.211

10. Altherr LC, Leise P, Pfetsch ME, Schmitt A (2019) Resilient layout, design and operation of energy-efficient water distribution networks for high-rise buildings using MINLP. Optim Eng 20(2):605–645. https://doi.org/10.1007/s11081-019-09423-8
11. Anderson BDO (1966) Algebraic description of bounded real matrixes. Electron Lett 2(12):464–465
12. Andersson P (1996) A process approach to robust design in early engineering design phases. PhD thesis, Lund Institute of Technology
13. Arlot S, Celisse A (2010) A survey of cross-validation procedures for model selection. Stat Surv 4:40–79. https://doi.org/10.1214/09-SS054
14. Avemann J, Calmano S, Schmitt S, Groche P (2014) Total flexibility in forming technology by servo presses. In: WGP Congress 2012, Advanced materials research, vol 907. Trans Tech Publications, pp 99–112. https://doi.org/10.4028/www.scientific.net/AMR.907.99
15. Ben-Tal A, El Ghaoui L, Nemirovski A (2009) Robust optimization. Princeton University Press, Princeton, NJ
16. Ben-Tal A, Jarre F, Kočvara M, Nemirovski A, Zowe J (2000) Optimal design of trusses under a nonconvex global buckling constraint. Optim Eng 1(2):189–213. https://doi.org/10.1006/jsvi.1999.2530
17. Ben-Tal A, Nemirovski A (1997) Robust truss topology design via semidefinite programming. SIAM J Optim 7(4):991–1016. https://doi.org/10.1137/S1052623495291951
18. Bendsøe MP, Sigmund O (2003) Topology optimization. Springer, Berlin
19. Bertsimas D, Brown DB, Caramanis C (2011) Theory and applications of robust optimization. SIAM Rev 53(3):464–501. https://doi.org/10.1137/080734510
20. Bertsimas D, Nohadani O, Teo KM (2010) Robust optimization for unconstrained simulation-based problems. Oper Res 58(1):161–178
21. Bhattacharyya O, Kapoor SG, DeVor RE (2006) Mechanistic model for the reaming process with emphasis on process faults. Int J Mach Tools Manuf 46(7–8):836–846
22. Birkhofer H (2011) From design practice to design science: The evolution of a career in design methodology research. J Eng Des22(5):333–359. https://doi.org/10.1080/09544828.2011.555392
23. Bölling C (2019) Simulationsbasierte Auslegung mehrstufiger Werkzeugsysteme zur Bohrungsfeinbearbeitung am Beispiel der Ventilführungs- und Ventilsitzbearbeitung. Dissertation, TU Darmstadt.http://tubiblio.ulb.tu-darmstadt.de/111353/
24. Bölling C, Abele E (2018) Simulation of multi-stage fine machining processes at the example of valve guide and valve seat. In: Pelz PF, Groche P (eds) Uncertainty in mechanical engineering III, applied mechanics and materials, vol 885. Trans Tech Publications, pp 255–266. https://doi.org/10.4028/www.scientific.net/AMM.885.255
25. lling C, Güth S, Abele E (2015) Control of uncertainty in high precision cutting processes: reaming of valve guides in a cylinder head of a combustion engine. In: Pelz PF, Groche P (eds) uncertainty in mechanical engineering II, applied mechanics and materials, vol 807. Trans Tech Publications, pp 153–161. https://doi.org/10.4028/www.scientific.net/AMM.807.153
26. Bölling C, Hoppe F, Geßner F, Knoll M, Abele E, Groche P (2018) Fortpflanzung von Unsicherheit in Prozessketten. wt Werkstatttechnik online 108(1–2):82–88
27. Bongard J, Zykov V, Lipson H (2006) Resilient machines through continuous self-modeling. Science 314(5802):1118–1121
28. Bretz A, Geßner F, Öztürk T, Rinn C, Abele E (2018) Adjustment of axis offset errors during reaming. In: Pelz PF, Groche P (eds) Uncertainty in mechanical engineering III, applied mechanics and materials, vol 885. Trans Tech Publications, pp. 267–275. https://doi.org/10.4028/www.scientific.net/amm.885.267
29. Brockwell PJ, Davis RA, Calder MV (2016) Introduction to time series and forecasting. Springer, Berlin
30. Bruneau M, Chang SE, Eguchi RT, Lee GC, O'Rourke TD, Reinhorn AM, Shinozuka M, Tierney K, Wallace WA, Von Winterfeldt D (2003) A framework to quantitatively assess and enhance the seismic resilience of communities. Earthquake Spectra 19(4):733–752

31. Calmano S, Hesse D, Hoppe F, Traidl P, Sinz J, Groche P (2015) Orbital forming of flange parts under uncertainty. In: Pelz PF, Groche P (eds) Uncertainty in mechanical engineering II, applied mechanics and materials, vol 807. Trans Tech Publications, pp 121–129. https://doi.org/10.4028/www.scientific.net/AMM.807.121
32. Carmichael MG, Liu D, Waldron KJ (2017) A framework for singularity-robust manipulator control during physical human-robot interaction. Int J Robot Res 36(5–7):861–876. https://doi.org/10.1177/0278364917698748
33. Chen RL, Cohn A, Pinar A (2011) An implicit optimization approach for survivable network design. In: 2011 IEEE network science workshop, pp 180–187. IEEE (2011). https://doi.org/10.1109/NSW.2011.6004644
34. Cheng G, Guo X, Olhoff N (2000) New formulation for truss topology optimization problems under buckling constraints. In: Topology optimization of structures and composite continua, pp 115–131. Kluwer Academic Publishers
35. Chiou EK, Lee JD (2016) Cooperation in human-agent systems to support resilience: A microworld experiment. Human Factors 58(6):846–863. https://doi.org/10.1177/0018720816649094
36. Conn AR, Vicente LN (2012) Bilevel derivative-free optimization and its application to robust optimization. Optim Methods Softw 27(3):561–577
37. Correa R, Ramirez CH (2004) A global algorithm for nonlinear semidefinite programming. SIAM J Optim **15**(1), 303–318. https://doi.org/10.1137/S1052623402417298
38. Curtis FE, Mitchell T, Overton ML (2017) A BFGS-SQP method for nonsmooth, nonconvex, constrained optimization and its evaluation using relative minimization profiles. Optim Methods Softw 32(1):148–181. https://doi.org/10.1080/10556788.2016.1208749
39. Curtis FE, Overton ML (2012) A sequential quadratic programming algorithm for nonconvex, nonsmooth constrained optimization. SIAM J Optim 22(2):474–500. https://doi.org/10.1137/090780201
40. Dağdeviren M, Yüksel İ (2008) Developing a fuzzy analytic hierarchy process (AHP) model for behavior-based safety management. Inf Sci 178(6):1717–1733. https://doi.org/10.1016/j.ins.2007.10.016
41. D'Ambrosio C, Lodi A, Wiese S, Bragalli C (2015) Mathematical programming techniques in water network optimization. Eur J Oper Res 243(3):774–788. https://doi.org/10.1016/j.ejor.2014.12.039
42. Deutsches Institut für Normung: DIN 8589-2:2003-09. Manufacturing processes chip removal – Part 2: Drilling, countersinking and counterboring, reaming; Classification, subdivision, terms and definitions. Beuth, Berlin (2003)
43. Diehl M, Bock HG, Kostina E (2006) An approximation technique for robust nonlinear optimization. Math Program 107(1):213–230. https://doi.org/10.1007/s10107-005-0685-1
44. Dieter G (1961) Mechanical metallurgy. McGraw-Hill
45. Dietrich I, Hedrich P, Bölling C, Brötz N, Geßner F, Pelz PF (2018) Concept of a resilient process chain to control uncertainty of a hydraulic actuator. In: Pelz PF, Groche P (eds) Uncertainty in mechanical engineering III, applied mechanics and materials, vol 885. Trans Tech Publications, pp 156–169. https://doi.org/10.4028/www.scientific.net/AMM.885.156
46. Dulger LC, Das MT, Halicioglu R, Kapucu S, Topalbekiroglu M (2016) Robotics and servo press control applications: experimental implementations. In: International conference on control, decision and information technologies. IEEE, Piscataway, NJ, pp 102–107. https://doi.org/10.1109/CoDIT.2016.7593543
47. Ederer T, Hartisch M, Lorenz U, Opfer T, Wolf J (2017) Yasol: an open source solver for quantified mixed integer programs. In: Winands MH, van den Herik HJ, Kosters WA (eds) Advances in computer games. Springer, Berlin, pp 224–233. https://doi.org/10.1007/978-3-319-71649-7_19
48. Ederer T, Lorenz U, Martin A, Wolf J (2011) Quantified linear programs: a computational study. In: Demetrescu C, Halldórsson MM (eds) Algorithms—ESA 2011. Springer, Berlin, pp 203–214. https://doi.org/10.1007/978-3-642-23719-5_18

49. Ederer T, Lorenz U, Opfer T (2013) Quantified combinatorial optimization. In: Huisman D, Louwerse I, Wagelmans AP (eds) Operations research proceedings. Springer International Publishing, pp 121–128. https://doi.org/10.1007/978-3-319-07001-8_17
50. Ederer T, Lorenz U, Opfer T, Wolf J (2012) Modeling games with the help of quantified integer linear programs. In: van den Herik HJ, Plaat A (eds) Advances in computer games. Springer, Berlin, pp 270–281. https://doi.org/10.1007/978-3-642-31866-5_23
51. Ederer T, Lorenz U, Opfer T, Wolf J (2014) Multistage optimization with the help of quantified linear programming. In: Lübbecke M, Koster A, Letmathe P, Madlener R, Peis B, Walther G (eds) Operations research proceedings 2014. Springer, Berlin, pp 369–375. https://doi.org/10.1007/978-3-319-28697-6_52
52. Eifler T, Engelhardt R, Mathias J, Kloberdanz H, Birkhofer H (2010) An assignment of methods to analyze uncertainty in different stages of the development process. In: Proceedings of the international mechanical engineering congress and exposition. Vancouver, pp 303–313. https://doi.org/10.1115/IMECE2010-39126
53. Fischetti M, Ljubić I, Monaci M, Sinnl M (2017) A new general-purpose algorithm for mixed-integer bilevel linear programs. Oper Res 65(6):1615–1637. https://doi.org/10.1287/opre.2017.1650
54. Florin MV, Linkov I (2016) IRGC resource guide on resilience. Technical Report, International Risk Governance Center (IRGC)
55. Folke C, Carpenter S, Walker B, Scheffer M, Chapin T, Rockström J (2010) Resilience thinking: integrating resilience, adaptability and transformability. Ecol Soc 15(4)
56. Frangioni A, Gentile C (2006) Perspective cuts for a class of convex 0–1 mixed integer programs. Math Program 106(2):225–236. https://doi.org/10.1007/s10107-005-0594-3
57. Freund T (2018) Konstruktionshinweise zur Beherrschung von Unsicherheit in technischen Systemen. Dissertation, TU Darmstadt
58. Freund T, Würtenberger J, Calmano S, Hesse D, Kloberdanz H (2014) Robust design of active systems: an approach to considering disturbances within the selection of sensors. In: Howards TJ, Eifler T (eds) Proceedings of the international symposium on robust design, pp 147–157
59. Freund T, Würtenberger J, Kloberdanz H, Blakaj P (2015) An approach to using elemental interfaces to assess design clarity. In: Uncertainty in mechanical engineering II, applied mechanics and materials, vol 807. Trans Tech Publications, pp 109–117. https://doi.org/10.4028/www.scientific.net/AMM.807.109
60. Freund T, Würtenberger J, Lotz J, Rommel C, Kirchner E (2017) Design for robustness—systematic application of design guidelines to control uncertainty. In: Maier A, Škec S, Kim H, Kokkolaras M, Oehmen J, Fadel G, Salustri F, Van der Loos M (eds) Proceedings of the 21st international conference on engineering design (ICED17), pp 277–286
61. Fritzson P, Engelson V (1998) Modelica—a unified object-oriented language for system modeling and simulation. In: European conference on object-oriented programming. Springer, Berlin, pp 67–90
62. Gally T (2019) Computational mixed-integer semidefinite programming. Dissertation, TU Darmstadt
63. Gally T, Gehb CM, Kolvenbach P, Kuttich A, Pfetsch ME, Ulbrich S (2015) Robust truss topology design with beam elements via mixed integer nonlinear semidefinite programming. In: Pelz PF, Groche P (eds) Uncertainty in mechanical engineering II, applied mechanics and materials, vol 807. Trans Tech Publications, pp 229–238
64. Gally T, Kuttich A, Pfetsch ME, Schaeffner M, Ulbrich S (2018) Optimal placement of active bars for buckling control in truss structures under bar failures. In: Pelz PF, Groche P (eds) Uncertainty in mechanical engineering III, applied mechanics and materials, vol 885. Trans Tech Publications, pp 119–130. https://doi.org/10.4028/www.scientific.net/AMM.885.119
65. Gally T, Pfetsch ME, Ulbrich S (2018) A framework for solving mixed-integer semidefinite programs. Optim Methods Softw 33(3):594–632
66. Gerwin D (1993) Manufacturing flexibility: a strategic perspective. Manag Sci **39**(4), 395–410. http://www.jstor.org/stable/2632407

67. Gleixner A, Bastubbe M, Eifler L, Gally T, Gamrath G, Gottwald RL, Hendel G, Hojny C, Koch T, Lübbecke ME, Maher SJ, Miltenberger M, Müller B, Pfetsch ME, Puchert C, Rehfeldt D, Schlösser F, Schubert C, Serrano F, Shinano Y, Viernickel JM, Walter M, Wegscheider F, Witt JT, Witzig J (2018) The SCIP optimization suite 6.0. Technical report, Optimization Online. http://www.optimization-online.org/DB_HTML/2018/07/6692.html
68. Goerger SR, Madni AM, Eslinger OJ (2014) Engineered resilient systems: a DoD perspective. In: Conference on systems engineering resarch, pp 865–872
69. Gorissen BL, Yanıkoğlu İ, den Hertog D (2015) A practical guide to robust optimization. Omega 53:124–137
70. Groche P, Hoppe F, Hesse D, Calmano S (2016) Blanking-bending process chain with disturbance feed-forward and closed-loop control. J Manuf Process 24:62–70
71. Groche P, Hoppe F, Sinz J (2017) Stiffness of multipoint servo presses: mechanics versus control. CIRP Ann **66**(1):373–376. https://doi.org/10.1016/j.cirp.2017.04.053
72. Groche P, Scheitza M, Kraft M, Schmitt S (2010) Increased total flexibility by 3D Servo Presses. CIRP Ann 59(1):267–270
73. Groche P, Schneider R (2004) Method for the optimization of forming presses for the manufacturing of micro parts. CIRP Ann 53(1):281–284
74. Güth S, Bretz A, Bölling C, Baron A, Abele E (2015) Control of uncertainty based on machining strategies during reaming. In: Pelz PF, Groche P (eds) Uncertainty in mechanical engineering II, applied mechanics and materials, vol 807. Trans Tech Publications, pp 162–168. https://doi.org/10.4028/www.scientific.net/AMM.807.162
75. Habermehl K (2014) Robust optimization of active trusses via mixed-integer semidefinite programming. Dissertation, TU Darmstadt
76. Hanselka H, Platz R (2010) Ansätze und Maßnahmen zur Beherrschung von Unsicherheit in lasttragenden Systemen des Maschinenbaus. Konstruktion 11–12:55–62
77. Hartisch M, Ederer T, Lorenz U, Wolf J (2016) Quantified integer programs with polyhedral uncertainty set. In: Plaat A, Kosters W, van den Herik J (eds) Computers and games. Springer, Berlin, pp 156–166. https://doi.org/10.1007/978-3-319-50935-8_15
78. Hartisch M, Herbst A, Lorenz U, Weber JB (2018) Towards resilient process networks-designing booster stations via quantified programming. In: Pelz PF, Groche P (eds) Uncertainty in mechanical engineering III, applied mechanics and materials, vol 885. Trans Tech Publications, pp 199–210. https://doi.org/10.4028/www.scientific.net/AMM.885.199
79. Hartisch M, Lorenz U (2018) Game tree search in a robust multistage optimization framework: Exploiting pruning mechanisms. Technical Report. arXiv:1811.12146
80. Hartisch M, Lorenz U (2019) Mastering uncertainty: towards robust multistage optimization with decision dependent uncertainty. In: Nayak AC, Sharma A (eds) PRICAI 2019: trends in artificial intelligence. Springer, Berlin, pp 446–458. https://doi.org/10.1007/978-3-030-29908-8_36
81. Hashimoto D, Kanno Y (2015) A semidefinite programming approach to robust truss topology optimization under uncertainty in locations of nodes. Struct Multidiscip Optim 51(2):439–461
82. Hauer T (2012) Modellierung der Werkzeugabdrängung beim Reiben – Ableitung von Empfehlungen für die Gestaltung von Mehrschneidenreibahlen. Dissertation, TU Darmstadt
83. Hedrich P (2018) Konzeptvalidierung einer aktiven Luftfederung im Kontext autonomer Fahrzeuge. Dissertation, TU Darmstadt. https://tuprints.ulb.tu-darmstadt.de/8469/
84. Hedrich P, Johe M, Pelz PF (2016) Design and realization of an adjustable fluid powered piston for an active air spring. In: 10th international fluid power conference. Dresden, Germany, pp 571–582. http://tubiblio.ulb.tu-darmstadt.de/78284/
85. Herrera M, Abraham E, Stoianov I (2016) A graph-theoretic framework for assessing the resilience of sectorised water distribution networks. Water Res Manag 30(5):1685–1699
86. Hiramoto K, Doki H, Obinata G (2000) Optimal sensor/actuator placement for active vibration control using explicit solution of algebraic Riccati equation. J Sound Vib 229(5):1057–1075. https://doi.org/10.1006/jsvi.1999.2530
87. Hollnagel E (2011) Epilogue: RAG – the resilience analysis grid. In: Hollnagel E, Pariès J, Woods DD, Wreathall J (eds) Resilience engineering in practice. a guidebook. CRC Press, Boca Raton, FL, pp 275–296

88. Hollnagel E (2011) Prologue: The scope of resilience engineering. In: Hollnagel E, Pariès J, Woods DD, Wreathall J (eds) Resilience engineering in practice: a guidebook. CRC Press, Boca Raton, FL, pp xxix–xxxix
89. Hollnagel E (2012) FRAM, The functional resonance analysis method. modelling complex socio-technical systems. Ashgate, Farnham
90. Hollnagel E (2014) Resilience engineering and the built environment. Build Res Inf 42(2):221–228. https://doi.org/10.1080/09613218.2014.862607
91. Hollnagel E (2019) Resilience engineering. http://erikhollnagel.com/ideas/resilience-engineering.html. Accessed 11 Nov 2019
92. Hoppe F, Pihan C, Groche P (2019) Closed-loop control of eccentric presses based on inverse kinematic models. Procedia Manuf 9:240–247. https://doi.org/10.1016/j.promfg.2019.02.132
93. Houska B, Diehl M (2013) Nonlinear robust optimization via sequential convex bilevel programming. Math Program 142(1–2):539–577
94. Hrycej T (2018) Robuste Regelung. Springer, Berlin, Heidelberg. https://doi.org/10.1007/978-3-662-54168-5
95. Isermann R, Münchhof M (2010) Identification of dynamic systems: an introduction with applications. Springer, Berlin
96. Jackson S (2010) Architecting resilient systems: accident avoidance and survival and recovery from disruptions. Wiley series in systems engineering and management. Wiley, Hoboken, NJ. https://doi.org/10.1002/9780470544013
97. Jackson S, Ferris TL (2013) Resilience principles for engineered systems. Syst Eng 16(2):152–164
98. Jackson WS (2016) Evaluation of resilience principles for engineered systems. PhD thesis, University of South Australia
99. Jansen M, Lombaert G, Schevenels M, Sigmund O (2014) Topology optimization of fail-safe structures using a simplified local damage model. Struct Multidiscip Optim 49(4):657–666. https://doi.org/10.1007/s00158-013-1001-y
100. Kanno Y (2017) Redundancy optimization of finite-dimensional structures: Concept and derivative-free algorithm. J Struct Eng 143(1):04016151–1–04016151–10 (2017). https://doi.org/10.1061/(ASCE)ST.1943-541X.0001630
101. Kanno Y (2018) Robust truss topology optimization via semidefinite programming with complementarity constraints: a difference-of-convex programming approach. Comput Optim Appl 71(2):403–433
102. Khandekar R, Kortsarz G, Mirrokni V, Salavatipour MR (2008) Two-stage robust network design with exponential scenarios. In: Halperin D, Mehlhorn K (eds) European symposium on algorithms. Springer, Berlin, pp 589–600. https://doi.org/10.1007/978-3-540-87744-8_49
103. Kirchner E (2020) Grundregeln der Gestaltung–Einfach, Eindeutig. In: Bender B, Gericke K (eds) Pahl/Beitz Konstruktionslehre: Methoden und Anwendung erfolgreicher Produktentwicklung. Springer, Berlin
104. Kirchner H, Pierer A, Putz M, Blau P (2016) Entwicklung einer Kippregelung für servoelektrische Exzenterpressen mit mechanisch entkoppelten Hauptantrieben. Automation 2016, vol 2284. VDI-Berichte. VDI Verlag, Düsseldorf, pp 1–12
105. Kleniati PM, Adjiman CS (2015) A generalization of the Branch-and-Sandwich algorithm: From continuous to mixed-integer nonlinear bilevel problems. Comput Chem Eng 72:373–386. https://doi.org/10.1016/j.compchemeng.2014.06.004
106. Kočvara M (2010) Truss topology design with integer variables made easy. Technical Report, Optimization Online
107. Kolvenbach P (2018) Robust optimization of PDE-constrained problems using second-order models and nonsmooth approaches. Dissertation, TU Darmstadt. https://www.dr.hut-verlag.de/9783843939690.html
108. Kolvenbach P, Lass O, Ulbrich S (2018) An approach for robust PDE-constrained optimization with application to shape optimization of electrical engines and of dynamic elastic structures under uncertainty. Optim Eng 19(3):697–731. https://doi.org/10.1007/s11081-018-9388-3

109. Kolvenbach P, Ulbrich S, Krech M, Groche P (2018) Robust design of a smart structure under manufacturing uncertainty via nonsmooth PDE-constrained optimization. In: Pelz PF, Groche P (eds) Uncertainty in mechanical engineering III, applied mechanics and materials, vol 885. Trans Tech Publications, pp 131–144. https://doi.org/10.4028/www.scientific.net/AMM.885.131
110. Koppka F (2009) A contribution to the maximization of productivity and workpiece quality of the reaming process by analyzing its static and dynamic behavior. Dissertation, TU Darmstadt
111. Kuttich A (2018) Robust topology design of mechanical systems under uncertain dynamic loads via nonlinear semidefinite programming. Dissertation, TU Darmstadt
112. Kuttich A, Götz B, Ulbrich S (2017) Robust optimization of shunted piezoelectric transducers for vibration attenuation considering different values of electromechanical coupling. In: Barthorpe R, Platz R, Lopez I, Moaveni B, Papadimitriou C (eds) Model validation and uncertainty quantification, vol 3. proceedings of the 35th IMAC, a conference and exposition on structural dynamics. Springer, Berlin, pp 51–59. https://doi.org/10.1007/978-3-319-54858-6_6
113. Kuttich A, Ulbrich S (2017) Feedback controller design and topology optimization for truss structures under uncertain dynamic loads. In: von Scheven M, Keip MA, Karajan N (eds) 7th GACM colloquium on computational mechanics for young scientists from academia and industry
114. Lass O, Ulbrich S (2017) Model order reduction techniques with a posteriori error control for nonlinear robust optimization governed by partial differential equations. SIAM J Sci Comput 39(5):S112–S139. https://doi.org/10.1137/16M108269X
115. Leise P, Breuer T, Altherr LC, Pelz PF, Development, validation and assessment of a resilient pumping system. In: Proceedings of the JIRC2020 (In press)
116. Lemaitre C (2008) Topologieoptimierung von adaptiven Stabwerken. Dissertation, Univeristy of Stuttgart. https://doi.org/10.18419/opus-300
117. Lipshitz R, Strauss O (1997) Coping with uncertainty: A naturalistic decision-making analysis. Organ Behav Hum Decis Process 69(2):149–163. https://doi.org/10.1006/obhd.1997.2679
118. Lorenz U, Martin A, Wolf J (2010) Polyhedral and algorithmic properties of quantified linear programs. In: de Berg M, Meyer U (eds) Algorithms—ESA 2010. Springer, Berlin, pp 512–523. https://doi.org/10.1007/978-3-642-15775-2_44
119. Lorenz U, Opfer T, Wolf J (2014) Solution techniques for quantified linear programs and the links to gaming. In: van den Herik HJ, Iida H, Plaat A (eds) Computers and games. Springer, Berlin, pp 110–124. https://doi.org/10.1007/978-3-319-09165-5_10
120. Lorenz U, Wolf J (2015) Solving multistage quantified linear optimization problems with the alpha-beta nested Benders decomposition. EURO J Comput Optim 3(4):349–370. https://doi.org/10.1007/s13675-015-0038-7
121. Ma DL, Braatz RD (2001) Worst-case analysis of finite-time control policies. IEEE Trans Control Syst Technol 9(5):766–774
122. Mars S (2013) Mixed-integer semidefinite programming with an application to truss topology design. Dissertation, FAU Erlangen-Nürnberg
123. Mars S, Schewe L (2012) An SDP-package for SCIP. Technical Report, TU Darmstadt and FAU Erlangen-Nürnberg
124. Mathias J (2016) Auf dem Weg zu robusten Lösungen. Dissertation, TU Darmstadt
125. Mathias J, Kloberdanz H, Eifler T, Engelhardt R, Wiebel M, Birkhofer H, Bohn A (2011) Selection of physical effects based on disturbances and robustness ratios in the early phases of robust design. In: DS 68-5: proceedings of the 18th international conference on engineering design (ICED 11), Impacting Society through Engineering Design, Vol 5. Design for X/Design to X, pp 324–335
126. Mathias J, Kloberdanz H, Engelhardt R, Birhofer H (2010) Strategies and principles to design robust products. In: Marjanović D (ed) Proceedings of design. Design Society, Zagreb, pp 341–350
127. Meng F, Fu G, Farmani R, Sweetapple C, Butler D (2018) Topological attributes of network resilience: a study in water distribution systems. Water Res 143:376–386

128. Mitschke M, Wallentowitz H (2004) Dynamik der Kraftfahrzeuge. Springer, Berlin, Heidelberg
129. Mitsos A (2010) Global solution of nonlinear mixed-integer bilevel programs. J Global Optim 47:557–582. https://doi.org/10.1007/s10898-009-9479-y
130. Mogul F (2013) GLYCODUR dry bearings. Data sheet, Federal Mogul
131. Mohr DP, Stein I, Matzies T, Knapek CA (2014) Redundant robust topology optimization of truss. Optim Eng 15(4):945–972. https://doi.org/10.1007/s11081-013-9241-7
132. Montgomery DC, Peck EA, Vining GG (2012) Introduction to linear regression analysis, vol 821. Wiley, New York
133. Müller TM, Leise P, Lorenz IS, Altherr LC, Pelz PF (2020) Optimization and validation of pumping system design and operation for water supply in high-rise buildings. Optim Eng 1–44. https://doi.org/10.1007/s11081-020-09553-4
134. Ouedraogo KA, Simon E, Vanderhaegen F (2013) How to learn from the resilience of human-machine systems? Eng Appl Artif Intell 26(1):24–34. https://doi.org/10.1016/j.engappai.2012.03.007
135. Ouyang M, Dueñas-Osorio L, Min X (2012) A three-stage resilience analysis framework for urban infrastructure systems. Struct Saf 36:23–31
136. Pahl G, Beitz W (1996) Engineering design: a systematic approach, 2nd edn. Springer, London
137. Pfetsch ME, Schmitt A (2020) Exploiting partial convexity of pump characteristics in water network design. In: Neufeld JS, Buscher U, Lasch R, Möst D, Schönberger J (eds) Operations research proceedings. Springer, Berlin, pp 497–503. https://doi.org/10.1007/978-3-030-48439-2_60
138. Righi A, Weber S, Tarcisio A, Wachs P (2015) A systematic literature review of resilience engineering: Research areas and a research agenda proposal. Reliab Eng Syst Saf 141:142–152. https://doi.org/10.1016/j.ress.2015.03.007
139. Rozvany GIN (1996) Difficulties in truss topology optimization with stress, local buckling and system stability constraints. Struct Optim 11(3–4):213–217. https://doi.org/10.1007/BF01197036
140. Schäfer C (2015) Optimization approaches for actuator and sensor placement and its application to model predictive control of dynamical systems. Dissertation, TU Darmstadt
141. Schänzle C, Altherr LC, Ederer T, Lorenz U, Pelz PF (2015) As good as it can be—ventilation system design by a combined scaling and discrete optimization method. In: Proceedings of the FAN
142. Schlemmer PD, Kloberdanz H, Gehb CM, Kirchner E (2018) Adaptivity as a property to achieve resilience of load-carrying systems. In: Pelz PF, Groche P (eds) Uncertainty in mechanical engineering III, applied mechanics and materials, vol 885. Trans Tech Publications, pp 77–87. https://doi.org/10.4028/www.scientific.net/AMM.885.77
143. Schmalz K (1970) Reibahle für hohe Kreisformgenauigkeit. WB Werkstatt + Betrieb, pp 313–318
144. Schmoeckel D (1991) Developments in automation, flexibilization and control of forming machinery. CIRP Ann 40(2):615–622. https://doi.org/10.1016/s0007-8506(07)61137-8
145. Schulte F, Engelhardt R, Kirchner E, Kloberdanz H (2019) Beitrag zur Entwicklungsmethodik für resiliente Systeme des Maschinenbaus. In: Krause D, Paetzold K, Wartzack S (eds) DFX 2019: proceedings of the 30th symposium design for X. The Design Society. https://doi.org/10.35199/dfx2019.1
146. Schulte F, Kirchner E, Kloberdanz H (2019) Analysis and synthesis of resilient load-carrying systems. Proc Des Soc Int Conf Eng Des 1(1):1403–1412. https://doi.org/10.1017/dsi.2019.146
147. SCIP-SDP—a mixed integer semidefinite programming plugin for SCIP (2019). http://www.opt.tu-darmstadt.de/scipsdp/
148. Shin S, Lee S, Judi DR, Parvania M, Goharian E, McPherson T, Burian SJ (2018) A systematic review of quantitative resilience measures for water infrastructure systems. Water 10(2):164. https://doi.org/10.3390/w10020164

149. Shinano Y, Rehfeldt D, Gally T (2019) An easy way to build parallel state-of-the-art combinatorial optimization problem solvers: A computational study on solving Steiner tree problems and mixed integer semidefinite programs by using ug[SCIP-*,*]-libraries. In: 2019 IEEE international parallel and distributed processing symposium workshops (IPDPSW), pp 530–541
150. Sichau A (2014) Robust nonlinear programming with discretized PDE constraints using second-order approximations. Dissertation, TU Darmstadt
151. Sichau A, Ulbrich S (2012) A second order approximation technique for robust shape optimization. In: Hanselka H, Groche P, Platz R (eds) Uncertainty in mechanical engineering, applied mechanics and materials, vol 104. Trans Tech Publications, pp 13–22. https://doi.org/10.4028/www.scientific.net/AMM.104.13
152. Siciliano B (2009) Robotics: Modelling, planning and control. advanced textbooks in control and signal processing. Springer, London
153. Slack N (1988) Manufacturing systems flexibility - an assessment procedure. Comput Integrated Manuf Syst 1(1):25–31. https://doi.org/10.1016/0951-5240(88)90007-9
154. Son YK, Park CS (1987) Economic measure of productivity, quality and flexibility in advanced manufacturing systems. J Manuf Syst 6(3):193–207. https://doi.org/10.1016/0278-6125(87)90018-5
155. Stolpe M (2019) Fail-safe truss topology optimization. Struct Multidiscip Optim 60(4):1605–1618. https://doi.org/10.1007/s00158-019-02295-7
156. Subramani K (2003) An analysis of quantified linear programs. In: Calude CS, Dinneen MJ, Vajnovszki V (eds) International conference on discrete mathematics and theoretical computer science. Springer, Berlin, pp 265–277. https://doi.org/10.1007/3-540-45066-1_21
157. Subramani K (2004) Analyzing selected quantified integer programs. In: Basin D, Rusinowitch M (eds) International joint conference on automated reasoning. Springer, Berlin, pp 342–356. https://doi.org/10.1007/978-3-540-25984-8_26
158. Sun PF, Arora J, Haug E Jr (1976) Fail-safe optimal design of structures. Eng Optim 2(1):43–53. https://doi.org/10.1080/03052157608960596
159. Taguchi G, Chowdhury S, Taguchi S (2000) Robust engineering, vol 224. McGraw-Hill, New York
160. Taguchi G, Chowdhury S, Wu Y (2005) Taguchi's quality engineering handbook. Wiley, New York
161. Taguchi G, Elsayed EA, Hsiang TC (1989) Quality engineering in production systems, vol 173. McGraw-Hill, New York
162. Taguchi G, Yano H, Chowdhury S, Taguchi S (2005) Taguchi's quality engineering handbook. Wiley, Hoboken, NJ. https://doi.org/10.1002/9780470258354
163. Thoma K, Scharte B, Hiller D, Leismann T (2016) Resilience engineering as part of security research: definitions, concepts and science approaches. Eur J Secur Res 1(1):3–19
164. Thomas JE, Eisenberg DA, Seager TP, Fisher E (2019) A resilience engineering approach to integrating human and socio-technical system capacities and processes for national infrastructure resilience. J Homeland Secur Emergency Manag 16(2):425. https://doi.org/10.1515/jhsem-2017-0019
165. Tierney K, Bruneau M (2007) Conceptualizing and measuring resilience: a key to disaster loss reduction. TR news 250
166. Tikhonov AN, Arsenin VY (1978) Solutions of ill-posed problems. Math Comput 32(144):1320. https://doi.org/10.2307/2006360
167. Tiller M (2012) Introduction to physical modeling with modelica, vol 615. Springer, Berlin
168. Timoshenko SP, Gere JM (1961) Theory of elastic stability. McGraw-Hill
169. Todini E (2000) Looped water distribution networks design using a resilience index based heuristic approach. Urban Water 2(2):115–122. https://doi.org/10.1016/S1462-0758(00)00049-2
170. Ulrich KT, Eppinger SD (2006) Product design and development, 3rd edn. McGraw-Hill, Boston, MA

171. Ulusoy A, Pecci F, Stoianov I (2020) An MINLP-based approach for the design-for-control of resilient water supply systems. IEEE Syst J 14(3):4579–4590. https://doi.org/10.1109/JSYST.2019.2961104
172. Verein Deutscher Ingenieure (2019) VDI 2221 Blatt 1:2019–11 Entwicklung technischer Produkte und Systeme - Modell der Produktentwicklung [Design of technical products and systems - Model of product design]. Beuth, Berlin
173. Walker B, Holling CS, Carpenter S, Kinzig A (2004) Resilience, adaptability and transformability in social-ecological systems. Ecol Soc 9(2)
174. Wolf J (2015) Quantified linear programming. Dissertation, TU Darmstadt
175. Woods DD (2017) Essential characteristics of resilience. In: Woods DD, Leveson N, Hollnagel E (eds) Resilience engineering. CRC Press, Boca Raton, pp 21–34
176. Yonekura K, Kanno Y (2010) Global optimization of robust truss topology via mixed integer semidefinite programming. Optim Eng 11(3):355–379
177. Zeng B, Zhao L (2013) Solving two-stage robust optimization problems using a column-and-constraint generation method. Oper Res Lett 41(5):457–461. https://doi.org/10.1016/j.orl.2013.05.003
178. Zhang Y (2007) General robust-optimization formulation for nonlinear programming. J Optim Theory Appl 132(1):111–124
179. Zhou M, Fleury R (2016) Fail-safe topology optimization. Struct Multidiscip Optim 54(5):1225–1243. https://doi.org/10.1007/s00158-016-1507-1

Chapter 7
Outlook

Peter F. Pelz, Peter Groche, Marc E. Pfetsch, and Maximilian Schaeffner

Bertolt Brecht once closed a text with the words "We are disappointed to see the curtain close and all questions are left unanswered" [1]. In this book, it has become clear that uncertainty is immanent in the product life cycle of technical systems in mechanical engineering from (B) production, (C) usage, (D) reuse to (E) sourcing. The latter is the starting phase of the following sequence B, C, D, E. Uncertainty has been relevant since the beginning of the industrialisation, cf. Theodor Fontane's ballad 'The Tay Bridge' quoted in Chap. 1 and this will continue to be so. Hence, we will never see "the curtain close", but a perpetual contribution of engineering science, applied mathematics, law and further branches of science to master uncertainty in mechanical engineering.

7.1 Towards the Complete Picture

The product life sequence B, C, D, E spans the temporal dimension. The spatial dimensions are captured by the system boundary. With further increasing system boundaries, we go from material to component and from techno-economic to socio-technical systems. In this outermost system boundary, market forces, social impact and regulatory rules become prominent.

In the presented book, we focused on (A) product design and the two phases (B) production and (C) usage of the product life cycle, cf. Fig. 1.6. Of course, this is not the complete picture: mastering economic uncertainty and uncertainty in acceptability, inevitably needs a holistic view on the product life sequence on the one hand and the extended system boundaries on the other hand, cf. Fig. 7.1.

P. F. Pelz (✉) · P. Groche · M. Schaeffner
Department of Mechanical Engineering, TU Darmstadt, Darmstadt, Germany
e-mail: peter.pelz@fst.tu-darmstadt.de

M. E. Pfetsch
Department of Mathematics, TU Darmstadt, Darmstadt, Germany

P. F. Pelz et al. (eds.), *Mastering Uncertainty in Mechanical Engineering*,
Springer Tracts in Mechanical Engineering,
https://doi.org/10.1007/978-3-030-78354-9_7

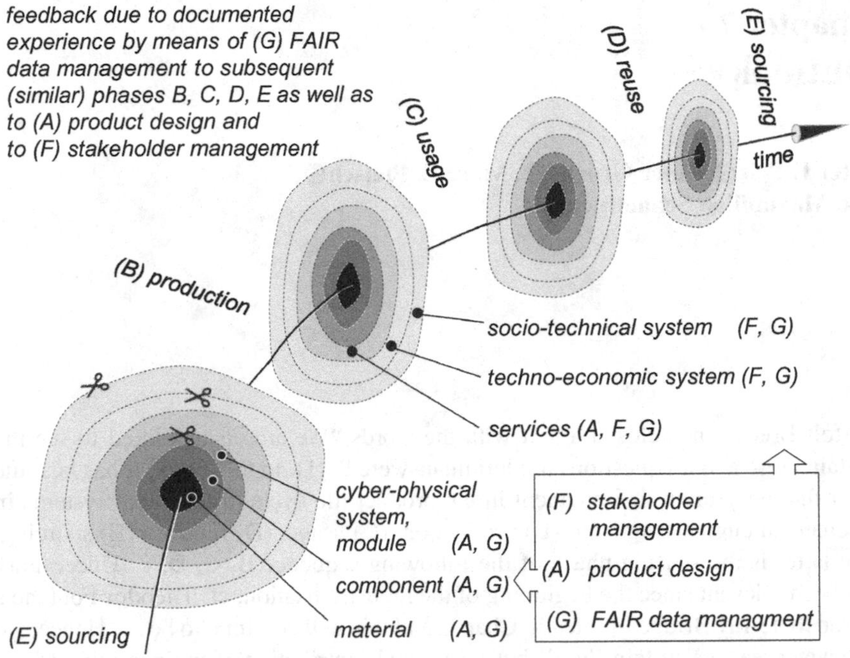

Fig. 7.1 The product life sequence B, C, D, E—rather than a cycle—is represented by the four phases (B) production, (C) usage, (D) reuse and (E) sourcing. The spatial dimensions are captured by the system boundaries extended from material to the socio-technical system. The phases (A) product design, (F) stakeholder interaction and market regulation as well as (G) FAIR data management address all temporal phases and system boundaries as indicated. The trajectory of the system in a Lagrangian representation shows the individuality of each system composed of individual components. The cloud symbolises the Eulerian observer fixed in space. This change of reference enables the feedback to subsequent similar phases as well as to (A) product design and (F) stakeholder management. Hence, (G) FAIR data management enables on the one hand learning from previous similar events; on the other hand it enables transparent quality KPI

It is evident that the relevant time period, the product life sequence, includes the phases (D) reuse/recycling and (E) sourcing. The second law of thermodynamics teaches us that there are no real systems without impact exceeding the phase (D) reuse/recycling [4]. Hence, it is indeed better to speak of product life sequences B, C, D, E, B, C, D, E … rather than a product life cycle.

The spatial extension of the system boundary from material, cyber-physical components, systems and services towards techno-economic or socio-technical systems needs not only contributions from (A) product design. Understanding and possibly control of (F) stakeholder management is as relevant as the (G) FAIR data management, where FAIR is the acronym for Findable, Accessible, Interoperable and Reusable. Stakeholder management includes analysis and control of stakeholder interaction as well as instruments for market regulation and market surveillance. As such, negotiating contracts is part of stakeholder management; the analysis of stakeholder interaction typically is a field of interest for sociology but also for economics and political science.

Some aspects of the extended view were indeed addressed in this book. Chapter 5 is exemplary for this. Mastering uncertainty in the assignment of functional requirement specifications and quality objectives needs the understanding of the stakeholder interaction combined with market regulation. The same is true for existing and emerging legal constraints. It is expected that digital humanities will influence this field in the future even more.

We are aware that Sustainable Systems Design, as discussed e.g. in Sects. 1.6 and 5.1.1, requires a holistic approach, i.e. the extension of system boundaries to socio-technical systems. Therefore Fig. 7.1 complements Figs. 1.10, 5.2 and 5.3. In short, sustainability can only be assessed from a combined socio-economic and technical perspective. Integration of these perspectives is essential for future research.

(F) stakeholder management, combining stakeholder analysis and market regulation, named side by side with (A) product design is part of economics, sociology and law. The scientific methods in that field stem e.g. from cybernetics or applied mathematics. Game theory is one branch of applied mathematics being beneficially applied to stakeholder analysis [3].

(G) FAIR data management [2] is an enabler for transparency in quality key performance indicators (KPI) to foster acceptability. Hence, the quality dimensions (i) effort F_1 measured in economic and social cost, (ii) availability F_2 and (iii) acceptability F_3 need to be further developed as enablers for Corporate Social Responsibility (CSR). These objectives apply to data, software, but also to already existing conclusions. FAIR data management requires (i) data competence, (ii) information technologies and (iii) data governance and curation. Therefore, FAIR data management is the prerequisite for the process from data to wisdom; it leads to a living digital twin being represented by a graph with persistently identified subjects and objects as nodes. The edges represent the predicate, i.e. functions mapping the data from the subject to the object. The graph in combination with a consistent ontology allows accessibility and reusability of the data.

With regard to the interaction between the interest groups, the consumer market differs from the capital goods market in the actors involved. For the former, these are manufacturers, planners, owners/operators and society. In the case of infrastructure systems, the number of stakeholders is further increased because owner and operator are usually not identical and different infrastructure systems are usually coupled. This considerably increases the complexity of stakeholder interactions. In the case of consumer goods, the acting stakeholders are usually limited to manufacturers, retailers, digital matchmakers, customers and society. In both cases, with or without online platform markets, it is clear that emerging block-chain based digital currencies will change the interplay between markets and stakeholders.

Components of production systems or fluid systems are traded on the capital goods market, a typical business to business market. The composed systems enable functions, such as producing, transport, heating, and many others. Mostly, these are typical infrastructure systems with (i) complex stakeholder interaction and market regulations, (ii) frequently unclear functional requirements and quality KPI as well as (iii) an only beginning smart modularisation.

7.2 Future of Mastering Uncertainty

This book builds on the tradition of Taguchi's robust design method, which has been used since the 1960s. At the same time, the world has continued to develop over the past 60 years, and the past 10 years in particular have seen significant new contributions to the mastering of uncertainty. Many of them are presented in this book, to name only some keywords:

(i) Rigorous classification of uncertainty,
(ii) extension of the system boundary towards socio-technical systems,
(iii) validated methods for mastering data, model and structural uncertainty and
(iv) active components serving mastering uncertainty in load-bearing systems.

It took two decades for Taguchi's methods to spread. The dissemination time of the presented newer concepts will be shorter for several reasons: First, the needs of society and the emergence of CSR are becoming powerful drivers for mastering uncertainty; second, digitalisation and computer power enable new methods, technologies, and strategies for quantifying and mastering uncertainty as presented in this book.

Our main focus has always been on mastering uncertainty. The three strategies to be most important in mastering uncertainty are

(i) design and operate robustly,
(ii) gain flexibility and
(iii) enable resilience.

There is still much to do for gaining robust, flexible, or resilient technical systems. In the following three sections, we anticipate the future regarding (i) to (iii).

7.2.1 Robustness

Section 6.1 illustrates that a wide range of methods and technologies is now available to master uncertainty through sufficient robustness. For both aspects, first uncertainty quantification, and second robust optimisation of components and systems, there is a need for multi-purpose, easy-to-use software frameworks. First and in more detail regarding uncertainty quantification: a software framework is needed supporting a consistent workflow from the quantification of uncertainty within the product design phase, to the propagation of uncertainty in the production phase and to the prediction of the system's reliability in the usage phase. Within this framework, efficient probabilistic parameter calibration methods, e.g. in a Bayesian framework, shall be available to cope with the increasingly complex and computationally intensive models used in the further virtualised product design. Second and regarding robust optimisation of components and systems: available mathematical methods are currently not supported by general purpose software, and prior modelling based on human experience is needed before using it for practical problems. Both facts still inhibit Sustainable Systems Design.

We expect software technologies to close this gap in the near future. Yet, there are still open research topics. The first addresses mathematical research, the second engineering research. Above all, the mathematical tools that enable robust optimisation, as described in Chap. 6, all exploit the underlying structure of the problem in one way or the other. Extending these methods to systems for which the corresponding mathematical structure is different, lacks refined methods and therefore requires mathematical research. Thus, although robust optimisation has developed into a relatively mature field with many contributions, there are still many open research questions, in particular for problems of practical interest, such as dynamical, i.e. transient problems.

As has been seen in Chaps. 1, 3 and 6, Robust Design and the related Sustainable Systems Design as it is understood here, can be seen as solving a constrained optimisation problem with the specified systems function as one constraint to be solved for a given design space. The objective has three dimensions: effort, availability and acceptability. To master uncertainty in the customer expectation, in material or component properties, usually an increase of effort, e.g. regarding material consumption, is needed. Section 3.5 listed seven inherent Robust Design and operating concepts that potentially offer additional freedom of design without additional cost and weight. Thus, tailored material or component behaviour could be one promising approach with graded material or component stiffness. Also the deliberate use of residual stresses potentially offers additional freedom of design without additional cost and weight.

When it comes to systems, the robustness of individual modules or components is strongly influenced by the behaviour of other components of the system. Thus, the application of Taguchi's DoE method as one tool of the Robust Design methodology can become quite expensive. Therefore, a further important research area is the efficient and comprehensive validation of a module's robustness under simulated realistically detuned operation and installation conditions.

In contrast to resilience, robust systems do not show recovery phases. Hence, robustness is achieved mainly by "smart" decisions made in the (A) product design prior to the product life sequence B, C, D, E. In the future, merging of data gained from experiences made in the product usage or a physical or cyber-physical product validation test with prior knowledge will become much more important. This merge will result in grey-box models enabling the mentioned "smart" decisions. Indeed, there is a need for integrating Bayesian methods with Robust Design and risk assessment in product design.

7.2.2 Flexibility

Section 3.5 introduced the concept of flexibility and Sect. 6.2 depicts promising approaches for mastering uncertainty by increased flexibility. At the same time, it also reveals the additional costs and complexity of design and production processes concomitant with higher flexibility. Further research work is necessary to provide either (i) smart modules or (ii) smart modularisation, cf. Sect. 3.5, which allows for the highest flexibility at a specific minimised cost. One promising approach for future developments could be the application of lean design engineering principles to obtain flexible systems.

First, smart modularisation is an interesting application field for optimisation methods presented in this book. Second, smart modules usually incorporate semi-active or active components within complex technical systems. They offer a freedom in usage and by this cover different customer needs or expectations. However, the reliable mastering of uncertainty, e.g. by the methods presented within this book, is necessary to, on the one hand, legitimate the increased effort associated with the semi-active and active components and, on the other hand, increase the acceptability for the customer and within society.

The current driver for modularisation is the speed when scaling up as well as satisfying customer demands. Functional units are integrated into modular type packages fulfilling a functional requirement specification. Further open questions are automated documentation as well as the approval processes. From a Sustainable Systems Design perspective, as defined here, it is clear that the specified function will be a constraint, whereas the minimisation of social costs measured in energy or material consumption will be an objective. This demands the definition of metrics and the aggregation of quality KPI from the component level up to the business level. Thus, commissioning, approval and learning will be enabled by FAIR data management as specified above.

7.2.3 Resilience

As discussed in Sect. 6.3, robust systems do not show a recovery phase after a severe impact, whereas resilient systems do. Besides seldom exceptions, only smart agents, humans or cyber-physical modules enable a recovery phase being characteristic for dynamically resilient systems. Those agents heal severely experienced damage by having the ability to measure, react, learn and anticipate.

We can imagine that in a composed system, agents interact in such a way that each agent measures its surrounding and all agents together react in a self-organised manner. This vision can be seen as a biologicalisation of products and processes. The driving potential for the agents to act is the loss of functional quality, cf. Chap. 1. From this perspective, the recovery phase of the resilience triangle is a Continuous Improvement Process (CIP) of products or the product design phase. Only now, the latter takes place within the usage phase. Hence, in this picture of dynamic resilience (A) product design becomes integrated into the product life sequence B, C, D, E mentioned above.

There are few examples of self-healing materials, components or systems without cyber or real agents. In Sect. 3.5, liquid sealant added to the inside of a tire was mentioned as one example: a puncture is self-healed by this sealant; a wooden boat seals itself by swelling the wood; a leather boot automatically seals small holes. In the named examples swelling is the basis for self-adaption or self-healing. In nature we observe stress-induced shape optimisation inline and online integrated in the life sequence B, C, D, E. Today, we use such shape optimisations offline in the (A) product design. Also here, the future task is to integrate (A) product design in the (C) usage phase as nature does. From this we conclude that the design of self-healing materials or self-repairing machines could be stimulated from nature. Their integration into technical systems could pave the way to the so far difficult to achieve recovery of structures.

In Sect. 6.3, static resilience is distinct from dynamic resilience: static resilience is the property of a system predefined by the system's design; dynamic resilience is the skill to react to a loss of functionality. The degree of static resilience e.g. of a water supply network is established in the (A) product design. The future will focus on the trade-off between static resilience and the costs achieving this static resilience. Our current research shows that there is a saturation of gained static resilience versus costs as one would expect. Still, there are open questions regarding the resilience of networks.

Mathematical tools to optimally design resilient systems have been developed for a long time, often under different names like network survivability, etc. The corresponding problems are inherently multi-level and an exploitation of the particular structure is necessary in order to be able to solve the corresponding optimisation problems. Similar to robust optimisation, the future is likely to see a refinement and extension of the available tools and hopefully software support. Moreover, incorporating learning into the systems poses interesting mathematical challenges.

7.3 Final Remarks

Our approach is mainly based on the creation of white-, grey- and black-box models and the use of those models for algorithmic supported systems design. The composition takes place for a known design space. We fully acknowledge that uncertainty sometimes can also be mastered by out-of-the-box thinking, cf. Chap. 1, where out-of-the-box means outside the known design space. Improvisations leading to processes and designs not foreseen in the originally created design space can definitely be stimulating and often help to create new break-through technologies and designs.

Although this book is based on Ratio and Reason, Intuition and Inspiration are the most important drivers. In that respect we are in agreement with the British empiricist

David Hume and others, cf. Sect. 1.3. The systematic development of the necessary creativity could be an important topic for future engineering education.

References

1. Brecht B (2012) Der gute Mensch von Sezuan: Parabelstück. Edition Suhrkamp. Suhrkamp, Frankfurt a. M
2. Commission European (2018) Turning FAIR data into reality: Final report and action plan from the European Commission expert group on FAIR data. Publications Office of the European Union, Luxembourg
3. Nash JF (1950) The bargaining problem. Econometrica 18(2):155–162. https://doi.org/10.2307/1907266
4. Prigogine I, Stengers I (1997) the end of certainty: time's flow and the laws of nature. Free Press, New York

Correction to: Mastering Uncertainty in Mechanical Engineering

Peter F. Pelz, Peter Groche, Marc E. Pfetsch, and Maximilian Schaeffner

Correction to: P. Pelz et al. (eds.), *Mastering Uncertainty in Mechanical Engineering*, Springer Tracts in Mechanical Engineering, https://doi.org/10.1007/978-3-030-78354-9

In the original version of the book, the following corrections have been incorporated:

On page 49, the last equation in the text has been changed.
On page 95, '#' has been replaced with ϑ in Eqs. 3.28 and 3.29.
In Chap. 5, the author name has been changed from Nasssr AlBaradoni to Nassr Al-Baradoni.

The book and the chapters have been updated with the changes.

The updated version of this chapters can be found at
https://doi.org/10.1007/978-3-030-78354-9_3 and
https://doi.org/10.1007/978-3-030-78354-9_5

P. Pelz et al. (eds.), *Mastering Uncertainty in Mechanical Engineering*, Springer Tracts in Mechanical Engineering, https://doi.org/10.1007/978-3-030-78354-9_8

Glossary

Acceptability One of three dimensions of a *system's quality* besides *effort* and *availability*. Formal acceptability is reached by conformity with explicit legal *constraints* or any implicit conventions; informal acceptability is fostered by functional quality and minimal social costs.

Active system → *system, active*

Actuator An energy converter generating potential influence on a *process*.

Adaptive system → *system, adaptive*

Aleatoric uncertainty → *uncertainty, aleatoric*

Algorithm Finite sequence of (computer-)instructions to solve a problem.

Anticipating Predictive *process* (and system) change with the aim of reducing *uncertainty*. Anticipating is one of four abilities/functions of a *resilient system.*

availability One of three dimensions of a *system's quality* besides *effort* and *acceptability*. Availability measures the relative usability of a technical *system* in time.

Black-box model → *model, black-box*

Buffering capacity Metric for evaluating the *resilience* of a technical *system*. The buffering capacity of a technical *system* measures the amount of structural change for which the fulfilment of a predetermined required minimum of functional performance can still be guaranteed. Depending on the context, the buffering capacity can attain continuous or integer values. (Example: In case of integer values, it describes the maximum number of components that can fail while still maintaining the required minimum of functional performance.) A system has a buffering capacity of *k* if it guarantees the required minimum of functional performance within a predetermined range of influencing factors for all possible failure scenarios of up to *k* components.

Component Synonym for an assembly or single item.

Conflict, data-induced A data-induced conflict exists when the interpretation and *usage* of uncertain *data* from more than one source leads to contradictory statements about the appropriate design of *processes* or *products*.

P. F. Pelz et al. (eds.), *Mastering Uncertainty in Mechanical Engineering*,
Springer Tracts in Mechanical Engineering, https://doi.org/10.1007/978-3-030-78354-9

Constraint Requirement for the *design* and *usage/operation* of a *socio-technical system*.

Data Generic term for a quantifiable system value.

Data management *Data* shall be findable, accessible, interoperable, reusable (FAIR). As such, FAIR data management fosters transparency and hence *acceptability*. It is the prerequisite for quality key performance indicators (quality KPI).

Data uncertainty The nature of data uncertainty depends on the form in which *data* are available. If data are stochastically distributed, there is *stochastic uncertainty*. If they are known to be within limits, but not stochastically distributed, there is *incertitude*. Unnoticed or ignored uncertainty occurs when there is neither stochastic uncertainty nor incertitude.

Data-induced conflict → *conflict, data-induced*

Design Methodical procedure from the first idea through planning, conception and development to the virtual elaboration of a (load-bearing) *product*.

Design point A technical *system* is designed for a specific design point. If the system designer strives for a *robust* design, considerations not only comprise one design point, but also an uncertainty area around it.

Design space The function of a *system* can only be realised within a certain design space. The design space is limited by physical laws as well as by the available resources or resource materials, components and technologies. The design space can be expanded by innovations or restricted by banning technologies, e.g. by the requirement for carbon-free energy supply. Systems that enable the same *function* usually differ in *quality*. The task of *Sustainable Systems Design* is to select from these competing systems one with an optimal quality within the design space.

Diagnosis A diagnosis is used to find the cause of *disturbances*. If a disturbance is not directly observable but only its effect on the system, only the symptoms of the disturbance can be observed. The diagnosis allows conclusions from these symptoms on the cause. In particular, it serves to find causes in *data-induced conflicts*.

Disturbance A disturbance leads to unexpected, unauthorised deviations of at least one system value. This can lead to a malfunction or failure of the system.

Effort One of three dimensions of a *system's quality* besides *acceptability* and *availability*. Effort measures the investment costs and social costs given by energy and material consumption to achieve a desired system *function*.

Epistemic uncertainty → *uncertainty, epistemic*

Flexibility A flexible *system* is characterised by the fact that it fulfils multiple predefined *functions* with accepted functional quality. Flexibility is used as a strategy to master *uncertainty* during the *product life cycle* of technical systems.

Function Desired relationship between a *system's* input and output with the aim of fulfilling a task.

Functional requirement Between stakeholders agreed and predefined *function* of a *system*.

Gracefulness Metric for evaluating the *resilience* of a technical *system*. Its behaviour may be described as "graceful degradation" at the boundary of its performance range towards the loss of the required minimum degree of the functional performance. Mathematically, it is defined by the directional derivative of the functional performance curve in the direction of a given influencing factor (or a vector of multiple influencing factors). In the case of non-differentiability, it is given by the limit from the direction of the design point.

Grey-box model → *model, grey-box*

Hardware-in-the-Loop Hardware-in-the-Loop (HiL) tests investigate the behaviour of real components connected to real-time simulated *systems* and allow the stepwise integration of a technical module or component into a real system by combining cyber world and real world.

Ignorance Disregarded but relevant reality. The effect of *uncertainty* is unknown or only suspected. Ignorance is associated with *model uncertainty* and with ignored possible manifestations of a *product*, *system* or *process*. No statement can be made about the probability distributions of an unfolding uncertain property.

Incertitude Limit values of an emerging uncertain *product* characteristic can be assumed. Furthermore, no probability distributions have to be presumed. There are known or estimated membership functions in fuzzy analysis or intervals in interval analysis. The variability is uncertain.

Information Information is derived from the interpretation of *data* and serves as the basis for decisions. Interpretation may be performed within *models*.

Irrelevant reality → *reality, irrelevant*

Learning Reduction of *model uncertainty* and *data uncertainty* through permanent model identification and model adaptation during the *product life cycle*. Learning is one of four abilities/functions of a *resilient system*.

Margin Metric for evaluating the *resilience* of a technical *system*. The margin of a technical system is the distance of the actual functional performance to the system's required minimum of functional performance.

Model Abstract image of an *object* in form of a mathematical model or other, such as imaginary on the basis of intuition. A mathematical model is substantiated in axiomatic (*white-box model*) or empirical (*black-box model*) terms or both (*grey-box model*). Mathematical models represent a functional relationship between input and output data, model parameters and internal variables, like states.

Model, black-box *Model* derived from measurements of a *process* or the experience of a user. In the first case, these models are called data-driven models today. In the second case, the deposited model is part of an expert system.

Model, grey-box *Model* that combines axiomatic and empirically derived relationships as well as (expert) user experience.

Model, white-box *Model* derived by deduction from axioms. *Model uncertainty* in white-box models arises from an impermissible model structure or impermissible simplifications, e.g. an assumption of quasi-stationary system behaviour, inadmissible constitutive equations as well as inadmissible initial and boundary conditions.

Model horizon Boundary of the *relevant reality* represented by the *model*.

Model uncertainty Model uncertainty arises from an incomplete mapping of the *object*. Parts of the *relevant reality* are ignored. In the case of model uncertainty, the functional relationship is suspected, unknown, incomplete or ignored—*ignorance* prevails in all cases.

Module Function-oriented group of *components* of a technical entity or *algorithm*; each with clear interfaces.

Monitoring Sensing a *process* by means of *data* acquisition and data analysis via *models* to obtain *information*. Monitoring is the one of four abilities/functions of a *resilient system*.

Object Generic term for *product*, *system* or *process*.

Objective Target for the *design* and *usage/operation* of a *socio-technical system*.

Operator An operator provides an effective quantity to be able to carry out a *process*. The effective quantity is the purpose of the operator and thus causes the desired change of state. In *production*, for example, the operator comprises machines and the necessary auxiliary material for the production process of the load-bearing structure. In the process of *usage*, however, the operator refers, for example, to the used load-bearing *system*.

Parameter (model) Model parameters are brought into a functional relationship in a *model*. They are a *data* component. Model parameters are derived from empirical data, literature or model analysis.

Passive system → *system, passive*

Performance range Basis for assessing the resilience of a technical system. The performance range describes the range of influencing factors in which a technical system is able to achieve a predefined required minimum functional performance. The performance range can be mathematically expressed by the so-called "super-level set" of the functional performance curve at the level of the required minimum functional performance.

Process A process transforms a primary state into a final state. The process is associated with an individual process or a *process chain*.

Process, time-invariant A *process* started at a time always shows the same behaviour. It can start at any time without a change of result. The *parameters* of its mathematical description and transfer functions of a controller are for example invariable in time (invariant).

Process, time-variant A *process* started at a time shows different behaviour over time, see *time-invariant process*.

Process chain A process chain is the combination of individual *processes*. They transform a primary state into a final state, with the operand going through various intermediate states. Process chains can be modelled across life cycle phases. A process chain can also be used to represent a component structure.

Process chain, resilient In a resilient process chain, *monitoring*, *responding*, *learning* and *anticipating* internal and external *disturbances* (for example machine failure, manufacturing uncertainty, slump in demand, and uncertainty in product usage) can be used to address *ignorance*.

Process model Common and applicable mode of communication for the various areas of expertise to define a *process chain* incorporating systematically and transparently the individual *processes* and the resulting *uncertainty*.

Product A product is an *object* that did not originate naturally, but is produced by man himself for other people, and that is used or consumed in the context of purpose-oriented action in *usage processes*.

Product life cycle The product life cycle describes the *process chain*: sourcing, *production*, *usage* and reuse/recycling.

Product properties Properties of *products* or *systems* are divided into *function* and *quality*.

Production The process of making products, components or systems.

Quality Measures—in the tradition of Taguchi—the *effort* with which a *function* is achieved. The effort is measured in economic and social costs. In addition, there is the *availability* and the *acceptability*.

Radius of performance Metric for evaluating the *resilience* of a technical *system*. It is connected to the technical system's performance range. The radius of performance measures the minimum distance of the design point to the specific value of an influencing variable for which the required minimum level of functional performance is no longer reached.

Reality, irrelevant The part of reality that is not necessary to answer a question.

Reality, relevant The part of reality necessary to answer a question.

Reliability The feature of a *product* to not fail with a certain probability under stated functional and environmental conditions during a specified period of time.

Resilient process chain → *process chain, resilient*

Resilient system → *system, resilient*

Responding *Process* intervention based on *information* with the aim of reducing uncertainty. Responding is one of four abilities/functions of a *resilient system*.

Risk analysis Specific operational measures to deal with uncertainty at the real *process* and *product* level. Risk analysis is limited to the identification and description of risks. The definition and application of specific measures are covered by risk management.

Robust Design Robust Design is an engineering design methodology also known as Taguchi methods. In Robust Design (i) *uncertainty* is replaced by *stochastic uncertainty* using the concept of quality loss functions; (ii) sourcing, design and production phases are hollistically treated by the concept of off-line quality control. The basis of off-line quality control is firstly the Design of Experiments (DoE) and secondly the robust optimisation in the so-called parameter design. Taguchi developed the methodology based on previous works of Ronald Fisher on DoE. In modern Robust Design the concept of perceived quality, i.e. customer experience, and social costs as quality measures are already anticipated.

Robust optimisation A *product* is designed and optimised in such a way that, even with unavoidable influence of disturbances and variations of input variables within the *model horizon*, the user expectations are completely fulfilled.

Robustness A robust *system* proves to be insensitive or only insignificantly sensitive to deviations in system properties or varying *usage*. Robustness is used as a strategy to master uncertainty from the different perspectives of mathematical optimisation, product or system *design* and *production*.

Semi-active system → *system, semi-active*

Socio-technical system In contrast to technical *systems* or *techno-economic systems* that only include technical components and their interaction regarding the flux of energy, material or information including money, socio-technical *systems* take into account the human being and the technical components. The effects of the interaction of humans and technology play a decisive role in the analysis of socio-technical systems.

Soft sensor Model-based acquisition of information at the *component* and/or *system* level. The target value is not measured directly, but determined based on a model. As such, soft sensors are familiar with state observers and Kalman filters based on Bayesian methods.

State variable The state variables, according to the state space representation of system theory, describe the current state of a system, regardless of its origin, e.g. force, speed, etc.

Stochastic uncertainty → *uncertainty, stochastic*

Structural uncertainty Only a part of all possible *structures* of the *design space* is evaluated, i.e. the remaining part of the *design space* is ignored.

Structure Combination of *functions* (functional structure), *components* (component structure) or *process* chains to fulfil a function.

Sustainable Systems Design Engineering design methodology representing the design process as a *constraint* optimisation process: *functional requirements* and *design space* form the constraints, *quality* dimensions give the *objectives*. The three quality dimensions are minimal *effort*, maximal *availability* and maximal *acceptability*.

System The system describes the totality of all elements considered. Setting a system boundary defines the object or *product*, respectively, the objects or products.

System, active An active *system* is characterised by the supply of external energy to influence a *process*. The external energy always influences the process via the *operator*. The term external energy does not include energy that is available to fundamentally necessary operations within the process, in particular no supply energy.

System, adaptive A technical system that can be adjusted to the particularities of various situations, due to its technical characteristics. Adaptivity is the prerequisite for a *resilient system*.

System, passive A passive *system* is characterised by the fact that external energy is only provided for the processes that are fundamentally necessary in the process, i.e. in particular as supply energy.

System, resilient A resilient technical *system* guarantees a predetermined minimum of functional performance even in the event of disturbances or failure of system components and a subsequent possibility of recovering at least the setpoint function. Resilience can be increased by adjusting the system state via *monitoring*, *responding*, *learning* and/or *anticipating*, as well as by systematically designing the system topology.

System, semi-active A semi-active *system* is characterised by the supply of external energy to influence the *operator*. In this case, any properties of the *operator* can be influenced by the external energy. The *process* itself is only affected indirectly by the external energy. The term external energy does not include any energy that is available to fundamentally necessary operations within the process, in particular no supply energy.

Techno-economic system In contrast to technical *systems* that only include technical components and their interaction, techno-economic systems take into account the flux of money and economic measures such as profit.

Time-invariant process → *process, time-invariant*

Time-variant process → *process, time-variant*

Uncertainty Uncertainty occurs when the *usage* properties and *process* characteristics of a *system* cannot, or can only be partially determined.

Uncertainty, aleatoric Natural, random and irreducible uncertainty.

Uncertainty, epistemic *Uncertainty* due to incomplete scientific knowledge. Epistemic uncertainty can be reduced by new insights.

Uncertainty, stochastic Partial to complete details on probability distributions of an emerging uncertain *product* characteristic are available. There are known or estimated probability density functions; the variability is always determined.

Usage Usage or operation of a component, *product*, *system* or *process*.

Validation Analysis to what extent a *model* after calibration is suitable for the description of a relevant functional relationship by comparison of reality and *model*. Furthermore, evaluation to what extent a *product* meets the predefined *quality* and functional *constraints* and to what extent a product is accepted by the customer and the society.

Verification Review whether the *model* is consistent and has been correctly solved. Furthermore, evaluation to what extent the design and production methods and technologies are selected correctly.

Vulnerability A *system's* vulnerability or violability.

White-box model → *model, white-box*

Printed in the United States
by Baker & Taylor Publisher Services

Printed in the United States
by Baker & Taylor Publisher Services